Standard Colorimetry

Current and future titles in the Society of Dyers and Colourists – John Wiley Series

Published

Standard Colorimetry: Definitions, Algorithms and Software
Claudio Oleari

The Coloration of Wool and Other Keratin Fibres
David M. Lewis and John A. Rippon (Eds)

Forthcoming

Theoretical Aspects of Textile Coloration
Stephen M. Burkinshaw

Natural Dyeing for Textiles: A Guide Book for Professionals
Debanjali Banerjee

Colour for Textiles: A User's Handbook, Second Edition
Roger H. Wardman and Matthew Clark

Giles's Laboratory Course in Dyeing, Fifth Edition
Uzma Syed

Standard Colorimetry

Definitions, Algorithms and Software

Claudio Oleari

Dipartimento di Fisica e Scienze della Terra, Università degli Studi di Parma, Italy

Software developed by **Gabriele Simone**-*Dipartimento di Informatica,
Università degli Studi di Milano, Italy*

society of dyers
and colourists

WILEY

This edition first published 2016
© 2016 John Wiley & Sons, Ltd

Registered office
John Wiley & Sons Ltd, The Atrium, Southern Gate, Chichester, West Sussex, PO19 8SQ, United Kingdom

For details of our global editorial offices, for customer services and for information about how to apply for permission to reuse the copyright material in this book please see our website at www.wiley.com.

Library of Congress Cataloging-in-Publication Data applied for.

A catalogue record for this book is available from the British Library.

ISBN: 9781118894446

Set in 10/12 pt in Times LT Std by Aptara, India
Printed and bound in Singapore by Markono Print Media Pte Ltd

1   2016

To those who strive to be just

Contents

Society of Dyers and Colourists

Society of Dyers and Colourists (SDC) is the world's leading independent, educational charity dedicated to advancing the science and technology of colour. Our mission is to educate the changing world in the science of colour.

SDC was established in 1884 and became a registered educational charity in 1962. SDC was granted a Royal Charter in 1963 and is the only organization in the world that can award the Chartered Colourist status, which remains the pinnacle of achievement for coloration professionals.

We are a global organization. With our Head Office and trading company based in Bradford, UK, we have members worldwide and regions in the UK, China, Hong Kong, India and Pakistan.

Membership: To become a member of the leading educational charity dedicated to colour, please email members@sdc.org.uk for details.

Coloration Qualifications: SDC's accredited qualifications are recognized worldwide. Please email edu@sdc.org.uk for further information.

Colour Index: The unique and definitive classification system for dyes and pigments used globally by manufacturers, researchers and users of dyes and pigments (www.colour-index.com).

Publications: SDC is a global provider of content, helping people to become more effective in the work-place and in their careers by educating them about colour. This includes text books covering a range of dyeing and finishing topics with an ongoing programme of new titles. In addition, we publish *Coloration Technology*, the world's leading peer-reviewed journal dealing with the application of colour, providing access to the latest coloration research globally.

For further information please email: info@sdc.org.uk, or visit www.sdc.org.uk.

Preface

"Standard Colorimetry" is an ambitious title that comes from the project of a small book, already fully written and never published, entitled *Concise Handbook of Standard Colorimetry*. The reviewers, who certainly knew my scientific production, suggested to broaden the content of the book, pointing me to chapters and contents. The book has become bigger, but more personal. This produced the change of the title, which contracted as *Standard Colorimetry*.

The books published in recent years on colorimetry are all excellent, comprehensive and authoritative, and written by authors and experts, and surely many readers have not felt the need for the publication of a further book. However, the differences between these books, including this one, are obvious.

Each book highlights the author's knowledge, expertise and experience, which are made of reliefs, accents that make the various points otherwise important and in this sense reveal the views of the author. These important features differentiate the various books.

I do not like to take possession of the sentences of others, so the text is full of quotations in inverted commas, indicating clearly the source. This is a way to go to the source and respect the authors.

A software that accompanies this book has the function of giving visual concreteness to the numbers that specify the colour and is a tool for all colorimetric calculations.

Today this book is the book I wish I had read in a sequential way, starting from the first row, when, at the age of about 45 years, the case led me to passionately study human colour vision.

I thank the unknown reviewers. I appreciate the quality of their work and their competence.

I thank the many colleagues that through dialogue, often with very short conversations, e-mail exchanges, or simply the seminars I attended, helped me to understand and know, led me to get a varied overview of colour science. I cannot cite everyone. I feel obliged to mention one name among them all, Robert M. Boynton, because in 2003 in a very short workshop in La Jolla he made us understand that every formula is obtained by engineering, but its value lies in its capacity to explain the phenomena and not simply to fit the phenomena. He had a high conception of science. Today there are too many formulae in colorimetry that have only a practical value but are unsatisfactory and do not help us to understand the phenomena.

Thanks to the readers who want to tell me the darkness and the errors encountered in reading the book or just want to comment. Send me suggestions and questions through e-mail: claudio.oleari@fis.unipr.it.

Claudio Oleari
2015

1

Generalities on Colour and Colorimetry

The *Commission Internationale de l'Éclairage* (CIE) is the official institution devoted to worldwide cooperation and the exchange of information on all matters relating to the science and art of light and lighting, colour and vision, photobiology and image technology.

CIE publications are the main reference for this book.[1–3] This book is about colorimetry and has the definitions of colour and colorimetry as its starting point.

1.1 Colour

In non-specialist language, the word 'colour' is ambiguous, because it is used to describe the quality of the objects, self-luminous and non-luminous, and to describe a quality of the viewing experience. These meanings of the same word 'colour' are different but they are not disjoint, because the first one is the stimulation of the visual experience and the other the visual experience itself. Between these two meanings there is a correspondence and colorimetry quantitatively describes this correspondence.

The colour of self-luminous and non-luminous objects is associated with a physical quantity, which is properly called *colour stimulus* and is measurable because it is external to the body of the observer:

> "*Colour stimulus* – visible radiation entering the eye and producing a sensation of colour, either chromatic or achromatic."[1]

The definition of colour as an effect of the colour stimulus is given by the *Optical Society of America* (OSA) in the 1952 report:

> "Color consists of the characteristics of light other than spatial and temporal inhomogeneities; light being the aspect of radiant energy of which a human being is aware through the visual sensations which arise from the stimulation of the retina of the eye."[4]

Standard Colorimetry: Definitions, Algorithms and Software, First Edition. Claudio Oleari.
© 2016 John Wiley & Sons, Ltd. Published 2016 by John Wiley & Sons, Ltd.

Among the many definitions of colour, the most comprehensive, albeit in its brevity, is given by the *American Society for Testing and Materials* (ASTM),[5] which with the definitions opens highly technical discussions, which are clarified later in the book:

1.   *"Colour of an object* – aspect of object appearance distinct from form, shape, size, position, or gloss that depends upon the spectral composition of the incident light, the spectral reflectance or transmittance of the object, and the spectral response of the observer, as well as the illuminating and viewing geometry."[5]

2.   *"Perceived colour* – attribute of visual perception that can be described by colour names such as white, grey, black, yellow, brown, vivid red, deep reddish purple, or by combinations of such names. *Discussion* – perceived colour depends greatly on the spectral power distribution of the colour stimulus, but also on the size, shape, structure, and surround of the stimulus area, the state of adaptation of the observer's visual system, and the observer's experience with similar observations."[5]

The 'perceived colour' is defined using the names of the colours. This means that the names of the colours represent fundamental concepts, which are not definable in other words. The perceived colour is incommunicable. Humans evoke the perceived colour in the interlocutors with conventional words – red, yellow, green, blue, black, grey, white, so on –.

1.2   Colorimetry

Robert W. Hunt[6,7] distinguishes between:

"Psychophysical colour terms – terms denoting objective measures of physical variables that are evaluated so as to relate to the magnitudes of important attributes of light and colour. These measures identify stimuli that produce equal responses in a visual process in specified viewing conditions."

and

"Psychometric colour terms – terms denoting objective measures of physical variables that are evaluated so as to relate to differences between magnitudes of important attributes of light and colour and such that equal scale intervals represent approximately equal perceived differences in the attribute considered. These measures identify pairs of stimuli that produce equally perceptible differences in response in a visual process in specified viewing conditions."[6]

Psychophysical colour terms regard *Psychophysical colorimetry* and psychometric colour terms regard *Psychometric colorimetry*. Both definitions of *psychophysical* and *psychometric colour* refer to colour stimuli, whose measurement and processing are same as those in the human visual system. The human visual system is a tool that measures the colour stimulus, as a camera, (psychophysics) and processes the signals produced quantifying the colour attributes according to a perceptive scale (psychometrics).

Psychophysical colorimetry is limited to the measurement of colour stimuli, attributing the same specification to different colour stimuli which induce equal colour sensations. This is exactly what happens in a photographic camera.

The human eye, unlike the camera, has a sensor – the retina – that has not the same optical properties in all its parts. The central part, for acute vision, is different from the surrounding parts, for which, according to a simplified diagram, there are two different colorimetries. In 1931 the CIE defined a colorimetry for

Table 1.1   *Scheme of the colour specification according to historical steps, stages of vision, visual fields and referred to the CIE standard systems.*

		Historical steps/stages of vision/systems		
		First stage of vision: transduction	Second stage of vision: colour difference and illuminant discounting	Third stage of vision: colour appearance and adaptation
		Psychophysics (Chapter 9)	Psychometrics (Chapter 11)	Colour appearance
v i s u a l	Visual field < 4°	CIE 1931 standard observer (X, Y, Z),	CIELUV system (L^*, u^*, v^*) CIELAB system (L^*, a^*, b^*)	CIECAM97 CIECAM02 Retinex
		Vos observer 2° CIE fundamental observer		
f i e l d	10° visual field	CIE 1964 supplementary standard observer (X_{10}, Y_{10}, Z_{10})	CIELUV system $(L^*_{10}, u^*_{10}, v^*_{10})$ CIELAB system $(L^*_{10}, a^*_{10}, b^*_{10})$	
		10° CIE fundamental observer		

acute vision – observer with a visual field of 2° described in Section 9.2 – and in 1964 a colorimetry for non-acute vision – observer with the field of view of 10° described in Section 9.3 –.

The distinctions between psychophysical and psychometric colorimetries, and between the 2° and 10° visual fields, have led to four different colorimetries, as summarized in Table 1.1.

Over time, the study of colour-vision has led to improving the standard observers by adding new cases within the schema of Table 1.1, that is, Vos and fundamental observers described in Sections 9.5 and 9.6. These improvements are considered so small that the industries and laboratories continue to use the standard CIE 1931 and CIE 1964.

This distinction among different colorimetries corresponds to a distinction among the historical phases of colorimetry[7]:

A first phase concerned with 'which colours match' and is termed *psychophysical* in strict sense (Chapter 9) and also *classical colorimetry* or *tristimulus colorimetry*.

A second phase concerned with 'whether colour differences are equal' and is termed *psychometric* (Chapter 11).

A third phase concerned with 'what colours look like' and is termed "*colour appearance*:

i. aspect of visual perception by which things are recognized and
ii. in psychophysical studies, visual perception in which the spectral and geometric aspects of a visual stimulus are integrated with its illuminating and viewing environment."[1]

The third phase of colour-appearance colorimetry is in rapid progress and is a subject of debate; therefore, it is not considered in this book.

References

1. ILV CIE S 017/E:2011, *ILV: International Lighting Vocabulary*, Commission Internationale de l'Éclairage, Vienna (2011). Available at: www.eilv.cie.co.at/free access (accessed on 3 June 2015)
2. CIE 15:2004, *Colorimetry*, 3rd edn, Commission Internationale de l'Éclairage, Vienna (2004).
3. CIE Supplement No. 2 to publication No. 15, *Colorimetry, Recommendations on Uniform Color Spaces, Color Difference Equations, Psychometric Color Terms*, Commission Internationale de l'Éclairage, Vienna (1978).
4. Committee on Colorimetry of the O.S.A., *The Science of Color*, OSA – The Optical Society of America, Washington, DC (1963).
5. ASTM E284 – 13b, *Standard Terminology of Appearance*, ASTM International, West Conshohocken, PA (2013).
6. Hunt RWG, The specification of colour appearance. I. Concepts and terms, *Color Res. Appl.,* **2**, 55–68 (1977).
7. Hunt RWG, Colour terminology, *Color Res. Appl.*, **3**, 79–87 (1978).

Bibliography

Berns RS, *Billmeyer and Saltzman's Principles of Color Technology,* 3rd edn, John Wiley & Sons, New York (2000).
Billmeyer F Jr and Saltzman M, *Principles of Color Technology*, 2nd edn, John Wiley & Sons, New York (1981).
Brainard D, Colorimetry, in *OSA Handbook of Optics*, 2nd edn, Bass M, ed., Vol. 1, Chap. 26, McGraw-Hill, New York (1995).
Committee on Colorimetry of the O.S.A., *The Science of Color*, OSA – The Optical Society of America, Washington, DC (1963).
Gegenfurtner KR and Lindsay TS, eds., *Color Vision, From Genes To Perception*, Cambridge University Press, Cambridge, (1999).
Gescheider GA, *Psychophysics – The Fundamentals*, 3rd edn, Lawrence Erlbaum Associates, London, (1997).
Hunt RWG and Pointer MR, *Measuring Colour*, 4th edn, Wiley-IS&T Series in Imaging Science and Technology, John Wiley & Sons (2011).
Kaiser PK and Boynton RM, *Human Color Vision*, Optical Society of America, Washington, DC (1996).
Kuehni RG, *Color, An Introduction to Practice and Principles*, John Wiley & Sons, New York (1997).
Kuehni RG, *Color Space, and Its Division*, John Wiley & Sons, New York (2003).
Kuehni RG and Schwarz A, *Color Ordered, A Survey of Color Order Systems from Antiquity to the Present*, Oxford University Press, New York (2008).
Le Grand Y, *Light, Colour and Vision*, 2nd edn, Chapman and Hall, London (1968).
MacAdam DL, *Color Measurement*, Springer, Berlin (1985).
MacAdam DL, ed., *Colorimetry – Fundamentals*, SPIE MS 77, SPIE – The International Society for Optical Engineering, Washington, DC (1993).
Malacara D, *Color Vision and Colorimetry: Theory and Applications, Second Edition*, SPIE Press (2011)
McDonald R, eds., *Colour Physics for Industry*, Society of Dyers and Colourist, Bradford (1987).
Ohta N and Robertson AR, *Colorimetry – Fundamentals and Applications*, John Wiley & Sons, New York (2005).
Schanda J, ed., *Colorimetry: Understanding the CIE System*, John Wiley & Sons, New York (2007).
Sève R, *Physique de la couleur*, Masson, Paris (1996).
Shevell S, ed., *The Science of Color*, 2nd edn, OSA-Elsevir, Amsterdam (2003).
Valberg A, *Light Vision Color*, John Wiley & Sons, Chichester (2005).
Wyszecky G and Judd DB, *Color in Business, Science and Industry*, John Wiley & Sons, New York (1975).
Wyszecky G and Stiles WS *Color Science*, John Wiley & Sons, New York (1982).

2

Optics for Colour Stimulus

2.1 Introduction

This chapter has solely educational purposes and only recalls classical optical phenomena, therefore its bibliography consist of textbooks, manuals and tutorials.

This chapter considers the physical nature of colour stimuli.

> "*Colour stimulus* – visible radiation entering the eye and producing a sensation of colour, either chromatic or achromatic."[1-3]

> "*Colour stimulus function* ϕ_λ – description of a colour stimulus by the spectral concentration of a radiometric quantity, such as radiance or radiant power, as a function of wavelength."[1-3]

> "*Relative colour stimulus function* $\phi(\lambda)$ – relative spectral power distribution of the colour stimulus function."[1-3]

> "The psychophysical specification of a colour stimulus is termed again *colour stimulus* and is denoted by a symbol in square brackets, for example, $[\phi]$"[1-3] (the context avoids ambiguity). Since the psycophysical colour stimulus is mathematically a vector, in this book the colour stimulus is always indicated in bold roman letters as scientific convention requires.

The colour stimulus function is a physical quantity measured in W/nm if it is the spectral distribution of the radiant power – also *spectral flux* –, or in W/(m² sterad nm) if it is the spectral distribution of the radiance. The relative colour stimulus function is dimensionless and the spectral plots of this function and that of the colour stimulus function are proportional, that is, $\phi_\lambda = k\,\phi(\lambda)$, where k is a constant with a suitable physical dimension. The colour stimulus functions must be used for the computations of absolute

Standard Colorimetry: Definitions, Algorithms and Software, First Edition. Claudio Oleari.
© 2016 John Wiley & Sons, Ltd. Published 2016 by John Wiley & Sons, Ltd.

quantities as, for example, the illuminance and the luminance, but generally the use of the relative colour stimulus function is more convenient and almost all the colour specification is made on a relative scale. [In colorimetry and photometry the absolute spectral distribution functions have dependence on the wavelength λ written as a subscript while the relative spectral distribution functions have dependence written in brackets (λ).]

These definitions of colour stimulus are merely radiometric (Section 2.4) and only physical operations on these quantities are considered, for example, addition of many colour stimuli.

Colour stimuli produce colour sensations, that usually continue to be called colour stimuli, which should not be confused with the previous colour stimuli (the context avoids ambiguity). Psychophysical operations can be made on these last colour stimuli, for example, addition, comparison and colour matching, and in this case colour stimuli, which represent colour sensations, are written in square brackets $[\psi]$ or as vectors.

The colour stimulus function characterized by a spectral power distribution constant is termed *equal-energy* (or *equi-energy*) *stimulus function* and denoted by the relative spectral radiance $E_E(\lambda) = 1$. The equal energy stimulus function often plays a role in the normalization of the tristimulus space, used for the psychophysical specification of colour stimuli (Section 6.13).

Colour vision is a phenomenon triggered by colour stimulus, and it happens by the interaction between visible electromagnetic radiation (light) and the visual system. Activation consists of the absorption of radiation by typical molecules located inside photosensitive cells of the retina on the fundus of the eye. Everything that happens before the activation itself concerns the interaction between the electromagnetic radiation and matter. It is therefore necessary to have a theoretical framework that describes the generation, propagation and interaction of light with matter in order:

1.   to have a basis for the understanding of the meaning of colour stimulus function and of colour perception and
2.   to quantify and forecast the light phenomena, and connect them with the physical nature of matter.

Point (1) is considered in this introductory section. The physical phenomena related to the colours, often called *colour physics*, are described in Chapter 3, while the fundamental optical phenomena used in colour science are concisely recalled where necessary.

It is known that visible light is an electromagnetic radiation constituted by mutually orthogonal electric and magnetic fields that propagate in space and time, in a vacuum and in matter, with a velocity dependent on the nature of the medium.

Electromagnetic radiation shows a dual nature, *corpuscular* and *wave nature*: the processes in which light is generated or absorbed can be explained assuming that it consists of quanta, postulated by Einstein and called *photons*, while light propagation in space and time is well explained by a wave-like behaviour, fully described by Maxwell's equations.

Everyday experience suggests we consider light as consisting of rays. *Geometrical optics* is based on this assumption and is derivable from wave optics by an approximation, termed *eikonal approximation*. Geometrical optics is valid only when light interacts with objects, which are much larger than its wavelength.

All these descriptions of light radiation enter the representation of vision phenomena and the lab equipment:

•   The image formation in the eye is well described by geometrical optics.
•   The transduction of light radiation into an electrical phenomenon in the photosensitive cells of the eye is a photochemical process and regards the quantum nature of light, that is, photons.
•   The photo-detection of solid state devices, generally used in radiometric, photometric and colorimetric instruments, is often a quantum phenomenon and regards the quantum nature of light.

- The emission of light radiation from a body – light source – is a quantum process, albeit divided into two different processes:
    1. transitions between electronic states of matter at different energy – responsible for the coloration of many bodies and
    2. blackbody emission, a phenomenon dependent only on the temperature of the body (Section 3.3).
- The interaction of the electromagnetic radiation with matter is describable with one of the three descriptions (geometrical, wave and quantum optics) according to the phenomena involved.

Here we describe briefly the characteristics of light according to the wave and quantum model. (It is assumed that the reader is familiar with geometrical optics.)

2.2 Electromagnetic Waves

The wave hypothesis dates back to Huygens (1629–1695), but then there are the Maxwell equations (1873) that fully describe the light as electromagnetic waves. The electromagnetic properties of the matter, in which there are waves, are represented by the *dielectric constant ε* and *magnetic permeability μ*. In fact, these are precisely the electromagnetic properties of matter that determine the light velocity $v = (\varepsilon\mu)^{1/2}$ and consequently the optical properties. The light velocity in the vacuum, denoted by c, is the maximum possible velocity and is an universal constant of physics. The *refractive index* $n \equiv c/v$, known in geometric optics, is a quantity that summarizes these properties. The refractive index is a function of the wavelength and is typical for any material.

The waves, solutions of Maxwell's equations in an optically homogeneous medium, are represented by sinusoidal functions called *wave functions*, whose argument, the *phase* of the wave, is defined at the point x of the space along the direction of propagation and the time instant t by

$$2\pi\left(\frac{x}{\lambda} - \frac{t}{\tau}\right) - \vartheta_0 = \kappa x - \omega t - \vartheta_0 \tag{2.1}$$

τ is the period;
$\nu = 1/\tau$ is the frequency;
$\omega = 2\pi\nu$ is the pulsation;
λ is the wavelength;
$\kappa = 2\pi/\lambda$ is the wave number (sometimes the wave number is simply the reciprocal of λ); and
ϑ_0 is the initial phase.

The electromagnetic waves have the following properties:

- The quantities described by wave functions are the electric field vector $\vec{E}(\kappa x - \omega t - \vartheta_0)$ and the magnetic field vector $\vec{H}(\kappa x - \omega t - \vartheta_0)$. A qualitative representation of a wave in space and time is given in Figure 2.1.
- $\vec{E}$ and $\vec{H}$ are perpendicular to each other and to the direction of propagation – *transverse waves* – (Figure 2.1).
- The waves associated with the electric filed are *plane waves* if the vectors $\vec{E}$ at any points of the propagation line belong to a plane. Since electric and magnetic fields are linked together by Maxwell's equations, the vectors $\vec{H}$ of the magnetic field belong to a plane orthogonal to the plane of the electric field. The surfaces on which the phase of the wave is constant are called *wave fronts*, and the planes tangent to these surfaces are perpendicular to the direction of propagation.

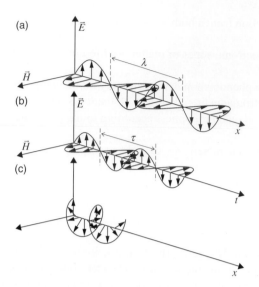

Figure 2.1 *(a) Plane electromagnetic waves represented in space at fixed time and (b) in the time in a fixed point in space. This graphical representation shows the wavelength and the frequency in space and time, respectively. (c) The third graph from the top represents a wave with 'circular polarization'; here only the electric field is represented and the magnetic field is orthogonal to the electric field. The electric field and magnetic field are mutually perpendicular and together orthogonal to the propagation direction.*

- A light wave is *linearly polarized* if the wave is a plane wave (Figure 2.1). Linear polarization and *plane polarization* are synonyms.
- Light *polarization*, represented in waves by the direction of the electric field, is a property with a role in the processes of reflection and refraction, and is therefore in part responsible for the surface appearance of bodies. Therefore, it has an important role in colorimetry and should not be ignored, especially in colorimetric measurements, in which the radiation reflected by a body is measured and the optical components of the instrument reflect light.
- An example of linearly polarized light is the light emitted by an LCD monitor. It can be easily checked with Polaroid sun glasses rotated on the screen, which operates as a polarization analyser.
- In the majority of cases, the light is not polarized and it is not possible to define a single plane for the oscillation of the electric field, but thanks to the superposition principle, it is possible to consider it as a sum of linearly polarized waves. The decomposition of a light beam in waves of different polarization facilitates the discussion of the phenomena of reflection and refraction. It is also of practical utility since many everyday devices polarize light as, for example, polaroid filters used in sunglasses.
- The flow of power ϕ_e of the electromagnetic radiation (energy per unit of time and area, measured in W/m^2) proceeds in the direction orthogonal to the fields $\vec{E}$ and $\vec{H}$.
- The electric and magnetic field magnitudes are proportional to each other $\left| \vec{H} \right| = \sqrt{(\varepsilon / \mu)} \left| \vec{E} \right|$.
- The wave is *coherent*, which means when the phase is known at a point in space and at an instant in time, then it can be determined at any other point and instant.
- The velocity at which the plane waves propagate in space is implicitly defined by the argument of the wave functions $(\kappa x - \omega t - \vartheta_0)$, which takes the same value for different pairs of variables x and t. If we consider the wave function at the instants t_1 and t_2, it follows that this function has the same values at the points x_1 and x_2 such that $(\kappa x_1 - \omega t_1) = (\kappa x_2 - \omega t_2)$, from which $\omega (t_1 - t_2) = \kappa (x_1 - x_2)$, and then the

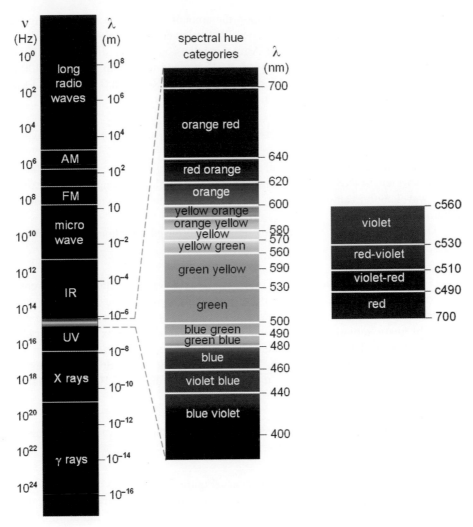

Figure 2.2 *Complete spectrum of electromagnetic radiation characterized by wavelength and frequency. The part of the spectrum related to visible light is expanded and a hue name is associated to each wavelength range. The colours printed here are only representative and approximate for many reasons, which will become clear on reading the book. For completeness, the non-spectral hues (purple and magenta hues) are added, which are specified by the complementary wavelengths (Section 4.3). 'c' before the wavelength numbers in the extraspectral region means 'complementary of'.*

wave advances rigidly with a velocity $v = (x_1 - x_2)/(t_1 - t_2) = \omega/\kappa = \lambda/\tau = \lambda v$. The velocity in the vacuum is $c = 2.997925 \times 10^8$ m/s.

• The waves are classified by their wavelength λ or frequency $v = c/\lambda$. Since the velocity depends on the medium, by convention the wavelength is considered in a vacuum and is very close to that in air. A classification according to decreasing wavelength (Figure 2.2) distinguishes between radio waves (used in telecommunications), microwaves (typical of radar and microwave ovens), infrared radiation (IR), visible radiation (VIS), ultraviolet radiation (UV), X-rays and γ-rays. There are no precise limits for the spectral range of visible radiation since they depend upon the amount of radiant power reaching the retina and the responsivity of the photosensitive cells. Generally, the lower limit is taken between 360 and 400 nm and the upper limit between 760 and 830 nm. A finer subdivision of UV radiation distinguishes between UVC with $100 < \lambda < 280$ nm, UVB with $280 < \lambda < 315$ nm and UVA with $315 < \lambda < 400$ nm, while in IR radiation we have NIR (near-IR) with $0.8 < \lambda < 2.5$ μm, and IR-A with $0.78 < \lambda < 1.4$ μm, IR-B with $1.4 < \lambda < 3.0$ μm and IR-C with 3 μm $< \lambda < 1$ mm.

"*Monochromatic radiation* – radiation characterized by a single frequency or a single wavelength.

NOTE 1 – In practice, radiation of a very small range of frequencies can be described by stating a single frequency or wavelength (NB: In reality a radiation with a single wavelength cannot exist).

NOTE 2 – The wavelength in air or in vacuum is also used to characterize a monochromatic radiation. The medium must be stated.

NOTE 3 – The wavelength in standard air is normally used in photometry, radiometry and colorimetry."[1-3]

A clarification is required: the radiation of a single wavelength is not selectable instrumentally (Section 12.3.1) and is theoretically definable only in an infinite optically homogeneous space. In a strict sense, the word 'monochromatic' is improperly used; anyway, it is used. This does not condition the science of colour, and the monochromatic radiations can be considered as a language approximation. A radiation, which is non-monochromatic but is defined within a narrow spectral range, is a *spectral radiation* defined within a band specified by its width (Section 12.3.1). In colorimetry, the bandwidth should be 1 nm (Section 6.10). In practice, often the spectral radiations used have a bandwidth up to 10 nm and Gaussian spectral distribution.

"*Monochromatic stimulus* – colour stimulus consisting of monochromatic radiation. Equivalent term: *spectral stimulus*."[1]

Anticipating the description of the colour sensations, Figure 2.2 proposes the visible electromagnetic radiation associated with the various spectral hues (albeit approximated by the printing process) and the corresponding names. There are extra-spectral hues provided by a mixture of radiation of short and long wavelengths with the intensity in a variable ratio. These extra-spectral hues cannot be associated with a wavelength; however, the convention is to associate them with the complementary wavelengths, denoted by the letter 'c' followed by the value of the complementary wavelength.

Complementary wavelength to a light radiation is the wavelength of the spectral radiation, which when mixed with this in an appropriate ratio of intensity generates a light-looking *achromatic* – without hue –.

The plane wave functions with different wavelengths are simultaneously solutions of Maxwell's equations because these differential equations are linear. It follows that also a linear combination of these waves is a wave function solution of Maxwell's equations. This is summarized in the *superposition principle*, so the sum of many solutions is still a solution. This latter property enables treating any light beam as a sum of monochromatic beams, which are defined by a single wavelength. This property is of particular importance in colour science.

Radiation defined by a superposition of different monochromatic waves is termed *heterochromatic*.

The refractive index of materials depends on the wavelength of the radiation. Since the velocity of light in a medium is smaller than that in a vacuum, the refractive index is greater than 1. Furthermore, it decreases with increasing wavelength. This is the basis of the phenomenon of *dispersion* and is of great importance in optics and in colour analysis (Section 4.3). All this is true in the absence of absorption, which leads to *anomalous dispersion*, the treatment of which is beyond the scope of this book.

2.3   Photons

The quantum nature of light is evident in the phenomena of light absorption, emission, photoelectric effect and in the blackbody emission. Emission or absorption of a light quantum is present in all these phenomena considered elementarily. The concept of wave, considered as a continuum, is not suitable to explain these phenomena. The hypothesis of light quantum, called *photon*, was given by Einstein to explain the photoelectric effect.

> The *photon* is a quantum of electromagnetic radiation, which is an indivisible entity with an energy $U_p = h\nu$ that is proportional to the frequency of the associated monochromatic wave $\nu = c/\lambda$. The proportionality constant $h = 6.62559 \times 10^{-34}$ Js, known as *Planck's constant,* is a universal constant of physics.

In atomic spectroscopy, the emission or absorption of a photon happens in the electronic transitions between two electronic states with different energies and the photon energy is equal to the difference in energy between the two electronic states.

The *photoelectric effect* is a process of interaction of electromagnetic radiation with matter in which the energy of the photons is transferred to as many electrons. The absorption of energy by the electrons changes the state, and it is customary to distinguish the *external* photoelectric effect from the *internal* one:

- In the *external* photoelectric effect, the irradiation of a conductor with photons can produce an emission of electrons with a kinetic energy equal to the difference between the energy of the absorbed photon and the binding energy of the extracted electron.
- In the *internal* photoelectric effect, the irradiation of a dielectric with photons can produce the passage of electrons from one electronic state, in which the medium is an isolator, to another, in which the medium is a conductor, and the energy of photons is equal to the difference between the two electronic states, the initial and final.

> *Quantum efficiency* is defined as the ratio of the number of elementary events contributing to the detector output to the number of incident photons. It represents the probability that a photon is absorbed by the illuminated body.

These phenomena are the base of functioning for many optoelectronic devices, primarily used as quantum photodetectors (photomultiplier tubes, photodiodes, phototransistors, CCDs, CMOS, etc.) (Section 12.3.1).

2.4   Radiometric and Actinometric Quantities

Electromagnetic radiation, regarded as a wave which propagates in space or as a set of photons travelling in space, is associated with a power ϕ_e that crosses an ideal surface in the space. This power is measured in watts [W]. Ideally, the radiation is spectrally decomposable into waves of defined wavelength, and then of defined energy and also in sets of photons of defined energy, so the radiation can be described analytically by the following:

- *Radiometric unit – "radiant spectral power distribution*[$\phi_{e,\lambda}$; ϕ_λ; P_λ] that is spectral power emitted, transmitted or received in the form of spectral radiation; measured in W/nm (the wavelength is measured in nm)." The subscript 'e' means 'energy'.

- *Actinometric unit* – spectral distribution of the number of photons $\phi_{p,\lambda}$, measured in the number of photons per nm. The subscript 'p' means 'photon'.

There is a distinction between spectral *radiometric* quantities, measured in watts, and spectral *actinometric* quantities, measured in number of photons. The energy of a photon $U_p = h\nu = h(c/\lambda)$ is defined by its frequency ν, as well as by the wavelength λ; therefore, the spectral power distribution can be written as the spectral distribution of the number of photons and vice versa: $\phi_{e,\lambda} = U_p\,\phi_{p,\lambda}$. The spectral specification of the radiation is generally used in spectrophotometric analysis and the actinometric specification is required by the absorption process of the molecules of the light sensitive cells of the eye, which is a quantum photochemical process. The activation of the colour vision system is defined by the actinometric spectral specification of the electromagnetic radiation entering the eye and absobed by the photosensitive molecules (Rushton's univariance principle, Sections 6.12 and 6.13.1).

All the actinometric quantities are analogous to the spectral radiometric ones with the substitution of $\phi_{e,\lambda}$ with $\phi_{p,\lambda}$ (the subscript 'p', which means 'photon', substitutes 'e') and of the word 'energy' with 'photon number'.

Consider the spectral radiometric quantities.

The luminous flux must be considered distinctly out of the eye and in the eye. The following quantities are considered especially outside the eye:

- the spectral density of the flux through a surface, named *spectral exitance* if emitted and *spectral irradiance* if incoming, measured in W/(m² nm)

"Spectral irradiance [$E_{e,\lambda}$; E_λ] / spectral radiant exitance [$M_{e,\lambda}$; M_λ] (at a point of a surface) – quotient of the spectral radiant flux, $d\phi_{e,\lambda}$, incident on/leaving an element of the surface containing the point, by the area, dA, of that element"[1]

$$E_{e,\lambda} = \frac{d\phi_{e,\lambda}}{dA} \quad / \quad M_{e,\lambda} = \frac{d\phi_{e,\lambda}}{dA} \quad W\,/\,(m^2 nm) \tag{2.2}$$

- the spectral density of the flow within a solid angle Ω whose vertex is considerable as the point at which the light is generated or concentrates, named *radiant intensity* and measured in W/(sterad nm)

"Spectral radiant intensity (of a source, in a given direction) [$I_{e,\lambda}$; I_λ] – quotient of the spectral radiant flux, $d\phi_{e,\lambda}$, leaving the source and propagated in the element of solid angle, $d\Omega$, containing the given direction, by the element of solid angle Ω"[1]

$$I_{e,\lambda} = \frac{d\phi_{e,\lambda}}{d\Omega} \quad W\,/\,(sterad\ nm) \tag{2.3}$$

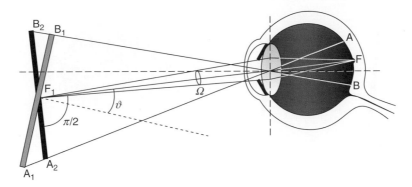

Figure 2.3 *Section of the ocular bulb and geometrical aspects of the radiation in relation with the definition of radiance. In this case, the apparent surface is shown.*

Inside the eye the activation of the colour vision phenomenon is produced by the spectral irradiance on the surface of the fundus of the eye on which an image is formed. This irradiance has its origin in the object observed and is limited by the opening of the pupil of the eye. For this purpose, consider the eye as a spherical optical chamber in which the various media, constituting the eye, with their refractive indices and their shapes, combine to build on the fundus of the eye an image of the scene in front of the eye (Section 5.2).

The spectral power distribution of the radiant flux $\phi_{e,\lambda}$ [W] that crosses the retina is measured by considering the path of light rays from an observed object point to their image point on the retina. Suppose the observer focuses on a region with a section A_1B_1 (Figure 2.3). Each point F_1 of this region can be considered as the source of rays that go in different directions and of these only those belonging to a solid angle Ω, having F_1F as axis, enter the eye and contribute to the image formation. The radiant flux must be related to the apparent section A_2B_2 orthogonal to F_1F axis, whose area, compared to that of the section A_1B_1, is area(A_2B_2) = area(A_1B_1) $\cos\vartheta$, where ϑ is the angle between the normals to the two sections and crossing F_1.

The radiant flux that crosses the area with section AB is obtained by integrating the radiance $L_{e,\lambda}$ on the solid angle Ω and on the apparent surface with section A_2B_2.

"*Spectral radiance* (for a wavelength interval dλ, in a given direction, at a given point of a real or imaginary surface) [$L_{e,\lambda}$; L_λ] – ratio of the spectral radiant power, d$\phi_{e,\lambda}$, passing through that point and propagating within the solid angle, dΩ, in the given direction, to the product of the area of a section of that beam on a plane perpendicular to this direction (dA $\cos\vartheta$) containing the given point and the solid angle dΩ"[1]

$$L_{e,\lambda} = \frac{d^2\phi_{e,\lambda}}{d\Omega\ dA\cos\vartheta} \quad \text{W} / (\text{sterad m}^2\text{nm}) \tag{2.4}$$

It is remarkable that the flux does not change with the distance d of the observed object A_1B_1 if its points have equal radiance (Figure 2.4). In fact, as the distance d changes, the area of section A_1B_1, corresponding to the image of section AB, varies proportionally to d^2, and the solid angle Ω varies proportionally to d^{-2}, resulting in no flux variation in AB. This implies that the *objects, both illuminated bodies and light sources, with the same spectral radiance, appear equal, and so independent of the distance.*

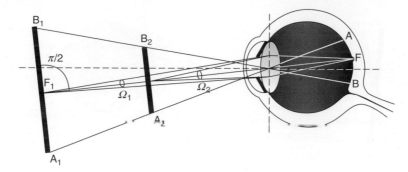

Figure 2.4   *Section of the ocular bulb and geometrical aspects of the radiation in relation with the definition of radiance. In this case, the solid angles and the distances of the observed objects is considered.*

This property favours the radiance compared to all other radiometric quantities, for which the various colorimetric quantities are calculated starting from the spectral radiance $L_{e,\lambda}$.

Furthermore, the relative spectral radiance $E_E(\lambda) \equiv L_e(\lambda) = 1$, due to the equal-energy source, called *equal-energy radiance*, has an important conventional role in colorimetry (Sections 6.13.1 and 8.2).

All the radiometric (and actinometric) quantities defined above differ from each other by a factor that takes into account the different geometries. Therefore, if the interest is for relative quantities, as almost always happens, it is enough to consider the relative spectral power distribution.

Conventionally, the spectral radiant flux is measured externally to the eye, on the corneal plane, where the light enters the eye.

2.5   Inverse Square Law

The *inverse square law*, also known as the *basic law of radiometry*, defines the relationship between the irradiance E_e produced by a point light source, viewed at a distance such that its dimensions are negligible, and the same distance. The law states that the irradiance E_e of the luminous flux through the element of area dA is proportional to the intensity and radiant I_e and varies inversely proportional to the square of the distance r:

$$E_e = \frac{I_e}{r^2} \tag{2.5}$$

This law, already advanced by Kepler in 1604, may be regarded as contained in the definitions of irradiance $E_e = d\phi_e/dA$ and radiant intensity $I_e = d\phi_e/d\Omega$, since $dA = r^2\, d\Omega$, and this can be checked with a simple substitution (Figure 2.5).

2.6   Photometric Quantities

The various monochromatic radiations have a different effect on the luminous sensation, which is quantified by the *relative luminous efficiency function* $V(\lambda)$ (Section 6.5 and Chapter 7). The luminous effect of a radiance, called *luminance* and denoted with L_v, is obtained by multiplying the spectral distribution of radiance by the luminous efficiency function and integrating on the wavelengths of the visible spectrum:

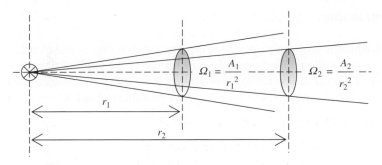

Figure 2.5   *Equal surface elements A_1 and A_2, located, respectively, at a distance r_1 and r_2 from a point source, underlying solid angles Ω_1 and Ω_2, which are inversely proportional to the square of the distance. Since the radiant intensity is uniform within these solid angles, the flow across the elements of the surface A_1 and A_2 is proportional to the subtended solid angles and therefore varies with the distance r in the same way to the change of the solid angle, that is, inversely proportional to the square of the distance r.*

$$\text{luminance } L_v = K_m \int\limits_{\substack{\text{visible} \\ \text{spectrum}}} V(\lambda) L_{e,\lambda}\, d\lambda \quad \text{cd/m}^2 \quad \text{with } K_m = 683 \text{ lm/W} \tag{2.6}$$

Analogously are defined all the other photometric quantities, directly related to corresponding radiometric quantities:

$$\text{luminous flux } \phi_v = K_m \int\limits_{\substack{\text{visiblel} \\ \text{spectrum}}} V(\lambda) \phi_{e,\lambda}\, d\lambda \text{ lumen} \tag{2.7}$$

$$\text{luminous exitance } \quad M_v = K_m \int\limits_{\substack{\text{visible} \\ \text{spectrum}}} V(\lambda) M_{e,\lambda}\, d\lambda \text{ lumen/m}^2 \tag{2.8}$$

$$\text{illuminance } \quad E_v = K_m \int\limits_{\substack{\text{visible} \\ \text{spectrum}}} V(\lambda) E_{e,\lambda}\, d\lambda \text{ lux, lumen/m}^2 \tag{2.9}$$

$$\text{luminous intensity } \quad I_v = K_m \int\limits_{\substack{\text{visible} \\ \text{spectrum}}} V(\lambda) I_{e,\lambda}\, d\lambda \text{ candela} \tag{2.10}$$

These photometric quantities are defined for the photopic vision, that is, colour vision (Sections 6.5 and 6.13.6). All the photometric quantities are defined in a more accomplished way in Chapter 7.

2.7   Retinal Illumination

Radiometric and photometric quantities are defined and measured on the corneal plane, where the light enters the eye. The illumination of the retina depends on the aperture of the pupil of the eye, and then the *troland* (named after Leonard T. Troland) denoted by Td is defined as the unit for measuring the *conventional retinal illuminance*. Typically, the troland refers to the ordinary or photopic troland, which is defined in terms of the photopic luminance:

$$I_v = L_v \times p \text{ Td} \tag{2.11}$$

where L_v is the photopic luminance in cd/m^2 and p is the pupil area in mm^2.

The *scotopic troland* is defined analogously.

References

1. CIE Publication S 017/E:2011, *ILV: International Lighting Vocabulary*, Commission Internationale de l'Éclairage, Vienna (2011). Available at: www.eilv.cie.co.at/ (accessed on 3 June 2015).
2. CIE Publication 15:2004, *Colorimetry*, 3rd edn, Commission Internationale de l'Éclairage, Vienna (2004).
3. CIE Supplement No. 2 to publication No. 15, *Colorimetry, Recommendations on Uniform Color Spaces, Color Difference Equations, Psychometric Color Terms*, Commission Internationale de l'Éclairage, Vienna (1978).

Bibliography

Bass M and DeCusatis C, *Handbook of Optics*, 3rd edn, OSA-MacGraw-Hill Companies, Inc. (2010).
Born M and Wolf E, *Principles of Optics: Electromagnetic Theory of Propagation, Interference and Diffraction of Light*, Cambridge University Press, Cambridge (1999).
Hecht E, *Optics*, Addison-Wesley (2002).

3

Colour and Light-Matter Interaction

3.1 Introduction

In this chapter, we apply the general principles of optics and those models related to the interaction between light and matter capable of justifying the colour and surface appearance of non-luminous bodies and the general characteristics of light-emitting bodies.

According to the CIE, the light radiation that enters the eye and activates the process of colour vision has several origins and their colours are classified as luminous colours and non-luminous colours.

3.1.1 Luminous Colours

> *"Luminous colour* – colour perceived to belong to an area that appears to be emitting light as a *primary light source* or that appears to be specularly reflecting such light.
> NOTE – Primary light sources seen in their natural surroundings normally exhibit the appearance of luminous colours in this sense."[1]

> *"Primary light source* – surface or object emitting light by a transformation of energy."[1]

The light *emission*, as an elementary event, consists of the creation of photons by atoms, molecules or crystals, which, being in an energetically excited state, de-energize. This phenomenon must satisfy the selection rules typical of the atom or of the molecule. For example, one of the selection rules is the global conservation of energy. Since every atom or molecule has states with defined energy, the emitted photons have defined energies.

The photon emission may be spontaneous or induced by another photon of suitable energy crossing the exited system. The LASER sources, that is, Light Amplification by Stimulated Emission of Radiation, are based on the phenomenon of induced emission.

Standard Colorimetry: Definitions, Algorithms and Software, First Edition. Claudio Oleari.
© 2016 John Wiley & Sons, Ltd. Published 2016 by John Wiley & Sons, Ltd.

3.1.2   Non-luminous Colours

"Non-luminous colour – colour perceived to belong to an area that appears to be transmitting or diffusely reflecting light as a *secondary light source.*
NOTE – Secondary light sources seen in their natural surroundings normally exhibit the appearance of non-luminous colours in this sense."[1]

"Secondary light source – surface or object which is not self-emitting but receives light and re-directs it, at least in part, by reflection or transmission."[1]

The non-luminous bodies are those that do not emit light and can only be seen if illuminated. It is the interaction between the electromagnetic radiation illuminating bodies that makes the bodies visible and classified according to the colour and geometric configuration of the light.

Traditionally, non-luminous colours are called

- *chemical colours* if caused by the colorant's absorption. The *colorants* can be schematically classified with some exceptions into
  1. *dyes* – substances which dissolve in the medium or react chemically with this constituting a homogeneous body that does not create light diffusion and
  2. *pigments* – which mixed in a medium, do not dissolve in them, and constitute a body optically inhomogeneous and therefore are light scatterers;

- *structural colours* if produced by other elemental optical processes such as reflection, refraction, interference, diffraction and scattering.[2] Particularly, structural colours are called
  1. *prismatic colours* if caused by refraction of light through a prism and
  2. *interference colour* if caused by interference phenomena.

Generally, *structural colour* is produced by elemental optical processes, which are present in microscopically structured surfaces, sometimes also called *schemochromes*:

"Schemochrome – any one of many colourless, sub-microscopic structures in organisms that serve as a source of colour by the manner in which they reflect light. Among those physical structures in organisms that fractionate light into its colour components are ridges, striations, facets, successive layers and multiple fine, randomly dispersed light-scattering bodies."[3]

3.1.3   Light Phenomena and Body Appearance

Among all the phenomena regarding the interaction of electromagnetic radiation with matter, only those with a role in colour science are considered. Limited to these phenomena, the interaction between light and matter can be classified as follows:

- Pure change in the direction of the light, as in the phenomena of
  1. reflection;
  2. refraction; and
  3. elastic diffusion.

- *Absorption* of electromagnetic radiation, which is done by subtracting part of the illuminating radiation in a wavelength selective way (the energy subtracted in the absorption is re-emitted generally in a non-radiative way, i.e. in a non-visible form).

- *Photoluminescence*, a phenomenon which consists of the excitement due to absorption of radiation belonging to a certain range of wavelengths, in general ultraviolet radiation, followed by subsequent emission of visible radiation, that is, of a longer wavelength. Photoluminescence is distinguished in *fluorescence* and *phosphorescence*. In colorimetry, fluorescence has a much more significant role.

- *Interference* and *diffraction* phenomena due to the wave propagation of light; in nature these phenomena are responsible for the colour of insects, of some feathers, fish scales, shells, so on, which are often iridescent.

The colour of objects is in relation to the spectral power distribution of the light, and the shape of objects is in relation to the geometric configuration of the light. This leads us to classify bodies as follows:

- *Transparent* bodies are those that can be traversed by light without diffusion (Section 3.4).

- *Translucent* bodies are those that allow some light to pass through (Section 3.5).

- *Opaque* bodies are those that transmit no light, and therefore reflect, scatter or absorb all of it (Section 3.5).

- *Gonio-apparent* are those bodies that are often iridescent and their colour changes as the direction of illumination and the observation point change.

3.2 Light Sources

The description of the physical principles on which the various light sources are based, although very interesting, is not considered here but an essential classification is given. The spectral distribution of the electromagnetic radiation emitted by each body depends on the nature of the body and on the temperature and is distinguished into two contributions:

1. Optical radiation emitted by *incandescence*, which is dependent only on the temperature; it is called *thermal radiation* or *blackbody radiation* and is independent of the nature of the emitting body (Section 3.3).

2. Optical radiation emitted by *luminescence*, which is caused by any non-thermal process, that is, by transition between electronic states of different defined energy in atoms, molecules, garnets, so on. These transitions with emission of light quanta happen after an excitation, which is produced in different ways and depends on the nature of the emitting body. Schematically, luminescence is divided into

- *Photoluminescence* if the excitation that precedes the emission is due to the absorption of electro-magnetic radiation; there is
  (a) *fluorescence* if the luminescence stands out with the cessation of illumination and
  (b) *phosphorescence* if the luminescence persists, with decreasing intensity, as well as the excitation ceases.

- *Electroluminescence* if the luminescence occurs by passage of current, for example, light-emitting diodes (LEDs).

- *Cathodoluminescence* if the luminescence follows an excitation produced by a collision with electrons, for example, cathode ray tubes (CRTs).

- *Chemiluminescence* if the luminescence occurs by chemical transformation, as in flames.

- *Gas-discharge luminescence* if the excitation is generated by the passage of an electrical current in a gas, where gas atoms and molecules are excited by collision with electrons and re-emit the excitation energy as photons.

The electric currents existing in the tungsten filament of a classic incandescent bulb, in an LED or in a gas discharge lead to an emission of light, which is based on different physical phenomena not considered here. A description of the light sources and illuminants for colorimetry is given in Chapter 8.

3.3   Planckian Radiator

"*Planckian radiator* (or *blackbody* or *full radiator*) – an ideal body that absorbs completely all incident electromagnetic radiation, whatever the wavelength, the polarization and the direction of incidence."[1]

If the body is in thermal equilibrium with its environment, that is, at equal constant temperature, it has four notable properties:

1. It is an ideal emitter that emits as much or more energy at any wavelength than any other body at the same temperature.
2. It is a diffuse isotropic emitter.
3. It emits electromagnetic radiation according to Planck's law, meaning that it has a *spectral exitance* that is determined by the temperature alone. The *Planck formula* is

$$M_{e,T,\lambda} = \frac{c_1}{\lambda^5} \frac{1}{\exp\left(\dfrac{c_2}{\lambda T}\right) - 1} \quad \text{W m}^{-3}, \tag{3.1}$$

where the wavelength λ is expressed in metres, the temperature T in kelvin and the constants $c_1 = 2\pi hc^2$ $= 3.741771 \times 10^{-16}$ [W m^2], $c_2 = h\,c/k. = 1.4388 \times 10^{-2}$ mK, c is the velocity of light in the vacuum, h is Planck's constant, $n \approx 1.00028$ the refractive index of standard air, k the Boltzmann constant and 'exp' is the exponential function with the Neperian base $e = 2.718282$ (Figure 3.1). The radiance corresponding to the exitance is $L_{e,T,\lambda} = M_{e,T,\lambda}/\pi$ W/(m^{3}sr) (3.5).

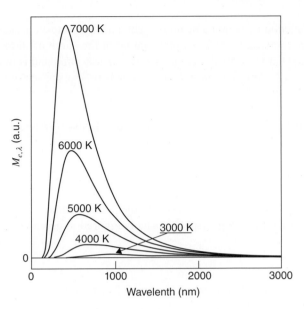

Figure 3.1   *Absolute spectral distribution of radiant power on the blackbody described by the Planck formula (3.1) at different absolute temperatures.*

4. The exitance described by the Planck's formula is the result of a theoretical model that is at the core of the quantum physics. This property makes blackbody radiation a reference in radiometry and then in photometry and colorimetry.

> *Grey body* is a body that has emissivity with the same relative spectral distribution as a blackbody at the same temperature but in a smaller amount.

The best practical realization of a body having blackbody emission is a cavity, the walls of which are maintained at a uniform temperature T. The radiation is emitted by a point-like hole – that is, with a negligible section compared to the size of the cavity – and communicating with the cavity has the spectral distribution of blackbody radiation at temperature T. Any light entering the hole is reflected indefinitely or absorbed inside and is unlikely to re-emerge, making the hole a nearly perfect absorber.

The radiation emitted by a body due to only thermal effect becomes visible as the temperature increases: first, the body appears red (~1000 K), then yellow (~2000 K), white (~6000 K) and finally blue (Section 8.3).

3.4   Light Regular Reflection and Refraction

Usually the reflection and refraction phenomena are represented in the approximation of geometrical optics (that is supposed to be known to the reader), in which light radiation is represented by light rays. These phenomena are shown here because they have importance in colour science and have a close connection with Fresnel's laws presented in Section 3.4.2.

Reflection is the phenomenon by which a beam of light, encountering a surface separating two media with different refractive indexes, is reflected propagating in the same medium from which it comes. The reflection considered here is termed *regular* or *specular reflection* and is that which is observed on glossy surfaces like glass. *Diffuse reflection*, which depends on the surface finish of materials, will be treated later on (Sections 3.5 and 3.9).

Refraction is the deviation of the path of a light beam crossing the surface separating two media with different refractive indexes.

These two phenomena are regulated by two sets of laws:

1. Snell's laws that describe the geometrical aspects of the phenomenon.
2. Fresnel's laws that describe the energy aspects (usually ignored in geometrical optics).

3.4.1   Snell's Laws

The three laws of Snell are given with reference to Figure 3.2 as follows:

1.   Directions of incidence, reflection and refraction all lie on a plane normal to the surface of separation between the media, the *incidence plane*.

2.   The angle of reflection ϑ_i' equals the angle of incidence, $\vartheta_i' = \vartheta_i$.

3.   The ratio between the sine of the angle of incidence ϑ_i and the sine of the angle of refraction ϑ_r in the transition from a material with refractive index n_1 to another with refractive index n_2 is constant $n_{21} = \sin(\vartheta_i)/\sin(\vartheta_r)$, where $n_{21} = n_2/n_1$ is said to be the *relative refractive index* between the two media.

Note that the reflection, at least for the geometrical aspects, does not depend on the two materials, that is, n_{21}; instead the refraction angle depends on n_{21} and we can distinguish between two situations:

*   passage from a medium with refractive index n_1 to a medium with a higher refractive index n_2 ($n_{21} > 1$) as, for example, in air-to-glass transition: in this case always $\vartheta_r \le \vartheta_i$ and

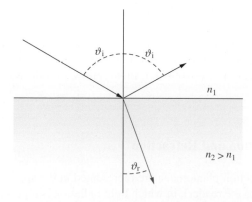

Figure 3.2   *Two media are characterized by two different refractive indices $n_2 > n_1$. A ray refracted, passing from a medium with a lower refractive index to a medium with a higher refractive index, and a reflected ray are represented.*

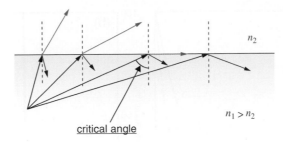

Figure 3.3   *Two media are characterized by two difference refractive indices $n_1 > n_2$. Reflection, refraction and critical angle are represented.*

- passage from a medium with a higher refractive index to a medium with a lower refractive index ($n_{21} < 1$): in this case always $\vartheta_r \geq \vartheta_i$.

In the second case, when $\sin(\vartheta_i) > n_{21}$, the third Snell law implies that $\sin(\vartheta_r) > 1$. This is impossible and therefore there is no refraction but only reflection (Figure 3.3). This phenomenon is called *total internal reflection*. The angle of incidence for which $\sin(\vartheta_i) = n_{21}$ has been termed *critical angle ϑ_c*.

As it is known, the phenomena of reflection and refraction exist even for metals, conductors of electricity, albeit with very different characteristics from the dielectrics. The metals are in fact opaque (unless they are very thin sheets as in the case of certain coatings for optical systems) and, if suitably smoothed, reflect light to a much greater extent than the dielectric. The high reflectance is due to the fact that the electrons, free to move within the metal, after absorbing a radiation, re-emit it following the laws of reflection.

Differently, the electrons in dielectrics are not free to move and when they absorb photons, they can re-emit photons of equal energy, but without following the laws of reflection, or re-emit photons with lower energy (*non-radiative* emission) and then to the observer the absorbed photons disappear definitely.

An *ideal (perfect) mirror* is a body that transmits or absorbs no light and reflects all the light according to the Snell reflection law.

3.4.2   Fresnel's Laws

The energy aspects of regular reflection and refraction are treated by the Fresnel laws expressing the relationship between the flux – energy in the unit of time – of the radiation reflected or refracted and the flux of the incident radiation as a function of the ratio $n_1(\lambda)/n_2(\lambda)$. There are several Fresnel laws, depending on the different situations of light polarization (Figure 3.4) and regarding the reflectance:

"*Reflectance ρ for an incident radiation of a given spectral composition, polarization and geometrical distribution, is defined as the ratio of the reflected radiant flux to the incident flux in the given conditions.*"[1]

If the light is polarized perpendicular to the incidence plane, the *reflectance $\rho_\perp(\lambda)$* is

$$\rho_\perp(\lambda) = \left[\frac{\sin(\vartheta_i - \vartheta_r)}{\sin(\vartheta_i + \vartheta_r)} \right]^2 \tag{3.2}$$

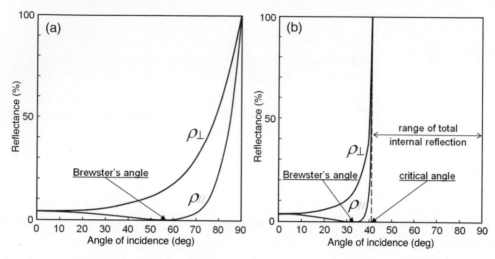

Figure 3.4   *Fresnel's reflectances, $\rho_\perp$ (black line) and $\rho_{//}$ (red line), represented as a function of the incidence angle ϑ_i. Figure (a) regards the case without a critical angle and figure (b) the case with a critical angle. The reflectance $\rho_{//}$ assumes the zero value at the Brewster angle, for which the reflected radiation can only be fully polarized with the electric field perpendicular to the incidence plane.*

If the polarization is parallel to the incidence plane, the reflectance $\rho_{//}(\lambda)$ is

$$\rho_{//}(\lambda) = \left[\frac{\tan(\theta_i - \theta_r)}{\tan(\theta_i + \theta_r)} \right]^2 \tag{3.3}$$

The dependence of the reflectance on the wavelength is obtainable by applying the Snell law and expressing the refraction angle ϑ_r as a function of the angle of incidence ϑ_i and of the refractive indices $n_1(\lambda)$ and $n_2(\lambda)$.

The reflectance $\rho_\perp(\lambda)$ monotonically increases with the incidence angle ϑ_i, while $\rho_{//}(\lambda)$ is equal to zero when $\tan(\vartheta_i) = n_2(\lambda)/n_1(\lambda)$. This angle is called the *Brewster angle* and is such that the reflected radiation has the polarization plane orthogonal to the incidence plane.

If the incident beam is not polarized, the reflectance is given by the average value of the two reflectances $\rho_\perp(\lambda)$ and $\rho_{//}(\lambda)$.

A very important case in colorimetry is when the angle of incidence is close to zero and the reflectance in this case is independent of the polarization and is

$$\rho(\lambda) = \frac{\left(n_2(\lambda) - n_1(\lambda)\right)^2}{\left(n_2(\lambda) + n_1(\lambda)\right)^2} \tag{3.4}$$

3.5   Light Scattering

"(Elastic) *Scattering* or *diffusion* – process by which the spatial distribution of a beam of radiation is changed when it is deviated in many directions by a surface or by a medium, without a change in frequency of its monochromatic components.

NOTE 1 – A distinction is made between

- selective scattering and
- non-selective scattering

according to whether or not the scattering properties vary with the wavelength of the incident radiation.
 NOTE 2 – The frequency is unchanged only if there is no Doppler effect due to the motion of the materials from which the radiation is returned."[1]

The illuminated material scattering light is as if it emits its own light. This emission is improper and the phenomenon is also called *improper luminescence*.

3.5.1 Lambertian Diffusion

A *Lambertian body*, or *uniform reflecting diffuser*, is an ideal body that exhibits *Lambertian reflectance* and obeys *Lambert's cosine law*, that is, reflects incident rays with an intensity proportional to the cosine of the reflection angle or, equivalently, reflects equal radiance in all directions.

This means that the Lambertian body appears to an observer with the same brightness (or luminance), independent of the viewing position.
 Light sources emitting equal radiance in all directions are Lambertian. The blackbody is a perfect Lambertian radiator.
 A surface emitting in a Lambertian way is obtainable as a port of an integration sphere. Since this kind of light emission is very important in colorimetric instrumentation (Section 12.2.2), the integration sphere is considered in a separate section (Section 3.10).

A Lambertian body that reflects all the incident light is a *perfect reflecting diffuser*. This body could be considered as a *white body*.

The perfect reflecting diffuser lit by an illuminance E_v (lm/m^2) has a luminous exitance M_v (lm/m^2) that satisfies the following relation with the reflected luminance L_v (cd/m^2):

$$M_v = E_v = \pi L_v \tag{3.5}$$

A uniform reflecting diffuser with a reflectance independent of the wavelength $\rho < 1$, lit by an illuminance E_v, has a luminous exitance M_v related to the emitted luminance L_v by the following relations:

$$M_v = \rho E_v, \; L_v = \frac{\rho E_v}{\pi} \tag{3.6}$$

3.5.2 Light Scattering on a Rough Surface

Refraction and reflection are phenomena occurring when the light crosses an optical discontinuity at the surface separating two media with different refractive indices.

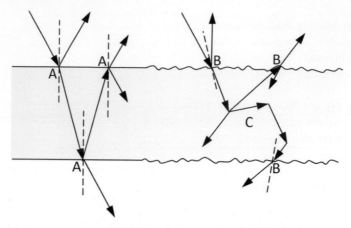

Figure 3.5   *Sketch of the path of a ray of light in interaction with a body: (A) regular or specular reflection, refraction and regular transmission according to Snell's laws on a smooth surface (Section 3.4); (B) diffuse reflection and diffuse transmission on a rough surface (Section 3.6); (C) internal diffusion due to optical heterogeneity; inside the body there may also be absorption and fluorescence (Section 3.5).*

Light scattering is present when the region, where the optical discontinuity exists, has a size comparable with the wavelength of the light, and light scattering's behaviour depends on the discontinuity size.

A *rough* surface can be considered as a sequence of micrometric elements of differently oriented flat surfaces, which produce reflection and refraction. A smooth surface has *regular* or *specular reflection*, and a rough surface has *diffuse reflection* (Figure 3.5). The surface appearance of the body depends on the roughness of the surface and can be described by the laws mentioned above. This quality is expressed by the *gloss* (Section 13.6). It depends on the amount of flux of radiation specularly reflected from the surface, which, in turn, depends

- on the smoothness of the surface and
- on the refractive indices of the two media.

There are standard methods for gloss measurement (Section 13.6).

The light scattered from the surface roughness and the light from optical heterogeneities inside the medium are not distinguishable. The global diffused light and that specularly reflected by illuminating in a directional way are distinguishable.

In analogy with specular and diffuse reflection, there is a regular, or specular, transmission and diffuse transmission, depending on whether or not there is any diffusion (Figure 3.5). The transmission depends on both the optical properties of the surface of separation between the media and the internal ones.

3.5.3   Light Scattering in an Optically Heterogeneous Medium

The optical heterogeneity of a medium causes diffusion for light moving through the medium.

A medium is optically heterogeneous when its refractive index changes from point to point, and this has various origins, as already said:

- The presence of particles dispersed in the medium, for example, pigment's granules, and then the body is chemically and optically heterogeneous.

- Local fluctuations of density of a chemically homogeneous medium due, for example, to thermal agitation.
- Electrical anisotropy of the molecules – in this case the diffusion of the light is called *molecular diffusion* –.

The laws of diffusion vary with the size of the optical heterogeneity of the media. If the optical heterogeneity has dimensions not exceeding a few tenths (~3/10) of the wavelength λ – in the case of visible light around 150 nm – diffusion occurs according to the following laws:

1. *Rayleigh's scattering* law – *Tyndall effect*, for which the fraction of the flux of the scattered radiation is proportional to λ^{-4}.
2. The angular distribution of the light intensity, expressed as a function of the angle ϑ between the incident and diffused rays, is proportional to $(1 + \cos^2\vartheta)$ (its graphical representation given in Figure 3.6 is called *scattering indicatrix*).
3. The scattered light is partially polarized and for $\vartheta = (\pi/2)$ it is entirely polarized on the plane containing the incident and diffuse beams.

In this case, the law of diffusion is summarized by the equation

$$I_e(\lambda,\vartheta) = \sigma I_{e,0}(\lambda)\lambda^{-4}(1 + \cos^2\vartheta) \ \text{W/sr} \tag{3.7}$$

where $I_e(\lambda,\vartheta)$ is the intensity of scattered light in the direction ϑ, $I_{e,0}(\lambda)$ is the intensity of the incident light and σ is a proportionality constant that depends on the particular physical system.

With the increasing size of the heterogeneity, the intensity of the scattered light decreases, the dependence with respect to the wavelength becomes λ^{-r} with $r < 4$, the forward scattering, that is for $\vartheta < (\pi/2)$, increases more and more and similarly the backward scattering for $\vartheta > (\pi/2)$ decreases and the polarization of light decreases.

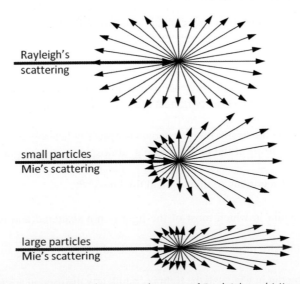

Figure 3.6   *Scattering indicatrix in the case of Rayleigh and Mie scattering.*

An example of Rayleigh's scattering is what happens in the earth's atmosphere. Light diffusion in air over short distances is barely appreciated, but for great distances the phenomenon is evident. The molecules (CO_2) and dust particles in the air, subjected to density fluctuations on a scale of distances much less than that of the wavelengths of visible radiation, diffuse the light in a selective manner in wavelength, giving the effect of the blue colour of the sky, the yellow of the sun at noon and red at sunset.

In 1908, G. Mie gave a mathematically rigorous and complete solution of the problem of the scattering of an electromagnetic wave on a sphere.[4] This kind of scattering is termed *Mie scattering*. The Mie-scattering model is valid for scatterers of any size and, in the limit in which these are much smaller than the incident wavelength, it regains the Rayleigh scattering. For this reason Mie scattering is used in the optical study of colloids and in meteorology; in fact the water droplets of clouds often have a greater or much greater size than the wavelength of visible light. The light scattered due to clouds is very variable and colour is from white to black. White means high backscattering, and black means almost complete transmittance.

3.6   Light Absorption and Colour Synthesis

Light absorption is the reverse phenomenon of light emission.

A body, after absorbing a photon and then being excited, can emit photons. If the emitted photons are visible, the emission is called *radiative*. In this case, if the energy of the emitted photon is equal to that of the absorbed photon, the process is called *elastic*; otherwise, according to *Stokes' law*, the process is termed *fluorescence* (Section 3.7).

The fluorescence effect is not to be confused with *Raman's effect*, which has no meaningful role in colorimetry.

3.6.1   Simple Subtractive Synthesis

Colouring is produced by simple subtractive synthesis if the light does not change its direction inside a coloured and transparent medium, and the colouring is due to wavelength selective absorption according to the Lambert–Bouguer and Beer laws (Section 3.8). Absorption is in general due to the presence of dyes and the body is optically homogeneous.

3.6.2   Complex Subtractive Synthesis

If the medium is coloured and translucent, wavelength-selective diffusion and absorption phenomena occur inside it simultaneously. As regards diffusion, schematically, two extreme cases exist, but all the intermediate cases are possible:

1.   The *turbid* medium is rendered inhomogeneous and optically discontinuous by a dispersion of particles of optically different materials, pigments (e.g. paints, fumes, emulsions) or gas (e.g. paper, natural textile fibres, foams).
2.   The medium, although chemically homogeneous, is optically heterogeneous due to density variations or electrical anisotropy of the molecules (e.g. atmosphere, certain plastics).

Turbid media are classified according to their optical thickness[5–6]:

- *Optically thin* are media in which most of the light is not scattered and has undergone a single diffusion event.
- *Optically intermediate* are media in which the light is almost completely diffused in a set of multiple scattering.
- *Optically thick* are media in which all the light is completely diffused in a set of multiple scattering.

Colouring is produced by *complex subtractive synthesis* when the phenomena, present in the case of the coloured opaque medium, are the same as for the coloured translucent medium but occur in an amount that does not depend on the thickness of the medium. This colouring is considered in Section 3.9 according to the Kubelka-Munk model.

3.7   Fluorescence

As stated before,

Fluorescence is a photoluminescence process, whereby materials absorb radiant power at one wavelength – *excitation* – and immediately, after within the time range of 0.5–20 ns, re-emit it in a *radiative* way, that is, non-thermally, at another wavelength:

1. *Stokes' shift* is the wavelength difference between positions of the band maxima of the absorption and emission spectra of the electronic transition of a molecule or an atom (Figure 3.7).
2. *Anti-Stokes' shift* is the wavelength difference if the emitted photon has a shorter wavelength (i.e. more energy) than that of the absorbed one. This extra energy of the emitted photon comes from the dissipation of thermal phonons in a crystal lattice, cooling the crystal in the process.

The whole emission band is present for every monochromatic radiation absorbed belonging to the band of the illuminating light.

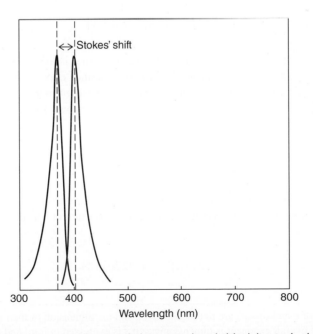

Figure 3.7   *Luminescence spectrum (red line) and absorption band (black line): the band of luminescence is independent of the monochromatic exciting radiation, provided that it belongs to the same absorption band. The absorption band is obtained as the envelope of the excitation lines.*

3.8   Transparent Media

Transparent medium – A medium in which light pass through without being scattered and has the property of transmitting rays of light in such a way that the human eye may see through the medium.

3.8.1   Internal Transmittance of a Medium

Consider electromagnetic radiation within an optically homogeneous medium, transparent and with dissolved dyes. The homogeneity consists in the centres of absorption due to the dye being uniformly distributed in the medium, without having any changes in the refractive index. The corpuscular representation, according to which the electromagnetic radiation is considered as a flux of photons, may be more effective to explain these laws, which are expressed by the following two elementary hypotheses:

1.   The generic photon, moving inside the medium, has a probability of colliding with a centre of absorption
   (a) proportional to the concentration c of the dye in the medium – *Beer's Law* – and
   (b) proportional to the element of path ds – *Bouguer-Lambert's Law* –.

2.   For each collision the photon has a probability of being absorbed, represented by the *absorption coefficient* $a(\lambda)$, which is typical of the dye and is a function of the photon wavelength λ.

Based on these assumptions, the probability that a photon is present after the path ds is reduced by the amount $dp = -ca(\lambda)\,ds$. This is a differential equation with the solution

$$p_\lambda(s) = p_\lambda(s = 0)\exp\left[-ca(\lambda)s\right] \tag{3.8}$$

where $p_\lambda(s = 0) = 1$ is the probability of having the initial photon, $p_\lambda(s)$ is the probability of finding the photon with wavelength λ at the point s without having been absorbed and 'exp' is the exponential function with Nepierian base $e = 2.71828$. Since electromagnetic radiation consists of many photons and the internal radiant flux $\Phi_{\downarrow i,\lambda}(s)$ at a defined wavelength associated with it is proportional to the number of photons, one can rewrite the relation (3.8) (subscript 'i' means 'internal') as

$$\Phi_{\downarrow i,\lambda}(s) = \Phi_{\downarrow i,\lambda}(s = 0)10^{-c\varepsilon(\lambda)s} \tag{3.9}$$

where $\Phi_{\downarrow i,\lambda}(s = 0)$ is the internal radiant flux at the beginning of the considered path. For practical reasons, logarithm with base 10 is considered, for which the absorption coefficient is replaced by *the linear extinction coefficient*:

$$\varepsilon(\lambda) = \log_{10}e\,a(\lambda) = 0.434229\,a(\lambda) \tag{3.10}$$

The units of measurement of c and of $\varepsilon(\lambda)$ are not independent, and the dimension of their product should be the reciprocal of that of s so that the overall exponent is dimensionless: if s is expressed in [cm] and the concentration in [mol/l], then $\varepsilon(\lambda)$ is expressed in [l/mol/cm], and again, as often happens, if the concentration is expressed in relative weight, $\varepsilon(\lambda)$ is expressed in [cm^{-1}].

The flux of a light beam within a transparent and coloured medium decreases exponentially and that medium of the thickness s is characterizable by one of the following three quantities derived from the laws of Bouguer-Lambert and of Beer:

- The *internal spectral transmittance* (Figure 3.8a):

$$\tau_i(\lambda,s) \equiv \frac{\Phi_{\downarrow i,\lambda}(s)}{\Phi_{\downarrow i,\lambda}(s=0)} = 10^{-c\varepsilon(\lambda)s} \qquad (3.11)$$

- The *internal optical density* or *absorbance* (Figure 3.8b):

$$\delta_i(\lambda,s) \equiv \log_{10}\left[\frac{1}{\tau_i(\lambda,s)}\right] = c\varepsilon(\lambda)s \qquad (3.12)$$

- The *logarithm of the internal optical density* (Figure 3.8c):

$$\log_{10}\delta_i(\lambda,s) = \log_{10}\varepsilon(\lambda) + \log_{10}(cs) \qquad (3.13)$$

which is a function of λ, and remains unchanged to vary the thickness and concentration, up to an additive constant, and therefore it is a characteristic function of the dye.

The total internal spectral transmittance and the total optical density, denoted with $\tau(\lambda)$ and $\delta(\lambda)$, respectively, if referring to a path through different media and characterized by the internal spectral transmittances $\tau_{i,1}(\lambda,s_1)$, $\tau_{i,2}(\lambda,s_2)$, $\tau_{i,3}(\lambda,s_3)$, …, or by optical density $\delta_{i,1}(\lambda,s_1)$, $\delta_{i,2}(\lambda,s_2)$, $\delta_{i,3}(\lambda,s_3)$, …, have the following properties:

$$\tau_i(\lambda,s) = \tau_{i,1}(\lambda,s_1)\tau_{i,2}(\lambda,s_2)\tau_{i,3}(\lambda,s_3)\cdots = \prod_j \tau_{i,j}(\lambda,s_j) \qquad (3.14)$$

$$\delta_i(\lambda,s) = \delta_{i,1}(\lambda,s_1) + \delta_{i,2}(\lambda,s_2) + \delta_{i,3}(\lambda,s_3)\cdots = \sum_j \delta_{i,j}(\lambda,s_j) = \sum_j c_j\varepsilon_j(\lambda)s_j \qquad (3.15)$$

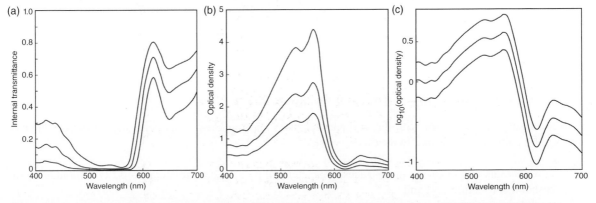

Figure 3.8 *(a) Internal Transmittance of three filters made with the same dye but with different thicknesses. (b) Internal optical densities of the three filters. (c) Logarithm of the internal optical densities, which reveals its modification in correspondence to the change of the thickness. The shape is invariant, independent of the concentration and thickness, as a characteristic property of the dye.*

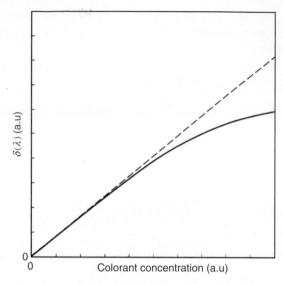

Figure 3.9 *Deviation of the actual behaviour (red line) from Beer's law (black line).*

These laws are also applicable to the media in which the simultaneous presence of more colorants exists distinguished by an index j, each of which is characterized by the linear absorption coefficient of $\varepsilon_j(\lambda)$ and with concentration c_j,

$$\Phi_{\downarrow i,\lambda}(s) = \Phi_{\downarrow i,\lambda}(s=0)10^{-\sum_j c_j \varepsilon_j(\lambda)s} \tag{3.16}$$

The Bouguer-Lambert law is considered exact, while Beer's law is sometimes not completely verified:

- The optical density grows below the linear growth with increasing concentration of the dye (Figure 3.9).
- The optical density may decrease with the increase beyond certain levels of light intensity.

The explanation of these phenomena is obtained by considering the molecular state of the solute as a function of concentration, but it, although known, is beyond the scope of this text.

3.8.2 Total Transmittance and Total Reflectance

Given a transparent medium with flat and parallel faces and thickness s, as generally an optical filter is, crossed by a flux perpendicular to the faces, consider

- the reflections that occur on the surfaces, characterized by a spectral reflectance $\rho_0(\lambda)$ computed by the Fresnel law (Section 3.4) and
- internal absorption to the medium, characterized by an internal transmittance $\tau_i(\lambda, s)$.

The phenomena of the total transmittance and total reflectance are represented in Figure 3.10 and are the sum of a series of elementary reflections, refractions and transmissions.

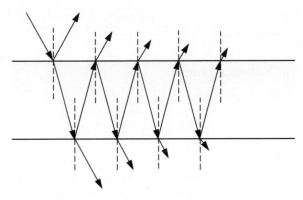

Figure 3.10   *Multiple reflections and refractions at the surfaces of a layer of material with refractive index different from that of the external medium, which contribute to the reflectance and the transmittance of the surface (the angle of incidence, which in reality is close to zero, is designed different from zero for a better graphical representation of the phenomenon).*

The *total spectral transmittance*, that is the inner plus the surface phenomenon, is

$$\tau(\lambda,s) = \frac{\Phi_{\downarrow,\lambda}(s)}{\Phi_{\downarrow,\lambda}(s=0)} = \frac{\tau_i(\lambda,s)\left[1-\rho_0(\lambda)\right]^2}{1-\left[\rho_0(\lambda)\tau_i(\lambda,s)\right]^2} \tag{3.17}$$

and the *total spectral reflectance* is

$$\rho(\lambda,s) = \frac{\Phi_{\uparrow,\lambda}(s=0)}{\Phi_{\downarrow,\lambda}(s=0)} = \rho_0(\lambda) + \frac{\rho_0(\lambda)\tau_i^2(\lambda,s)\left[1-\rho_0(\lambda)\right]^2}{1-\left[\rho_0(\lambda)\tau_i(\lambda,s)\right]^2} \tag{3.18}$$

The quantities $\tau(\lambda, \sigma)$ and $\rho(\lambda, \sigma)$ are the result of all the contributions due to the two surfaces of the filter (Figure 3.10):

$$\tau(\lambda,s) = \tau_i(\lambda,s)\left[1-\rho_0(\lambda)\right]^2 \left\{1+\left[\rho_0(\lambda)\tau_i(\lambda,s)\right]^2 + \left[\rho_0(\lambda)\tau_i(\lambda,s)\right]^4 + \cdots\right\}$$

$$\rho(\lambda,s) = \rho_0(\lambda) + \rho_0(\lambda)\left[1-\rho_0(\lambda)\right]^2 \tau_i^2(\lambda,s)\left\{1+\left[\rho_0(\lambda)\tau_i(\lambda,s)\right]^2 + \left[\rho_0(\lambda)\tau_i(\lambda,s)\right]^4 + \cdots\right\} \tag{3.19}$$

In the case of non-planar and rugged surfaces, the phenomenon is in general complex and must be analysed on a case-by-case basis by suitable models.

3.9   Turbid Media

The translucent and opaque media are turbid media and have internal optical heterogeneities.

The treatment of turbid media requires appropriate models depending on the optical thickness. Here the turbid media with high optical thickness are mainly considered, because of their greater interest in colorimetry.

Almost all paints can be considered optically thick turbid media. The model, which in this field has achieved the greatest success, is that of Kubelka and Munk.[7–10]

3.9.1   Two-Flux Model of Kubelka-Munk

Here an electromagnetic radiation is considered within a coloured, translucent or opaque medium, in interaction with the same medium, while the reflection due to the separation surface between the media is added later with the *Saunderson correction*, as it has already been done in the case of transparent bodies distinguishing the internal and surface transmittance and the reflectance.

The translucency or opacity is generated by the light diffusion, which may occur by internal optical discontinuities of the body due to the presence of pigment granules, or bubbles of air or the molecular complexity of the medium in which colorants are dissolved. Sometimes the turbid systems have abnormalities such as flocculation, sedimentation and the mutual alignment between the pigment particles, which are ignored here. Despite this simplification, the phenomenon is more complicated than the case considered by the law of Bouguer-Lambert and Beer, because the diffusion alters the direction of the light. A full description of the phenomenon (many flux theory) would have little practical value, because of its complexity. Here the simplified analysis advanced by Kubelka-Munk is proposed, according to which only the backward diffusion of light is considered, that is, in the direction opposite to that of the incident flux. This simplification is possible, because the size of the particles is small and produces turbidity, which have a section of approximately one order of magnitude higher than that of the wavelength of the visible radiation and have therefore diffusion prevailing along the direction of the incident ray and poorly selective in wavelength, far from the situation of Rayleigh's law. Referring to Figure 3.11, each elementary layer of thickness ds of the medium is seen as crossed by two fluxes, one descending $\Phi_{\downarrow i,\lambda}(s)$ and one ascending $\Phi_{\uparrow i,\lambda}(s)$ (subscript 'i' means 'internal'). For this simplification, the model is said to be a *two-flux model*. The laws of Bouguer-Lambert and Beer are generalized to this new situation:

- The generic photon moving inside the medium has a probability of colliding with a centre of absorption proportional to the concentration c of the dye or pigment in the medium and to the path element ds.

- The photon for each collision has a probability of being absorbed represented by the *absorption coefficient* $K(\lambda)$, which is typical of the dye or pigment and is a function of the photon wavelength λ.

- The photon for each collision has a probability of being backscattered, represented by the *scattering coefficient* $S(\lambda)$, which is typical of the pigment and is a function of the photon wavelength λ.

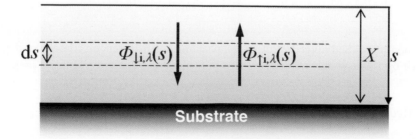

Figure 3.11   *Fluxes within a paint layer of thickness X crossing the elementary layer ds according to the model of Kubelka-Munk.*

The descending flux $\Phi_{\downarrow i,\lambda}(s)$, crossing the layer d$s$, has a variation d$\Phi_{\downarrow i,\lambda}(s)$ constituted by three distinct contributions:

1. a reduction due to absorption equal to $-K(\lambda)\,\Phi_{\downarrow i,\lambda}(s)\,\mathrm{d}s$;
2. a reduction due to backscattering equal to $-S(\lambda)\,\Phi_{\downarrow i,\lambda}(s)\,\mathrm{d}s$; and
3. an increment due to the backscattering of the ascending flux equal to $S(\lambda)\,\Phi_{\uparrow i,\lambda}(s)\,\mathrm{d}s$.

The total variation is

$$d\Phi_{\downarrow i,\lambda}(s) = -\left[K(\lambda) + S(\lambda)\right]\Phi_{\downarrow i,\lambda}(s)\mathrm{d}s + S(\lambda)\Phi_{\uparrow i,\lambda}(s)\mathrm{d}s \tag{3.20}$$

Analogously, the upward flux has a variation

$$-d\Phi_{\uparrow i,\lambda}(s) = -\left[K(\lambda) + S(\lambda)\right]\Phi_{\uparrow i,\lambda}(s)\mathrm{d}s + S(\lambda)\Phi_{\downarrow i,\lambda}(s)\mathrm{d}s \tag{3.21}$$

(The different signs in the two equations are due to the choice of measuring the flux with respect to the distance s, and in the second equation ds is negative.)

Together these relations constitute a set of two differential equations, the resolution of which is laborious. The magnitude of interest from the colorimetric point of view is the reflectance of the body and not the fluxes $\Phi_{\downarrow i,\lambda}(s)$ and $\Phi_{\uparrow i,\lambda}(s)$, of which the reflectance is a function. Here the solution is given for the situation of paintings and in general objects coated with a coloured layer. The *internal reflectance* is defined as the ratio between the reflected flux $\Phi_{\uparrow i,\lambda}(s=0)$ and the incident flux $\Phi_{\downarrow i,\lambda}(s=0)$ on the surface $s=0$, ignoring the Fresnel reflection on the surface of the coloured layer:

$$R_i(\lambda) = \frac{\Phi_{\uparrow i,\lambda}(s=0)}{\Phi_{\downarrow i,\lambda}(s=0)} \tag{3.22}$$

and depends

- on the thickness X of the coloured layer and
- on the reflectance $R_g(\lambda)$ of the surface of separation between the coloured layer and the substrate (subscript 'g' means 'ground').

The reflectance is obtained by the integration of equations (3.20) and (3.21) and is given by

$$R_i(\lambda) = \frac{(R_g - R_{i,\infty})/R_{i,\infty} - R_{i,\infty}(R_g - 1/R_{i,\infty})e^{SX(1/R_{i,\infty}-R_{i,\infty})}}{R_g - R_{i,\infty} - (R_g - 1/R_{i,\infty})e^{SX(1/R_{i,\infty}-R_{i,\infty})}} \tag{3.23}$$

where $R_{i,\infty}(\lambda)$ is the internal reflectance limit of the layer, which does not change for each increment of the thickness of the layer itself, and is independent of the thickness X and of the substrate reflectance $R_g(\lambda)$,

$$R_{i,\infty}(\lambda) = 1 + \frac{K(\lambda)}{S(\lambda)} - \left[\left(\frac{K(\lambda)}{S(\lambda)}\right)^2 + 2\frac{K(\lambda)}{S(\lambda)}\right]^{1/2} = \left\{1 + \frac{K(\lambda)}{S(\lambda)} + \left[\left(\frac{K(\lambda)}{S(\lambda)}\right)^2 + 2\frac{K(\lambda)}{S(\lambda)}\right]^{1/2}\right\}^{-1} \tag{3.24}$$

Equation (3.24) can be rewritten in the form

$$\frac{K(\lambda)}{S(\lambda)} = \frac{\left(1 - R_{i,\infty}(\lambda)\right)^2}{2R_{i,\infty}(\lambda)} \tag{3.25}$$

This relationship is also compatible with the solution $R_{i,\infty} = 1 + (K/S) + [(K/S)^2 + 2(K/S)]^{1/2}$, which, however, is not a physical solution because it corresponds to a reflectance $R_{i,\infty} > 1$.

In practice, the most interest is for the reflectance of totally opaque layers of paint and then the attention is for the solution $R_{i,\infty}(\lambda)$ (3.24). Then the ratio $K(\lambda)/S(\lambda)$ characterizes the totally opaque medium.

Kubelka-Munk Mixing Law

A mixture of pigments is characterized by an absorption coefficient $K_{mix}(\lambda)$ and a scattering coefficient $S_{mix}(\lambda)$, which are functions of the absorption and scattering coefficients of the individual pigments, $K_j(\lambda)$ and $S_j(\lambda)$, according to the *Kubelka-Munk mixing law*:

$$\frac{K_{mix}(\lambda)}{S_{mix}(\lambda)} = \frac{\sum_j c_j K_j(\lambda)}{\sum_j c_j S_j(\lambda)} \tag{3.26}$$

where c_j's indicate the concentrations of various pigments j. In the colorimetric practice, the concentrations are expressed in relative weight.

Figure 3.12 shows a commented example of application of equations (3.24) and (3.26), and Figure 3.13 gives the colorimetric result.

3.9.2 Saunderson's Equation

In the Kubelka-Munk analysis, Saunderson's equation[10] takes into account the reflection, described by the Fresnel laws (Section 3.4.2), which occurs at the separation surface between the air and the coloured layer, for example, a paint. This correction is limited to considering unpolarized light and the two cases of illumination geometry of greater practical importance, the ones most followed in reflectance measurements in industrial laboratories, implemented by an integrating sphere (Section 12.2.2 and Figure 12.2b and 12.2c) and denoted with

- (de:0°) if the specular reflection component is excluded and
- (di:0°) if the specular reflection component is included.

(These terms will become clearer after discussing the role of the integrating sphere (Sections 3.10 and 12.2.2).)

Let $R'(\lambda)$ be the measured reflectance, which, according to Saunderson, is a function of

- the reflectance $R_{i,\infty}(\lambda)$ evaluated by the Kubelka-Munk equations (3.24);
- the Fresnel reflectance $\rho_0(\lambda)$ – often denoted by k_1 – evaluated for normal incidence of the illuminating light (Section 3.4), which in the case of paint in the air is $\rho_0(\lambda) \cong (n(\lambda) - 1)^2 / (n(\lambda) + 1)^2$, where $n(\lambda)$ is the refractive index of the coloured layer at the surface; and
- the Fresnel reflectance of $\rho_i(\lambda)$ – often denoted with k_2 – evaluated for the diffused light that crosses the surface inside the coloured layer

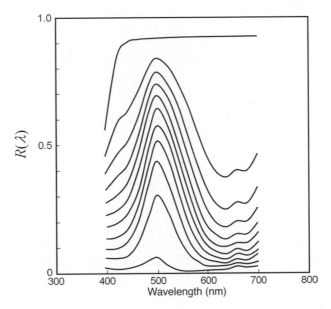

Figure 3.12   *Reflectances of a mixture of two pigments, one white and one green, according to mixing ratios with increasing step equal to 10% (the top curve represents the 100% white pigment, the bottom curve 100% green pigment) and computed by the mixing law of Kubelka-Munk corrected by Saunderson. It is observed that greatest variations in reflectance are at the extreme steps, 0–10% and 90–100%. This reveals that the formulation for a mixture of pigments, in which one pigment enters the mixture with a percentage lower than 10%, is difficult to make, because a very small percentage change involves an appreciable variation of reflectance.*

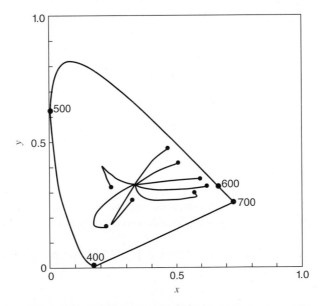

Figure 3.13   *Chromaticity of colours obtained by mixing different pigments in varying ratios with the white pigment (Section 6.13). It is observed that in the process of subtractive synthesis implemented with mixtures of pigments, the chromaticity moves away from its straight line of additive colour mixing (Section 6.13.8). Such behaviour is calculated with the Kubelka-Munk model.*

and is

$$R'(\lambda) = k\rho_0(\lambda) + \frac{\left(1 - \rho_0(\lambda)\right)\left(1 - \rho_i(\lambda)\right)R_{i,\infty}(\lambda)}{1 - \rho_i(\lambda)R_{i,\infty}(\lambda)} \tag{3.27}$$

where $k = 1$ in the case (di:0°) and $k = 0$ in the case (de:0°) (Section 12.2.2). The following equation gives the reflectance $R_{i,\infty}(\lambda)$ as a function of the measured $R'(\lambda)$:

$$R_{i,\infty}(\lambda) = \frac{R'(\lambda) - k\rho_0(\lambda)}{\left(1 - \rho_0(\lambda)\right)\left(1 - \rho_i(\lambda)\right) + \rho_i(\lambda)R'(\lambda) - k\rho_0(\lambda)\rho_i(\lambda)} \tag{3.28}$$

In the case of a paint placed in the air and constituted of pigments dispersed in a resin, the internal reflectance can be approximated to $\rho_i(\lambda) \cong 0.68 \langle n \rangle - 0.56$, where $\langle n \rangle$ is the mean refractive index of the resin, and sometimes is set $\rho_i(\lambda) \cong 0.4$.

3.9.3   Colorant Characterization and Formulation

The *colorant formulation* is the definition of the concentrations of pigments and dyes suitable to produce a paint, to dye a fabric, to colour a surface of paper, wood, so on, with colour equal to that of a given standard. In these cases, the colour is obtained by subtractive synthesis and is due to the absorption of radiation by colouring substances, which contribute to define the reflectance and the transmittance of the bodies. The problem of colour reproduction is mainly considered as a reproduction of the reflectance or transmittance, and colorimetry is considered in a second step:

- in the colorimetric specification (Section 12.4);
- in the computation of the colour difference between standard and reproduction (Section 11.4); and
- in the computation of the metamerism index (Section 11.4.7).

The colorant formulation takes place in two steps, the first is the physical reproduction of the reflectance and the second step concerns the colorimetric optimization of the reflectance.

To reproduce a reflectance, the colorant must be characterized, that is, the functions $K_j(\lambda)$ and $S_j(\lambda)$ must be known. So, first the colorant has to be characterized. The equations that relate the colorant formulation and the colorant characterization are the same.

Here, consider a coloured fully opaque layer. The reproduction of a spectral reflectance by using known pigments is obtained by applying the Kubelka-Munk mixing law (3.26) together with the solution (3.24) and (3.25) of the Kubelka-Munk equations and of the Saunderson correction (3.27) and (3.28):

1. By equation (3.28), the Kubelka-Munk reflectance $R_{i,\infty}(\lambda_j)$ is computed starting from the measured reflectance $R'(\lambda_j)$.
2. Equation (3.25) gives

$$\left(\frac{K_{\mathrm{mix}}(\lambda_j)}{S_{\mathrm{mix}}(\lambda_j)}\right) = \frac{\left(1 - R_{i,\infty}(\lambda_j)\right)^2}{2R_{i,\infty}(\lambda_j)} \tag{3.29}$$

3. The law of mixing of Kubelka-Munk consists of a set of equations (3.26), one equation for each wavelength λ_j:

$$\left(\frac{K_{\mathrm{mix}}(\lambda_j)}{S_{\mathrm{mix}}(\lambda_j)} \right) = \frac{\sum_i c_i K_i(\lambda_j)}{\sum_i c_i S_i(\lambda_j)} \quad \text{with} \quad \sum_i c_i = 1 \tag{3.30}$$

In the colorant characterization, the concentrations c_j are known and the functions $K_j(\lambda)$ and $S_j(\lambda)$ are unknown, while in the colorant formulation the functions $K_j(\lambda)$ and $S_j(\lambda)$ are known and the concentrations c_j are unknown.

Colorant Characterization

Once the concentrations are chosen and the paint is produced, the reflectances can be measured and the $K_j(\lambda)$ and $S_j(\lambda)$ are computed by equations (3.29) and (3.30). In this way, the colorant characterization is obtained at a given wavelength. The characterization goes on by considering one wavelength and a minimum number of pigments at a time. The minimum number of different pigments present in a mixture is two. In the case in which both pigments have to be characterized, the problem at each wavelength is so defined:

- the unknowns for each pigment are two and
- an equation of the type (3.30) exists for each ratio of concentrations of the two pigments.

It follows that for the characterization of two pigments, four equations are necessary, but, since the equations are homogeneous, only three equations are necessary to define K_i and S_i, up to one factor.

In practice, the characterization starts with the black and white pigments. Once these are characterized, the coloured pigments are considered, each of which is mixed in a known concentration ratio with the black and white pigments.

In the characterization of black and white pigments, the most significant experience indicates the following three mixtures (the mixture consisting of a single pigment, that is present at 100%, is called *mass tone*):

1. the mass–tone mixture of a white pigment, with a Kubelka-Munk reflectance $R_{i,\infty,\mathrm{W}}$;
2. the mass–tone mixture of a black pigment, with a Kubelka-Munk reflectance $R_{i,\infty,\mathrm{K}}$;
3. the mixture of a black pigment with concentration $c_\mathrm{K} = 3\%$ and of a white pigment with concentration $c_\mathrm{W} = 1 - c_\mathrm{K} = 97\%$, with a Kubelka-Munk reflectance $R_{i,\infty,\mathrm{KW}}$;

from which the three equations of the type (3.29) and (3.30) for each wavelength follow. The solution of this set of equations is obtained by requiring that the white pigment be a pure wavelength independent scatterer with diffusion coefficient $S_\mathrm{W} = 1$. The problem is represented by three linear equations with three unknowns, whose solutions at a fixed wavelength are

$$S_\mathrm{W} = 1$$
$$K_\mathrm{W} = \frac{M_\mathrm{W}}{S_\mathrm{W}} = M_\mathrm{W}$$
$$K_\mathrm{K} = -\frac{c_\mathrm{W}(M_\mathrm{W} - M_\mathrm{KW})}{(1 - c_\mathrm{W})(M_\mathrm{K} - M_\mathrm{KW})} M_\mathrm{K} \tag{3.31}$$
$$S_\mathrm{K} = \frac{K_\mathrm{K}}{M_\mathrm{K}} = -\frac{c_\mathrm{W}(M_\mathrm{W} - M_\mathrm{KW})}{(1 - c_\mathrm{W})(M_\mathrm{K} - M_\mathrm{KW})}$$

where

$$M_K = \frac{\left(1 - R_{i,\infty,K}\right)^2}{2R_{i,\infty,K}}, \; M_W = \frac{\left(1 - R_{i,\infty,W}\right)^2}{2R_{i,\infty,W}}, \; M_{KW} = \frac{\left(1 - R_{i,\infty,KW}\right)^2}{2R_{i,\infty,KW}} \tag{3.32}$$

and K_W and S_W, and K_K and S_K are the absorption and scattering coefficients of the white and black pigments, respectively.

Once K_W, S_W, K_K and S_K are known, the characterization of the other coloured pigments follows. Two mixtures are considered for each colour pigment, one mixture with a white pigment and the other with a black pigment.

If the coloured pigment is organic, the mixing ratios, which experience indicates as the most significant, are

- mix with $c_{CW} = 5\%$ and 95% of a coloured pigment in white, with a Kubelka-Munk internal reflectance $R_{i,\infty,CW}$;
- mix with $c_{CK} = 99\%$ of a coloured pigment and 1% black, with a Kubelka-Munk internal reflectance $R_{i,\infty,CK}$; which are followed by two linear equations, whose unknowns are the coefficients S_C and K_C characterizing the coloured pigment, which for each fixed wavelength are

$$S_C = \left\{ \frac{(1 - c_{CK})\left[M_{CK}S_K - K_K\right]}{c_{CK}} - \frac{(1 - c_{CW})\left[M_{CW}S_W - K_W\right]}{c_{CW}} \right\} \frac{1}{M_{CW} - M_{CK}}$$

$$K_C = \left\{ \frac{(1 - c_{CK})\left[M_{CK}S_K - K_K\right]}{c_{CK}M_{CK}} - \frac{(1 - c_{CW})\left[M_{CW}S_W - K_W\right]}{c_{CW}M_{CW}} \right\} \frac{1}{1/M_{CK} - 1/M_{CW}} \tag{3.33}$$

where

$$M_{CK} = \frac{\left(1 - R_{i,\infty,CK}\right)^2}{2R_{i,\infty,CK}}, \; M_{CW} = \frac{\left(1 - R_{i,\infty,CW}\right)^2}{2R_{i,\infty,CW}} \tag{3.34}$$

If the coloured pigment is inorganic, everything is as in the previous case, except for the optimal mixing ratios, which experience has shown to be

- mix with $c_{CW} = 25\%$ of the coloured pigment and 75% of the white and
- mix with $c_{CK} = 99\%$ of the coloured pigment and 1% of the black.

Reproduction of the Spectral Reflectance of the Turbid Media

Assuming known coefficients $K_i(\lambda)$ and $S_i(\lambda)$, the set of equations (3.30) is unusual because the number of equations – in general 22 or 43, depending on whether the reflectance is measured with a step of 20 or 10 nm – is much greater than the number of unknowns c_i, usually not more than six. The problem cannot have an exact solution, unless the case is very lucky, and requires a technique to obtain a solution and a criterion for evaluating the goodness of the solution. At this point of the analysis, approximations are involved, which have to be selected on a case-by-case basis.

Despite these simplifications, the search for the unknown concentrations that will produce the reflectance that best approximates the standard is still complicated. According to the classical technique, the problem is tackled in two stages:

1. searching for the best reflectance using the *Gall method*[11] and
2. optimizing the solution from the colorimetric point of view with the *Allen method*,[12-13] searching in the tristimulus space the solution closest to the standard.

These methods are tedious and we refer the interested readers to the original publications.[11–13] Here a technique is proposed that seems to be effective.

Equation (3.30) can be transformed easily into linear and homogeneous equations:

$$\sum_{i=1}^{n\,\text{pigments}} c_i \left[K_i(\lambda_j) - S_i(\lambda_j) \frac{K_{\text{mix}}(\lambda_j)}{S_{\text{mix}}(\lambda_j)} \right] \equiv \sum_{i=1}^{n\,\text{pigments}} c_i Q_i(\lambda_j) = 0 \tag{3.35}$$

and then rewritten in matrix form as a scalar product of two vectors:

$$\begin{pmatrix} c_1 & c_2 & \cdots & c_n \end{pmatrix} \begin{pmatrix} K_1(\lambda_j) - S_1(\lambda_j) \left[K_{\text{mix}}(\lambda_j) \middle/ S_{\text{mix}}(\lambda_j) \right] \\ K_2(\lambda_j) - S_2(\lambda_j) \left[K_{\text{mix}}(\lambda_j) \middle/ S_{\text{mix}}(\lambda_j) \right] \\ \vdots \\ K_n(\lambda_j) - S_n(\lambda_j) \left[K_{\text{mix}}(\lambda_j) \middle/ S_{\text{mix}}(\lambda_j) \right] \end{pmatrix} = \tilde{\mathbf{C}} \mathbf{Q}(\lambda_j) = 0 \tag{3.36}$$

where n is the number of pigments and the row vector unknown $\tilde{\mathbf{C}} = (c_1 \quad c_2 \quad \cdots \quad c_n)$ should be orthogonal to all vectors $\mathbf{Q}(\lambda_j)$. If an exact solution exists, the vectors $\mathbf{Q}(\lambda_j)$ belong to a hyper-plane in n-dimensional space orthogonal to the vector $\tilde{\mathbf{C}}$. The 'vector' obtained from the anti-symmetric product of $(n-1)$ vectors $\mathbf{Q}(\lambda_j)$ related to $(n-1)$ different wavelengths would be the solution of the problem. Since the exact solution usually does not exist, before evaluating the possible products of anti-symmetric vectors $\mathbf{Q}(\lambda_j)$, for each of these anti-symmetric products the spectral reflectance is calculated by equations (3.24), (3.25), (3.26) and (3.27), and finally the root mean square (RMS) value with respect to the reflectance of the standard is computed. The solution with the smallest RMS value is the best from the physical point of view, but often it is good also from the colorimetric point of view. This solution, readily available, in general already is a good solution to the problem and has the great advantage of indicating whether the selected pigments lead to acceptable colour reproduction. The colorimetric optimization can be obtained by the technique of *deepest descent*,[14] according to which variations in the concentrations c_i's are introduced that reduce the distance between the standard and imitated colours, specified in a uniform space of perceived colours (Section 11.4).

The software presented in Sections 16.11.1 and 16.11.2 allows us to understand the role that the pigments present in a mixture as a function of the percentage in which they are present.

3.10 Ulbricht's Integration Sphere

The integrating sphere is for its optical properties a basic device for measuring radiometric, photometric and colorimetric quantities. A detailed description of the theory of the integrating sphere can be found in the literature.[15–20]

The ideal integrating sphere is a sphere whose inner surface is a Lambertian diffuser that does not absorb light. In reality, the inner surface is not exactly Lambertian and absorbs a little light. Through ports it is possible

- to introduce light inside;
- to take light to measure it; and
- to take light to illuminate a sample in a hemispherical way.

Section 12.2.2 considers the use of the integrating sphere in radiometry, photometry and colorimetry. Here the characteristics of the integrating sphere useful for its understanding and for practical use are considered. Schematically, the properties are divided into geometrical and optical.

Geometrical Characteristics

Consider a sphere equipped with two circular ports, one for the entry of light and one for the output. Define:

R_s	radius of the sphere
$A_s = 4\pi R_s^2$	area of the internal surface of the sphere
r_i	radius of the input port
$A_i = \pi r_i^2$	area of the input port
r_e	radius of the exit port
$A_e = \pi r_e^2$	area of the exit port
$f_e = A_e/A_s$	ratio of the area of the exit port and the area of the inner surface of the sphere
$f = (A_i+A_e)/A_s$	ratio between the sum of the areas of all the ports and the area of the inner surface of the sphere
$A = (A_i+A_e)$	area of all the ports

In the case of multiple input and exit ports, areas A_i and A_e refer to the sum of the areas of all the input and exit ports, respectively.

The following considerations require that the total area of the ports is negligible compared to the area of the internal surface of the sphere: this occurs in practice if the total area of the ports is less than one-tenth of the area of the inner surface of the sphere.

A diffusing Lambertian baffle should be placed on the straight lines joining the input port with the port where there is a sample to be hemispherically illuminated, thus avoiding the sample from being illuminated directly by the source.

Optical Characteristics

A single quantity optically characterizes an integrating sphere and this is the spectral reflectance of the inner surface of the sphere $\rho(\lambda)$, which in the ideal sphere is equal to 1.

The main properties of the integrating sphere come from having an internal Lambertian surface, that is, with a radiance $L_{e,A}(\vartheta)$ emerging from the generic surface element dA equal for all the directions ϑ. By virtue of this, there exists an equal incident flux on the whole internal surface of the sphere and for all the directions of incidence (Figure 3.14). The calculation of the flux exiting the element dA and crossing the element dS is given by the product of the radiance $L_{e,A}(\vartheta)$ from dA for the solid angle subtended by dS, that is, $[\cos\vartheta \, dS/(2 \, r \cos\vartheta)^2]$, and for the element of apparent surface emitting ($dA \cos\vartheta$). After simplification we have $[L_{e,A}(\vartheta) \, dA \, dS/(4 \, r^2)]$, which is an independent magnitude of ϑ, because $L_{e,A}(\vartheta)$ is equal for all the angles ϑ, and independent of the distance between the two surface elements dA and dS. This implies that the flux crossing dS is the same for all the points of the sphere from which comes the flux and for this, if the element dS is an output port, this is a Lambertian light source.

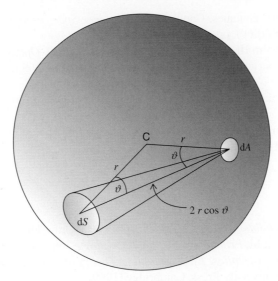

Figure 3.14   *Integrating sphere of radius r, on the inner surface of which are considered in perspective view the elements of the surface dS and dA with a mutual distance equal to (2 r cosϑ), where dS receives light from dA. The point C is the centre of the sphere.*

References

1. CIE Publication S 017/E:2011, *ILV: International Lighting Vocabulary*, Commission Internationale de l'Éclairage, Vienna, Austria (2011). Available at: www.eilv.cie.co.at/ (accessed on 3 June 2015).
2. Kinoshita S, Yoshioka S and Miyazaki J, Physics of structural colors, *Rep. Prog. Phys.*, **71**, 1–30 (2008).
3. Encyclopædia Britannica. Available at: http://www.britannica.com/EBchecked/topic/527235/schemochrome (accessed on 3 June 2015).
4. Mie G, Beiträge zur Optik Truber Medien, speziell kolloidaler Metallösungen, *Annal. Phys.*, **330**, 377–445 (1908).
5. Craker WE and Robinson PF, The effect of pigment volume concentration and film thickness on optical properties of surface coatings, *J. Oil Colour Chem. Assoc.*, **50**, 111–133 (1967).
6. Billmeyer FW Jr and Richards LW, Scattering and absorption of radiation by lighting materials, *J. Color Appearance*, **2**, 4–15 (1973).
7. Kubelka P and Munk F, Ein Beitrag zur Optik der Farbanstriche, *Z. Tech. Phys.*, **12**, 593–601 (1931).
8. Kubelka P, New contributions to the optics of intensely light-scattering materials, Part II, non-homogeneous layers, *J. Opt. Soc. Am.*, **44**, 330–355 (1954).
9. Völz HG, *Industrial Color Testing*, VCH, Weinheim (1994).
10. Saunderson JL, Calculation of the color of pigmented plastics, *J. Opt. Soc. Am.*, **32**, 727–736 (1942).
11. Gall L, Computer time-sharing aids color matching, *Paint Varnish Prod.*, **August**, 37–44 (1971).
12. Allen E, Basic equations used in computer color matching, *J. Opt. Soc. Am.*, **56**, 1256–1259 (1966).
13. Allen E, Basic equations used in computer color matching, II. Tristimulus match, two-constant theory, *J. Opt. Soc. Am.*, **64**, 991–993 (1973).
14. Press WH, Flannery BP, Teukolsky SA and Vetterling WT, *Numerical Recipes*, Cambridge University Press, New York (1992).
15. Wendlandt WW and Hecht HG, The integrating sphere, Cap. X, in *Reflectance Spectroscopy*, Interscience Publishers, New York (1966), pp. 253–273.

16. Kortüm G, Experimental techniques, Cap. VI, in *Reflectance Spectroscopy, Principles, Methods, Applications*, Springer, Berlin (1969), pp. 217–232.
17. Jacquez JA and Kuppenheim HF, Theory of the integrating sphere, *J. Opt. Soc. Am.*, **45**, 460–470 (1955).
18. Hisdal J, Reflectance of perfect diffuse and specular samples in the integrating sphere, *J. Opt. Soc. Am.*, **55**, 1122–1128 (1965).
19. Hisdal J, Reflectance of non-perfect surfaces in the integrating sphere, *J. Opt. Soc. Am.*, **55**, 1255–1260 (1965).
20. Clarke FJJ and Compton JA, Correction methods for integrating-sphere measurement of hemispherical reflectance, *Color Res. Appl.*, **11**, 253–262 (1986).

Bibliography

Bass M and DeCusatis C, *Handbook of Optics*, 3rd edn, OSA-MacGraw-Hill Companies Inc. (2010).
Billmeyer FW Jr and Saltzman M, *Principles of Color Technology*, John Wiley & Sons, New York (1981).
Born M and Wolf E, *Principles of Optics: Electromagnetic Theory of Propagation, Interference and Diffraction of Light*, Cambridge University Press, Cambridge (1999).
Elias M and Lafait J, *La Couleur - Lumière, Vision et Matèriaux*, Edition Belin, Paris (2006)
Hecht E, *Optics*, Addison-Wesley (2002).
Judd DB and Wyszecki G, *Color in Business, Science and Industry*, John Wiley & Sons, New York (1975).
McDonald R ed., *Colour Physics for Industry*, Society of Dyers and Colourists, Bradford (1987).
Sève R, *Physique De La Culeur*, Masson, Paris (1996)

4

Perceptual Phenomenology
of Light and Colour

4.1 Introduction

Often, visual experience has led humans to consider mistakenly colour as the exclusive property of objects and light sources, without thinking that the visual experience exists only because a light flux enters the eye and activates the visual system of the observer.

Isaac Newton[1] wrote:

"For the rays to speak properly are not coloured. In them there is nothing else than a certain power and disposition to stir up a sensation of this or that colour."

According to Newton, colours are sensations and colour sensations are not measurable physical quantities, but attributes of the visual sensations and therefore these entities are purely subjective and incommunicable. The sensation that every other observer experiences is unknown, although everybody calls that colour sensation with the same name. At the moment, science does not have any tools able to 'observe' and measure the last stage of the visual process that human beings experience as perception and regard consciousness. In addition, the human being communicates colour sensations with language and the colour specification is then made with the names that distinguish the colour sensations. This verbal specification concerns observer individuality and therefore does not have the objective value required by science, but certainly is a way for knowledge of the human visual system.

Only measurements on the colour stimuli are considered objective, while the attributes of perceived colours are subjective.

In this chapter perceived colours are considered in relation to verbal language and to perceptive and physical phenomena in order to understand mainly whether a colour ordering is possible.

Moreover, perceptive phenomena produced by different colour stimuli simultaneously present in the visual field are considered.

Standard Colorimetry: Definitions, Algorithms and Software, First Edition. Claudio Oleari.
© 2016 John Wiley & Sons, Ltd. Published 2016 by John Wiley & Sons, Ltd.

4.2 Perceived Colours, Categorization and Language

The distinction between colour sensation and colour perception is a debated item. Consider the following two definitions as a guide, not as an answer to the debate:

- *Colour sensation* – The sense organs register the colour stimulus, 'decode' it and transform it into a neural signal – colour sensation –, that is then transmitted to the brain.

- *Colour perception* – In the brain, the neural signal is organized and interpreted. Colour perception involves 'making sense' of colour sensations.

Perceived colour is a psychological activity following a retinal and neuronal stimulation produced by a visible light. Colour experiences cannot be shared directly and the only way for communicating perceived colour is by language. It means that an interconnection between the words used to specify colour perception, human thought and colour sensations exists. Human beings are able to distinguish between approximately 10 million different colours,[2] but do not use 10 million colour names. It means that the human brain attempts to distinguish among the colour perceptions using *similarities* and then collecting colour sensations in clusters and differentiating in a cluster with additional words. Conceptual clustering of the colour sensations is an attempt to explain how knowledge is represented. In this approach, clusters are generated by first formulating their conceptual descriptions and then classifying the entities according to the descriptions, that is, by words. This sentence is the definition of a process of *categorization*.

The process of categorization seems to lead us to know the deep rules of human thought and therefore directly related to the process of vision present in the visual system. For this reason, colour categorization has been carefully considered.

So many different colour sensations constitute a continuum and no visible borders exist in this colour continuum. The *categorical perception* is the experience of percept invariances in sensory phenomena that vary along a continuum.

The connection between language and categorization has led to the definition of

1.   "Basic colour term:
     (a) A colour word that is monolexemic (unlike 'reddish-yellow');
     (b) A colour word whose extension is not included in that of any other colour term (unlike 'scarlet', whose extension is included in 'red');
     (c) A colour word whose application is not restricted to a narrow class of objects (unlike 'roan');
     (d) A colour word that is psychologically salient (unlike 'puce')."

A basic colour term names a basic colour category.

2.   "*Basic colour names* (or *terms*) – a group of 11 colour names found in anthropological surveys to be in wide use in fully developed languages: white, black, red, green, yellow, blue, brown, grey, orange, purple, pink."[3,4]

With regard to the second definition, the anthropological investigation was conducted by Berlin-Kay[5] who originally thought that these 11 categories were universal, that is, independent of the culture of the people who had expressed them. Characteristics of the basic colour terms were to be general and salient[6]:

1.   A term is *general* if it applies to diverse classes of objects and its meaning is not subsumable under the meaning of another term.

2.   A term is *salient* if it is readily elicitable, occurs in the idiolects of most speakers, and is used consistently by individuals and with a high degree of consensus among individuals.

This book does not enter a philosophical and linguistic confrontation. The classification of Berlin and Kay[5] is not shared by everyone, but represented a significant step.

We must pay attention to a miscategorization, in which diverse colour sensations are grouped together considering common qualities which are meaningless with respect to the categorization. An example of miscategorization is the over-categorization based upon over-similar qualities that virtually all colours have in common. A kind of over-categorization is the result of a distinction of additional aspects related to the variability in lightness, as in the distinction among white, black and grey, that all belong to the same category of achromatic colours. A kind of over-categorization is related to the variability of the intensity of coloration as for the brown and pink colours that are not autonomous categories.

There exists a small set of six perceptual landmarks, instead of 11 foci of Berlin-Kay, that can be identified with the Hering *elementary* colours (Section 4.4) – black, white, red, yellow, green and blue – which individually or in combination form the basis of the denotation of most of the major colour terms of most of the languages in the world. These colours are termed *focal colours* and the hues of these focal colours are named *unique hues*[6] (Section 4.4), that is, pure hues that does not contain a hue of any other colour. For example, unique yellow is the yellow that is neither greenish nor reddish. Moreover, a classic set of studies by Eleanor Rosch Heider[7–9] found that these focal colours were also remembered more accurately than other colours, across speakers of languages with different colour naming systems. Focal colours are perceptually elemental. At the same time focal colour is defined as the best example of each colour name, the typical colour, universal all across different languages. These definitions of focal colours and unique hues, although different, are very similar up to be almost identical[10]. Focal colours seem to constitute a universal cognitive basis for both colour language and colour memory.

4.3   Light Dispersion and Light Mixing

The rainbow is a natural phenomenon that offers the viewer an ordered set of colours, which occurs with continuous passage without jumps from red to violet. These colours are naturally disposable on a segment. This colour order concerns the natural spectral decomposition of the electromagnetic radiation due to the refraction of light and is not based on perception. The rainbow colours produced by the spectral lights do not cover the whole set of colours seen. This is not surprising because in the rainbow there is a dispersion of coloured lights and not mixtures of them. The most obvious colour missing in the rainbow is white, but also purple hues are absent. To see the missing hues an operation opposite to dispersion is needed. It is necessary to combine the spectral lights in mixtures with variable mixing ratios.

Everyone knows that a beam of white light of the sun, when it passes through a prism in a dark room, is dispersed into its spectral light bands as in the rainbow, which, following Newton, are termed as *red, orange, yellow, green, blue, indigo* and *violet*. These names associated with the various spectral bands are the result of a commonly accepted categorization, although if within a band colours vary according to a continuous scale and no jump exists between contiguous bands. Physics associate an exact wavelength to any spectral light and psychophysics a dominant wavelength to any spectral band (Section 2.2). The observer associates a hue to any wavelength and to any band, that, as a perceived quantity, is a subjective association and, as a linguistic term, is a conventional association.

These colour terms represent hues, but not all the hues. The definition of hue given by CIE is:

"Hue – attribute of a visual perception according to which an area appears to be similar to one of the colours: *red, yellow, green* and *blue*, or to a combination of adjacent pairs of these colours considered in a closed ring"[3] (Figure 4.1).

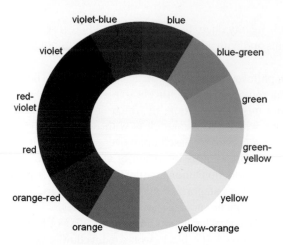

Figure 4.1   *Colour wheel.*

The definition of hue cannot be given without a verbal example. The colour order in the rainbow and in the CIE definition of hue is the same, although CIE considers only Hering's elementary chromatic colours (Section 4.4).

In a perceptual sense, colours are subdivided into:

1.   *"Chromatic colour* – perceived colour possessing hue."[3]

2.   *"Achromatic colour* – perceived colour devoid of hue, and the colour names *white*, *grey* and *black* are commonly used for colourless and neutral objects."[3]

The set of six colours – black, white, red, yellow, green and blue – is a perceptual landmark in the colour-categorisation process and in optical refraction. The coloured lights in scenes of daily life are perceived as mixtures of these six elementary colours (Section 4.4) and, at the same time, are mixtures of spectral lights with different intensities.

"Additive mixture of colour stimuli – stimulation that combines on the retina (Sections 5.2 and 5.4) the actions of various colour stimuli in such a manner that they cannot be perceived individually."[3]

4.3.1   Newton's Prism Experiment, Colour Wheel and Colour Attributes

Here the apparatus used by Isaac Newton to produce an additive mixture of lights is considered (Figure 4.2). The light source was the sun and Newton used a sunbeam entering in a dark room through a hole in the window. An equivalent beam of light is obtainable today from a source, for example, a lamp with electrical discharge in xenon. The light beam enters the first prism at the right in the Figure 4.2 and exits dispersed. A converging lens is placed equidistant between the first prism and a second prism, and the distance between the two prisms is four times the focal length of the lens so that dispersed rays are recomposed into a single beam that exits from the second prism. The colour of the recomposed beam is white, as the one that entered the first prism.

The recombined beam is visible on a diffusing screen, like a sheet of paper, crossing the beam after the second prism.

The amounts of spectral lights of the initial beam can be modulated by introducing a comb with teeth of different length and size on a plane close to the lens. After this modulation, the recombined beam appears coloured. This phenomenon is the *additive mixture of colour stimuli* already defined. In this way all the

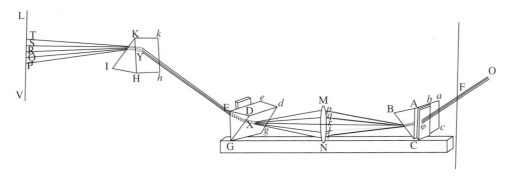

Figure 4.2   *Newton's experimental apparatus for studying the compound colours of spectral lights (from the original figure of Newton[1]).*

colours are obtainable. In particular, the continuous set of purple hues, from violet to red, is obtained by red and violet light mixings in variable ratio. The purple hues are absent in the spectral decomposition of white light. The set of colour-hues is now complete because the two extremities of the rainbow colours are connected by the continuous set of purple hues (Section 10.2). The set of colour-hues can be arranged with continuity and without jumps on a circle.

It is evident that the purple colours are not spectral colours, but come from the mixture of two spectral colours. Similarly, the hues of the various spectral colours are obtainable from the mixture of the rays adjacent in the spectral decomposition made with the prism, for example, the colour orange is obtained by mixing rays with the colours red and yellow. This is no longer true if the mixture regards suitable pairs of rays: for example, a mixture of a red ray with a green ray with suitable intensities and wavelengths produce an achromatic beam (although Newton was not able to see this kind of mixing). The mixture of a blue spectral ray with a suitable yellow spectral ray with intensity ratio ranging from zero and increasing, at the beginning the colour is blue and as the ratio increases the colour fades, becomes achromatic and then it begins to turn yellow. This experiment defines

"Complementary colour stimuli – color stimuli that produce a specified achromatic stimulus when they are suitably mixed in an additive manner."[4]

"Complementary wavelength of a colour stimulus [λ_c], wavelength of the monochromatic stimulus that, when additively mixed in suitable proportions with the colour stimulus considered, matches the specified achromatic stimulus."[3]

Now, the colour wheel can be defined (Figure 4.1):

Colour wheel or *colour circle* – The continuum of colours "arranged in a circle, where complementary colours (or their names) are arranged on opposite sides of a circle."[11]

Each hue of the colour wheel is in correspondence to a spectral light (Figure 2.2) with exclusion of the purple hues, which are obtained by mixing red and blue hues.

The hue is not constant in the varying mixture of two lights with complementary wavelength. This phenomenon is called *Abney hue shift* and is considered in Section 6.19.

The colour attribute that decreases in the colour fading is the

"Colourfulness – attribute of a visual perception according to which the perceived colour of an area appears to be more or less chromatic."[3]

The last colour attribute considered here is

> "*Brightness* – attribute of a visual perception according to which an area appears to emit, or reflect, more or less light."[3]

Now, the *basic attributes* used to specify the colour of a beam of light are known: hue, brightness and colourfulness.

4.3.2   Maxwell's Disk Experiment

Everyone knows the toy constituted by a multicolour spinning top. The rotating disc, on to which radially slit and overlapping coloured discs are mounted to divide the surface into sectors of different colours, produces a visual colour mixing. A defined light source illuminates the rotating disks. The reflected light is an *additive mixture* of the lights reflected by the sectors of the rotating disks, but with respect to the illuminating light, the disks absorb light and therefore this colour mixing is *subtractive*. As stated by Helmholtz, this phenomenon is defined by,

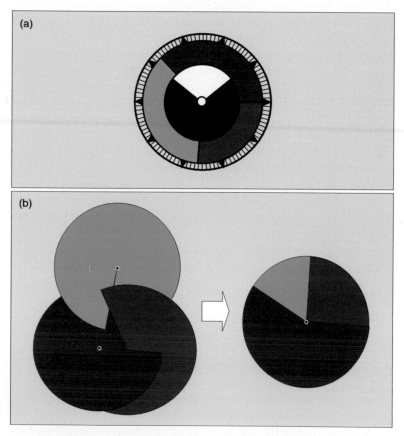

Figure 4.3   *(a) Normal view of the Maxwell disks. The inner circle, constituted by two disks slipped together, one black and one white, can be adjusted so that the relative proportions of black and white paper mixed by rotation form a neutral grey. (b) The outer disk with three coloured sectors is the visible part of three 'primary' colour disks, which are partially overlapped. The areas of the three sectors can be adjusted so that, when the discs are spun, the colour of the mixture of three 'primary' colours matches the neutral grey of the internal circle. The complete matching is obtained by adjusting the ratio between white and black colours of the central disks.*

Talbot's law: "If any part of the retina is excited with intermittent light, recurring periodically and regularly in the same way, and if the period is sufficiently short, a continuous impression will result, which is the same as that which would result if the total light received during each period were uniformly distributed throughout the whole period."[12]

This rotating disc is one of the first instruments for a quantitative measurement of colour matching, which is defined as

Disk colorimeter – additive colorimeter in which components are produced by sectors of chromatic samples occupying various proportions of a circular disk.

This was used by J. K. Maxwell for investigating colour mixing. Previously, both Claudius Ptolemy, ~150 A.C., and Ibn Al-Haytham Alhazen, 1000 A.C., had experimented with rotating disks with coloured sectors.

Maxwell demonstrated that Newton's centre of gravity rule[13] (Section 10.2) is correct and that almost all colour stimuli could be synthesized by different combinations of the three 'primary' lights, red (vermillion), green and blue (ultramarine). Figures 4.3 and 4.4 illustrate this instrument: three coloured paper disks could

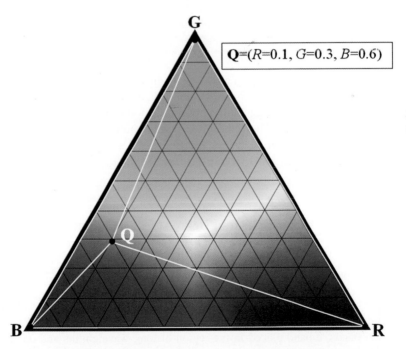

Figure 4.4 *The complete gamut of colours obtainable by mixing the three colours of Figure 4.3 together (vermillion, emerald green and ultramarine colour stimuli) represented by a triangle, called Maxwell's triangle. This triangle is a generalization of the bipolar diagram considered by Hering (Figure 4.6), although the quantities involved are different. A set of three coordinates (R, G, B), termed barycentric coordinates, are defined on this diagram for specifying each colour **Q**. These coordinates represent the areas of the three triangles with a vertex at the point **Q** and one side common with the **RGB** triangle. The sum of the three coordinates R+R+B=1. These coordinates are also proportional to the angles of the three sectors of the rotating disk of Figure 4.3. Maxwell's triangle shows the quality aspect of psychophysical colour, which is termed chromaticity. This quality does not indicate the quantity aspect of psychophysical colour, which represents the amount of light.*

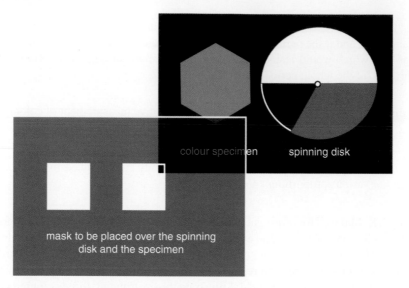

Figure 4.5   *Sketch of the Maxwell disk used as a colorimeter. Two, three or more disks can be slipped together partially overlapping and showing portions of each. The colours of the non-overlapping portions of the disks are mixed additively by spinning the disks and the resulting colour is matched to a standard colour on the left. A suitable mask provides a comparative view of the two colours in isolation.*

be clamped together to the disk of a top in such a way as to allow different amounts of the primary colours to be mixed when the top is spun. In the original Maxwell disk, a smaller central disc contained the colour sample which was to be matched by adding together different amounts of the primary colours. To eliminate the effect of the different brightness's of the central disc and the outer colours, different amounts of black could be added to the central disc. After Maxwell, Munsell and Ostwald used the disk for the construction of their atlases (Sections 14.3 and 14.4). Figure 4.5 illustrates the use of the Maxwell disk as a visual colorimeter.

4.4   Unique Hues, Colour Opponencies and Degree of Resemblance

Leonardo da Vinci wrote in his 'A treatise of painting'[14] that in nature there are six particularly simple colours: four elementary chromatic hues – red, yellow, green and blue – and two achromatic colours – white and black –.

Moreover, the colours of the colour wheel (Figure 4.1) are put in relation to the considered spectral bands, each one characterized by its proper wavelength, with the exclusion of purple hues that are provided by a mixture of two spectral bands, a band placed in the region of long wavelengths and one in the short wavelengths. Figure 14.5 shows the correspondence between wavelengths and colours of the wheel as proposed by Ostwald. The correspondence between spectral bands and wavelength is measurable and objective, while the correspondence between perceived hues and spectral bands is subjective.

This categorical classification of hues is strongly related to appearance and has the following characteristics:

- There is the presence of four hues defined by only one term: red, yellow, green and blue. These hues are those of the focal colours (Section 4.2).

- All other hues are given by the ordered mixture of two contiguous one-term hues, for example, yellow-red and reddish yellow.

- No one hue is defined by the union of two one-word non-contiguous term, that is, red-green, green-red, blue-yellow and yellow-blue.

This hue classification represented by this colour wheel follows the theory of Ewald Hering,[15] who called the red, yellow, green and blue hues *elementary hues*. Today these hues are called *unique* (or *unitary*) "because they cannot be further described by the use of hue names other than its own",[3] that is, they have the fundamental property that no presence of other known hues is perceived, for example, in the yellow hue no other unique hues are perceived. Also, if we mix unique hues opposite on the colour wheel, such as yellow and blue or red and green hues, the resulting hues are not termed with the names of the two colours entering the mixture. A greenish-red or a reddish-green does not exist, both in the perception and in the colour imagination.

Then Hering formulated a theory that two opponent colour systems are present in the human visual system, red-green and blue-yellow. Perceptual opponency of red-green and of yellow-blue forms the conceptual basis for quantifying the redness, greenness, yellowness and blueness of monochromatic lights. The perceptions of colour would be encoded using these opponents mechanisms.

In addition, Hering analysed the perception of the contrast between light and dark, and postulated a black-white mechanism. The perceptions of black and white are considered fundamental as those of red, yellow, green and blue.

The three mechanisms proposed by Hering are in agreement with the phenomenology of achromatic or neutral colours, dark colours, pastel colours, ..., that is,

- *grey colours* obtainable by suitable mixings of white with black,

- *brown colours*, arise from the mixing of red, yellow and black,

- *light* and *pastel* colours, such as *pink*, are obtainable by appropriate mixings of red, blue and white.

Although black is inversely related to white, it is not fully incompatible with it (as red is incompatible with green and yellow with blue): a particular grey can always be described in terms of its percentage of either whiteness or blackness.[16] The achromatic colours are a bit different set up in comparison with the antagonistic fashion of red opposed to green and yellow opposed to blue. Hering realized that black and grey are not produced by absence of light coming from the scene but arise when and only when the light from the object is less than the average of the light coming from the surrounding regions. Similarly, white arises only when the surround is darker and when no hue is present. Hering was conscious of the mutual interaction in the perception of contiguous colours. This phenomenon is typical of *related colours* and is considered in *colour appearance*.

As the green colour mixed with the red one fades the colour red to make an achromatic mixture, so, for example, this red hue present in the red-yellow colours can be cancelled by mixing this with a suitable amount of green. In this way we get a faded yellow colour. The theory of Hering has found an important confirmation in psychophysical experiments made by Jameson and Hurvich (Section 6.16), by a chromatic cancellation method.

Hering introduced the concept of *degree of resemblance*, useful to define a colour by its degree of resemblance to the six elementary colours. This concept of resemblance had a basic role in the definition and construction of the Natural Colour order System (NCS) (Section 14.7) and is an important contribution

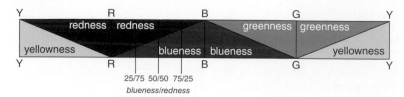

Figure 4.6   *Hering's bipolar diagrams related to the mixture of contiguous unique hues. Each vertical section of the diagram consists of two segments belonging to the two regions of the ingredient colours, respectively. The lengths of the two segments represent the degree of resemblance of the colour considered with respect to two ingredient colours and can be related to the two arms of the yoke of a balance with a moving equilibrium point on which two plates are placed for the ingredients of the mixture (Figure 10.2). As an example three colours are considered resulting from the resemblance with the blue unique hue (blueness degree) and red unique hue (redness degree), whose redness/blueness ratio is equal to 25/75, 50/50 and 75/25 (these numbers are percentage numbers), respectively.*

to colour scaling. Hering represented the pure chromatic colours of the colour wheel by four bipolar diagrams between the elementary colours yellow, red, blue and green (Figure 4.6): for example, a pure chromatic purple colour is characterized by the relationship between its redness and blueness, b/r or r/b, where r and b are the degree of resemblance to the elementary colours red and blue. Analogous for greenness, blueness, blackness and whiteness. Of course, no bipolar diagram for red-green and blue-yellow is possible. Hering inserted these four bipolar diagrams bent on the hue circle. As an example in Figure 4.7,

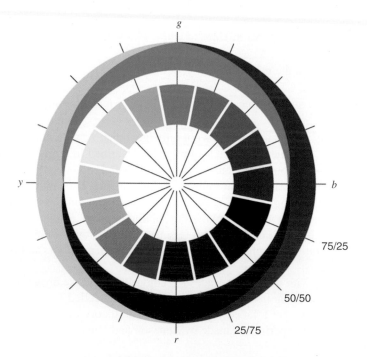

Figure 4.7   *Hering's hue wheel surrounded by a circle representing the succession of four bipolar diagrams associated with the mixing of the four pairs of contiguous unique hues. The four bipolar diagrams are bent and adapted to a circle for graphic reasons. In particular the same three colours are considered as in Figure 4.6.*

three pure chromatic colours in the region between blue and red elementary colours are specified by their redness and blueness.

The bipolar diagram is a way of representing the mixing of two ingredients. The principle is the same used by Maxwell for the mixture of three ingredients and represented by Maxwell's triangle. So Maxwell's triangle is a generalization of the bipolar diagram with three ingredients. The difference between the systems of representation of Hering and Maxwell lies in the meaning of the ingredients: Maxwell considered the amount by which each colour stimulus (the extent of the circular sector of the rotating disc dedicated to each colour) enters the mixture, while Hering considered the amount of resemblance entering the mixture (evaluation of a magnitude perceived by the observer and expressing the degree of similarity of the colour considered against the colours ingredients). This concept of resemblance is an important contribution to colour scaling.

4.5   Colour Similitude

In 1962 R. N. Shepard used a technique, called *observers' judgments of similarity*, for quantifying the perceived similarity between all the pairs of colour filters with half-band widths of about 12 nm, whose dominant wavelengths were almost randomly chosen between 434 and 674 nm. Then, Shepard, by using a multidimensional scaling algorithm, translated these similarity judgments into a two dimensional diagram (Figure 4.8), where dissimilarity is represented as distance.[17,18] Shepard excluded extra-spectral colours

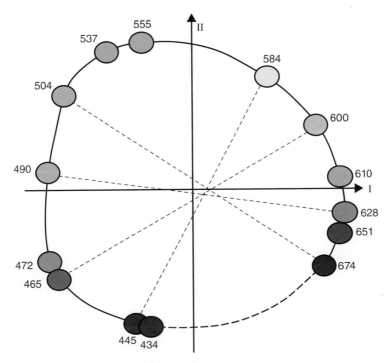

Figure 4.8   *Bi-dimensional diagram drawn by Shepard[17–19] for representing the perceptual dissimilarity between 91 pairs of colours. The numbers associated to colours represent their dominant wavelengths.*

(purple and magenta) to avoid an automatic convergence to a circle. The task recalls the ordering as it is in the Farnsworth Dichotomous D-15 Test (Section 16.3), where the observer has to judge the smallest colour difference between at most 14-colour pairs. Here the observers judged in turn all the 91 possible pairs of lights and the task was to rate the qualitative similarity of each colour pair on a five-step ordinal rating scale.

The simplest diagram for geometrically representing the similarities between colour pairs is to represent the 14 colours with points on a line. It appears evident that it is impossible to arrange the distances between the colour pairs satisfying the judgments. This task is easier if the line with the 14 points is bent on a plane and the distances between colour pairs are measured on the plane. This computation of optimization has been made with a multidimensional scaling algorithm. The result was obtained for a graphical representation on a plane, but the number of dimensions was free. This procedure accurately reproduced the ordering in the hue circle but also the perceptual spacing and complementarity between hues. An increment of the dimension certainly gives a more accurate fitting, but probably an over-fitting is possible. Particularly, the Euclidean distance represents the perceived similarities only qualitatively and a refinement of the result could be possible with the definition of suitable metrics.[19] Since the considered colours are not of spectral lights, the diagram has not to be confused with a spectrum locus.

The result is very interesting, because with a simple task it was possible to obtain a highly meaningful diagram.

4.6   Unrelated and Related Colours

The colour attributes considered so far – brightness, hue, colourfulness – are defined by coloured lights seen in isolation in a dark room. The colours seen in this contest are termed

> *"Unrelated colour* – colour perceived to belong to an area seen in isolation from other colours."[3]

Unrelated colours are studied in laboratories. In the natural world, a single isolated light is rare and the colours experienced in daily life are perceived in the presence of others and together the colours are interacting in the visual processing and mutually conditioning their colour appearance. These colours are:

> *"Related colour* – colour perceived to belong to an area seen in relation to other colours."[3]

All this is well known to embroiderers who choose the colour of a wire not seeing the colour on the spool, but on the embroidered fabric; so it is for painters who do not judge the splash of colour on the palette but on the painted canvas.

These perceived colour visual effects are a consequence of the visual organization of the receptive fields in the retina (Section 5.5) and on the spatial organization of the colour stimuli in the scene.

The quantification of colour differences is very important in practice. It is the result of comparisons of perceived colours (excluding the memory colours), which are necessarily in the same visual field and therefore are related colours.

4.6.1   Relative Attributes

In this subsection we collect in a very schematic way almost all the different kinds of interactions among related colours.

Recalling the *relative attributes* of the related colour, consider that related colours exhibit not only the relative attribute, but also the absolute attributes of hue, brightness and colourfulness. Here the classical definition of relative attribute as a function of the absolute attribute, based on the interaction with a reference white, are given:

1. "*Lightness* (of a related colour) – brightness of an area judged relative to the brightness of a similarly illuminated area that appears to be white or highly transmitting"[3]

$$\text{Lightness} = \frac{\text{Brightness}}{\text{White Brightness}}$$

The distinction between brightness and lightness, points to a fundamental property of an isolated achromatic stimulus: an isolated stimulus can vary in brightness but not lightness because lightness, by definition, is perceived in relation to a second stimulus that appears white. The perceptual dimension of the lightness ranges from black to white.

A single isolated colour stimulus cannot vary in lightness, and, if achromatic of any radiance, appears white, never grey or black. Reducing the luminance of the isolated achromatic stimulus causes it to appear dimmer while still white. An observer sees that varying the radiance of an isolated achromatic stimulus does not alter its appearance along the black-white perceptual dimension.

2. "*Chroma*, colourfulness of an area judged as a proportion of the brightness of a similarly illuminated area that appears white or highly transmitting"[3]

$$\text{Chroma} = \frac{\text{Colourfulness}}{\text{White Brightness}}$$

3. "*Saturation*, colourfulness of an area judged in proportion to its brightness"[3]

$$\text{Saturation} = \frac{\text{Colourfulness}}{\text{Brightness}} = \frac{\text{Colourfulness}/\text{White Brightness}}{\text{Brightness}/\text{White Brightness}} = \frac{\text{Chroma}}{\text{Lightness}}$$

then *saturation* is the chroma of an area judged in proportion to its lightness. (The definition of saturation, given for unrelated colours, is easily rewritten for related colours).

4.7   Colour Interactions

Colour interaction –

1. mutual influence of nervous visual signals producing a mutual dependence of the perceived colour attributes;
2. effect exploited by artists – first Paul Signac, George Seurat and Henri Matisse –, where uniform dashes of bright contrast colours are juxtaposed on the canvas producing modified contrast effects between the contiguous patches.

These two definitions of colour interaction are based on physiology and on perception, but regard the same phenomenon. The relative colour attributes are dependent on the colour stimuli present in the field of view and their values are related to the spatial organization of the colour stimuli.

Often, freely speaking, the locution 'colour-stimuli interaction' is used for colour interaction, although the colour stimuli, as physical quantities, are not interacting, but the interaction happens in the visual process, as stated in the definition.

The most noticeable phenomena produced by the interaction are due to the simple structure of few stimuli viewed in temporal succession or simultaneously in a still situation. The phenomena are very dependent on the complexity of the scene, although constituted by few stimuli.

1.   *Successive contrast* happens when a viewer sees an imaginary coloured field (*after-image*) in the area of view after having intently viewed an object for a while and then changed the gaze to a different, contrasting area or closed the eyes.

Physiologically, the photoreceptors of a given type, after exposure to strong light in their sensitivity range, become desensitized. For a few seconds after the light ceases, they will continue to signal less strongly than they otherwise would. Colours observed during that period will appear to lack the colour component detected by the desensitized photoreceptors.

For completeness the classification of after-images, made according to their appearance compared with the original response, is reported:
- *complementary after-image*: after-image of which the hue is approximately the complementary of the hue of the sensation produced by the original stimulus; (Section 16.4.4)
- *homochromatic after-image*: brief after-image with approximately the same hues as the original;
- *negative after-image*: after-image with lightness relations the reverse of the original image;
- *positive after-image*: fleeting after-image with lightness relations like the original;
- *Swindle ghost*: variety of positive after-image that may last a minute or more;
- *recurrent vision*: succession of after-images resulting from a flash stimulus of a fraction of a second duration.

2.   *Simultaneous contrast* happens when the appearance of a colour stimulus moves away from the colour of the inducing stimulus produced by contiguity with other colour stimuli. Generally only two colours are present and in mutual contact. Simultaneous contrast is a change in colour appearance active within the lightness domain as well as within the chromatic domain (coloured shadows, colour induction).[20–28] This interaction is distinguished between:

- *Simultaneous achromatic contrast* (or *brightness induction* or *brightness contrast*). Consider a colour stimulus surrounded by another colour stimulus, as the luminance of the surround approaches and surpasses the luminance of the colour stimulus, its appearance falls rapidly toward dark colour. An achromatic unrelated colour, when its visual field is modified by introducing another colour stimulus, can appear white, brighter white, grey or black (Figure 4.9).
- *Simultaneous chromatic contrast*, when the appearance of a colour stimulus moves away from the colour of the inducing stimulus[20–22] (Figure 4.10).
- *Simultaneous saturation contrast (induced saturation)* happens when a colour stimulus is surrounded by a field that is of the same colour but of stronger saturation, the colour stimulus appears to be tinted less vivid (less saturated). On the other hand, when a colour stimulus is surrounded by a field that is of the same colour but of weaker saturation, the colour stimulus appears to be tinted more vivid (more saturated) (Figure 4.11).

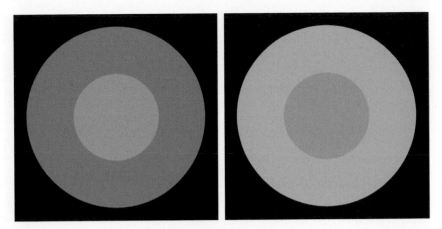

Figure 4.9   *Example of simultaneous achromatic contrast (or brightness induction or brightness contrast). The same grey disk with a luminance factor Y displayed internal to two grey backgrounds with luminance factor lower and greater than Y, respectively, appears differently bright: brighter if the background is darker and darker if the background is lighter.*

- *Crispenin.* A very little difference between two colour stimuli, which cannot be perceived if the two stimuli are seen on a very different surround stimulus, can appear enhanced if the two colour stimuli have a surround that appears between them. This phenomenon is a direct consequence of the simultaneous contrast.

  Brightness cispening happens if the difference between the stimuli regards only the brightness (Figures 4.12 and 4.13).
  Chroma crispening happens if the difference between the stimuli regards only the chroma[29] (Figure 4.14).

Figure 4.10   *Example of simultaneous chromatic contrast. Two equal pink disks are shown on a red and on a green background, respectively. The disk on the green background appears redder.*

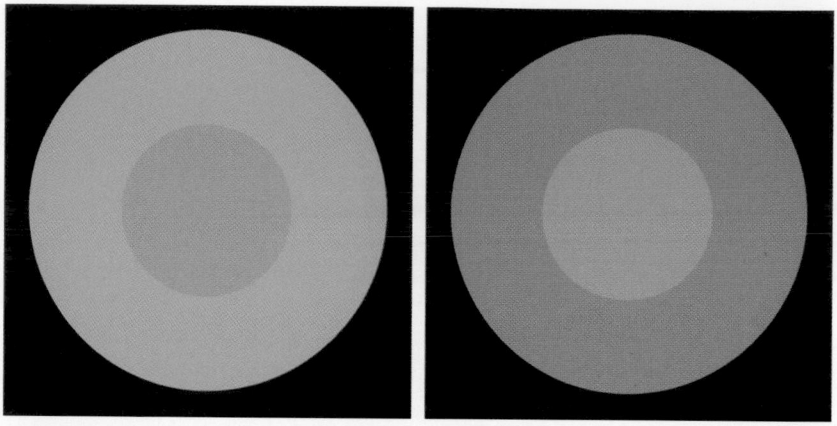

Figure 4.11 *Example of simultaneous saturation contrast. Two equal dim cyan disks are presented on a bright cyan and a grey background, respectively. The two disks appear differently: more saturated if the background has a weaker saturation and less saturated if the background has a stronger saturation, that is, the disk on the grey background appears more saturated.*

3.   *Spreading* or *assimilation*[30–32]. In related colours, an assimilation appears when a perceived colour shifts toward the adjacent inducing colour. Such an interaction is reciprocal. The change in colour appearance of any stimulus depends on the chromaticities of the interacting stimuli and on their spatial configuration. In general, assimilation occurs with fields of relatively high spatial frequency. Assimilation is distinguished into:

- *Achromatic assimilation* – also *von Bezold's spreading effect* –. A perceptual phenomenon in which the colour of an area is perceived as closer to the colour of the surroundings than it would if viewed in isolation. Assimilation occurs with stimuli with fine spatial structure, for example, a thin strip of colour between two black bars will look darker than it would between two white bars (Figure 4.15).

Figure 4.12 *First example of brightness crispening effect. Two grey disks are shown on three backgrounds different in lightness. The two disks are almost indistinguishable from each other on a background with a high brightness contrast (white and black backgrounds). The lightness difference appears if the disks are on a background with a lightness between that of the two disks.*

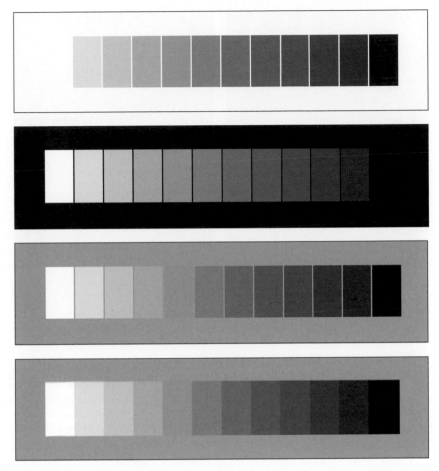

Figure 4.13 *Second example of brightness crispening effect. A grey scale almost uniform in lightness (CIE 1976 L* scale; section 11.2), ranging from black and white, is shown on three different backgrounds – white, black and grey –. The appearance of the scale presents no particular jump if the background is white or black, but if the background is grey the samples lighter than the background appear lighter and those darker appear darker with respect of the appearance of the same samples on white or black background. Particularly the samples of the grey appear uniform if the samples are separated, while appear bent with a lightness gradient appearing lighter in the side with contiguity with a darker sample and darker on the other side with contiguity with a lighter sample. This last phenomenon is that of the simultaneous lightness contrast.*

- *Chromatic assimilation.* Consider the example of Figure 4.16: two identical squares of equal grey stimulus have a superimposed grid red, yellow, respectively. The two backgrounds appear of different perceived colours. The grey background with red grid appears to be reddish while the grey background with yellow grid appears to be yellowish. Figure 4.17 shows two sets of identical red strips superimposed on a white and a black background. The red strips interlaced with white strips appear lighter, while the red strips interlaced with black strips appear darker. In Figure 4.18 four sets of strips of red, yellow, green and blue colours, respectively, are shown in interlaced combination with two colours: from the left, yellow-green, green-blue, yellow-red and red-blue. The four colours appear differently in the different combinations.

Figure 4.14   *Example of chroma crispening effect.[29] Two pink disks are shown on three different backgrounds. They look generally almost identical but their difference looks greatest on the background of most similar chroma. The grey scale image obtained by a transformation made by a customary program for imaging is shown in order to prove that the difference is of the chroma, because the imaging program leaves in the transformation the luminance factor unvaried, that looks equal for the two disks. (After a similar figure in Reference[29].)*

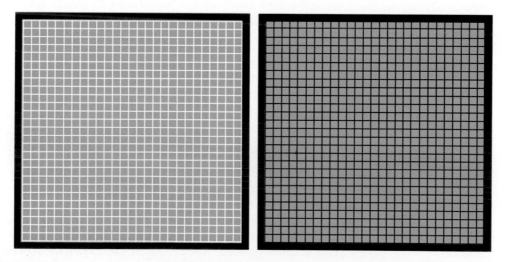

Figure 4.15   *Example of achromatic assimilation. Two grey squares with an overlying grid, white in the left square and black in the right one, appear lighter and darker, respectively. (After a similar figure in Reference[26])*

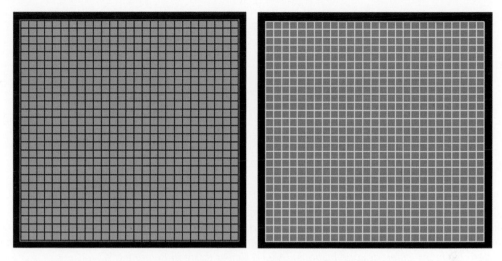

Figure 4.16   *Example of chromatic assimilation. Two grey squares with an overlying grid, red in the left square and yellow in the right one, appear reddish and yellowish, respectively. (After a similar figure in Reference[26].)*

4.   *White's effect*[33] and *Munker's illusion*[34]. White's effect is a brightness illusion where certain stripes of a black and white grating arc partially replaced by a grey rectangle (Figure 4.19). Both of the grey bars are identical. The appearance of the grey stimuli appears to shift toward the appearance of the top and bottom bordering stripes. White effect is called *Munker's illusion* when the achromatic colours are substituted by chromatic colours (Figure 4.19).

Figure 4.17   *Example of chromatic assimilation. Two sets of identical red strips appear lighter if on white background (left) and darker if on black background (right). (After a similar figure in Reference[26].)*

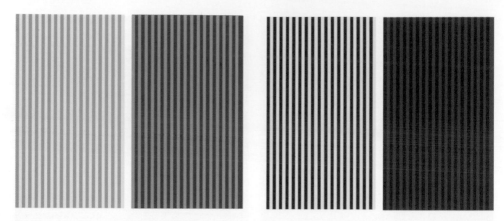

Figure 4.18   *Four sets of strips of red, yellow, green and blue colours, respectively, are shown in interlaced combination with two colours: from the left, yellow-green, green-blue, yellow-red and red-blue. The four colours appear differently in the different combinations. (After a similar figure in Reference[26].)*

5.   *Greyness* or *dark colour*. The adjective 'dark' is generally used to describe perceived colours with low levels of lightness in comparison with other colour stimuli in the scene. Dark colours are percepts that cannot be achieved as unrelated colours and are an example of how one colour stimulus is conditioning the appearance of another stimulus.

   The percepts of grey and black are examples of dark colours, which occur only as related colours seen simultaneously with other colours of higher luminance and in general are percepts that include a component of perceptual blackness or greyness.

   Blackness/greyness – Attribute of a visual sensation according to which an area appears to have more or less black/grey content.

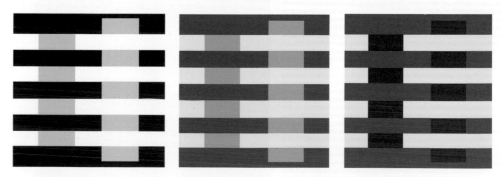

Figure 4.19   *The same geometrical pattern is shown in different coloured ways. In the left figure two equal grey strips appear unlike (White effect[33]). In the central figure two equal green strips appear unlike and similarly in the right figure two equal red strips appear unlike (Munker illusion[34]). The different extension and shape of the contiguity of these green (centre) and red (right) strips with the blue and yellow induce different appearances.*

The set of basic colour names given by Berlin and Kay is constituted by 11 names (Section 4.2). We can divide this set of colours into two sets, one corresponding to spectral lights and the other apparently not corresponding to spectral lights. The colours of this second set are white, black, grey and maroon. These colours exist only as related colours:

- An achromatic stimulus that appears white as unrelated colour, when seen with a surround of a colour stimulus with higher luminance appears grey and eventually black.
- A colour stimulus that appears yellow-orange as unrelated colour, appears brown when surrounded by an achromatic light of significantly higher luminance.

These colours are termed *dark colours* and their names, in addition to the black, grey, brown and maroon, are those of the objects that have those colours: midnight blue, ebony, taupe, Davy's grey, charcoal, café noir, black bean, black olive, onyx, phthalic green, navy blue, etc. The perceptual experience that regards dark colours is called *greyness* and was studied by Evans[35].

References

1. Newton I, *Opticks, or Treatise of the Reflections, Refractions, Inflections & Colours of Light*, Smith & Walford, London (1704), reprinted by Dover Publication Inc, NY (1952).
2. Brown R and Lenneberg E, A study in language and cognition, *J. Abnorm. Soc. Psychol.*, **49**, 454–462 (1954).
3. CIE Publication CIE S 017/E:2011, *ILV: International Lighting Vocabulary*, Commission Internationale de l'Éclairage, Vienna, Austria (2011). Available at: www.eilv.cie.co.at/ (accessed on 8 June 2015)
4. ASTM E284:13b, *Standard Terminology of Appearance*, ASTM International, West Conshohocken, PA (2013).
5. Berlin B and Kay P, *Basic Color Terms: Their Universality and Evolution*, University of California Press, Berkeley (1969).
6. Kay P and McDaniel CK The linguistic significance of the meanings of basic colors terms, *Language*, **54**, 610–646 (1978).
7. Heider E, Focal color areas and the development of color names, *Dev. Psychol.*, **4**, 447–455 (1971).
8. Heider E, Probabilities, sampling, and ethnographic method: the case of dani color names, *MAN*, **7**, 448–466 (1972).
9. Rosch E, Natural categories, *Cognit. Psychol.*, **4**, 328–350 (1973).
10. McDaniel CK, *Hue Perception and Hue Naming*. Unpublished BA thesis, Harvard University, Cambridge (1972).
11. wordnet, Princeton, Available at: www.wordnet.princeton.edu/ (accessed on 8 June 2015).
12. Hyde EP, Talbot's law as applied to the rotating sector disk, *Bull. Bureau Stand.* Bulletin 2, No. (1), 1–32 (1906).
13. Maxwell JK, Experimens on colour, as perceived by the eye, with remarks on colour-blidness, *Trans. R. Soc. Edinb.*, XXI, Part II, 126–156 (1857).
14. da Vinci Leonardo, *A Treatise of Painting*, translated to English by Rigaud JF and Bell G, London (1906), original Italian print (1651).
15. Hering E, *Grundzüge der Lehre vom Lichtsinn*, Springer, Berlin (1920); *Outlines of a Theory of the Light Sense*, Hurvich LM and Jameson D, eds., Trans., Harvard University Press, Cambridge (1964).
16. Sivik L, Color systems for cognitive research, in *Color Categories in Thought and Language*, HardinLC and MaffiL, eds., Cambridge University Press, New York (1997).

17. Shepard RN, The analysis of proximities: multidimensional scaling with an unknown distance function, I. *Psychometrika*, **27**, 125–140 (1962).
18. Shepard RN, The analysis of proximities: multidimensional scaling with an unknown distance function, II. *Psychometrika* **27**, 219–246 (1962).
19. Schiffman SS, Reynolds ML and Young FW, *Introduction to Multidimensional Scaling*, Academic Press, New York (1981).
20. Albers J, *Interaction of Color*, Yale University Press (1963).
21. Leo M. Hurvich, *Color Vision*, Sinauer Associates Inc., Sunderland (1981).
22. Evans RM, *Introduction to Color*, John Wiley & Sons, New York (1948).
23. Cao D and Shevell SK, Chromatic assimilation: spread light or neural mechanism? *Vision Res.*, **45**, 1031–1045 (2005).
24. Devinck F, Hardy JL, Delahunt PB, Spillmann L and Werner JS, Illusory spreading of watercolour, *J. Vision*, **6**, 625–633 (2006).
25. Logvinenko AD and Hutchinson SJ, Evidence for the existence of colour mechanisms producing unique hues as derived from a colour illusion based on spatio-chromatic interactions. *Vision Res.*, **47**, 1315–1334 (2007).
26. Shevell SK, Color appearance, in *The Science of Color*, 2nd ed., Shevell SK Ed., Optical Society of America and Elsevier Science, 149–187 (2003).
27. Smith VC, Jin PQ and Pokorny J, Color appearance: neutral surrounds and spatial contrast, *Vision Res.*, **38**, 3265–3269 (1998).
28. Walraven J, Spatial characteristics of chromatic induction: the segregation of lateral effects from stray-light artefacts, *Vision Res.*, **13**, 1739–1753 (1973).
29. Johnson GM and Fairchild MD, Chapter on visual psychophysics and color appearance, in *Digital Color Imaging Handbook*, Sharma G, ed., CRC Press, Washington, DC (2003).
30. Cornelissen FW and Brenner E, On the role and nature of adaptation in chromatic induction, in *Channels in the Visual Nervous System; Neuro-Physiology, Psychophysics and Models*, Blum B, ed., Freund Publishing House, London and Tel Aviv, pp. 109–123 (1991).
31. Blackwell KT and Buchsbaum G, The effect of spatial and chromatic parameters on chromatic induction, *Color Res. Appl.*, **13**, 166–173 (1988).
32. De Valois RL and De Valois KK, *Spatial Vision*, Oxford University Press, New York (1988).
33. White M, The effect of the nature of the surround on the perceived lightness of gray bars within square-wave test gratings, *Perception*, **10**, str. 215–230 (1981).
34. Munker H, *Farbige Gitter, Abbildung Auf Der Netzhaut und Ubertragungstheoretische Beschreibung Der Farbwahr Nehmung Habilitationschrift*, Ludwig-Maximilians Universität Munchen (1970).
35. Evans RM, *The Perception of Color*, John Wiley & Sons, New York (1974).

5

Visual System

5.1 Introduction

This chapter has solely an introductory purpose and, as it only considers generalities on the human visual system, the bibliography consists of textbooks, manuals and tutorials.[1-10] Only a few specialized articles are considered.

Colour vision is a phenomenon consisting of a sequence of events, the disclosures of which require different tools and methodologies, which are typical of physics, bio-chemistry, physiology and psychophysics. The science of colour vision investigates all the stages of the vision phenomenon, while it is not able to measure how brain activity gives rise to the conscious experience behind colour sensations.

The phenomenon of vision is activated by the colour stimulus that is an electromagnetic radiation, which is measured on the outside of the eye. The *eye* is the organ of vision with two functions:

1. One function is to create on the retina an optical image of the scene external to the eye. This phenomenon is optical and regards:
    i. geometrical optics for the construction of the image on the retina and its chromatic aberrations and
    ii. wave optics to define the resolution of the retinal image.
2. The other function is retinal and concerns the generation of nerve signals, which give rise to the visual sensations in the brain. The retina is a light sensitive layer of *tissues*, the outside part of the nervous system, in which occurs the transduction of the luminous flux into a nerve signal. Transduction is a photo-chemical phenomenon and, as such, is a quantum phenomenon and the nervous signal, generated in this process, is due to the number of photons (light quanta) absorbed by the photosensitive cells of the retina. In the retina, nerve cells with a complex processing of the nerve signals resulting from the transduction of the photoreceptors, produce the signal to be transmitted to the brain through the optic nerve.

The signals pass from the retina to the brain, where there are different places intended to treat the various aspects of vision. The correspondence between the colour sensations and colour stimuli is obtained by applying the tools of physiology and psycho-physical analysis.

This chapter

1. invokes those concepts of optics that are required to describe the activation of the visual system and
2. describes the anatomy and physiology of the visual system.

All that is treated concisely and only for the purposes of this book.

5.2   Eye Anatomy and Optical Image Formation

The eye is the organ of vision that has the structure of a camera, with an objective lens and a surface sensitive to the light, the retina.

The role of the eye is comparable with that of a photographic camera, even if with many fundamental differences (Figure 5.1). The features of the eye emerge enhanced from this comparison and make us better grasp the exceptional properties of the eye.

The elements that characterize a camera are the lens, the diaphragm, the shutter and the photosensitive device, which may be a film or a solid-state sensor with subsequent electronic processing and a storage memory.

The optical system of the eye, that replaces the camera lens, consists of four different media at mutual contact, placed in succession, with different geometric shapes and different refractive indices (Figure 5.1):

1. the *cornea*, which is the outer part of the eye with an approximately parabolic shape, refractive index $n = 1.37$, is responsible for most of the refraction and therefore of the optical power;
2. the *aqueous humour*, with refractive index $n = 1.33$;
3. the *crystalline lens*, with $n = 1.42$ and lenticular shape; and
4. the *vitreous humour* with $n = 1.33$.

(The refractive indices are wavelength dependent and are conventionally defined for the wavelength $\lambda = 589.3$ nm corresponding to the average of the yellow lines of sodium.)

Before the crystalline lens is the *iris*, a membrane with a central hole, termed *pupil*. The pupil diameter ranges from 1 or 2 mm to about 8 mm and automatically varies with the variation of luminous flux with a function similar to that of the diaphragm of the camera. The variations of the pupil have only a minimal effect on the visual process.

Inside the back of the eye is situated the *retina*.

The optical properties of the eye are:

1. "*Accommodation* – adjustment of the dioptric power of the crystalline lens by which the image of an object, at a given distance, is focused on the retina."[1]

    Visual accommodation, which is the focus of the scene in front of the eye on the retina and particularly on the fovea, is implemented by changing the thickness and curvature of the crystalline-lens. The state of vision where an object at infinity is in sharp focus with the crystalline-lens in a neutral or relaxed state is termed *emmetropic*. Otherwise, the focus in the camera occurs by moving the objective lens along the optical axis.

2. Visual acuity or visual resolution:
    i.   "Qualitatively – capacity for seeing distinctly fine details that have very small angular separation."[1]

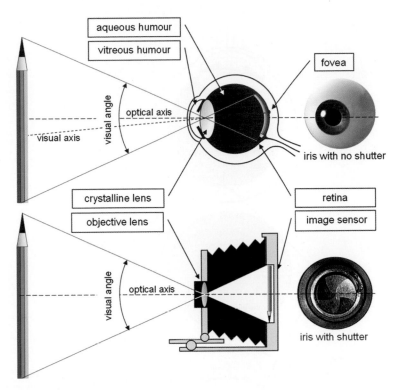

Figure 5.1 *Comparison of the eye with a camera showing similitudes and differences. Mainly, a shutter distinguishes the camera from the eye.*

ii. "Quantitatively – any of a number of measures of spatial discrimination such as the reciprocal of the value of the angular separation in minutes of arc of two neighbouring objects (points or lines or other specified stimuli) which the observer can just perceive to be separate."[1]

Visual acuity depends mainly on the size and shape of the photosensitive cells of the retina and is highest on the fovea (Section 5.4).

3. *Optical* and *visual axes* (Figures 5.1, 5.2 and 5.3)
   i. The *optical axis* of the visual system, analogous to that of the objective lens of the camera, passes through the centres of curvature of the cornea and of the lens, but not through the centre of the fovea.
   ii. The *visual axis*, which passes through the image nodal point of the optical system and the centre of the fovea.

4. *Field of view, visual angle* and *visual field*
   i. The *field of view* is the extent of the observable scene that is seen at any given moment. Since the eye is inserted into an orbit, the field of view is limited upward by the upper eyebrow, downwards by the cheekbone and side-wards by the nose. The field of view is represented on the retina by radial angles from the centre of the fovea, that is, from the centre of the field of view. Only towards the temple the useful range is about 90°, while it is on average 60° towards the nose, 70° upwards and 80° downwards. (Figure 5.2)

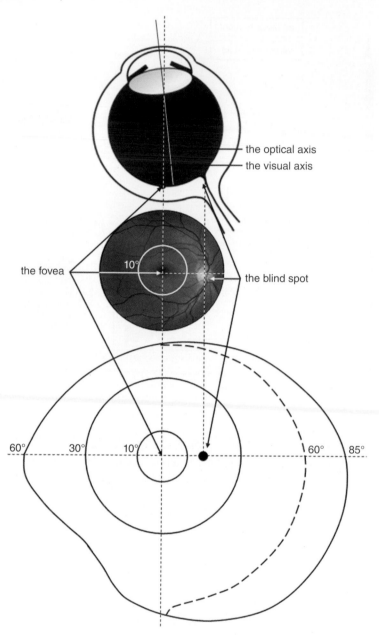

the optical axis
the visual axis
the fovea
10°
the blind spot
60°   30°   10°   60°   85°

Figure 5.2   *Interrelation between the eye, the retina and the visual field. At the top there is a section of the eye, in the middle a picture of the fundus of the eye and at the bottom a representation of the visual field of the right eye. The section of the eye is aligned on the photograph of the fundus of the right eye and the visual field is as it appears to the observer. The right and left fields of view (here non-represented) are overlapping and centred on the attention point, that is represented by the angular origin coincident with the fovea. (The optic nerve or blind spot is approximately 7.5° high and 5.5° wide.)*

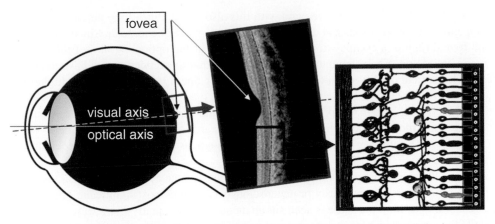

Figure 5.3   *Two steps of magnification of the retina in relation with the eye. A section of the human eye shows its various structures (left), also described in Figure 5.1. A thin piece of retina is enlarged in a picture made by the optical-coherent-tomography technique (OCT) (centre), and a sketch of a section of the retina revealing its layers (right). This sketch is enlarged in Figure 5.6.*

For optical instruments or sensors, the field of view is the *acceptance angle* through which a detector is sensitive to light (a solid angle measured in ste-radiant and often a plane angle measured in degrees, which is a section of a circular symmetric solid angle).

---

   ii.   When the eye observes a scene, at the same time all the object points of the scene, that constitute the field of view, are visible and their image is formed on the retina. The set of these points define the *visual field*. The field of view is the input of the visual system, which generates a picture in the brain called the visual field. (This distinction between field of view and visual field is important in neurology, psychology and visual neuroscience,[11] but in this book can be considered as synonymous.) The fovea subtends an angle in the visual field of about 0.087 rad (5°).[1]

   iii.   The *visual angle* – also object's *angular subtense* or object's *angular seize* – is the angle that a viewed object subtends at the object-nodal point of the eye, usually measured in degrees of arc (Figure 5.1).

   The eye does not take a rigid position, but it can turn into the orbit around a centre of rotation, located about 15 mm from the apex of the cornea. When we want to observe a point, with attention such as a star, the eye rotates so that the image of the star is formed in the fovea.

Reduced Eye

The anatomy of the eye is actually very complex, but for the evaluation of its optical performance it is sufficient to consider a *reduced eye* model.[4] Such a model has become useful in the design of optical instruments for use in ophthalmology and eye surgery, and in the calculation of the locus of the object points, that have the image on the retina, the chromatic aberrations and the diffraction patterns on the retina.

Acuity decreases rapidly outside the fovea and with this also the quality of the perceived image. The low visual acuity, chromatic aberrations and diffraction are overlapping phenomena. Since the region for acute vision is the foveal region, a computation of the aberrations of the optical system of the eye is meaningful only for foveal vision. These computations have shown that these phenomena have almost negligible visual effect. The optical quality of an objective lens of the camera is generally much higher than that of the optical system of the eye.

As regards to the visual system, the properties of vision in the visual field are not uniform, but vary greatly when the image point moves away from the visual axis, towards the periphery of the field. The quality of vision decreases for the optical properties of the optical system, as said before, and especially for the non-uniformity of the retina.

The mobility of the eye compensates for this limitation, because the eye scans in foveal vision the regions of the scene, that the observer consciously or unconsciously considers most significant. The resulting behaviour of the observer is as if all the scene is seen simultaneously in an equally sharp way.

To conclude the comparison between camera and eye, the sensor of the camera corresponds to the retina of the eye, and huge differences exist between these:

- The camera sensor generates a signal representing exclusively the exposure, which is determined by the aperture of the lens and the integration time.
- The retina generates a nerve signal
  - little affected by the opening of the pupil;
  - generated in a process without defined integration over time, that is, no shutter defines an exposure time;
  - dependent on the activations of the photoreceptors belonging to a receptive field (not individual photoreceptors), which interact, producing visual phenomena and particularly the adaptation phenomenon, for which
    - the retina operates in a huge range of luminous flux, whose extreme thresholds are in the ratio of $1:10^{14}$ and
    - vision depends weakly on the light level and on the spectral quality of the light that illuminates the scene.

5.3   Eye and Pre-retina Physics

This section considers the radiometric aspects of luminous radiation from its measurement out of the eye until its crossing with the photoreceptors of the retina.

The light collection of the optical system of the eye is considered in Section 2.4. The description of this phenomenon has been anticipated for introducing radiometric quantities. Spectral radiance is the magnitude most significant to the phenomenon of vision.

The activation of the photoreceptors is given by the number of photons absorbed by them and therefore we need to know the spectral distribution of the photon flux $\phi_{p,\lambda}$ arriving on the photoreceptors themselves (Section 5.5). Since the energy of a photon of wavelength λ is given by $U_p = h\nu = hc/\lambda$ (Section 2.3), it is possible to consider the spectral distribution of the power flux $\phi_{e,\lambda}$, that is $\phi_{e,\lambda} = \phi_{p,\lambda} \times h\nu$, and then use spectral radiometric units [W/nm], typical of the measuring instruments (Section 2.4), instead of the actinometric units required by physiology.

The light entering the eye, before getting to collide with the photoreceptors, travels through various media characterized by their optical properties. Moreover, the light is collected according to geometrical rules.

The phenomena that reduce the luminous flux are the absorption within a medium or a tissue and the reflection in the transition between media having different refractive indices.

Loss of Light Due to Reflection

The greater reflection occurs on the outer surface of the cornea, where, by assuming a refractive index $n_{air} = 1$ for air and $n_{cornea} = 1.37$ for the cornea and applying the Fresnel formula (Section 3.4.2), the reflectance for orthogonal incidence is approximately

$$\rho_{cornea} = \left(\frac{n_{cornea} - n_{air}}{n_{cornea} + n_{air}} \right)^2 \cong 2.5\% \tag{5.1}$$

over the entire visible spectral range, which corresponds to a spectral transmittance $\tau_{cornea}(\lambda) \cong 97.5\%$.

Between other media, the difference in refractive index is so low that an overall loss by reflection is in the order of 0.3%.

Loss of Light Due to Absorption

Three of the optical media of the eye – the cornea, the aqueous humour and the vitreous humour – have refractive index similar to that of water and very low optical density, for which, in the range of wavelengths between 400 and 600 nm, the aqueous and vitreous humour together absorb less than 10%, the cornea between 10% and 20% depending on the wavelength. For colorimetry, these absorptions are almost negligible, especially since they are almost constant over the entire visible spectral range.

The absorption of the crystalline lens is rather important and very dependent on the wavelength. The optical density of the crystalline δ_{lens} (Figure 5.4) is very high, greater than 3 units below 390 nm, but is less than 10% between 450 and 700 nm. Also the lens changes with age: it yellows, grows in thickness and increases its ability to spread light. These phenomena are not the same for all individuals.

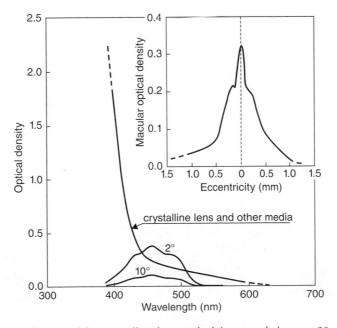

Figure 5.4 *Spectral optical density of the crystalline lens and of the macula lutea at 2° and 10° from the fovea centre. In the top right corner plot of macular optical density as a function of the distance from the centre of the fovea to the periphery. (After Reference 12.)*

5.4   Anatomy of the Retina

Before entering a more detailed description of the retina, two basic definitions should be given:

- *Cell* – the basic structural, functional and biological unit of all known living organisms.
- *Nerve cell* or *neuron* – an electrically excitable cell that processes and transmits information through electrical and chemical signals (*neurotransmitters*). These signals between neurons occur via specialized connections with other cells, termed *synapses*, small areas between axons and dendrites, which the nerve signals are transmitted through. A typical neuron possesses a cell body, dendrites and one axon. *Dendrites* are thin structures that arise from the cell body, often extending for hundreds of microns and branching multiple times, giving rise to a complex *dendritic tree*. An *axon* or *nerve fibre* is a special cellular extension that arises from the cell body and travels for a distance, from microns to as far as 1 m in humans.

"*Retina* is a membrane sensitive to light stimuli, barely half a millimetre thick (few tenths of a millimetre), that is situated inside the back of the eye and lines the inside of the eyeball."

Figure 5.5 is a picture of the optical fundus shot through the pupil of the eye. The back of the eye, seen through the pupil, is the retina and appears differently coloured and textured:

- A transparent red colouring of the blood covers the entire retina.
- A brown colouring is added to the red coloration in the central region of the retina gradually degrading up to fade. This region is called the *macula lutea*.

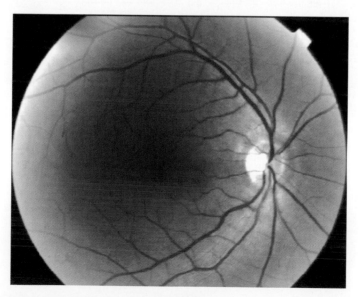

Figure 5.5   *Image of the retina (fundus) photographed through the eye pupil. The fovea clearly appears in the centre, the macula lutea in the central region, the blind spot at the right, the blood vessels everywhere except in the foveal region, and soft light streaks, axons of the ganglion cells, converging from all regions of the retina onto the blind spot.*

- The retina is covered by blood vessels, which emerge from a large region of a light colour, known as *optic disc*, thin and branch out everywhere excluding the central region, named *fovea*. The optic disc is also called the *papilla* and the *blind spot* because this region is without photoreceptors. The blind spot is located about 15° from the fovea, measured as the angle of vision on the retina, towards the nose.
- Light streaks converge onto the *optic disc*. They are the axons, which originate from the retinal ganglion cells, lie on the inner surface of the retina, converge into the optic disc and together constitute the *optic nerve*. The optic nerve is made up of about 1.3 million nerve fibres. The optic nerve passes through the retina at the optic disc, and exits from the eyeball to the brain.

Figure 5.3 shows different magnifications of a section of the retina. Figure 5.6 represents schematically the structure of a section of the retina showing four cell layers separated by two intermediate plexiform layers, which are evident also in the OCT picture of Figure 5.3.

The retina is considered a part of the brain because

- the retina includes the sensory neurons that respond to light;
- the retina includes intricate neural circuits that perform the first stages of visual processing; and
- the ganglion cells of the retina generate electrical signals, that travel down the optic nerve – set of axons of the ganglion cells – into the brain for further processing and visual perception.

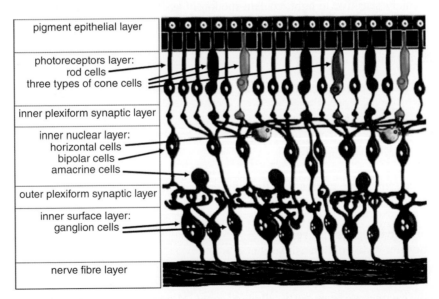

Figure 5.6 *Magnified sketch of a section of the retina. From the top: the photoreceptors close to a dark row of cells called the pigment epithelium; then four layers of different kinds of nervous cells: horizontal cells, bipolar cells, amacrine cells and finally the ganglion cells. The axons of the ganglion cells constitute the optical nerve, which carries the visual nerve signal from the eye to the brain. The light that enters the eye through the pupil crosses the ganglion cell layer (bottom in the figure) and the other layers of cells before crossing the photoreceptors and getting adsorbed activating the visual process. (After Reference 8.)*

5.4.1   Retina Layers

Starting from the outside towards the inside, the six layers, represented in Figures 5.3 and 5.6 are the following:

1. The eye *pigment epithelial layer* – a very dark tissue constituted by flattened cells is full of melanin granules, which provides *retinal*, or vitamin A, to the pigment-bearing membranes of the photoreceptors, that they have to be in contact with.
2. The *photoreceptors layer* against the back of the eyeball in contact with the pigment epithelial layer. The photoreceptor cells are *cones* and *rods*. The existence of two kinds of photoreceptive cells is at the basis of the *duplicity theory*, that states that two transduction mechanisms exist, related to the two different kinds of photosensitive cells (Section 6.2). The outermost part of the photoreceptors, facing the pigmented epithelium, is called the *outer segment* and is constituted by a membrane containing a chemical substance (*photopigment*) that transforms – it isomerizes – when it absorbs electromagnetic radiation of appropriate wavelength. Since this transformation is in the photopigment, an electrical signal is generated in the photoreceptive cell, which is the first neural response to colour stimuli, initiating the sequence of neural events that constitute the visual process. The two types of photoreceptors, rods and cones, have different functions and different characteristics:

Rods

Rods are responsible for night vision, or *scotopic*, that is colourless vision at low levels of illumination (Sections 6.2.1 and 6.2.2).

Rods are 40–70 μm long and thin with a diameter of 1–2.5 μm.

Rods are extremely sensitive to light, contain rhodopsin as the light absorbing pigment responsible for transduction, and provide achromatic vision.

Cones

Cones are responsible for daily or *photopic vision*, that is the colour vision at high illumination levels.

Cones have a squat shape with a conical tip.
Cones are 40–70 μm long with a diameter 1–9 μm.
Cones are less light sensitive than rods.

Cones are divided into three classes with three different photopigments sensitive to light, responsible for transduction:

  i. L-cones, long wavelength sensitive cones
 ii. M-cones, medium wavelength sensitive cones
iii. S-cones, short wavelength sensitive cones

(Sometimes the L-cones are improperly called red cones, M-cones green and S-cones blue.) A comparison of the signals generated by the three types of cones implies that the vision is in colour (Section 6.13).

3. The *outer plexiform layer*, region containing synapses linking the photoreceptors with the dendrites of the bipolar and horizontal cells.
4. The *inner nuclear layer*, region with one to four types of *horizontal* cells, 11 types of *bipolar* cells and 22–30 types of *amacrine* cells.[8] The *horizontal* cells establish contacts between photoreceptors. The

bipolar cells receive nerve signals through their dendrites from the photoreceptors, with whose terminals they form synapses. *Amacrine* cells, localized in the proximity of the ganglion cells, are of many different types and have different functions. The horizontal cells and amacrine regulate the transmission of information from the receptors to the bipolar cells and from bipolar to ganglion cells, respectively.

5.   The *inner plexiform layer*, the region where the bipolar and amacrine cells connect to the ganglion cells.
6.   The *inner surface layer* of the retina with about 20 types of *ganglion* cells.
7.   The *nerve fibre layer*. Impulses from the ganglion cells travel to the brain via more than a million optic nerve fibres per eye. As already said, these fibres are the axons of the ganglion cells, are very long, lie on the inner surface of the retina and converge in the *optic disc*, where they meet to form the optic nerve (Figure 5.5).

Optical radiation must pass through the cells of these layers before reaching the photoreceptor cells and exciting the pigment molecules.

The structure of the retina is not uniform and is divided into several parts.

5.4.2   Fovea

Fovea or *fovea centralis* is the central part of the retina, thin and depressed, which contains almost exclusively cones and constitute the site of most distinct vision (Figures 5.7a and 5.7b). The fovea has a diameter of a few tenths of a millimetre and subtends an angle at a maximum of about 0.087 rad – that is, 5° – in the visual field.[1]

Foveal cones are particularly long and thin. The bipolar and ganglion cells associated with these foveal cones are not in front of them, as is the case in other parts of the retina, but are shifted around towards the edges of the fovea. Then this is the only region of the retina where the photoreceptors receive direct radiation coming from external objects and focused by the optical system of the eye. Moreover in the fovea for each foveal cone, there is a direct line with at least one bipolar cell and one ganglion cell. For all these properties, the fovea is the region of the retina that allows the best view of fine details, that is, the best *visual acuity*. (Section 5.2).[8] Maximum discrimination of the optical system of the eye is of six pairs of lines per millimetre viewed at a distance of 25 cm, corresponding to 26 pairs of lines to the degree (Figure 5.8).

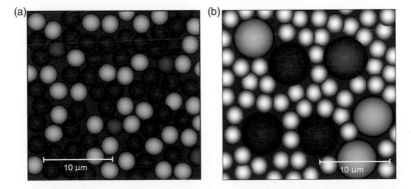

Figure 5.7   *(a) Sketch of the cone mosaic in the fovea, where only cones are present. The colours of the cones are indicative and not realistic: blue for the S cones, green for the M cones and red for the L cones. The distribution of the different types of cones is not regular and is very different between different individuals. Often the cone mosaic presents clusters of cones of the same type. (b) Sketch of the mosaic of cones and rods outside but near the fovea, where rods and cones are both present. The rods are thin while the cones are squat. The colours of the cells are indicative and not realistic: grey for the rods, blue for the S cones, green for the M cones and red for the L cones. (After Reference 13.)*

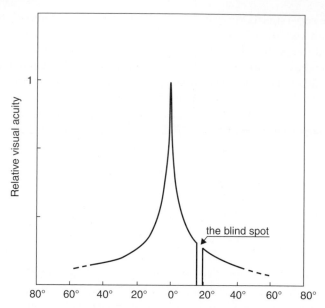

Figure 5.8 *Diagram of the relative acuity of the right human eye in degrees from the fovea, along a line passing through the fovea and the blind spot of an eye from the nasal (right) to the temporal side (left) of the retina. The visual acuity is the reciprocal of the smallest angle α that can be resolved between two adjacent black bars of a grating: $VA = 1/\alpha$ [arcmin^{-1}]. The value $VA \approx 1$ is considered normal for an illuminance of 200–300 lx. This value corresponds approximately to six line pairs per millimetre at a distance of 25 cm, or equivalently to 26 line pairs per degree. (After References 14–16.)*

5.4.3 Foveola

"*Foveola* is the central region of the fovea which contains no blood vessels."[1] The foveola subtends an angle of about 0.017 rad – that is, 1° – in the visual field. The central area of the foveola subtends an angle of about 0.003 rad – that is, 0.2°– and contains no S-cones.

5.4.4 Extra Fovea

Outside the fovea the photoreceptor layer contains rods, in addition to cones. Rods are few in number in the region near the fovea, but increase in number in regions progressively more peripheral (Figures 5.7b and 5.9).

Outside the fovea also there are changes in the relationship between photoreceptors and bipolar cells and between bipolar and ganglion cells. More photoreceptors are in contact with each bipolar cell, and more bipolar cells are in contact with each ganglion cell. This convergence of a growing number of photoreceptors on every single ganglionic cell grows gradually away from the fovea.

The total number of ganglion cells, that is equal to the number of fibres of the optic nerve, is approximately only 1.3 million, compared to the many millions of photoreceptor cells.

The retina is a vascularised tissue and the vascularization acts as a filter for the light that reaches the photoreceptor. This filter is similar to a 3-μm-thick layer of blood, with the exclusion of the foveal region, where vascularization is absent. Its absorbance has a bandwidth of about 40 nm with the maximum at about 430 nm. The major blood vessels create a shadow on the photoreceptor, which is imperceptible to the viewer.

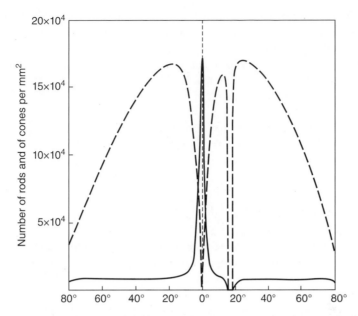

Figure 5.9 *Angular distribution of rods (dashed line) and cones (continuous line) at the bottom of the retina, along a line passing through the fovea and the blind spot of an eye from the nasal (right) to the temporal side (left) of the retina. The angle 0° corresponds to the centre of the fovea. In the fovea the rods are completely absent while the cones are mainly concentrated in the fovea. Their distribution is such as to allow colour vision over the entire visual field.[14–16]*

5.4.5   Macula Lutea

Macula lutea or *yellow spot* is a layer of inert photo-stable pigment covering a region of the retina having the fovea at the centre, whose optical density decreases from the centre to the periphery.[1] The extension of the macula lutea is about 10° horizontally and 6° vertically (Figure 5.5).

The optical density of the macular pigment is plotted as the spectral density and as a function of the radial angle centred in the fovea in Figure 5.4. This pigment acts as a non-uniform filter modifying the spectral composition of the radiation that enters the photoreceptors leading to different photoreceptor activations. This implies that macular vision is different from extramacular vision and therefore two colorimetric observers have been defined for the various retinal region, one for the foveal vision – 2°–4° visual field – and a second for extramacular vision – 10° visual field – (Chapter 9, Sections 9.2, 9.3 and 9.6).

5.4.6   Rod and Cone Distribution

In each retina, there are approximately between 75 and 150 millions of rods and 6–7 millions of cone cells.[15] The distributions of cones and rods on the retina are very different and very non-uniform, moreover great differences exist between individual observers. Recent studies[13] proved large individual variations, not only in the relative numbers of cone types, but also in their spatial distributions on the retina. Often clusters of one cone type surround areas of another type. Figure 5.7a is a reproduction of such a distribution 1° away from the central fovea. (For a long time it was assumed wrongly that the relative number of cones in the retina is in the proportion of 32:16:1 between L-, M- and S-cones.)

The distribution of the various types of photoreceptors in the retina has a particular structure that reveals different visual functions in various parts of the retina. (Figure 5.9)

The L-cones and M-cones are much more numerous than S-cones. S-cones are absent in the centre of the foveola, an area of about 20', and throughout the retina does not exceed in total 10% of the total number of cones.

Such retina heterogeneity and these differences between individuals do not cause appreciable differences in colour vision. The chromatic sensation for macular vision is substantially equal to that of extramacular vision, notwithstanding that the pigment of the macula filters modifies the spectral distribution of the radiation and therefore entails a different illumination of the photoreceptors.

In the blind spot and behind the blood vessels, the photoreceptors are enlightened little or not at all. This involves visual gaps which are integrated in the visual process creating a post-receptor visual field without interruption. This phenomenon is purely illusory, because the image that is formed on the blind spot is not perceived, and is absent from the visual field.

5.5   From the Retina to the Brain

In the retina, there are various types of photoreceptor cells that are differentiated by their type of photosensitive pigment and their different sensitivity to light. The visual situations are distinguished in correspondence to the different sensitivities of different kinds of photoreceptors:

1.   "*Scotopic vision*: vision by the normal eye in which rods are the principal active photoreceptors; it normally occurs when the eye is adapted to levels of luminance of less than ~10^{-3} cd/m^2. The rods are active at levels of illumination ranging from 10^{-6} to 10 cd/m^2, typical of night and twilight. Scotopic vision is characterized by the lack of colour perception and by a shift in visual sensitivity towards shorter wavelengths."[1]

2.   "*Photopic vision*: vision by the normal eye in which the cones are the principal active photoreceptors; it normally occurs when the eye is adapted to levels of luminance of at least 10 cd/m^2. The cones are active at levels of illumination ranging from 0.01 to 10^8 cd/m^2, typical of daylight. Colour perception is typical of photopic vision."[1]

3.   "*Mesopic vision*: vision by the normal eye intermediate between photopic and scotopic vision; both cones and rods are active. Mesopic vision is between two luminance levels ranging from 0.01 to 10 cd/m^2."[1]

These definitions use the adjective 'adapted', which refers to the adaptation phenomenon with obvious effects in the perception of brightness and colour appearance. This is considered in Sections 6.2 and 6.20.

5.5.1   Scotopic Vision

In scotopic vision operate only rods, which are very sensitive to light and are active at a flux of light at which the cones are insensitive.

Rhodopsin is the photopigment of the rods, that is a protein, opsin, reversibly bound to retinal, one of the many forms of vitamin A. Rhodopsin, if exposed to light with spectral power distribution in the range between 400 and 600 nm, immediately has a conformational change in opsin – *cis*-retinal isomerizes into *trans*-retinal and rhodopsin bleaches – converting light into metabolic energy. Bleached rhodopsin takes about 45 minutes for complete regeneration.[17] Rhodopsin absorbs green-blue light and appears reddish-purple, therefore it is called 'visual purple'. The absorbance spectrum of rhodopsin depends on opsin protein to which it is bound and visual perception is directly dependent on such a spectrum (Figure 5.10). In Section 6.12, the absorbance

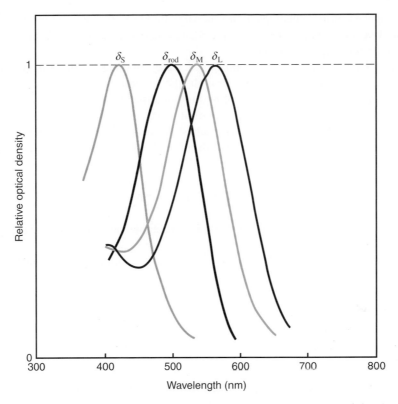

Figure 5.10   *Relative spectral absorbance of human rhodopsin δ_{rod} (dashed line) and the three photopsins δ_S, δ_M and δ_L normalised at their maximum. The quantum efficiencies of the photoreceptors considered as photodetectors are $\eta_S \propto \delta_S$, $\eta_{rod} \propto \delta_{rod}$, $\eta_M \propto \delta_M$, $\eta_L \propto \delta_L$. (After References 19–21.)*

of the photopigment is considered in relation to the spectral sensitivity of scotopic vision, assessed with psychophysical technique, and applying Rushton's univariance principle.

5.5.2   Photopic Trichromatic Vision

In photopic vision, with daytime luminous flux, only the cones operate, while the rod pigment is completely bleached and transparent to light. The cone-type photoreceptors are of three classes with photopigments with three different absorption spectra.

The cone photopigments are *iodopsins* – or *cone opsins* –. These pigments are very close analogues of the rod rhodopsin and consist of a protein termed *photopsin* bound with a retinal. Three different *photopsins* characterize the three classes of cones[18]:

1. photopsin I, *protan*, *erythrolabe* (absorption maxima for red light), present in the long wavelength sensitive cones (L-cones);
2. photopsin II, *deutan*, *chlorolabe* (absorption maxima for green light) present in the medium wavelength sensitive cones (M-cones); and
3. photopsin III, *tritan*, *cyanolabe*, (absorption maxima for bluish-violet light) present in the short wavelength sensitive cones (S-cones).

The absorbance spectra of these photopsins are shown in Figure 5.10. The absorption curves of the L- and M-cones are partially overlapping, while that of the S-cones is largely separated from the other two. Colour vision is dependent on these absorption spectra.

More recent data from human photoreceptors[19–21] were obtained by measuring the light absorbed by the photopigment molecules directly through the photoreceptors. This procedure is called micro-spectro-photometry. The wavelengths of maximum absorbance of each curve are 420 nm for the short wavelength sensitive cones, 534 nm for the middle and 564 nm for the long ones.

Over the past decades, research has disclosed much of the genetic basis of the production of visual pigments[22–23] and has led to the finding that people with 'normal' trichromatic vision possess genetic variants of the cone-pigment types[24]:

- Among L-cone pigment genes, approximately 56% have serine amino acid and 44% have alanine amino acid at the same genetic position in DNA.
- Among M-pigment genes, approximately 6% have serine amino acid and 94% have alanine amino acid.

These polymorphisms in the L- and M-pigment genes produce variations in colour vision,[25–26] thus, the one who is called a standard observer, defined in psychophysical way, has spectral cone absorbances that could be considered as a weighted average of the different genetic absorbances.

Colour vision based on three types of photoreceptors and three consequent channels for conveying and processing colour information is termed *trichromatic*. The condition of having trichromatic vision is termed *trichromacy*.

5.5.3   Rushton's Univariance Principle and Photoreceptor Activation

The visual process starts with the transduction of energy of absorbed light radiation into nerve signals. The absorption by the photopigment molecules is a photochemical process according to *Stark-Einstein's photochemical equivalence law*. This law says that every photon that is absorbed will cause a (primary) chemical or physical reaction. The application of this law has been made by W. A. H. Rushton to quantify the activation of the rods in visual processing. This is stated in *Rushton's univariance principle*:

> "*Principle of Univariance* – The visual effect of a light depends upon the quantum catch but not upon what quanta are caught."[27]

The absorbance of each type of photo-pigment changes with the wavelength and thus the probability that a photon is absorbed varies analogously with the wavelength. Rod and cone visions result entirely from the catch of quanta by photoreceptor pigments. The *activation process* of a photoreceptor consists in the absorption of light quanta by photopigment molecules and their isomerization.[28] The *activation of a photoreceptor* is defined by the number of absorbed photons per unit time and gives no information on the spectral quality of the absorbed radiation (Figures 5.10 and 5.11). The activation of the L-, M- and S-cones is specified by a set of three numbers (L, M, S). The spectral absorbances of the three kinds of cones are so overlapping that no radiation can activate only one kind of cones at a time and generally all three kinds of cones are simultaneously activated, although differently. The possibility of giving different visual sensations to several different spectral radiations is due to the three types of cones, L-, M- and S-cones, that absorb differently and then are activated differently. Photoreceptor activation (L, M, S) is the *psychophysical specification of a colour sensation* (Sections 6.13, 7.8, 9.5 and 9.6).

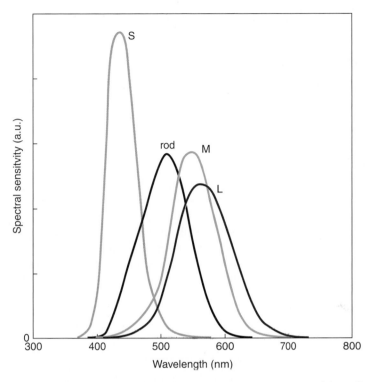

Figure 5.11   *The spectral sensitivity (derived from absorption spectrum measured by reflexion densitometry on the living eye) of the three cone pigments erythrolabe, chlorolabe, cyanolabe and of the rods scaled to give equal areas under each curve. (After References 18, 29–30.)*

5.5.4   Horizontal Cells

Horizontal cells link together many photoreceptor cells and bipolar cells and have different functions[8]:

(i)   Horizontal cells influence cone-photoreceptor activity with feedback signals from the inner plexiform layer.

(ii)   Horizontal cells modulate the photoreceptor signal under different lighting conditions, allowing signalling to become less sensitive in bright light and more sensitive in dim light (adaptation process).

(iii)   Horizontal cells modulate the photoreceptor signal shaping the receptive field of the bipolar cells. (This complicated circuit from horizontal cell to photoreceptor to bipolar cells is still not completely understood).

(iv)   Horizontal cells can make the response of the bipolar cells colour-coded.

5.5.5   Bipolar Cells

Bipolar cells receive signals from the photoreceptors, either rods or cones, but not both, or from horizontal cells: bipolar cells are designated *rod bipolar* or *cone bipolar* cells depending on whether they receive synaptic input from either rods or cones. There are roughly 10 distinct forms of cone bipolar cells. Bipolar cells are so-named because they have a central body from which two sets of processes arise: they receive input signal from the photoreceptors or the horizontal cell and pass on to the ganglion cells, directly or indirectly via amacrine cells.

Unlike most neurons, bipolar cells communicate via graded potentials.

5.5.6 Amacrine Cells

The functional roles of the amacrine cells is almost unknown. They have extensive dendritic branches with bipolar and ganglion cells and give most input to *phasic magnocellular* ganglion cells (MC) – also *parasol* cell – than to *tonic parvocellular* ganglion cells (PC) – also *midget* cell –.[8]

5.5.7 Ganglion Cells and Visual Pathways

There are about 100 times more photoreceptors in the retina than there are ganglion cells and nerve fibres. Therefore, on average, signals from many rods and cones converge onto one ganglion cell (pp, 123–124 of Reference 6).

Every single ganglion cell is activated only if a specific area of the retina is illuminated, which is called the *receptive field* of the cell. Typically, the receptive field of ganglion cells consists of two concentric regions, which have opposite effects on the activation of the cell: the *centre* and the *surround* of the receptive field.[8]

The structure of the net of the neural cells is different in the fovea with respect to the other regions. H. Kolb[8] wrote:

"In contrast to the rest of the retina, the human fovea contains *midget* ganglion cells, which have minute dendritic trees connected in a one-to-one ratio with midget bipolar cells. The channel from midget bipolar to midget ganglion cell carries information from a single cone, thus relaying a point-to point image from the fovea to the brain. Each red or green cone in the central fovea connects to two midget ganglion cells, so at all times each cone can either transmit a dark-on-light (OFF) signal or a light-on-dark (ON) message. The message that goes to the brain carries both spatial and spectral information of the finest resolution."

The ganglion cells actually are classified in three types specified in Table 5.1, to which correspond proper visual pathways (pp. 127–128 in Reference 6).

Table 5.1 *Classification of the visual pathways.*

Visual pathway	Tonic[a] opponent parvo-cellular	Tonic[a] opponent konio-cellular	Phasic[b] non-opponent magno-cellular
Ganglion cell	Parvocellurar *midged*	Koniocellurar	Magnocellurar *parasol*
Percentage of all the ganglion cells	70–80%	10–20%	~10%
Receptive field		Relatively large	Large
Input	Input from L- and M-cones	Input from S cone	Input from L- and M-cones
Opponency	Opponent	Opponent	Non-opponent, with spectral response $V(\lambda)$-like

[a]A tonic cell responds with a sustained response to a change in the stimulus.
[b]A phasic cell responds transiently to a change in stimulus.

At first sight, the three visual pathways could be associated to the three opponencies hypothesized by Hering, but many years of research showed that these opponencies are not tightly linked to unique hues, as supposed by Hering[31] (Sections 4.4 and 6.16) (pp 280, 395 in Reference 6).

5.5.8 From the Ganglion Cells to the Visual Cortex

Figure 5.12 shows the pathway of the visual nerve signals from the retina to the primary visual cortex.

The set of the axons of the ganglion cells, approximately 1.3 million fibres, constitutes the optic nerve. Most of these axons send ganglion signals to the lateral geniculate nucleus and then to the brain. A very small portion of the axons terminate in the *pretactal nucleus*, from which depart nerve fibres directed to the midbrain, where, through another two types of fibres[32-33]:

1. they innervate the constrictor muscle of the pupil (*miosis*), depending on the brightness of the scene view;
2. they innervate the muscle of visual accommodation reflex, that is, eye adaptation for the near objects vision (corresponding to the autofocus system in a camera).

The optic nerves from the two eyes cross at the *optic chiasm* and from here the axons are reorganized to constitute the *optic tracts* (Figure 5.12 and 10.5): axons of the ganglion cells of the right side of each of the two retinas come together to constitute the right optic tract, and those of the left side the left optic tract. The axons of the two optic tracts terminate in the lateral geniculate nuclei (LGN), left and right, respectively. In this way, the information coming from the two eyes is organized according to the two half-visual fields: right visual field signals converge into the left LGN and left visual field signals converge into the right LGN.

The LGN does not perform visual signal processing and preserves the separation of information from the two eyes. The neurons of the LGN have properties similar to those of the ganglion cells of the retina. LGN reproduces in an orderly way the distribution of retinal cells: in each layer of the NGL there is a map of the visual semifield reproducing the order present in the retina.

The axons of the cells of the two right and left LGN constitute two bundles called *optic radiations*, which end in the *primary visual cortex* of the left and right occipital hemisphere, respectively. In the right and left parts of the primary visual cortex are reproduced the maps of the left and right visual semi fields. Such a reproduction on the cortex has different scales to those on the retina. The fovea has a very small area on the retina while the corresponding area on the cortex is very large. On the contrary, the areas of the retina outside the fovea are represented on the visual cortex by a scale reducing as one moves away from the centre of the visual field. The cortical areas are proportional to the amount of information, which are greater in numbers in the foveal region compared to the eccentric regions of the retina.

There are three channels of information from retina to primary cortex:

1. a luminance channel, *phasic*, that is, it responds transiently to a change in stimulus, and
2. two cone-opponent channels, *tonic*, that is, responds with a sustained response to a change in the stimulus.

The distinction between the two kinds of opponent colour neurons, with different spectral properties, is present also in the visual cortex, although the neurons respond in part differently from the neurons of the retina and LGN.

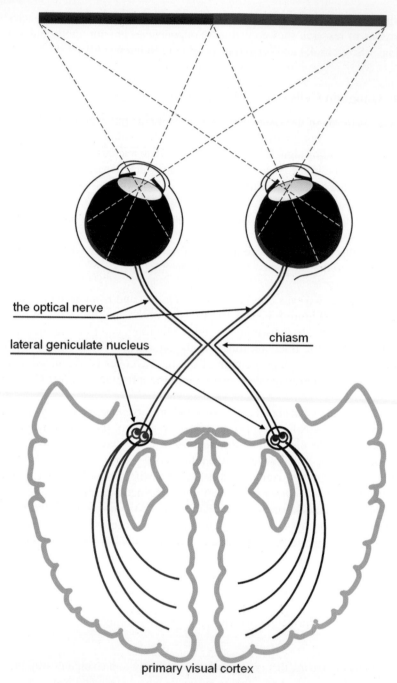

Figure 5.12 *Scheme of visual neural pathways, from the retinas of the two eyes to the primary visual cortex in the right and left hemisphere of the brain, through the mid-station of the lateral geniculate nuclei (Figure 10.5). The visual field is subdivided into the right and left parts and represented by red and blue, respectively. Right and left parts of the scene are seen by both eyes and by a proper separation are separated and recombined separately in the left and right parts of the visual cortex. This achieves a mixing of the nerve signals coming from the two eyes but related to equal visual semi-fields.*

A great deal of information arrives at the primary visual cortex – also called *striate cortex* or *V1 area* – and regards the colour and geometry of the scene, separately:

- A one to one correspondence exists of the points on the cortex with the points of the scene projected onto the retina. A visual field is associated with each neuron. The neurons within their field are active in correspondence to the orientations of the contrast lines between light and dark areas.[7]
- The majority of the other cells has properties of chromatic response, with excitatory response to the radiation of a part of the spectrum and inhibitory for other. The chromatic properties of cortical cells are more varied than the ganglion cells and the neurons of the LGN, and cover not only red-green and blue-yellow opponencics but present opponencies with other orientations (p. 395 in Reference 6).
- "Double opponency color cells exist, which are both spatially and chromatically opponent, therefore are able to signal chromatic boundaries. That was impossible for single opponent cells in the retina and in LGN."[30]

In the brain, there are visual areas with different functions, which are not considered in detail here. Here only important aspects of colour vision are considered. The neurons of the primary visual cortex (area V1), in turn, send their axons to a neighbouring area of the visual cortex (V2), and from there to other areas.

Area V4 contains an unusually large proportion of cells with sharp chromatic selectivity. The cells in area V4 of the visual cortex respond to the colour and not to the wavelengths of a stimulus. In particular the response to a coloured stimulus in the middle of its receptive field depends on the colour of light falling in surrounding regions, in a manner that such a response is correlated with the coloured appearance of the colour stimulus to a human observer.[34–37]

5.6   Visual System and Colorimetry

Most of the phenomena presented in this chapter concern the physiology of the primate visual system. It is believed that human physiology is very similar. At the moment, the phenomena described here seem relevant for colour vision. The study is ongoing and this experimental information, albeit limited, provides a basis of comparison for each theoretical model of the human visual system.

After this extremely synthetic view of the visual system, how brain activity gives rise to conscious experience remains an enigma, as enigmatic is the processes behind colour qualities, for example, the redness of red (pp. 127–128 in Reference 6).

Colorimetry should be defined on the greatest physiological evidence. Four main physiological processes are evident in this synthetic view of the colour visual system:

1. the activation of three kinds of photosensitive cells;
2. a mixing of the signals of the photosensitive cells;
3. a processing of the visual signals (tonic channel) that manifests different opponencies at two different levels – in ganglion cells and in visual cortex –; and
4. a processing of the visual signals (phasic channel) that manifests a spectral response $V(\lambda)$-like, probably to be associated to the luminosity (Section 6.5).

Any attempt to define a theory/model for colorimetry should consider this evidence. Other important inputs for such a theory are given by psychophysics (Chapter 6).

Bibliography

1. CIE Publication 017.2/E:2009, *ILV: International Lighting Vocabulary*, Commission Internationale de l'Éclairage, Vienna (2009).
2. Artal P, Optics of the eye and its impact in vision: a tutorial. *Adv. Opt. Photon.* **6**, 340–367 (2014).
3. Wyszecki G and Stiles WS, *ColorScience, Concepts and Methods, Quantitative Data and Formulae*, John Wiley & Sons, New York (1982).
4. http://paperzz.com/doc/1498054/how-to-model-the-human-eye-in-zemax—contrast-optical; http://paperzz .com/download/1498054#pdf (accessed 8 June 2015)
5. Kaiser PK and Boynton RM, *Human Color Vision*, Optical Society of America, Washington (1996).
6. Valberg A, *Light Vision Color*, John Wiley & Sons, Chichester (2005).
7. Hubel D, *Eye, Brain, and Vision*, Henry Holt and Company (1995).
8. Kolb Helga, How the retina works, *Am. Sci.*, **91**, 28–35 (2003).
9. Werner JS, Human colour vision: 1. Colour mixture and retino-geniculate processing, *Neuron. Coding Percept. Syst. World Scient.* **79**, 101–124 (2001).
10. Packer O and Williams DR, Light, the retinal image, and photoreceptors, in *The Science of Color*, 2nd ed., Chap. 2, OSA-Elsevir, Amsterdam (2003).

References

11. Smythies J, A notes on the concept of the visual field in neurology, psychology, and visual neuroscience, *Perception*, **25**, 369–371 (1996).
12. van Norren D and Vos JJ, Spectral transmission of the human ocular media, *Vision Res.*, **14**, 1237–1244 (1974).
13. Roorda A and Williams DR, The arrangement of the three cone classes in the living human eye, *Nature*, **397**, 520–522 (1999).
14. Curcio CA, Sloan KR Jr, Packer O, Hendrickson AE and Kalina RE, Distribution of cones in human and monkey retina: individual variability and radial symmetry, *Science*, **236**, 579–582 (1987).
15. Riggs L A, Vision, in *Woodworth and Schlosberg's Experimental Psychology*, 3rd ed., Chap. **9**, Kling, JW and Riggs, LA, eds., Holt, Rinehart, and Winston, New York, pp. 273–314 (1971).
16. Hadjikhani N, Projection of rods and cones within human Visual cortex, *Hum. Brain Mapp.*, **9**, 55–63 (2000).
17. Stuart JA and Brige RR, Characterization of the primary photochemical events in bacteriorhodopsin and rhodopsin, in *Rhodopsin and G-Protein Linked Receptors*, Lee AG, ed., Part A, Vol. 2, JAI Press, Greenwich, CT, pp. 33–140 (1996).
18. Wald G, *The Molecular Basis of Visual Excitation*, Nobel Lecture (1967). Available at: http://www. nobelprize.org/nobel_prizes/medicine/laureates/1967/wald-lecture.pdf (accessed on 12 December 1967).
19. Bowmaker JK and Dartnall HJA, Visual pigments of rods and cones in a human retina, *J. Physiol.*, **298**, 501–511 (1980).
20. Bowmaker JK, Dartnall HJ and Mollon JD, Microspectrophotometric demonstration of four classes of photoreceptor in an old world primate, Macaca fascicularis, *J. Physiol.* **298**, 131–43 (1980).
21. Dartnall HJA, Bowmaker JK and Mollon JD, Human visual pigments: microspectrophotometric resuls from the eyes of seven persons. *Proc. R. Soc. Lond. B Biol. Sci.*, **220** (1218), 115–130 (1983).
22. Nathans J, Piantanida TP, Eddy RL, Shows TB and Hogness DS, Molecular genetics of inherited variation in human color vision, *Science*, **232**, 203–210 (1986).
23. Neitz M and Neitz J, Molecular genetics and the biological basis of color vision, in *Human Color Vision: Perspectives from Different Disciplines*, Backhaus GK, Kliegl R, and Werner JS, eds., Walter de Gruyter Berlin, pp. 101–119 (1998).

24. Sharpe LT, Stockman A, Jägle H and Nathans J, Opsin genes, cone photopigments, color vision and color blindness, in *Color Vision. From Genes to Perception*, Gegenfurtner KR, and Sharpe LT, eds., Cambridge University Press, Cambridge, 3–51 (1999).
25. Nathans J, Thomas D and Hogness DS, Molecular genetics of human color vision: the genes encoding blue, green and red pigments, *Science*, **232**, 193–202 (1986).
26. Neitz J, Neitz M and Jacobs GH, More than three different cone pigments among people with normal color vision, *Vision Res.*, **33**, 117–122 (1993).
27. Rushton WAH, Color vision: an approach through the cone pigments, *Invest. Ophthalmol. Visual Sci.*, **10**(5), 311–322 (1971).
28. Wald G, The molecular basis of visual excitation, *Nature*, **219** (5156), 800–807 (1968).
29. Marks WB, Dobelle WH and MacNichol EF, Visual pigments of single primate cones. *Science*, **143**, 1181–1183 (1964).
30. Baker HD and Rushton WAH, The red-sensitive pigment in normal cones. *J. Physiol.* **176**, 56–72 (1965).
31. Gegenfurtner KR, Cortical mechanisms of color vision, in *CGIV 2010/MCD '10 rth European Conference on Colour in Graphics, Imaging, and Vision* (CD-ROM), Joensuu, Finland, June 2010, pp. 1–4 (2010).
32. Buettner-Ennever JA, ed., *Neuroanatomy of the Oculomotor System*, Elsevier, Amsterdam (1988).
33. Carpenter MB and Sutin J, *Human Neuroanatomy*, 8th edn., Williams & Wilkins, Baltimore, CT (1983).
34. Zeki SM, Colour coding in the cerebral cortex: the responses of wavelength-selective and colourcoded cells in monkey visual cortex to changes in wavelength composition. *Neuroscience*, **9**, 767–781 (1983).
35. Lennie P and Movshon JA, Coding of color and form in the geniculostriate visual pathway (invited review), *J. Opt. Soc. Amer.* A, **22**, 2013–2033 (2005).
36. Roe AW and Ts'o DY, Visual topography in primate V2: multiple representation across functional stripes, *J. Neurosci.*, **15**, 3689–3715 (1995).
37. Xiao Y, Wang YI and Fellman J, A spatial organized representation of colour in macaque cortical area V2. *Nature*, **421**, 535–539 (2003).

6

Colour-Vision Psychophysics

6.1 Introduction

Psychophysics is a very vast discipline that covers all the senses and as such is defined on rules, methods and criteria used with typical specificities in the study of the different senses. Here only the psychophysics of colour vision, which is the basis for colorimetry, is considered.

> *Psychophysics* is the discipline that studies the relationship between physical stimuli and their subjective correlates, the sensations.

The term 'psychophysics' was coined by G. T. Fechner[1], whose main idea was that two independent dimensions exist in the relationship between the physical and mental quantities:

1. the stimulus dimension, ϕ (in colour vision ϕ is a luminous radiation entering the eye) and
2. the sensation dimension, Ψ (in colour vision Ψ is the corresponding colour sensation or an attribute of the colour sensation).

In his project, Fechner attempted to establish a rigorous relationship between the physical and mental quantities, that is, a mathematical equation to describe the relationship between physical events and conscious experience.

6.1.1 Psychophysics and Physiology

J. K. Maxwell[2] wrote:

> "If the sensation that we call colour has any laws, it must be something in our own nature which determines the forms of these laws".

Standard Colorimetry: Definitions, Algorithms and Software, First Edition. Claudio Oleari.
© 2016 John Wiley & Sons, Ltd. Published 2016 by John Wiley & Sons, Ltd.

This sentence recalls *physiology* – the scientific study of functions, organ systems, organs, cells and bio-molecules, which carry out the chemical or physical functions in a living system.

A physiological sensory system is between physical stimuli and sensations, and the correlation between physical stimuli and sensations provides knowledge on the physiological system. There is

1. *Descriptive psychophysics*, if the psychophysical measurements provide a quantitative description of the capacities of the sensory system, making a descriptive correspondence between sensations and physical stimuli.

2. *Analytical psychophysics*, if a new hypothesis enters the physiological structure of the visual system for relating sensations and physiological response. In this way, psychophysical measurements enable the testing of hypotheses about the nature of biological mechanisms that underlie sensory perception.

Principle of Nomination and the Analytical Psychophysics of Colour Vision

Analytical psychophysics needs a correlation of visual sensory experience with the physiological structure of the visual system and the physiological response. This correlation is based on a fundamental hypothesis known as

the *principle of nomination*[3] – also known as the *psycho-physical linking hypothesis*, declaring that identical neural events give rise to identical psychological events: that is, two physical stimuli producing the same neural response yield the same sensory experience.

To the principle of nomination corresponds

the *reflexive form of the principle of nomination* stating that when two different physical stimuli produce the same sensory experience, they produce the same neural response.

The reflexive principle of nomination in a psychophysical experiment makes it possible to test a physiological hypothesis.

On the principle of nomination is based

the *method of response invariance*, also known as *criterion-response technique*, according to which are sought the physical stimuli that generate identical sensations and identical neural responses.

In colour vision this method regards the metamerism phenomenon (Section 6.13.2) and is used to determine the *spectral sensitivity* of the human visual system.

6.1.2   Visual Judgement

The colour sensations are characterized by *colour attributes* – such as hue, brightness, lightness and colourfulness (Chapter 4) –, which depend on the colour stimuli, visual situation and adaptation (Sections 6.2 and 6.20). Psychophysical experiments consider these attributes together or separately.

All psychophysical experiments are based on the visual judgement made by an observer, who has to judge the equality of the perceived attributes of two visual sensations, or to detect the discrimination threshold or to quantify their perceived difference. A practical distinction of the colour attributes is given with the experiments shown in this chapter. These comparisons need

1. a defined *visual situation* with an *observer adaptation* – mode of colour appearance –;
2. a *method* for visual judging;
3. a *criterion* to detect the discrimination threshold or the match of the colour attributes; and
4. a *scaling technique* for quantifying the perceived differences.

6.1.3 Modes of Colour Appearance and Viewing Situations

Human vision is a complex phenomenon in which colour is only a quality of visual perception, which is dependent on the other colour stimuli present in the scene, on the configuration of the scene and, in the case of non-luminous colours, on the illuminant. Therefore, the perceived colours are *related colours* (Section 4.6) and depend on the illuminant adaptation. This complexity is studied by

"*Colour appearance*:

1. aspect of visual perception by which things are recognized by their colour
2. in psychophysical studies: visual perception in which the spectral aspects of a colour stimulus are integrated with its illuminating and viewing environment",[4] to which the observer is considered adapted.

The definition of perceived colour given in reference[4] presents the several modes of colour appearance in a very concise way. The names for various modes of appearance are intended to distinguish among qualitative and geometric differences of colour perceptions.[5, 6]

First distinguish between

1. "*primary light source* – surface or object emitting light by a transformation of energy"[4] and

2. "*secondary light source* – surface or object which is not self-emitting but receives light and re-directs it, at least in part, by reflection or transmission".[4]

It follows that the qualitative differences of colour appearance among colours perceived from these two light sources are

1. "*luminous colour* (often *self-luminous colour*) – colour perceived to belong to an area that appears to be emitting light as a *primary light source*, or that appears to be specularly reflecting such light"[4] and

2. "*non-luminous colour* (often *non-self-luminous colour*) – colour perceived to belong to an area that appears to be transmitting or diffusely reflecting light as a *secondary light source*".[4]

Modes of Colour Appearance

The most important modes of colour appearance are the following[4]:

- *Illuminant mode*, typically of direct sight of a light source, to which corresponds the *"Illuminant colour* mode – colour seen as ascribed to a source of illumination".[4]

- *Object mode*, just the vision of an illuminated object, characterized by diffuse reflection or transmission of diffuse radiation, to which corresponds the *"Object-colour* mode – colour seen as ascribed to an object".[4]

- *Aperture mode*, from which the term *"Aperture colour* mode – perceived colour for which there is no definite spatial localisation in depth, such as that perceived as filling a hole in a screen".[4]

Synonymous with *film colour*. Perceived colour typical of the radiation that, before entering the eye, passes through the aperture of a diaphragm and the eye is focused on the edge of the diaphragm in order to avoid associating the radiation with an object or a source. In this way, the observer loses any information that denotes the object (or source), such as surface non-uniformity, polish, texture, type of lighting (grazing, diffused, etc.), and the light that enters the eye on the retina has at any point the same spectral decomposition. In this case, the field of vision is totally dark with the exclusion of the part within the aperture of the diaphragm.

- *"Surface-colour* mode – colour perceived as belonging to a surface from which the light appears to be diffusely reflected or radiated".[4]

- *"Surface-colour* mode – colour perceived as if belonging to the surface of an object – without necessarily this being true in a physical sense, and that the colour is not perceived as luminous".[7]

Katz's definition of surface colour[7] considers only non-luminous colour and, if not specified, generally this is the definition used, also in this book. Often 'surface colour' and 'object colour' are considered synonymous.

- *"Volume-colour* mode – colour perceived as belonging to the bulk of the substance".[4]

- *"Illumination-colour* mode – colour perceived as belonging to the light falling on objects".[4]

- *Ganzfeld-colour* mode ('Ganzfeld', from German for 'complete field') – perceptual deprivation caused by exposure to an unstructured, uniform stimulation field.

Psychophysical colorimetry, in a strict sense, studies colour as an isolated phenomenon and therefore requires that the observation takes place in a controlled and repeatable mode. The most used modes are the 'aperture mode' and the 'surface mode'. A pair of colour stimuli for a comparison in the same visual field are considered related colours, although the phenomenon studied is limited to only two colour stimuli.

Viewing Situations

The overall visual situation has a fundamental role in judging a colour stimulus and should be defined in every detail in order to take account of all the interactions produced by the colours present in a scene. This description of the scene is too complicated and practically unattainable. It is necessary to define a minimum number of important components of the viewing field. Following the proposal of R. W. G. Hunt[8] (p. 294 of Reference 8), the viewing field is divided into four components:

- "*Proximal field* – the immediate environment of the colour element considered, extending typically for about 2° from the edge of that colour element in all or most directions."

- "*Background* – the environment of the colour element considered, extending typically for about 10° from the edge of the proximal field in all or most directions. When the proximal field is of the same colour as the background, the latter is regarded as extending from the edge of the colour element considered."

- "*Surround* – a field outside the background. Typically, in the view of an image display, it is the area outside the display that is filling the rest of the visual field."

- "*Adapting field* – the total environment of the colour element considered, including the proximal field, the background and the surround, and extending to the limit of vision in all directions."

This subdivision of the visual field shows a correspondence with the structure of the retina.

6.1.4 Colour Stimuli

Colour stimuli are proposed to the observer in different ways, in space and/or in time. The main property of colour stimuli is spatial uniformity and this can be obtained only by optical devices. The simplest view is in a booth of light where the observer sees uniform colour specimens uniformly lit, but in this case only wide spectral bands can be considered. For narrow band lights with high intensity, other techniques are required, generally obtained in the Maxwellian view.

Maxwellian View

Often the psychophysical experiments are based on a comparison between two colour stimuli, one termed *standard*, generally fixed, and the other *test* (or *comparison*), changeable by the experimenter or adjustable by the observer.

Often the view is monocular Maxwellian on two concentric circular fields, in which the internal circle is or is not a bipartite field (Figure 6.1), but many other views exist.[9]

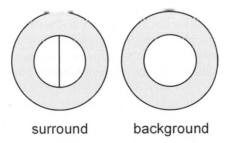

surround   background

Figure 6.1 *Different configurations of stimuli used in colour-vision psychophysics. The stimuli are compared either for colour-matching, or for colour-difference detection or quantification. The comparison is made in bipartite fields on a background with the two stimuli juxtaposed (left) – the background stimulus can also be darkness – or between circular fields inside a circular background (right).*

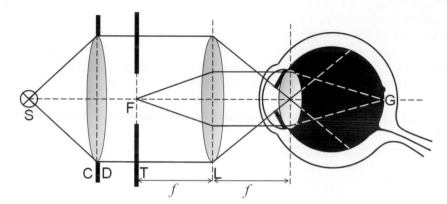

Figure 6.2   *Optical configuration of a Maxwellian view system: light source S, collimating lens C, diaphragm D, and target T situated in the focal plane of lens L, first focal point F of lens L and conjugate of G on the retina of an emmetropic eye (Section 5.2), whose crystalline lens is at the centre of the entrance pupil of the eye and is the conjugate of S. The focal length of the lens is f.*

In the case of the bipartite field, one half is for the standard stimulus, the other half for the test stimulus and the surround is dark or with an *adaptation stimulus*. In the case of two concentric fields, the central field is for the test stimulus, and the background for the standard stimulus. In psychophysical experiments with these fields of view, the surrounds or background stimuli have the important effect of controlling the state of adaptation, although in a very simple visual situation.

The bipartite field used for colour-matching, also with a dark surround, presents two colours in the surface mode, which becomes the aperture mode when the match is obtained.

The Maxwellian view is a method of observation, in which a converging lens forms an image of the light source on the plane of the entrance pupil of the observer's eye. If the observer's eye is focused on the lens, the lens appears as a disc filled with light of uniform intensity[10] (Figure 6.2).

In a bipartite field the *minimum distinct border* criterion is satisfied when the distinctness of the straight line border between the two contiguous parts of the bipartite field is minimal with respect to a variation in either sense of a physical parameter controlling the test stimulus. This criterion is especially used in the brightness matching of different colour lights, which forms the basis of *heterochromatic photometry by the direct comparison method* (Section 7.3).

Flicker Effect

A different way for comparing standard and test fields is the *minimum Flicker* method developed by Frederick Ives[11]. The comparison is made in time, not in space. Standard and test stimuli – generally circular – are presented to the observer alternately in time at a variable temporal frequency (Figure 6.3). Both fields must have an identical shape and size, and must be exactly overlapping. Generally the view is monocular Maxwellian. If the dominant wavelengths of the two fields are too different, then the chromatic aberration of the eye focuses the corresponding images on different surfaces on the retina and presents the two stimuli with the outer edge vibrating (which has to be avoided).

Three different visual situations exist:

1. If the two fields alternate slowly, around 2 Hz, the observer first sees one stimulus and then the other in a distinct way. This is the typical view for complete colour-matching. If the alternation frequency increases, the two stimuli begin to fuse (e.g., red and yellow fields fuse to a pulsating orange).

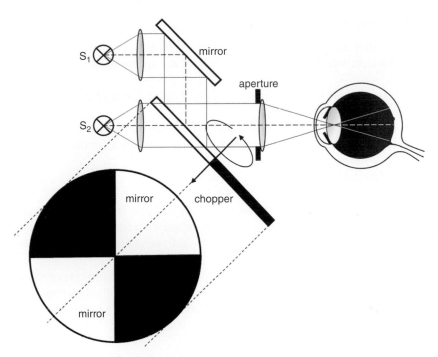

Figure 6.3 *Optical configuration of a Maxwellian view system (Figure 6.2) organized for a dynamic flicker, that is, with time alternating stimuli with a selectable alternation frequency (few Hz to 100 Hz). A chopper, constituted by a rotating disk subdivided in sectors alternately transparent or reflecting specularly, produces the alternate sequence of the two stimuli in temporal comparison.*

2. If the alternation frequency between the test and reference fields is high, around 35–40 Hz, the pulsating flicker disappears, unless the radiance differences between the test and reference fields is very large.

3. If the alternation frequency between the test and reference fields is about 15–20 Hz, the flicker perception disappears for a very narrow range of chromatic radiance.

The perceptual criterion in *heterochromatic flicker photometry* is a *just noticeable flicker* in the situation (3). In this case, the observer is asked to adjust the intensity of the test stimulus until the perceived flicker is minimum and in this case the two stimuli are different but with equal brightness (Section 7.3). If the flicker disappears completely and is independent of the frequency, the colours of the two stimuli are matching.

Whether an intermittent colour stimulus appears as a flicker or as a continuous sensation depends on the frequency of light pulse presentation. The frequency at which the flicker just disappears is termed the critical flicker frequency or *critical fusion frequency*. Here the alternation is between a test bright field and a reference dark field set at zero radiance and the critical flicker frequency is a measure of retinal sensitivity.[12]

The appearance mode of vision given by the visual judgement apparatuses described here is the aperture mode.

Also *binocular vision* apparatuses and *asymmetric vision* apparatuses – different and suitable defined fields of view for the two eyes – are used.

6.1.5   Colour-Attribute Matching

Visual sensations can differ in colour attributes – for example, hue, brightness and colourfulness – as a function of the visual adaptation (Sections 6.2 and 6.20), of the spatial extension and of the time duration (Sections 4.6 and 4.7).

The point at which an attribute of a test stimulus is judged equal to that of the standard stimulus is called the *point of subjective equality*.

In a *matching experiment*, two stimuli are presented to the observer, who is asked to adjust one to match the other, or to adjust one to match a particular attribute of the other, for example, the brightness. This is an application of the *method of response invariance*.

Brindley[13] distinguishes the matches into two classes:

1.   *Class A* match, if two different colour stimuli are indistinguishable in a *complete colour match*, that is, the test colour stimulus appears the same in colour as the reference colour stimulus, and for the reflexive principle of nomination, different colour stimuli with equal sensations activate equal neural responses;

2.   *Class B* match, if the observer is asked to match only a particular attribute of the stimuli, which are obviously different, there is an *incomplete match* between standard and test, that is, different stimuli activate different neural responses and different sensations with the exclusion of one attribute of the colour sensation.

An *incomplete match* relevant in colorimetry is the *heterochromatic brightness match*, when a colour stimulus appears in brightness equal to a given different colour stimulus (e.g. flicker photometry and minimum distinct border method). Other incomplete matches are *hue match, saturation match, brightness and hue match*, and *saturation and brightness match*.

Conclusions based on Class B matches may be less secure than those based on Class A matches because the incomplete matching is limited to particular attributes of colour sensation.

In visual matching, the physical stimuli are measured while no scale of perceived attributes is measured. Other forms of colour-matching are (Section 6.20):

- *Dichoptic* or *haploscopic colour-matching* – asymmetric matching made between the retinal images of the two eyes, which are seeing two different scenes with different adaptation lights.

Asymmetric matching is important for studying chromatic adaptation (Section 6.20) and brightness adaptation. In this method, the two scenes are lit with lights differing in brightness and/or in chromatic content,[14,15] and once the matching is obtained information of the adaptation on the perceived colour is obtained (Section 6.20). The two colours in asymmetric matching are termed *corresponding colours*.

- *Memory matching* – mainly used to form a pair of corresponding colours related to two different illuminants (Section 6.20).

This technique is not widely used because it requires a long training period. Generally, memory matching is carried out without the interposition of any optical devices using both eyes and seeing object-colour stimuli. This technique provides a steady state of adaptation with free eye movement. First, observers have to be trained using a colour order system until they are very familiar with its scales – for example, Value, Chroma and Hue in the case of the Munsell system (Section 14.4) – to be able to describe the colour of any object. To form a pair of corresponding colours related to two different illuminants, the observer has to adjust slightly the tristimulus values of a colour sample under a test illuminant to match a remembered colour stimulus under the reference illuminant. The tristimulus values of both colours are measured. Memory matching has a lower precision than the haploscopic technique and suffers from memory distortion.

6.1.6 Visual Detection Threshold and Sensitivity

Sensitivity is defined as the ratio of the response of the visual system and the magnitude of the physical stimulus that provokes this response. The quantification of the subjective response to a physical stimulus is generally difficult, while it is easier to measure the magnitude of a physical stimulus variation ΔI that is barely detected by the observer. This magnitude of the stimulus variation ΔI is called the *threshold* and is quantified as one unit of the response of the visual system; therefore, the *psychophysical sensitivity* is

$$s = 1 / \Delta I \tag{6.1}$$

The sensitivity is related to the threshold judgements and is of two kinds:

1. *Absolute sensitivity*: the judged quantity is the *absolute threshold* or *stimulus threshold*, the smallest amount of physical stimulus necessary to produce a sensation. Stimuli with intensities below the threshold are considered not detectable.

2. *Differential sensitivity*: the judged quantity is the *difference threshold*, the amount of change in a stimulus required to produce a *just noticeable difference* (jnd) in the sensation when the stimulus above the absolute threshold is applied to the visual system.

Traditionally, there are three (four) methods for testing observer perception in stimulus-detection experiments – absolute sensitivity – and in stimulus-difference-detection experiments – differential sensitivity –:

1. *Method of adjustment*: The observer is active and is asked to alter the level of the stimulus until it is just barely detectable against the background noise. This is repeated many times and at the end the mean value is calculated.[16]

2. *Method of limits*: The levels of a considered property of the stimulus are presented in descending order – descending method of limits– or in ascending order – ascending method of limits – to a passive observer, who has to say if the stimulus is detectable against the background noise. The ascending and descending methods are used alternately and the thresholds are averaged.[16]

3. *Method of constant stimuli*: The levels of a considered property of the stimulus are presented repeatedly and randomly to a passive observer, who has to say whether the stimulus is detectable against the background noise or not.[16,17] The result is represented by the graph of a 'frequency of detection' function (percentage of 'yes' responses against the stimulus intensity), termed *psychometric function*.[16] (Figure 6.4) Generally, the threshold is the stimulus intensity that would be detected 50% of the times.

A fourth method exists, in which the observer is asked to perform only colour-matching:

4. *Colour-matching distribution*: The observer performs repeated operations of colour-matching. The distribution of the stimuli of these colour matches has a mean value and a standard deviation, which, if the distribution is normal, takes on a relevant statistical significance: 68.26% of the observations are within one standard deviation, 95.44% within two and 99.74% within three. Since 99.74 is very close to 100, three standard deviations are considered representing well the threshold of discrimination and are equal to 1 jnd.

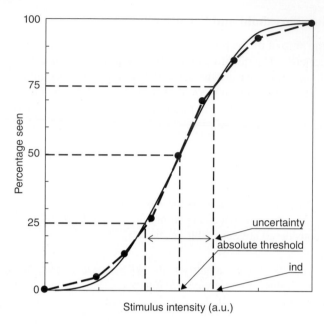

Figure 6.4 *Typical plot of a psychometric function, which shows the relationship between the percentage of times that a stimulus is perceived and the corresponding stimulus intensity (empirical points represented by dots and an interpolating line). Generally, the absolute threshold is defined as the intensity at which the stimulus is detected 50% of the times with an uncertainty ranging between 25% and 75% and the jnd as the intensity at which the stimulus is detected 75% of times.*

This method has been used in the measurements of Macadam's chromatic discrimination ellipses (Section 6.17) and on the wavelength discrimination (Section 6.10).

The absolute and difference thresholds change as some aspect of the stimulus (wavelength, spatial frequency, adaptation time, intensity level, etc.) is systematically varied. The resulting relations are generally called *stimulus critical value functions*, since they describe how the threshold (critical stimulus value) changes as a function of other aspects of the stimulus.

6.1.7   Scaling of Colour Attributes

In a general sense,

"*psychometrics* – any branch of psychology concerned with psychological measurements".[18]

In the strict sense of colour vision, psychometrics is the section of psychophysics studying colour-difference evaluation and colour attribute scaling. In this book only the following definition, given by Hunt[19,20] and used by the CIE, is considered[21] (Chapter 1):

"*Psychometric terms* – terms denoting objective measures of physical variables that are evaluated so as to relate to differences between magnitudes of important attributes of light and colour and such that equal scale intervals represent approximately equal perceived differences in the attribute considered. These measures identify pairs of stimuli that produce equally perceptible differences in response in a visual process in specified viewing conditions."

Colour scale is a scale in which the perceived colour stimuli of objects change in a systematic manner usually in one attribute, that is, as having equal differences of some attribute of colour sensation.

Psychometric is the colorimetry with colour specification made on uniform scales of colour attributes and with rules for evaluating colour differences.

Psychometric is the visual adaptation used for defining uniform scales of colour attributes and for evaluating colour differences (Section 14.6).

The experimental problem of the scaling is to discover the colour stimuli corresponding to a series of colours separated by equally perceived distances.

Generally, a scale of colour attribute has an origin, to which the number 0 corresponds.

Different scales have been defined in colour psychometrics and are object of a debate not yet closed.

The *scaling method* is a procedure that attempts to give a quantitative specification of a perceptual attribute. Different scaling methods produce different types of scales, each with different mathematical properties and practical use.

The scales important in colorimetry regard the hue, lightness, brightness, chroma, saturation, etc. These magnitudes are in part absolute and in part relative; therefore, they have to be judged as unrelated and related colours, respectively. The separate judgement of colour components, which are fused in one colour sensation, is particularly difficult. Burns and Sheep[22] have shown that observers respond to the overall lack of similarity between stimuli rather than to similarity itself, or lack of it, between the separate colour components. This difficulty suggests directly reading the section where the scaling problem is considered (Sections 6.7, 6.8, 6.9, 6.10, 6.17, 6.18 and Chapter 14). Particularly important is the comparison between the different methods employed in the definition of the scales in the atlases considered in Chapter 14. Here only four classical scaling methods are reported for completeness: nominal, ordinal, interval and ratio method (Figure 6.5).

Nominal scales – These are the simplest form of numerical scales. Numbers are used as names for objects. The values of the numbers have no meaning other than to identify the different players. Any mathematical operation performed on this type of scale is arbitrary. In colour appearance, a nominal scale can be given to colour names, such as reds, greens, yellows and blues. This scale can then be used for determining the category of a given colour stimulus.

Ordinal scales – These scales have a magnitude of order associated with them. Objects can be ranked in ascending or descending order based on the magnitude of a certain trait. A colour-appearance example of this type of scale might be the sorting of a series of paint chips in order of lightness. The resulting scale would reveal only that one paint chip was lighter than the others, and there would be no information as to how much lighter. The only mathematical operation that is valid for an ordinal scale is the greater than/less than operator. Any other operation should be considered arbitrary.

Interval scales – An interval scale is any scale that has equally spaced units, or intervals. For example, in this type of scale, if one sample is judged to be one unit away from a reference stimulus, considered as

Scale	Intervals	Zero	
Nominal	No intervals	Without zero	
Ordinal	Unequal intervals	Without zero	
Interval	Equal intervals	Without zero	
Ratio	Equal intervals	With zero	

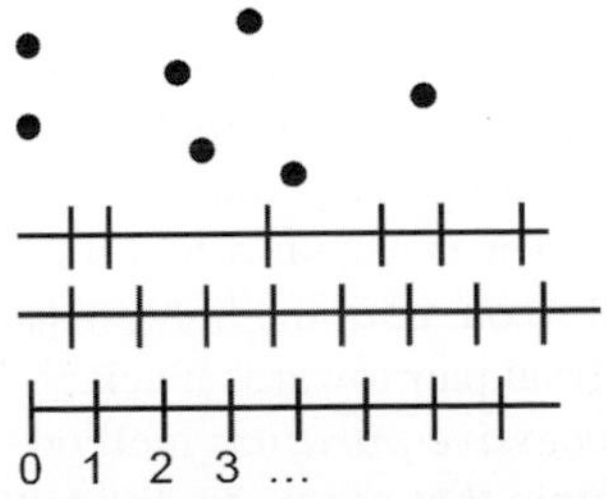

Figure 6.5 *Classification of psychometric scales according to Stevens.*

an 'anchor' in the judgement, and a second sample is judged to also be one unit away, though in a different direction, the differences between the anchor and the first or second sample is still said to be perceptually equal. Again, given a unit of difference as anchor, copies of this unit positioned at different points on the scale are judged to also be one unit and are still said to be perceptually equal. There is no meaningful zero in an interval scale, that is, the zero value is arbitrary.

Interval scales are obtainable also with the four following methods:

1. *Partition scaling method*, for constructing the scale of a psychophysical attribute directly from the observer's judgement, who is asked to subdivide the psychological continuum into equal sensory intervals.
2. *Equisection scaling method*, a particular partition scaling method, where the partition is made in equal parts.
3. *Bisection scaling method*, a particular equisection scaling method, where the equisection is a bisection into equal perceived parts, that is, where the observer has to choose the intermediate stimulus between two given stimuli.
4. *Ratio scaling*, as one of the previous scalings with the addition of a meaningful zero point.

    Ratio scales are obtained with different ratio-scaling methods – among which the difference is often very small:

    • *Ratio estimation* – The experimenter presents two stimuli to the observer whose task is to estimate the ratio between them.
    • *Ratio production*, often called *fractionation* – The experimenter presents a stimulus called *standard* and the observer has to adjust a variable stimulus to a prescribed ratio, for example, $\frac{1}{2}$, $\frac{1}{3}$, $\frac{2}{3}$, $\frac{1}{4}$ and $\frac{3}{4}$.
    • *Magnitude estimation method with standard* and *without standard* – Experimenter presents a *standard* stimulus with a number assigned called the *modulus*, then, for the subsequent stimuli, the observer reports numerically their estimation relative to the standard so as to preserve the ratio between the sensations and the numerical estimates. If the standard is chosen by the observer, the method is without standard.
    • *Magnitude production method* – the inverse of magnitude estimation – the experimenter tells the observer a numerical value of a visual reference attribute and asks him or her to adjust a stimulus to have a perceptual attribute value as that number times the reference.
5. *Resemblance-scale method* – This kind of scaling based on the resemblance of the colours has been introduced by E. Hering (Section 4.4, 4.5) and used for the definition of the NCS colour ordering system (see *NCS's axioms* in Section 14.7). (This method could be classified as a 'partition scale method', but the use of resemblances induces us to consider it separately.)

Typically, the lightness scale has been defined in different adaptations with different methods, different tasks and different criteria, and generally with different results. Two examples of lightness scales are given below:

1. *Magnitude production* method – Given a series of many achromatic samples, from the lightest to the darkest possible, from the mid-region of this series choose two samples, L_1 and L_2, with an arbitrary small lightness interval. The observer's task is first to choose one, L_3, from all the lighter samples so that the lightness interval L_3–L_2 would appear equal to the interval L_2–L_1. Then L_1 is laid aside and the observer is asked to find a new lighter sample, L_4, and match the L_3–L_4 lightness interval with L_3–L_2, and so on, until the lightest sample is reached. Then the same assessment is performed from the original interval pair towards black.[23]
2. *Successive partition* method – a kind of ratio production. Here the observer has to choose the grey sample that seems to divide the greyscale between the white and the black samples into two halves

and then further divide each of these two parts into two halves, and so on. This method yields practically identical results to the magnitude production method, but the standard deviations for this *successive partitions* procedure is more than double the previous one. Black sample is assumed as zero.[23]

6.2 Adaptation

"*Adaptation* – process by which the state of the visual system is modified by previous and present exposure to stimuli that may have various luminance values, spectral distributions and angular subtenses.
  NOTE – Adaptation to specific spatial frequencies, orientations, sizes, etc. is recognized as being included in this definition."[4]

The adaptation is a response modification of the visual system to light stimulation. The human visual system changes its sensitivity as a function of the time evolution of the observed scene and therefore has a time evolution itself. Once the observed scene has no more change, the adaptation process continues until it becomes complete and then stops. The notation 'adaptation' is used both for the process of adjustment and the end state of complete adaptation. The CIE ILV[4] defines:

"*State of adaptation* – state of the visual system after an adaptation process has been completed.
  NOTE – The terms *light adaptation* and *dark adaptation* are also used, the former when the luminances of the stimuli are of at least 10 cd m^{-2}, and the latter when the luminances are of less than some hundredths of a cd·m^{-2}."[4] (Sections 2.6 and 6.5)

The human visual system, after being adapted to a bright light, may need some time of the order of more than half an hour to become completely dark adapted, while, after being adapted to darkness, may need only a few minutes to become completely day-light adapted. These two processes are not symmetrical and are considered separately (Sections 6.2.1 and 6.2.2).

The *duplicity theory* (Sections 5.4 and 5.5) states that two transduction mechanisms exist, which are related to two different kinds of photosensitive cells: the rods and the cones.

Therefore, three kinds of vision exist: photopic vision, scotopic vision and mesopic vision, with which different adaptations are associated.
The state of adaptations enters the definition of:

"*Perceived colour* – characteristic of visual perception that can be described by attributes of hue, brightness (or lightness) and colourfulness (or saturation or chroma)."[4]
  "Perceived colour may appear in several modes of colour appearance", which are in correspondence with different adaptations (Section 6.1.3).

The adaptation phenomenon depends on the adapting light, both on the luminous level and on the chromatic quality. The CIE ILV[4] defines the following other phenomena:

- "*Chromatic adaptation* – visual process whereby approximate compensation is made for changes in the colours of stimuli, especially in the case of changes in illuminants" (Section 6.20).

- "*Adaptive colour shift* – change in the perceived colour of an object caused solely by change of *chromatic adaptation*" (Section 6.20).

- "*Illuminant colour shift* – change in the perceived colour of an object caused solely by change of illuminant in the absence of any change in the observer's state of *chromatic adaptation*" (Section 6.20).

- "*Resultant colour shift* – combined *illuminant colour shift* and *adaptive colour shift*" (Section 6.20).

The following quantities, useful to describe the adaptation phenomena, are also defined:

- "*Contrast sensitivity* – reciprocal of the least perceptible (physical) contrast, usually expressed as $S_c \equiv L/\Delta L$, where L is the average luminance and ΔL is the luminance difference threshold [dimensionless].
  NOTE – The value of S_c depends on a number of factors including the luminance, the viewing conditions and the *state of adaptation*" (Section 6.20).

- "*Luminance difference threshold* [ΔL], smallest perceptible difference in luminance of 2 adjacent fields. [cd/m^2 = lm/(m^2·sr)].
  NOTE – The value depends on the methodology, luminance, and on the viewing conditions, including the *state of adaptation*" (Section 6.20).

- "*Luminance threshold* – lowest luminance of a stimulus which enables it to be perceived.
  NOTE – The value depends on field size, surround, *state of adaptation*, methodology, and other viewing conditions". (Section 6.20)

[often, instead of the luminance, is considered the retinal illuminance measured in troland (Td) (Section 2.7)].

The mechanisms behind the different adaptation processes are still not completely known; therefore, this chapter describes psychophysically the main phenomenological aspects of the adaptation.

The effect of the stimulation on the photoreceptors is a modification of its own effectiveness. Whenever there is a change in retinal illuminance and/or spectral power distribution of the stimulus, the visual mechanism starts re-adapting to the changing stimulus. Thus, in the real situation, when the light flux crossing the retina is continuously changing, the visual sensitivity at any particular time and place on the retina is a resultant of the effects of the stimulations due to the light flux in the time, previous and actual, and on the considered place in relation with the contiguous places. The term *local adaptation* refers to the effect of a stimulus, which has been confined to a specific region of the retina.

In psychophysical experiments with only few colour stimuli in the field of view, the surround stimuli have the important effect of controlling the state of adaptation.

The time change in sensitivity depends on the duration and also on the degree of stimulation. If the adapting light level changes by a relatively small amount, the visual system compensates for the change almost immediately, but if the light level changes by a lot, it takes a long time to reach the complete adaptation. If a new stimulus remains the same for a long enough time, the adaptation level reaches an equilibrium and the adaptation is *complete*. The time required for a complete adaptation depends on the starting level and on the new stimulus. The sensitivity change is very sudden in the initial phase, and the initial phase has a duration that is a small fraction of the whole time required for a complete adaptation.

The term *adaptation level* defines the kind and degree of steady stimulation that would produce the same state of sensitivity as it exists at any moment and place on the retina. The adaptation level determines the range of responsiveness. The fully dark-adapted visual system cannot discriminate any luminances below an *absolute threshold*, nor can the fully light-adapted visual system discriminate any luminances above an upper *terminal threshold*. The overall range within which the optimally adapted visual system is effective is about from 10^{-5} to 10^5 cd/m^2. The retina is bound to be adapted to some level at any time and only a small part of the full range is available at any time. The ratio of maximum luminance over minimum luminance detectable for the full range is in the order 10 billion to 1, while the momentary range for ordinary levels of luminance is in the order 1000 to 1.

The adaptation mechanisms produce changes in threshold visibility, colour appearance, visual acuity and sensitivity over time.

6.2.1   Brightness Adaptation

The adaptation process described above, which allows this great extension of sensitivity range of the retina, is called *brightness adaptation.*

Four mechanisms underlie the adaptation in such a wide range of luminances:

1.   Pupil size
2.   Switchover from rods to cones in the passage from scotopic to photopic vision
3.   Bleaching/regeneration of the photopigments in the photosensitive cells – rods and cones –
4.   Feedback from the horizontal cells to control the responsiveness of the photosensitive cells[24]

These four mechanisms have the combined effect of making the retina more sensitive at low light levels and less sensitive at high light levels, with important consequences for perception.

The pupil size has only a small part in the adaptation process: the luminous flux entering the eye is proportional to the pupil area; therefore, since the pupil diameter ranges from 1 or 2 mm to about 8 mm, the luminous flux is approximately modulated by a factor of 16–64. A change of 10 log units of luminance induces the pupil to change in diameter from approximately 7–8 mm down to about 1–2 mm.[25] This range of variation produces a little more than one log unit change in retinal illuminance, so pupillary action alone is not sufficient to completely account for visual adaptation.[26]

The neuron net of the retina has a very limited response range: from −80 to +50 mV of graded potential in the non-spiking cells of the retina – rods, cones and horizontal cells –, and 0 to about 200 spikes per second for ganglion cells.

The main parts of the adaptation process are due to the mechanisms (3) and (4) producing changes in the retina sensitivity.

The effect of light on photopigments is their bleaching or depletion (photochemical effect) and, after the bleaching, a regeneration of photopigments exists (chemical effect). Pigment bleaching makes the receptors less sensitive to light and at high light flux produces a compression of their response. However, pigment bleaching cannot completely account for adaptation because the time-courses of the early phases of dark and light adaptation are too rapid to be explained by pigment bleaching alone.[27]

Adaptive processes sited in the neural network of the retina (horizontal cells) have a multiplicative process effectively scaling the input by a constant related to the background luminance.[24] This process acts very rapidly and accounts for changes in sensitivity over the first few seconds of adaptation. A slower acting subtractive process reduces the base level of activity in the system caused by a constant background. This last process accounts for the slow improvement in sensitivity measured over minutes of adaptation.[28] Sensitivity increases in dim light and decreases in bright light, inducing a more or less constant range of response in the visual system.

For a given set of visual conditions, the current sensitivity level of the visual system is called the *brightness-adaptation level*. These combined phenomena have the effect that the perceived brightness is approximately constant in a wide range of brightness adaptation levels.

The response of photosensitive cells may no longer increase if the light flux is so intense that the regeneration of the photopigments is not able to counterbalance the bleaching completely. This situation is known as *photopigment saturation*.

The term brightness adaptation is not defined in the CIE ILV[4]; it is seldom used and the phenomena described here are generally considered in the light-adaptation process.

6.2.2   Threshold in Dark Adaptation

The human visual system is not static, but adapts to the situation in which it is located, and the time of adjustment when switching between different visual situations is short, almost instantaneous, if the difference between the visual situations is small, but it is tens of minutes in the case of large differences, for example, in the passage from outside, with the light of day, to an interior illuminated by a candle light. The study of the visual system must start from the knowledge of its behaviour as a function of the environmental illumination. Relevant to colour perception are the adaptation phenomena; particularly relevant are the mechanisms of dark, light, and chromatic adaptation. In this section, the time-course of detection threshold in dark adaptation is investigated by the psychophysical method.

The Time-Course of Detection Threshold of Dark Adaptation

The visual system adapts to any change in light level and becomes more or less sensitive, depending on whether the light level decreases or increases. Dark adaptation refers to the change in visual sensitivity that occurs when the level of illumination is decreased. Visually, dark adaptation is experienced as the temporary blindness that occurs when we go rapidly from photopic to scotopic levels of illumination. Dark adaptation is a progressive decrease in the visual detection threshold as a function of how long the observer has been in the dark after prior stimulation by light. The threshold decreases over a time of approximately 5–8 minutes for cone stimulation and over 40–60 minutes for rod stimulation. During dark adaptation the increment-threshold detection of a light well represents the dark-adaptation process. Before the experiment the observer is completely adapted to a uniform background intensity. After the background light is removed, in darkness the test observer threshold is measured periodically.

For the threshold measurement, the stimulus is a uniform spot of light large enough to stimulate cones and rods for central fixation. In determining such adaptation curves, the experimenter changes the stimulus magnitude (flux of photons), and the only task of the subject is to say 'yes' or 'no', meaning 'seen' or 'not seen', respectively. The threshold criterion is defined as that intensity of the stimulus that results in 'yes' more frequently than 'no', for instance, say, in 60% of all the presentations. Threshold is the inverse of sensitivity. In the darkness the detection threshold is measured continuously over more than 30 minutes. Let us consider the time-course of dark adaptation given by Hecht et al.[29–31] (Figure 6.6) where a violet light stimulus was used.

In the first 5 minutes, after the adapting field is switched off, the threshold drops rapidly, but then it levels off at a relatively high level because the cone system has reached its greatest sensitivity, and the rod system is still not significantly regenerated. After about 7 minutes rod system sensitivity takes over that of the cone system and the threshold begins to drop again. Changes in the threshold can be measured out to about 35 minutes, at which point the visual system has reached its absolute levels of sensitivity, and the threshold has dropped nearly 4 log units. A change of slope separates the two curve branches that represent the rod and cone systems, respectively. This separation point is known as the Purkinje break[32–34] and indicates the transition from detection by the cone system to detection by the rod system. The biphasic curve is a confirmation of the *duplicity theory* (Sections 5.4, 5.5 and 6.2), which states that two transduction mechanisms related to rods and cones exist.

The course of dark adaptation is influenced by the intensity and duration of light pre-adaptation, and different but analogous curves are measured in correspondence to different pre-adaptations.

The dark adaptation curves are different if stimuli of different wavelengths are used (Figure 6.7). The curve is no longer biphasic in the case that the stimulus is of a wavelength of 650 nm, which shows that L-cones

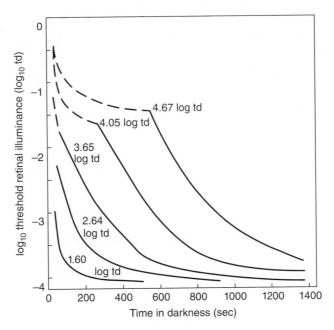

Figure 6.6   *The time-course of dark adaptation as measured by using violet light and following different levels of pre-adapting retinal illuminance.*[29–32]

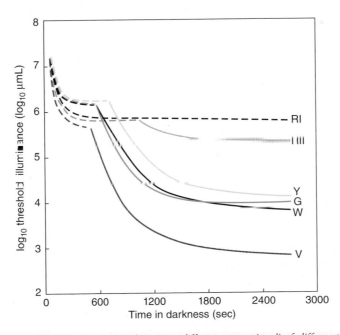

Figure 6.7   *The time-course of dark adaptation by using different test stimuli of different wavelengths. Observers were pre-adapted to 2000 mL for 5 minutes (1 L = 10⁴/π cd/m²). A 3° test stimulus was presented 7° on the nasal retina. The colours were: RI (extreme red) λ = 680 nm; RII (red) λ = 635 nm; Y (yellow) λ = 573 nm; G (green) λ = 520 nm; V (violet) λ = 485 nm and W (white).*[32,33]

are equally or more sensitive than rods at this wavelength. Moreover, the Purkinje break is most prominent if light of short wavelength is used, because the rods are much more sensitive than the cones to short wavelengths, once the rods are dark adapted.

6.3   Absolute Thresholds in Human Vision

The absolute threshold for vision is represented by the minimum number of photons that the human visual system can detect. It was assessed first by Hecht, Shlaer and Pirenne[35] in 1942.

The phenomenon is not confined to a single photoreceptor cell, but regards the receptive field of a ganglion cell, to which more photoreceptor cells belong. In addition, activation due to the absorbed photons takes effect only if the temporal distance between the individual events of absorption is less than one's own time of the visual system (camera sensors do not have this limitation, so the exposure time can be freely chosen). Therefore, the experiment was conducted according to the following rules:

- The experiment room was completely dark and the observer was completely dark adapted after an adaptation time of 40 minutes.
- Since the visual situation was scotopic, the physical stimulus was a light beam with a wavelength of 510 nm, to which corresponds the maximum sensitivity of the rods, and focused on the retina at 20° to the right of the fovea, where the rod-cell density is high.
- A test field 10 minutes in diameter (1 minute = 1/60°) was employed to be sure that it is within a receptive field.
- The physical stimulus was a flash with a duration of 0.001 second, lower of the proper temporal summation of the photoreceptors.

No light was seen by the observer excluding the flash test.

Thresholds were defined as the stimulus energy resulting in a sensation 60% of the times according to the *method of limits*. They were measured over a period of months for seven observers. Stimulus energy was measured by a thermopile at the cornea of the eye. To specify the corresponding number of photons absorbed at the threshold by the photopigment of the rods (rhodopsin), the threshold values measured at the cornea were reduced by loss factors due to the absorption and reflection of the light within the eye:

- Approximately 4% of the light is back-reflected by the cornea (Fresnel's reflectance, Section 3.4.2).
- Approximately 50% of the light at a wavelength of 510 nm entering the eye is absorbed by the ocular media before reaching the retina.[36]
- Approximately 20% of the light reaching the retina is absorbed by the rhodopsin of the receptors.

The measured threshold value is only 5–14 photons absorbed by rods, activating the scotopic visual process. "This small number of quanta, in comparison with the large number of rods involved – 500 –, precludes any significant two quantum absorptions per rod, and means that in order to produce a visual effect, one quantum must be absorbed by each of 5 to 14 rods in the retina".[35]

6.4   Absolute Threshold and Spectral Sensitivity in Scotopic and Photopic Visions

The human visual system is sensitive to light mainly in the wavelength range 400–750 nm, but is not equally sensitive to different spectral lights.

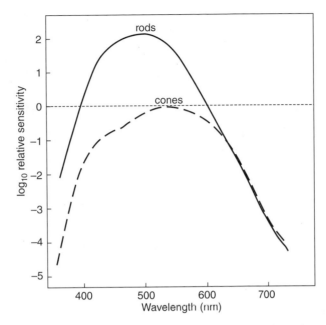

Figure 6.8 *Scotopic (rods) and photopic (cones) spectral sensitivity functions in logarithmic scale. Wald's data from Reference 34.*

Photopic and scotopic visions have to be considered separately, and it is possible because in the fovea only cones exist and in the periphery a large part of the sensitive cells is of rods and the few cones have sensitivity much lower than rods. G. Wald[37] measured the absolute thresholds of photopic and scotopic visual systems separately. Test stimuli of variable wavelength, size of 1° and duration time of 40 ms (i.e., 1/25 s) were presented to the observers either within the fovea or 8° above the fovea. Thresholds were defined according to the method of limits.

Two different threshold curves were obtained for photopic and scotopic visions (Figures 6.8 and 6.9). The photopic curve regards the three kinds of cones together. The overlapping of the curves in the long-wavelength region over 650 nm is interesting, which reveals almost equal sensitivity of the rods and the L-cones.[38]

Particularly, Figure 6.10 shows the threshold function for small stimuli presented only to the fovea, where only cone receptors are present.[38] The three parts of the curve reveal the wavelengths at which each of the S-, M- and L-cones have their lowest thresholds. Because the threshold curves for each type of cone overlap substantially, in this experiment, it is impossible to determine completely the three independent curves of the individual cone types.

6.4.1 Silent Substitution Method

Anyway, by applying a method known as the *selective adaptation method* or *silent substitution method*, it is possible to consider the three kinds of cones separately, although not completely. Consider three light bands, for which the three types of cones have maximum sensitivity. The exposition of the observer to one or two of these lights reduces the sensitivity of the corresponding cones, and then the three kinds of cones can be selected individually, although incompletely. Then, under selective adaptation, the spectral sensitivity curves

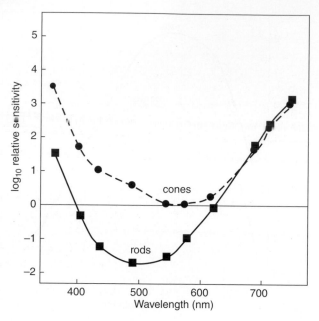

Figure 6.9   *Relative threshold for the detection of light on a logarithmic scale as a function of wavelength. Wald's data from References[37–40].*

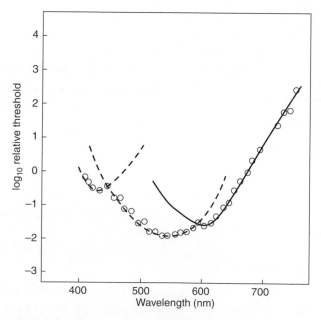

Figure 6.10   *Relative detection threshold in foveal vision on a logarithmic scale as a function of the wavelength (experimental points represented by circles; dashed and continuous lines represent the supposed functions related to different types of cones).[38]*

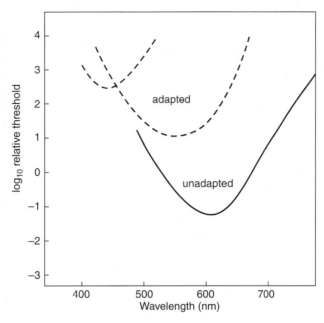

Figure 6.11   *Relative detection threshold after visual adaptation to short-wavelength light on a logarithmic scale (dashed line) and relative non-adapted detection threshold on a logarithmic scale (continuous line) related to cone vision.*[38]

were psychophysically measured.[37–44] As an example consider the L-cones of which the response is studied in isolation. The L-cones are equally sensitive to pairs of radiation of different wavelengths – for example, 580 and 620 nm –. These cones give the same response (silent substitution) to the alternation of these two radiation of equal intensity, while another type of cone, for example, M-cones, is stimulated alternately more or less, because they have different sensitivity for the two radiations. The S-cones have negligible sensitivity to these radiations.

In Wald's experiment (1964), the observer was asked to detect a circle of 1°, stimulating only the foveal cones with a radiation of variable wavelength, presented against a 3.5° background illuminated with fixed adaptation light (Figure 6.11):

- L-cones have been selected with a blue adaptation background, which causes substantial elevations in the thresholds of M- and S-cones.
- M-cones have been selected with a purple adaptation background, which causes substantial elevations in the thresholds of S- and L-cones.
- S-cones have been selected with a bright yellow adaptation background – wavelength longer than 550 nm –, which causes substantial elevations in the thresholds of L- and M-cones.

The measurements of light corresponding to the absolute threshold were made at the cornea of the eye and not inside at the eye receptors; consequently, the spectral curves in Figure 6.12 represent the spectral sensitivities of the cone visual system. The relative spectral sensitivity is expressed as the reciprocal of the measured threshold and the logarithm of sensitivity is plotted as a function of the wavelength of the test stimulus. Wald tried to obtain three spectral sensitivity curves of naked cones (Figure 6.13), taking into account

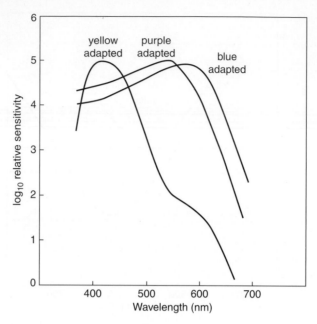

Figure 6.12 *Relative visual sensitivities on a logarithmic scale in adapted foveal vision ignoring the absorption of light by elements of the eye before the retina. The three curves are obtained after adaptation to yellow light, purple light and blue light, respectively.*[40]

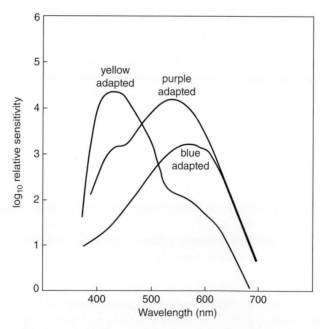

Figure 6.13 *Relative visual sensitivities on a logarithmic scale in adapted foveal vision considering the absorption of light by elements of the eye before the retina.*[40]

the absorbances of the different media of the eye, with exclusion of the cones – that is, cornea, lens and non-visual pigments of the fovea –. The results of these experiments are very important, because they give a deep knowledge of photoreceptor physiology, through the *reflexive principle of nomination*, although the separation produced by selective adaptation is incomplete. The effect of the adaptation is impressive and in any psychophysical measurement has to be considered.

6.5 Luminous Efficiency Function

Brightness is one of the attributes of a colour sensation. The comparison of brightness of different colours needs criteria for the brightness-discrimination threshold and criteria for brightness matching. Brightness matching is considered in this section. Different brightness-matching criteria have been defined over time. Two equally bright spectral lights, characterized by the wavelengths λ_s and λ_t, have a particular ratio of the corresponding radiances $L_{e,s}$ and $L_{e,t}$. The ratio of these two radiances, where the standard radiance $L_{e,s}$ is chosen with the lowest magnitude, that is, at the wavelength of 555 nm for photopic vision (507 nm for scotopic vision), is named

Spectral luminous efficiency function, and is defined for

1. rod scotopic vision (distinguished by a prime index) with a large field of view and low radiance stimulating only the rods:

$$V'(\lambda) = \frac{L_{e,\lambda}}{L_{e,\lambda=507}} \text{ for scotopic vision} \tag{6.2}$$

2. cone photopic vision with a field of view within to the fovea, excluding the rods:

$$V(\lambda) = \frac{L_{e,\lambda}}{L_{e,\lambda=555}} \text{ for photopic vision} \tag{6.3}$$

The criteria for judging the brightness matches are as follows:

- *Heterochromatic minimum flicker* (Section 6.1.4), which defines the *flicker brightness*
- *Direct heterochromatic brightness match* – in this technique the view is monocular Maxwellian on a bipartite field in foveal vision, whose two halves propose two monochromatic lights, mutually spaced by a few nanometres. The observer must match in brightness the two halves. The small colour difference between the two lights makes their brightness matching possible. All the visible spectrum is considered step by step.
- *Minimum distinct border*, already defined in Section 6.1.4 – In this case the comparison is made in a static visual situation between two contiguous regions of the retina.

In 1924 the CIE proposed the standard $V(\lambda)$ for photopic vision and in 1951 the standard $V'(\lambda)$ for scotopic vision (Sections 7.3, 7.4, 7.6) (Figure 6.14). Both proposals are obtained by a suitable weighted average of empirical data produced by different laboratories and different judging methods and criteria.

For the standard photopic photometric observer, the first three criteria were used to obtain empirical data. Particularly the test field was of 2° and in many cases with a large surround with a brightness of different amounts. The differences among different laboratories and observers are considerable. This

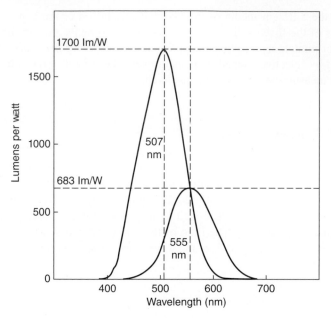

Figure 6.14   *Absolute luminous efficiency functions for the scotopic vision $K'_m V'(\lambda)$ (maximum at 507 nm) and photopic vision $K_m V(\lambda)$ (maximum at 555 nm) as standardized by the CIE (Chapter 7). The maximum for scotopic vision is at 507 nm and for photopic vision at 555 nm.*

variability of the empirical data is the subject of discussion and criticism, particularly because the function $V(\lambda)$ entered the definition of the CIE 1931 standard colorimetric photopic observer (Figure 6.14) (Sections 7.3 and 7.4).

For the standard scotopic observer $V'(\lambda)$ (Section 7.6), the empirical data were produced in two laboratories with direct comparison in a bipartite field of 20° with a standard field of white light. The observer was completely dark adapted for one hour before starting the experiment and had to fixate the top of the dividing line of the bipartite field.

A presentation of the standard photometric observers is given in Chapter 7.

6.5.1   Abney Additivity Law and Luminance

In this section, the colour stimuli are considered as psychophysical quantities, therefore are denoted by symbols in square brackets following the historical notation.

The *Abney additivity law* renders the knowledge of the luminous efficiency function very powerful in both disciplines, photometry and colorimetry.

> *Abney's law* is an "empirical law stating that if two colour stimuli, [A] and [B], are perceived to be of equal brightness and two other colour stimuli, [C] and [D], are perceived to be of equal brightness, then the additive mixtures of [A] with [C] and [B] with [D] will also be perceived to be of equal brightness.
>   NOTE – The validity of Abney's law depends strongly on the observing conditions."[4]

More generally, Abney's laws are the basic laws of the heterochromatic brightness matching[45,46] and formally say:

Given four psychophysical luminous stimuli, [A], [B], [C] and [D], the following linearity relations hold true:

1. Symmetry law: if [A] ⇔ [B], then [B] ⇔ [A].
2. Transitivity law: if [A] ⇔ [B], and [B] ⇔ [C] then [A] ⇔ [C].
3. Proportionality law: if [A] ⇔ [B], then a[B] ⇔ a[A] $\forall\ a > 0$ and real.
4. Additivity law: if [A] ⇔ [B] and [C] ⇔ [D], then ([A] + [C]) ⇔ ([B] + [D]).
5. Visual stimuli have a defined scale with common origin.

(The symbol '⇔' means that colour stimuli are *brightness matching*.)

It means also that the total brightness composed of a mixture of wavelengths λ_i is equal to the sum of the brightnesses of its monochromatic components:

$$\sum_{\lambda_i=380}^{780} L_{e,\lambda_i} V(\lambda_i)\Delta\lambda_i \tag{6.4}$$

Consequently, the brightness of a general spectral radiance $L_e(\lambda)$ is defined by the integral

$$\int_{380}^{780} L_{e,\lambda} V(\lambda)\,d\lambda \tag{6.5}$$

to which the absolute *photopic luminance* corresponds (Section 2.6)

$$L_v = K_m \int_{380}^{780} L_{e,\lambda} V(\lambda)\,d\lambda, \quad K_m = 683 \text{ lumen/watt} \tag{6.6}$$

Analogously for the *scotopic luminance*

$$L_v' = K_m' \int_{380}^{780} L_{e,\lambda} V'(\lambda)\,d\lambda, \quad K_m' = 1700 \text{ lumen/watt} \tag{6.7}$$

(The constants K_m and K_m' are discussed in Chapter 7.)

Violations of brightness additivity may occur as brightness enhancement or inhibition. In brightness enhancement/inhibition, the brightness of a mixture is higher/less than the sum of the brightnesses of the components viewed alone.

The criteria for judging the brightness matches do not produce an exact brightness efficiency. Figure 6.15 compares the luminous efficiency function $V(\lambda)$ used to define the luminance and $V_{10}(\lambda)$ (Section 7.7), with the *brightness efficiency functions* $V_{b;2}(\lambda)$ and $V_{b;10}(\lambda)$ for 2° and 10° visual fields proposed by Ikeda-Nakano [47–49] in 1986.

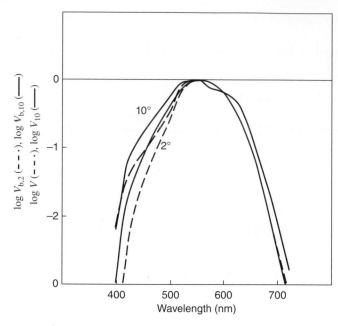

Figure 6.15   *Comparison of the luminous efficiency functions V(λ), used to define the luminance, and V₁₀(λ)* *(Chapter 7) with the brightness efficiency functions Vᵦ;₂(λ) and Vᵦ;₁₀(λ) for 2° and 10° visual fields given by Ikeda-Nakano[47,48] in 1986.*

6.6   Light Adaptation and Sensitivity

Visually, light adaptation is experienced as the temporary blindness that occurs when we go rapidly from scotopic to photopic levels of illumination. Instantaneously and momentarily the bright light appears to the observer as a white light because the sensitivity of the receptors is set to dim light and large amounts of the photopigment are broken down. During the light adaptation process, the sensitivity of the retina decreases dramatically, inhibiting the rod function and favouring the cone system. Rod sensitivity is lost during light adaptation to photopic vision. Within about 1 minute the cones are sufficiently excited by the bright light to take over. The almost complete process for light adaptation occurs over 5–10 minutes.

The human visual system operates over a wide range of luminance levels.

The state of brightness adaptation defines the sensitivity:

- Under low levels the human visual system has a very high sensitivity and can discriminate against a black background a luminous spot as little as 100 photons entering the eye.
- In the photopic range the human visual system requires a luminous spot of thousands or millions of photons to be seen against a background of higher illumination.

The sensitivity of the human visual system is represented by the reciprocal of the absolute intensity threshold, which is measured by a psychophysical experiment known as *threshold versus illuminance* experiment (TVI). The measured quantity is the illuminance at the cornea I_v and the unit of measure is troland (Section 2.7). The observer is adapted to a uniformly lit background with retinal illuminance I_B. This experiment

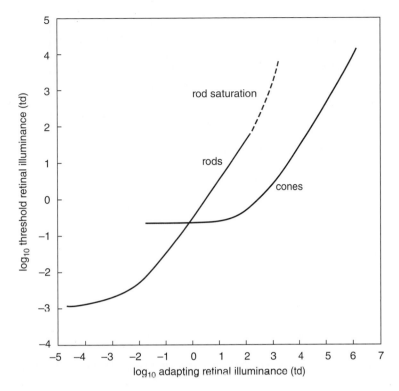

Figure 6.16 *Threshold versus retinal illuminance curve.*[53] *The detection threshold becomes higher as the background adapting retinal illuminance increases. Light of two different wavelengths is used (580 nm for the test and 500 nm for the background).*[51]

measures the minimum illuminance increment ΔI_v of a test spot – a sharp-edged circular target in the centre of the visual field – required to produce a visual sensation on the background. This is achieved by placing an observer in front of a background wall of a given adapting illuminance, and, once the adaptation is obtained, by increasing the retinal illuminance of the test spot ΔI_v from zero until it is just noticeable to the observer. The test phase has to be very quick avoiding any conditioning of the test spot on the adaptation.

Figure 6.16 shows the TVI: the threshold of detectable retinal illuminance increment increases as the background adapting retinal illuminance increases (abscissa). The TVI has two branched curves, one related to the rod vision and the other to the cone vision. Both curves have an analogous shape. Consider the rod curve that is represented by four sections:

1. The TVI below –4 log units is almost constant and the low background luminance does not significantly affect the threshold.
2. The TVI between –4 and –2 log units follows the square root law of *de Vries-Rose*.[50]
3. The TVI between –2 and +2 log units is proportional to the background adapting retinal illuminance and the slope $\Delta I/I_\mathrm{B}$ is constant. This section of the TVI is known as *Weber's Law*.
4. The TVI over +2 log units rises rapidly and the rod system starts to become incapable of detecting any stimulus. This is known as *saturation* and is represented in the figure with a dotted line.[51]

Consider the cone curve of a TVI [52] experiment as plotted in Figure 6.17, where the plot regards the threshold sensitivity $L_\mathrm{B}/\Delta L$ – that is, the inverse Weber ratio – where L_B is the background luminance.

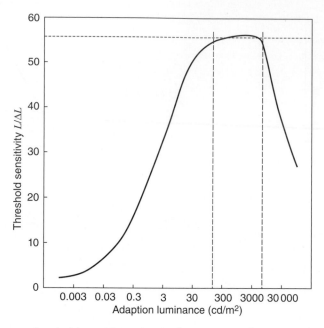

Figure 6.17   *Plot of the cone-threshold sensitivity, that is, the inverse Weber ratio, as a function of background adapting luminance L_B. The contrast sensitivity increases with the adaptation luminance up to approximately 50 cd/m². The Weber ratio is constant for a higher luminance, from approximately 100 cd/m² up to approximately 10 000 cd/m².[52]*

The threshold sensitivity for discriminating small light increments on a background increases as the background luminance increases up to about 50 cd/m². Between 50 and 10 000 cd/m² the threshold sensitivity is approximately constant. In this range of luminance Weber's law $\Delta L/L_B \cong 0.02$ holds true. This result refers to an experiment with a medium test field size – that is, a size non-disturbing the state of adaptation –. These results depend on the test size and retinal eccentricity because the distributions of the rods and cones on the retina are not uniform.

6.7   Weber's and Fechner's Laws

The light-adaptation phenomenon studied by the TVI experiment in Section 6.6 has shown that the discrimination threshold, at retinal illuminance between –2 and +2 log units, is proportional to the background adapting retinal illuminance and the slope $\Delta I/I_B$ is constant. The retinal illuminance is proportional to the luminance of the observed scene; therefore, $\Delta L_v/L_{vB} = \Delta I/I_B = constant$. What is significant is that this relationship appears to be strong enough to be considered a law. This relationship was discovered by E. H. Weber in 1834 and is known as *Weber's law*.[54] The change in stimulus intensity that can just be discriminated ΔL_v is a constant fraction k of the starting adaptation intensity of the stimulus L_v:

$$\Delta L_v = k\,L_v \ \text{ or } \ \frac{\Delta L_v}{L_v} = k \tag{6.8}$$

and this relationship is valid for a photopic luminance range from ~50 to ~10 000 cd/m². This law holds true with some approximation in a wider range from 28 to 100 000 cd/m². An improvement in this wider range is obtainable by adding a suitable constant $\Delta L_v/(L_v–L_{v,g})$.

6.7.1   Contrast Sensitivity

The constant fraction, denoted by $C_W \equiv \Delta L_v / L_v$, is known with different names: *Weber contrast*, *Weber ratio* and *Weber fraction*. Its reciprocal $S_c = 1/C_W$ is termed *contrast sensitivity*. The value of S_c depends on the luminance level, the viewing conditions and the state of adaptation.[4]

In daily life, sensitivity to contrast generally improves as the illumination level increases. This is a violation of Weber's law.

6.7.2   Fechner's Scaling

The empirical determination of the number of jnd above absolute threshold corresponding to a value of the physical stimulus requires a very long task, starting at absolute threshold and measuring successive values of ΔL_v along the physical continuum of L_v. Weber's law simplifies this task, because

Weber's law defines the jnd $\Delta \Psi = 1$ as a function of the stimulus and, by integration, the formal expression of *Fechner's law* – better referred to as *Fechner's scale* – follows:

$$\frac{\Delta L_v}{L_v} = k \Delta \Psi \quad \rightarrow \quad \int \frac{dL_v}{L_v} = k \int d\Psi \quad \rightarrow \quad \Psi = \frac{1}{k} \ln \frac{L_v}{L_{v,0}} \tag{6.9}$$

where $L_{v,0}$ is the luminance corresponding to a zero sensation and could represent the absolute threshold, below which no stimulus is perceived at all. This law states that the perception is proportional to the logarithm of the stimulus by a factor $1/k$.

The threshold has been widely investigated by Blackwell.[55]

Equation (6.9) is one of the hypotheses of Fechner's psychophysics.

6.8   Stevens' Law

"*Brightness* – attribute of a visual perception according to which an area appears to emit, or reflect, more or less light."[4]

Brightness as an attribute of a subjective sensation cannot be directly measured, while radiance and luminance can be measured. Psychophysics tries to relate brightness and luminance. In Section 6.7.2 the relation is given by Fechner's law; here it is given by Stevens' law.

6.8.1   Brightness Scaling and Stevens' Law

Brightness is an unrelated colour attribute; therefore, the observer has to judge the brightness seeing the stimulus as isolated in the darkness. Stevens defined the brightness in two steps:

1. first, a perceived scale of the brightness is made,
2. afterwards, its relation with the luminance is made.

Then, he gave:

"*Stevens' law* – A scaling of a sensory magnitude, Ψ, that follows a power law of stimulus magnitude ϕ, i.e., $\Psi = constant \times \phi^\beta$."[4,56,57]

The power β ranges from 0.44 to 0.33 and depends on the adaptation level and the surrounding light.

Moreover, in the 1963 experiment, Stevens considered brightness as a function of the adaptation level.[57] Here again Stevens defined brightness in two steps:

1. first, with the magnitude estimation method (Section 6.1.7), the experimenter showed to the observer a standard, arbitrarily called 10;
2. afterwards, the observer assigned to a set of stimuli numbers in proportion to their apparent brightness.

Then the *haploskopic matching* method (Section 6.1.5) was used and the comparison was made quickly, before a significant change of adaptation. The relation between brightness and luminance has been investigated with different adaptation levels.

"The observer estimates generated a pair of brightness functions, one for each eye. The validity of these functions was checked by interocular brightness matching. The results are described by the equation

$$\Psi = k\left(L_v - L_{v,0}\right)^{\beta} \tag{6.10}$$

when Ψ is brightness, L_v luminance and $L_{v,0}$ the absolute threshold. All the parameters, k, $L_{v,0}$ and β, change systematically with light adaptation. The exponent β increases from 0.33 for the dark-adapted eye to 0.44 for the eye adapted to 1 lambert."[57] (1 lambert = $10^4/\pi$ cd/m^2.)

The exponent $\beta < 1$ represents a compressive function.

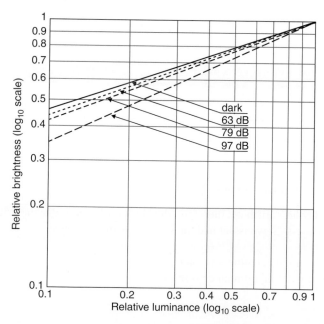

Figure 6.18   *Relative brightness versus relative luminance on logarithmic scales in four different adaptation-luminance levels according to Stevens' law.*[56, 57]

The logarithm of the two sides of equation (6.10) gives a linear equation where the variables are the logarithms of the psychological and the physical quantities, respectively

$$\log\left(\frac{\Psi}{\Psi_{max}}\right) = \beta \log\left(\frac{L_v - L_{v,0}}{L_{v,max} - L_{v,0}}\right) \tag{6.11}$$

Therefore, the power law, when plotted on log-log axes, plots as a straight line with a slope equal to the exponent β (Figure 6.18).

6.9   Fechner's and Stevens' Psychophysics

G. T. Fechner[1], in his attempt to construct *psychophysics*, wished to establish a mathematical equation to describe the relationship between physical events and conscious experience. Therefore he added the following two hypotheses to his general definitions of physical stimulus ϕ and sensation Ψ (Section 6.1):

1. Sensation magnitude could be quantified indirectly by relating the discrimination-threshold values $\Delta\phi$ on the physical scale to the corresponding values of the jnd in sensation on the psychological scale.
2. All values of jnd are equal psychological increments in sensation magnitude, regardless of the size of $\Delta\phi$ in physical units. The jnd is the smallest detectable increment in a sensation; therefore, it always has the same size. The jnd is assumed as the *base unit* of the metric scale. The scale of sensation magnitude is obtained by counting jnd, starting at absolute threshold. The intensity in physical units of a stimulus at absolute threshold, which represents the transition between sensation and no sensation, is assumed to correspond to the zero point on the psychological scale of sensation magnitude. (This is not the case of the visual attribute 'hue', where the zero point is very conventional.)
3. Sensation Ψ is proportional to the logarithm of the physical stimulus ϕ (*Fechner's law*), $\Psi = k \ln(\phi/\phi_0)$, and this function follows from the *Weber law*, $\Delta\phi = k \phi$, where k and ϕ_0 are empirical constants.

These hypotheses are limited to the case in which the visual attribute considered is well represented by an intensity, such as lightness. Attributes such as hue could follow different hypotheses.

In 1957 S. S. Stevens considered the Fechner procedure for defining the perceived scale as an *indirect method*, based on threshold discrimination, and proposed a *direct method*. "On this crucial point Stevens wrote: "On numerous perceptual continua, direct assessments of subjective magnitude seem to bear an orderly relation to the magnitude of the stimulus. To a fair first-order approximation, the ratio scales constructed by 'direct' methods (as opposed to the 'indirect' procedures of Fechner) are related to the stimulus by a power function of one degree or another (*Stevens' power law*)". Stevens wrote that he was giving "'new look' to psychophysics". With Stevens, the metric scale is modified with a different base unit and Fechner's law is substituted by *Stevens's law*.

These two different proposals for psychophysics should not be considered parties of a controversy (as it was) but as approaches to different psychophysics: that of Fechner is *local* on the scale and that of Stevens is *global*.[58]

6.10   Wavelength Discrimination

Hue is one of the attributes of a colour sensation. All three attributes of colour sensation – hue, colourfulness and brightness – for spectral stimuli depend on wavelength. Here the hue discrimination of spectral colour stimuli is considered. Psychophysical colour discrimination experiments allow us to know how much change in colour stimulus is required to detect one jnd in hue. The comparison of the hues of different colours needs criteria for hue-discrimination threshold and for hue matching.

The threshold of hue discrimination is represented by the amount of change in wavelength $\Delta\lambda$ for spectral stimuli that is required to detect a change of one jnd in hue. In the historical experiment,[59] the empirical apparatus presented a monocular Maxwellian view of bipartite field, one half filled with light of a standard wavelength λ and the other with light of a test wavelength $(\lambda + \Delta\lambda)$. Both lights of bipartite fields are narrow spectral bands of light that are varied in wavelength and radiance. Any measurement started with both bipartite fields with equal stimuli, and the observer was asked to change the wavelength of one half at constant brightness until one jnd of hue was detected. The judgement was made with increasing wavelength with a threshold in correspondence to $(\lambda + \Delta\lambda^+)$ and with decreasing wavelength with a threshold in correspondence to $(\lambda + \Delta\lambda^-)$. The average value $\langle\Delta\lambda\rangle = (\Delta\lambda^+ + \Delta\lambda^-)/2$ was assumed to represent the threshold of wavelength discrimination. This method is a combination of the method of limits and the method of adjustment. The result depends on the brightness level, surround of the bipartite field, field size and region of the retina used for the judgment. In Figure 6.19, the result of the classical experiment of Wright and Pitt is shown.[59]

Notice that near 500 nm and 600 nm observers can detect wavelength differences of about 1 nm. It is remarkable that the wavelength discrimination $\langle\Delta\lambda\rangle$ is between 1 and 2 nm in a wide range from 480 to 630 nm, and $\langle\Delta\lambda\rangle < 4$ nm between 440 and 650 nm. This discrimination capacity is very useful for defining the spectral characteristics of the instruments used in photometry and colorimetry: only light bands can be used, inside which the human visual system cannot discriminate; therefore, the bandpass function of the instrument has to be lower than 1 nm. If the bandpass function is higher, the results of any colorimetric computation have to be considered as approximate. The CIE chose 1 nm as the right step for any colorimetric and photometric computation.

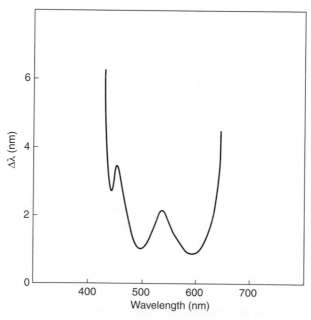

Figure 6.19 *Plot of the wavelength discrimination as a function of the wavelength, that is, the smallest wavelength difference that can be detected as a chromatic difference when luminance is constant.*[59]

A different way for measuring the wavelength discrimination is associated with the procedure of colour-matching, whose intrinsic uncertainty is quantified by its standard deviation. In the procedure of colour-matching, the two fields are initially different in wavelength. The observer adjusts the wavelength and radiance of the comparison field until it appears to be identical to the standard. The colour-matching data are represented by a distribution, with an average value and a standard deviation. Three standard deviations are assumed equal to the discrimination step, that is, to one jnd.[60] This technique is an alternative for evaluating the wavelength discrimination. The standard deviation procedure is more time-consuming than step by step, but may be more accurate as it shows less dependence on the application of the discrimination-threshold criterion.

6.11 Saturation Discrimination and Least Colorimetric Purity

Colorimetric purity and saturation are two mutually related quantities. Consider their definitions:

"*Colorimetric purity* – quantity defined by the relation

$$p_c \equiv \frac{L_d}{L_n + L_d} \tag{6.12}$$

where L_d and L_n are the respective luminances of the monochromatic stimulus and of the specified achromatic stimulus that match the colour stimulus considered in an additive mixture."[4]

"*Saturation* – colourfulness of an area judged in proportion to its brightness.
 NOTE – For given viewing conditions and at luminance levels within the range of photopic vision, a colour stimulus of a given chromaticity exhibits approximately constant saturation for all luminance levels, except when the brightness is very high."[4]

Colorimetric purity is measured by colour-matching with the considered stimulus matched with a mixture of a spectral stimulus L_d with an achromatic stimulus L_n. By definition the colorimetric purity of a spectral stimulus is equal to 1 and that of the neutral stimulus is 0.

In the saturation-discrimination experiment, two colour stimuli with equal luminance are compared in a bipartite field or in two concentric circular fields (Section 6.1.4). The standard colour is achromatic and the test is obtained by mixing the achromatic colour L_n with a spectral colour of defined wavelength $L_{v,\lambda}$. The observer defines $L_{v,\lambda}$ at the discrimination threshold as a function of the wavelength that can be quantified with one of the methods described in Section 6.1.6. The luminance $L_{v,\lambda}$ of the spectral light entering the mixture is related to the *saturation-discrimination threshold*, which is equal to the colorimetric purity threshold of the test stimulus at the discrimination threshold with the standard white:

$$p_c = \frac{L_{v,\lambda}}{L_n + L_{v,\lambda}} \tag{6.13}$$

The saturation discrimination describes the degree of paleness of the colour and is dependent on the luminance level L_n and on the wavelength (Figure 6.20). This figure shows that yellow light (~580 nm) has low saturating power, whereas blue (below 500 nm) and red (over 600 nm) have high saturation power because a big quantity of yellow light is required to make the white stimulus appear coloured, while a small amount of red and blue is required.[61–65]

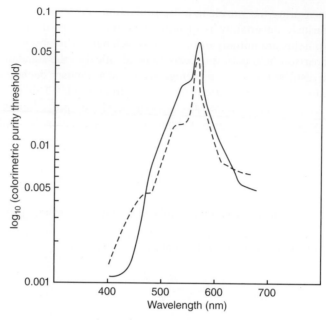

Figure 6.20   *Threshold of the colorimetric purity from white, as a function of the wavelength of the monochromatic radiation that added to the white modifies so barely perceptible saturation. The curves represent the logarithm of the colorimetric purity threshold of two observers.*[61–65]

6.12   Rushton's Univariance Principle and Scotopic Vision

In Section 6.5, scotopic sensitivity has been defined by psychophysical measurements. These kinds of measurements have led to the definition of the CIE 1951 standard photometric scotopic observer (Section 7.6). Here the scotopic sensitivity is considered in relation to the absorption of photons due to the rod's photopigment, rhodopsin. Photon absorption is a step in the physiological process of scotopic vision. The Rushton univariance principle[66] (Section 5.5.3) relates psychophysical with physiological quantities. This section is an interpretation of reference 67.

The photon flux crossing the rods is that measured in the corneal plane of the eye and reduced by the absorption of the crystalline lens and, albeit to a lesser extent, of other media which constitute the eye.

The known measured quantities are as follows:

- The relative scotopic luminous efficiency function $V'(\lambda)$ (Section 6.5);
- The crystalline lens transmittance $t_L(\lambda)$ (Section 5.3);
- The rodopsine absorption coefficient $\alpha(\lambda)$ (Section 5.5.3), which is proportional to the quantum efficiency of the scotopic visual system η_{rod} (recall that the quantum efficiency of a physical photodetector is the ability of the photons to be absorbed in the junction region of the detector);
- The radiance entering the eye measured in the corneal plane.

Rushton's univariance principle states that the spectral scotopic luminous efficiency function measured in actinometric units $[V'(\lambda)/\lambda]$ [recalling that $V'(\lambda) = L_e(\lambda = 500)/L_e(\lambda)$ and the actinometric units are obtained by dividing the spectral energy by the photon energy, $h\nu = hc/\lambda$ (Section 2.4); therefore $[V'(\lambda) 500/\lambda] = [L_e(\lambda = 500) 500]/[L_e(\lambda) \lambda]$ with λ in nm, and corrected by the crystalline lens absorption $t_L(\lambda)^{1.16}$ (the

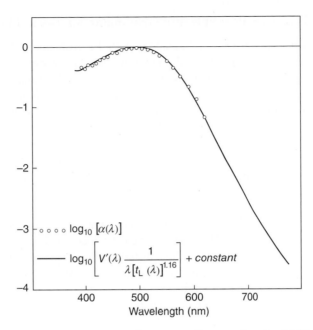

Figure 6.21 *Comparison of the standard scotopic luminous efficiency function V′(λ), measured in quantum units and corrected for crystalline lens absorption, with the relative absorption coefficient of the rod pigment on a logarithmic scale.*

exponent 1.16 takes into account the pupil aperture and the thickness of the crystalline lens)] is proportional to the quantum efficiency of the scotopic visual system:

$$\alpha(\lambda) \propto \frac{V'(\lambda)}{\lambda} \frac{1}{[t_L(\lambda)]^{1.16}} \tag{6.14}$$

The quantities in this equation are compared on a logarithmic scale in Figure 6.21. The result is an important qualification of the combination of the Rushton univariance principle with the principle of nominations of psychophysics applied to a simple physiological model of the transduction step of scotopic vision.[67]

From the Rushton univariance principle with the principle of nominations, it follows that the *rod activation* produced by a radiance $L_{e,\lambda}$ ($L_{p,\lambda}$ in actinometric unit) measured in the corneal plane, that is, the number of absorbed photon in the unit of time, is defined by

$$\int_{360}^{780} L_{p,\lambda} t_L^{1.16} \alpha(\lambda)\, d\lambda \propto \int_{360}^{780} L_{e,\lambda} V'(\lambda)\, d\lambda \tag{6.15}$$

and specifies the scotopic visual sensation.

6.13 Tristimulus Space

The first step in the color-vision process is represented by the activation of the three kinds of cones, which associates a vector in a linear vector space to any color stimulus. The vectors of this space, called tristimulus space, are a psychophysical color specification. The subsections of this section introduce the tristimulus space from a physiological point of view, assuming the Rushton univariance principle, and by a psychophysical point of view.

6.13.1 Rushton's Univariance Principle and Grassmann's Laws in Photopic Vision

In this section we consider the photopic vision in analogy with the treatment of the scotopic vision made in Section 6.12. The aim is to relate psychophysical quantities with the corresponding physiological ones of colour vision. Section 6.4 presented the spectral sensitivity curves of the three kinds of cones determined psychophysically under selective adaptation by Wald[39,40], and the same cone sensitivities corrected for absorption of light by elements of the eye before the retina. The comparison of these two sets of data (Figures 6.12 and 6.13) with the absorbances of the pigments of the cones shown in Figure 5.10 confirms the goodness of the Rushton univariance principle combined with the *principle of nomination* of psychophysics applied to a very simple model of eye physiology. We must pay attention to the fact that the psychophysical data have been obtained by insulating the three different kinds of cones by suitable adapting lights and the isolation was not complete; therefore, the three curves are non-completely independent.

Historically, colorimetry did not start from this point, but on spectral colour-matching functions and Grassman's laws. Let us recall how Rushton wrote on this important point: "Grassmann's Laws were important formulations of the facts of colour-matching. They include such statements as in a colour match any component may be replaced by one that matches it perfectly, without disturbing the exactness of the original match. All these laws follow as **trivialities** from the principle of univariance applied to the cone pigments".[68] As in the previous section, here we apply the Rushton univariance principle and the 'principle of nomination' of psychophysics to derive Grassmann laws. We assume that there are three kinds of cones, and that their activations are independent.

Rushton's principle states that visual sensation depends only on the number of photons absorbed by the cones and these numbers are the *cone activation values*, or simply *cone activations*:

$$\begin{cases} L = K_L \int_{360}^{780} L_{p,\lambda}\tau(\lambda)\eta_L(\lambda)\,d\lambda & \text{for L-cones} \\[2mm] M = K_M \int_{360}^{780} L_{p,\lambda}\tau(\lambda)\eta_M(\lambda)\,d\lambda & \text{for M-cones} \\[2mm] S = K_S \int_{360}^{780} L_{p,\lambda}\tau(\lambda)\eta_S(\lambda)\,d\lambda & \text{for S-cones} \end{cases} \tag{6.16}$$

where

- $L_{p,\lambda} = L_{e,\lambda}(\lambda/hc)$ is the radiance entering the eye measured on the cornea and measured in actinometric units,
- $\tau(\lambda)$ is the transmittance of all the media crossed by the photons in the eye before colliding with the cones, which depends on the region of the retina considered [different observers have been standardized, CIE 1931 for 2° visual field and CIE 1964 for 10° visual field[69] (Sections 9.2 and 9.3) and the Stockman and Sharpe fundamentals for both 2° and 10° visual fields (Section 9.6)],
- $\eta_L(\lambda)$, $\eta_M(\lambda)$ and $\eta_S(\lambda)$ are the probability that a photon with wavelength λ is absorbed by an L-, M- and S-cone, respectively,
- K_L, K_M and K_S are conventional normalization constants, which generally are chosen such that the activation values are $L = M = S = 1$ or 100 for the equal energy stimulus function $E_E(\lambda) = 1$ (Section 2.1).

Rushton's univariance principle states that the spectral sensitivity functions of the cones, referred to radiometric units, are proportional to their spectral quantum efficiencies corrected by the transmittance of all the media of the eye; therefore, the three *cone spectral sensitivity functions* – also termed *cone fundamentals* or *fundamental colour-matching functions* – are defined as follows:

$$l(\lambda) \equiv K_L\tau(\lambda)\eta_L(\lambda)\lambda, \ m(\lambda) \equiv K_M\tau(\lambda)\eta_M(\lambda)\lambda, \ s(\lambda) \equiv K_S\tau(\lambda)\eta_S(\lambda)\lambda \tag{6.17}$$

where the factor λ is required for the conversion from the actinometric to the radiometric units. The cone activations are

$$
\left(
\begin{array}{l}
L = \displaystyle\int_{360}^{780} L_{e,\lambda} l(\lambda)\, d\lambda \\[2mm]
M = \displaystyle\int_{360}^{780} L_{e,\lambda} m(\lambda)\, d\lambda \\[2mm]
S = \displaystyle\int_{360}^{780} L_{e,\lambda} s(\lambda)\, d\lambda
\end{array}
\right.
\tag{6.18}
$$

where the argument of these integrals is written as dependent on the spectral radiance $L_{e,\lambda}$. But the CIE uses a more general quantity, termed *spectral colour-stimulus function* and denoted by ϕ_λ, that is, a "description of a colour stimulus by the spectral concentration of a radiometric quantity, such as radiance or radiant power, as a function of wavelength".[4] All these spectral quantities are proportional to each other (Section 2.1).

The sets of cone activations (L, M, S) are the psychophysical specifications of colour stimuli, which are called simply *colour stimuli* and denoted by a symbol in square brackets, for example, [A] (the context loses its ambiguity).

With regard to the three kinds of cones – L, M and S – the property of linearity of the integrals (6.18), which define the whole set of activation values (L, M, S), is such that

- at any addition of radiances $L_{e,\lambda,1} + L_{e,\lambda,2}$, there is the addition of the corresponding cone activation values $(L_1 + L_2, M_1 + M_2, S_1 + S_2)$;
- at any multiplication of the radiances by a real positive number a, there is the product of the cone activation values by the same number (aL, aM, aS);
- the null element exists, represented by the cone activation values $(L = 0, M = 0, S = 0)$ and corresponding to a null radiance.

These properties of the set of three activation values (L, M, S) are the trichromatic generalization of Grassmann's laws.[70–73] Formally,

the *trichromatic generalization of Grassmann's laws* state that, given four psychophysical colour stimuli [A], [B], [C] and [D], the following linearity relations hold true:

1. Symmetry law: if $[A] \equiv [B]$, then $[B] \equiv [A]$
2. Transitivity law: if $[A] \equiv [B]$, and $[B] \equiv [C]$ then $[A] \equiv [C]$
3. Proportionality law: if $[A] \equiv [B]$, then $a[B] \equiv a[A]\ \forall\ a > 0$ and real
4. Additivity law: if $[A] \equiv [B]$ and $[C] \equiv [D]$, then $([A] + [C]) \equiv ([B] + [D])$

(where the symbol '$\equiv$' means 'matching'; this symbol has only historical meaning and is no longer used in the vector notation of colour stimuli.).

The *trichromatic generalization of Grassmann's laws* is the definition of a tridimensional linear vector space with a common origin in the vector $(0, 0, 0)$ defined on the real number field (Figure 6.22). Therefore, hereafter the following terminology is used: the psychophysical colour stimuli [Q] are represented by *tristimulus vectors* $\mathbf{Q} = (L, M, S)$ (denoted by bold roman letters, as mathematical convention requires) in the *tristimulus space* and the vector components (L, M, S) are termed *tristimulus values*.

In a linear vector space, an infinite set of reference frames exists and the reference frame defined by equations (6.16) is the *fundamental reference frame*, as conceived by König[74], where the three components of

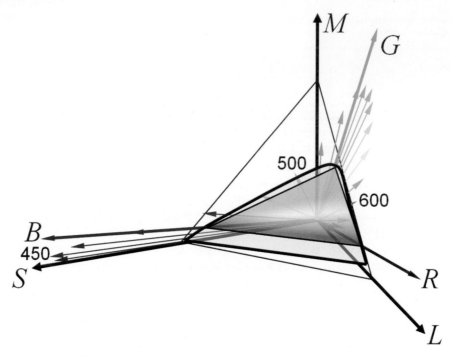

Figure 6.22   *Fundamental reference frame in the tristimulus space in which the components of the vectors, that is, the fundamental tristimulus values, are proportional to the cone activations. In the figure are plotted the coloured vectors* $\mathbf{U}_{LMS,\lambda}$ *that represent the activations due to monochromatic stimuli with a unity of power and whose components as a function of wavelength are the colour-matching functions in the fundamental reference frame. Three vectors* $\mathbf{R}$, $\mathbf{G}$ *and* $\mathbf{B}$ *related to three monochromatic lights – red, green and blue, respectively – are assumed as reference vectors for the laboratory reference frame, introduced below in Section 6.13.5. On the plane* $L + M + S = 1$ *is defined the chromaticity diagram (Section 6.13.3), distinguished as a grey region. Inside the chromaticity diagram is a coloured triangle (coloration only indicative) that represents all the chromaticities possible by mixing in a variable way the red, green and blue colours represented by the vectors* $\mathbf{R}$, $\mathbf{G}$ *and* $\mathbf{B}$.

a vector represent the activations of the three kinds of cones. In this reference frame, the basic vectors are $\hat{\mathbf{L}}$, $\hat{\mathbf{M}}$ and $\hat{\mathbf{S}}$. The origin is the same for all the reference frames.

The tristimulus space in the fundamental reference frame is also called *cone activation space*.

In the fundamental reference only a portion of the octant with positive components has psychophysical and physiological meaning and this is due to the following properties of the tristimulus values:

- Tristimulus values are positive because the negative activation of the cones is impossible.
- It is impossible to activate a single type of cone leaving the other inactivated, because never two of the three spectral quantum efficiencies of the photopigments of the three types of cones are simultaneously zero (Section 6.4.1), excluding the wavelength extremes of the visible range in which all three quantum efficiencies go to zero. Therefore, the vectors with a component that is not zero and the other two equal to zero, represent non-physical stimuli.

The tristimulus space is completely defined by the knowledge of the tristimulus vectors $\mathbf{U}_\lambda$ associated with the unit-power colour stimulus functions $\Phi_{U,\lambda}$, whose spectral power distribution $\Phi_U(\lambda'-\lambda)$ has the shape of an extremely narrow band centred in the wavelength λ ('extremely narrow' means that the cone sensitivity functions are considerable as constant inside the band).

By definition, the unit-power stimulus functions have the following property:

$$\int_{380}^{780} \varPhi_{\mathrm{U}}(\lambda' - \lambda)d\lambda' = 1 \qquad (6.19)$$

and, because their spectral bands are extremely narrow, it follows that

$$\begin{cases} \displaystyle\int_{380}^{780} \varPhi_{\mathrm{U}}(\lambda' - \lambda)l(\lambda')d\lambda' = l(\lambda) \\[2ex] \displaystyle\int_{380}^{780} \varPhi_{\mathrm{U}}(\lambda' - \lambda)m(\lambda')d\lambda' = m(\lambda) \\[2ex] \displaystyle\int_{380}^{780} \varPhi_{\mathrm{U}}(\lambda' - \lambda)s(\lambda')d\lambda' = s(\lambda). \end{cases} \qquad (6.20)$$

and

the stistimulus values associated to the unit-power monochromatic tristimulus vectors are the colour-matching functions (subindex LMS specifies the fundamental reference frame of the tristimulus space)

$$\mathbf{U}_{\mathrm{LMS},\lambda} = \big(l(\lambda), m(\lambda), s(\lambda)\big) \qquad (6.21)$$

The mathematics used in equations (6.19), (6.20) and (6.21) is approximate, and the mathematics of the distributions, or generalized functions, should be used. Anyway this representation is useful to represent conceptually the phenomena and is used only here. Figure 6.23 shows the colour-matching functions ($l(\lambda)$, $m(\lambda)$ and $s(\lambda)$), that is, the unit-power monochromatic tristimulus values, which are derived from the psychophysical measurement described below. The standardized colour-matching functions, obtained as an average from a panel of observers, are denoted with a bar $\big(\bar{l}(\lambda), \bar{m}(\lambda), \bar{s}(\lambda)\big)$ (Chapter 9).

Each colour stimulus can be considered as the sum of monochromatic components and the corresponding tristimulus vector, both for the univariance principle and for Grassmann's laws, can be considered as the sum of the tristimulus vectors associated with these monochromatic components.

Colour equation is termed the "vector representation of the match of 2 colour stimuli, of which, for instance, one may be an additive mixture of 3 reference colour stimuli."[4]

Here mathematics enters psychophysics in a very powerful way. A mathematical knowledge of the tristimulus space is necessary

- for an accurate definition of this space by colour-matching measurements (Section 6.13.4, 6.13.5, 6.13.6, Chapter 9),
- for colour management in connection with the instrument used and the laboratory,
- for colour coding in colour reproduction (Chapter 15).

The first property of a linear space is the possibility to choose a *reference frame* according to a specific convenience. Particularly important are the reference frames RGB and XYZ (Sections 9.2.1 and 9.2.2, Chapter 15).

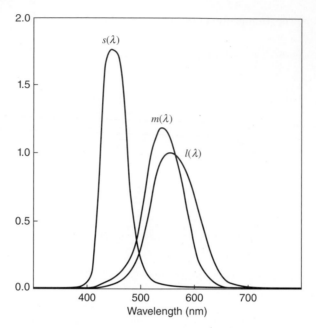

Figure 6.23   *Unit-power monochromatic tristimulus values, (l(λ), m(λ), s(λ)), also termed colour-matching functions in the fundamental reference frame, normalized to have the integral area in the visible spectral range equal to 1.*

A very important tool in tristimulus space is the chromaticity diagram, shown in Sections 6.13.3 and 6.13.8.

The activation process of the cones by an illuminated surface is visually represented in the software described in Section 16.7.

6.13.2   Metamerism

The metamerism phenomenon is evident in equation (6.18).

> "*Metameric colour stimuli* – spectrally different colour stimuli that have the same tristimulus values in a specified colorimetric system. Equivalent term: *metamers*.
>   NOTE – The corresponding property is called *metamerism*."[4]

The same set of tristimulus values, defined by equations (6.18), is obtainable from an infinite set of different radiances $L_{e,\lambda}$, for which the integrals (6.18) have the same result. The correspondence between spectral radiances and colour stimuli is many to one with the exclusion of the monochromatic or spectral radiances, which are in a one-to-one correspondence with the corresponding tristimulus vectors.

The term 'metamer' in the colour field has been introduced by Ostwald[75] and was defined as follows, although non-mathematically:

> "Colour metamers are those that, even though they are composed differently in regard to types of light or wavelengths, have identical appearance. The identical appearance is not limited to hue but is complete, that is, also in regard to white and black content." ['white and black content' regards the colour model of Ostwald (Section 14.3.3)]

Metamerism is a very important phenomenon from a scientific and a practical point of view. Reproductions of colours according to various techniques (television, photography, screen plate printing by inks, etc.) are metameric reproductions because the radiances reflected or emitted by the reproductions are different, often very different from the radiances reflected or emitted by the originals.

6.13.3 Chromaticity

The colour stimuli are represented in the tristimulus space by vectors, and a way for their specification is represented by length and direction. The chromatic sensations depend on the ratios of the cone activation, which depend on the direction of the tristimulus vector. Therefore, the directions of the tristimulus vectors represent the chromatic sensation called *chromaticity* of the colour stimuli and is specified by chromaticity coordinates. The technique used for representing the chromaticity is given by a *chromaticity diagram*: a plane figure constituted by the intersections of the tristimulus vectors, or their extensions, with a plane, usually the unit plane. This plane is represented by the equation $L + M + S = 1$ and intersects the three reference axes at the points $(1, 0, 0)$, $(0, 1, 0)$ and $(0, 0, 1)$ (Figures 6.22 and 6.24). This representation is possible because the directions of the vectors, and then the chromaticity, are in bijection (one-to-one correspondence) with the points of the plane. In particular, the tristimulus vectors related to monochromatic stimuli intersect the unit plane along a line called *spectrum locus*.

A coordinates system on the plane of the chromaticity diagram should be introduced in order to define the chromaticity of the colour stimuli. These coordinates are called *chromaticity coordinates*, and a point q on the chromaticity diagram defines the chromaticity of the corresponding tristimulus vector Q. The chromaticity diagram in the LMS reference frame is contained in an equilateral triangle, whose vertices, L, M and S, are given by the intersection of the reference axes with the plane $L + M + S = 1$. On this triangle each intersection point q of the generic vector $Q = (L, M, S)$ with the unit plane is defined by three *barycentric coordinates*

$$l = \frac{L}{L + M + S}, \quad m = \frac{M}{L + M + S}, \quad s = \frac{S}{L + M + S} \tag{6.22}$$

for which the identity equation $l + m + s = 1$ holds true (Figure 6.24). This identity equation constrains the three chromaticity coordinates among them; therefore, the chromaticity $q = (l, m)$ of the tristimulus vector Q is defined by two coordinates, the third being automatically defined $s = 1 - l - m$.

The triangle LMS is a triangular diagram (Figure 6.24), in which the ratios of the areas of the triangles MqS, SqL and LqM are equal to those of the chromaticity coordinates, and also to those of the tristimulus values

$$\text{area}(MqS) : \text{area}(SqL) : \text{area}(LqM) = l : m : s = L : M : S \tag{6.23}$$

The three triangles have the equal base and then the relationships between the areas (6.23) are also satisfied by the corresponding heights (Figure 6.24).

Two parallel tristimulus vectors $Q_1 = (L_1, M_1, S_1)$ and $Q_2 = (L_2, M_2, S_2)$ intersect the chromaticity diagram in the same point, that is, have equal chromaticity, and the corresponding cone activations are in the same ratios $L_1 : M_1 : S_1 = L_2 : M_2 : S_2$, or $L_1 : L_2 = M_1 : M_2 = S_1 : S_2$.

Conventionally, the equation of the plane for the chromaticity diagram is the unit plane, but it is not necessary. Particularly, the CIE (u, v) chromaticity diagram is defined on the plane $X + 15Y + 3Z = 1$ in the XYZ reference frame (Section 11.3.1); the chromaticity diagram of MacLeod-Boynton, conceived with a physiological aim, is defined on the equi-luminant plane $L + M = 1$ (Sections 9.5.2 and 9.6.2). [NB: The MacLeod-Boynton chromaticity coordinates are denoted with the same symbols ($l, m = 1 - l, s$), but are based on a different normalization (Sections 9.5.2 and 9.6.2).]

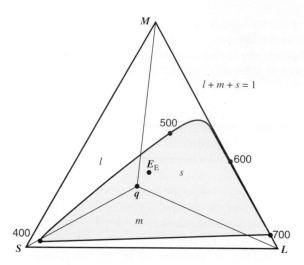

Figure 6.24 *Chromaticity diagram represented by the grey region within the triangle* **LMS** *on the plane with the equation L + M + S = 1. The curved border of the diagram is the 'spectrum locus' and its points represent the chromaticities of the monochromatic lights, while the points of the straight line represent the purple hues and is termed 'purple line'. The points* **q** *on this plane defined by three coordinates (l, m, s = 1 − l − m), termed barycentric coordinates, that are proportional to the areas of the triangles* **MqS**, **SqL** *and* **LqM**, *respectively. The chromaticity of the equal-energy stimulus* **E**$_E$ *is in the centre of the triangle* **LMS**.

6.13.4   Reference Frames in Tristimulus Space

In Section 6.13.1 the tristimulus space is represented in the fundamental reference system, in which the three components of the vectors represent the activations of the three kinds of cones. The reference frame is defined by three orthogonal axes. In this space any triplet of axes, provided independent, can be considered as a reference frame and the transformation between the two reference frames is linear and represented by a matrix. The new reference frame is represented by three orthogonal axes.

To implement a transformation of the reference frame, it is necessary to know the new basic vectors in the initial reference frame. For convenience, let us consider a generic set of coordinates (X, Y, Z) – this is the notation of the standard reference of the CIE (Sections 9.2.2 and 9.3) – for the initial reference frame and a set (R, G, B) for the final reference frame: $\hat{\mathbf{R}}, \hat{\mathbf{G}}$ and $\hat{\mathbf{B}}$ are the *reference stimuli* in the RGB reference frame, (R, G, B) the tristimulus vectors, the vector components R, G and B *tristimulus values*, and (r, g) the chromaticity coordinates, which are defined as

$$r = \frac{R}{R+G+B}, \; g = \frac{G}{R+G+B}, \; b = \frac{B}{R+G+B}, \; r+g+b=1 \tag{6.24}$$

and analogously, in the XYZ reference frame, $\hat{\mathbf{X}}, \hat{\mathbf{Y}}$ and $\hat{\mathbf{Z}}$ are the reference stimuli, (X, Y, Z) are the tristimulus vectors, the vector components X, Y and Z are tristimulus values, and (x, y) are the chromaticity coordinates, which are defined as

$$x = \frac{X}{X+Y+Z}, \; y = \frac{Y}{X+Y+Z}, \; z = \frac{Z}{X+Y+Z}, \; x+y+z=1 \tag{6.25}$$

The basic vectors $\hat{\mathbf{R}}, \hat{\mathbf{G}}$ and $\hat{\mathbf{B}}$ and the corresponding chromaticities in the XYZ reference frame are

$$\hat{\mathbf{R}} = (X_r, Y_r, Z_r) = c_r(x_r, y_r, z_r) \quad \text{with } c_r = X_r + Y_r + Z_r$$
$$\hat{\mathbf{G}} = (X_g, Y_g, Z_g) = c_g(x_g, y_g, z_g) \quad \text{with } c_g = X_g + Y_g + Z_g \qquad (6.26)$$
$$\hat{\mathbf{B}} = (X_b, Y_b, Z_b) = c_b(x_b, y_b, z_b) \quad \text{with } c_b = X_b + Y_b + Z_b$$

whose lengths are such that their sum is equal to a specified neutral or achromatic stimulus

$$\mathbf{W} = \hat{\mathbf{R}} + \hat{\mathbf{G}} + \hat{\mathbf{B}} = (X_n, Y_n = 1, Z_n) \text{ with chromaticity } (x_n, y_n) \qquad (6.27)$$

and $z_n = 1 - x_n - y_n$ (often $Y_n = 100$ instead of $Y_n = 1$). In general the *neutral stimulus* is the equal energy stimulus $\mathbf{E}_E$ with the stimulus function $E_{E,\lambda} \equiv L_{e,\lambda} = 1$, but in some cases is the stimulus of the considered illuminant (Sections 15.2 and 15.3).

The transformation matrices are

$$\begin{pmatrix} X \\ Y \\ Z \end{pmatrix} = \mathbf{T}_{RGB \to XYZ} \begin{pmatrix} R \\ G \\ B \end{pmatrix} = \begin{pmatrix} c_r x_r & c_g x_g & c_b x_b \\ c_r y_r & c_g y_g & c_b y_b \\ c_r z_r & c_g z_g & c_b z_b \end{pmatrix} \begin{pmatrix} R \\ G \\ B \end{pmatrix}$$

$$\begin{pmatrix} R \\ G \\ B \end{pmatrix} = \mathbf{T}_{XYZ \to RGB} \begin{pmatrix} X \\ Y \\ Z \end{pmatrix} = \frac{1}{\Delta} \begin{pmatrix} r_x / c_r & r_y / c_r & r_z / c_r \\ g_x / c_g & g_y / c_g & g_z / c_g \\ b_x / c_b & b_y / c_b & b_z / c_b \end{pmatrix} \begin{pmatrix} X \\ Y \\ Z \end{pmatrix} \qquad (6.28)$$

$$\mathbf{T}_{RGB \to XYZ} = \left(\mathbf{T}_{XYZ \to RGB} \right)^{-1}$$

with

$$\Delta = x_r(y_g z_b - y_b z_g) + x_g(y_b z_r - y_r z_b) + x_b(y_r z_g - y_g z_r) \qquad (6.29)$$

$$\left\{ \begin{array}{l} c_r = [X_n(y_g z_b - y_b z_g) - Y_n(x_g z_b - x_b z_g) + Z_n(x_g y_b - x_b y_g)] / \Delta \\ c_g = [-X_n(y_r z_b - y_b z_r) + Y_n(x_r z_b - x_b z_r) + Z_n(x_r y_b - x_b y_r)] / \Delta \\ c_b = [X_n(y_r z_g - y_g z_r) - Y_n(x_r z_g - x_g z_r) + Z_n(x_r y_g - x_g y_r)] / \Delta \end{array} \right. \qquad (6.30)$$

$$\left\{ \begin{array}{l} r_x = (y_g z_b - y_b z_g) \\ r_y = (-x_g z_b + x_b z_g) \\ r_z = (x_g y_b - x_b z_g) \end{array} \right. , \left\{ \begin{array}{l} g_x = (y_b z_r - y_r z_b) \\ g_y = (-x_b z_r + x_r z_b) \\ g_z = (x_b y_r - x_r z_b) \end{array} \right. , \left\{ \begin{array}{l} b_x = (y_r z_g - y_g z_r) \\ b_y = (-x_r z_g + x_g z_r) \\ b_z = (x_r y_g - x_g z_r) \end{array} \right. \qquad (6.31)$$

The linear transformations considered here can be computed with the software presented in Section 16.8.6.

6.13.5   Measurement of the Colour-Matching Functions in the RGB Reference Frame

Instrumental Reference Frame and Colour Matching Apparatus

In Section 6.13.1, the tristimulus space is obtained from the spectral quantum efficiencies $\eta_L(\lambda)$, $\eta_M(\lambda)$ and $\eta_S(\lambda)$ of the three types of cones and the univariance Rushton principle. Historically, the tristimulus space has been constructed in a different way. The reasons are historical and especially because the physiology of the visual system was not known as it is today. Moreover, the way followed has had advantages and led to a tristimulus space accurate enough to allow a standardization. The experiments were psychophysical and based on colour-matching.

Figure 6.25 shows a sketch of the colour-matching apparatus with a bipartite field. The bipartite field is shown also in Figure 6.1.

The physically significant vectors in tristimulus space represent tristimuli to which corresponds a radiance. Three vectors representing three colour stimuli with different radiances, physically produced by three light sources, constitute a possible reference frame in tristimulus space if the corresponding tristimulus vectors are independent, that is, one of the tristimulus vectors cannot be equated to a suitable mixture of the other two. Such a reference frame, called an *instrumental reference frame*, is constructed here. Figure 6.22 presents the LMS reference frame with the vectors associated with monochromatic unit radiances and the basic vectors $\hat{\mathbf{R}}, \hat{\mathbf{G}}, \hat{\mathbf{B}}$ (termed *reference stimuli* and often, colloquially, *primary stimuli*), corresponding to three light sources of a laboratory, one red, one green and one blue, respectively. Figure 6.26 is the transformation of

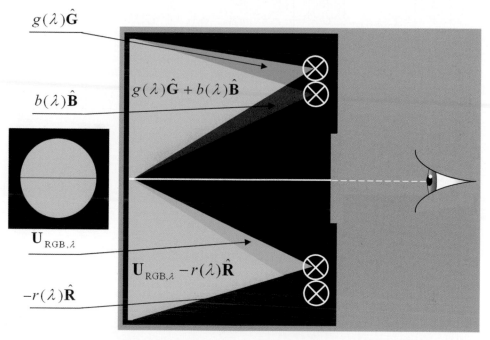

Figure 6.25   *Cross section of a colour-matching apparatus, constituted by a box with a hole, which the observer looks through. The observer sees a bipartite field as shown on the left of the figure. The box is internally subdivided into two parts by a baffle avoiding any mixture of the lights of the two parts. The ground of the box, seen by the observer, is a Lambertian non-wavelength selective diffuser that in the two parts is uniformly lit by different lights. The observer has to match the colours of the two reflected lights visually. This task consists of modifying the emissions of the four light sources in order to perform the colour-matching. The sum of the reference green and blue lights $g(\lambda)\hat{\mathbf{G}} + b(\lambda)\hat{\mathbf{B}}$ properly modulated matches the sum of a unit power monochromatic light $\mathbf{U}_{RGB,\lambda}$ of wavelength λ and the reference red light $-r(\lambda)\hat{\mathbf{R}}$ (in this case $r(\lambda)$ is a negative quantity). Analogous measurement can be made in a Maxwellian view.*

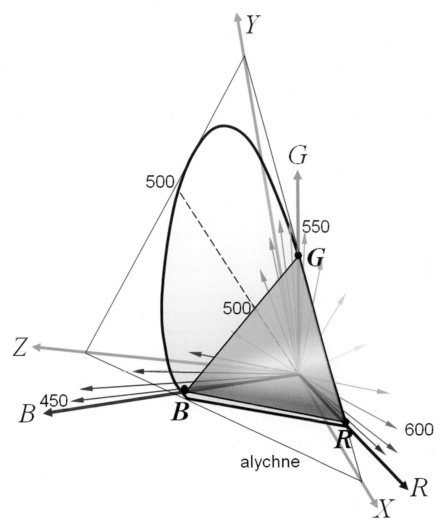

Figure 6.26 *Laboratory-reference frame in the tristimulus space in which the vectors **R**, **G** and **B** related to three monochromatic lights – red, green and blue, respectively – constitute a set of three orthogonal vectors. This reference frame is obtained from the fundamental one of Figure 6.22 by a linear transformation. A one-to-one correspondence exists between the elements of the two reference frames as exists between this figure and Figure 6.22. In the figure are plotted the coloured vectors $\mathbf{U}_{RGB,\lambda}$ that represent the monochromatic stimuli with a unity of power and whose components as a function of wavelength are the colour-matching functions in the RGB reference frame. Here the chromaticity diagram is on the plane $R + G + B = 1$, where (R, G, B) are the coordinates or tristimulus values in this reference frame. The XYZ reference frame is presented in Sections 6.13.6 and 9.2.2.*

Figure 6.22 to the RGB reference frame. The monochromatic colour stimuli of wavelength λ and unit radiance, specified in the LMS reference frame by $\mathbf{U}_{LMS,\lambda} = (l(\lambda), m(\lambda), s(\lambda))$, are specified in the RGB reference frame by

$$
\begin{pmatrix} r(\lambda) \\ g(\lambda) \\ b(\lambda) \end{pmatrix} = \mathbf{T}_{LMS \rightarrow RGB} \begin{pmatrix} l(\lambda) \\ m(\lambda) \\ s(\lambda) \end{pmatrix}
\tag{6.32}
$$

and the three components represent the intensity of colour stimuli associated with the radiances of the three RGB light sources, whose colour stimuli match the monochromatic unit power stimulus with wavelength λ. In the laboratory, an observer can make the experiment of matching a monochromatic colour stimulus with a mix of stimuli, produced with three light sources, one red, one green and one blue. This is an experiment of colour-matching and the functions $r(\lambda)$, $g(\lambda)$ and $b(\lambda)$ are called *colour-matching functions*. This is the way used by Maxwell[76] in the first construction of the tristimulus space. The RGB reference frame was used by Guild and Wright for producing the data used in the construction of the CIE 1931 standard observers.[77] (Section 9.2).

Measurement of the Colour Matching Functions

In the fundamental reference frame, the functions $l(\lambda)$, $m(\lambda)$ and $s(\lambda)$ represent for each wavelength the activations of the three corresponding types of cones due to the monochromatic colour stimulus with unit radiance at wavelength λ. As a consequence of the nature of the tristimulus vector space, the stimulus of a monochromatic unit radiance, represented by the vector $\mathbf{U}_{\mathrm{LMS},\lambda} = (l(\lambda), m(\lambda), s(\lambda))$, once transformed into the RGB reference frame by equation (6.28), is expressible as the sum of three independent vectors:

$$\mathbf{U}_{\mathrm{RGB},\lambda} = r(\lambda)\hat{\mathbf{R}} + g(\lambda)\hat{\mathbf{G}} + b(\lambda)\hat{\mathbf{B}} \qquad (6.33)$$

where $\hat{\mathbf{R}}, \hat{\mathbf{G}}, \hat{\mathbf{B}}$ are three vectors representing three monochromatic colour stimuli of unit radiance, and $r(\lambda)$, $g(\lambda)$ and $b(\lambda)$ are the corresponding radiances necessary to obtain the colour-matching represented by equation (6.33). Then these functions are the colour-matching functions in the reference frame defined by the $\hat{\mathbf{R}}, \hat{\mathbf{G}}, \hat{\mathbf{B}}$ reference stimuli. The requirement of these vectors to be chosen as three reference stimuli, suitable to define a reference frame, is to be independent, that is, no linear combination of them with not all zero coefficients may be equal to the null vector.

Colour-matching is obtained by equating visually different mixtures of lights belonging to the two parts of a bipartite field and is represented by a vector equation (termed *colour equation*), as derived from the univariance principle.

There are two methods to assess visually the colour-matching functions:

1. the *minimum saturation method* proposed by Maxwell[76],
2. the *maximum saturation method*, used for defining the CIE standard observers.

In both methods, the colour-matching criterion for the observer is that the line dividing the two parts of the bipartite field is made invisible.

Maximum Saturation Method

The maximum saturation method simply considers colour-matching as represented by equation (6.33). The reference frame is defined by three monochromatic reference stimuli $\hat{\mathbf{R}}, \hat{\mathbf{G}}, \hat{\mathbf{B}}$. The matching is done in a bipartite field, but, if the red, green and blue lights are put in the same part of the bipartite field, colour-matching is impossible. A negative light source should be used, that is, not natural and seemingly absurd. The solution was given by Maxwell, who said: "by transposing the negative term to the other side" of the colour-matching equation "it becomes positive, and then the equation may be verified."[65] Therefore, the mixture of the monochromatic stimulus $\mathbf{U}_{\mathrm{RGB},\lambda}$ with the reference stimulus that in equation (6.33) was negative, but now in the other side is positive, matches with a mixture of the other two reference stimuli. At the change in wavelength of the monochromatic stimulus, the chromaticity of the light proposed in the bipartite changes moving on the sides of the triangle whose vertices are the chromaticities of the reference stimuli. Since saturation of a stimulus is the distance between its chromaticity and that of a selected achromatic stimulus, the saturation on the sides of the triangle is the highest with that hue. Therefore, the name 'maximum saturation method'.

Maxwell's Minimum Saturation Method

The minimum saturation method was used by Maxwell first[76]. In this method it is considered a colour-matching in which the bipartite field presents white light, that is, saturation zero. Therefore, the name 'minimum saturation method'. Consider the monochromatic reference stimuli $\hat{\mathbf{R}}, \hat{\mathbf{G}}, \hat{\mathbf{B}}$, and the white stimulus $\mathbf{W} = r_W \hat{\mathbf{R}} + g_W \hat{\mathbf{G}} + b_W \hat{\mathbf{B}}$. The same stimulus $\mathbf{W}$ is also obtainable as a mixture of the three reference stimuli and of one monochromatic stimulus $\mathbf{W} = e_\lambda \mathbf{U}_{RGB,\lambda} + r_\lambda \hat{\mathbf{R}} + g_\lambda \hat{\mathbf{G}} + b_\lambda \hat{\mathbf{B}}$. In this case colour-matching occurs between the two stimuli $\mathbf{W}$:

$$\mathbf{W} = r_W \hat{\mathbf{R}} + g_W \hat{\mathbf{G}} + b_W \hat{\mathbf{B}} = e_\lambda \mathbf{U}_{RGB,\lambda} + r_\lambda \hat{\mathbf{R}} + g_\lambda \hat{\mathbf{G}} + b_\lambda \hat{\mathbf{B}} \tag{6.34}$$

and, once the colour match is obtained, the radiances $r_W, g_W, b_W, r_\lambda, g_\lambda, b_\lambda$. and e_λ are measured. In this equation, the substitution of the stimulus $\mathbf{U}_{RGB,\lambda}$ with that given by equation (6.33) and rearranging gives

$$\left[r(\lambda)e_\lambda + r_\lambda - r_W \right] \hat{\mathbf{R}} + \left[g(\lambda)e_\lambda + g_\lambda - g_W \right] \hat{\mathbf{G}} + \left[b(\lambda)e_\lambda + b_\lambda - b_W \right] \hat{\mathbf{B}} = 0 \tag{6.35}$$

from which the colour-matching functions are

$$r(\lambda) = \frac{r_W - r_\lambda}{e_\lambda}, \ g(\lambda) = \frac{g_W - g_\lambda}{e_\lambda}, \ b(\lambda) = \frac{b_W - b_\lambda}{e_\lambda} \tag{6.36}$$

Colour-Matching Functions and Tristimulus Vectors in the RGB Reference Frame

Different individual observers have different colour-matching functions, but the differences can be considered small. The differences are due to the age (yellowish crystalline lens), to different macular pigmentation and to genetic differences. A suitable average of matches of many observers is used to define the standard observer (Section 9.2.1).

The colour-matching functions are dependent on the region of the retina considered and this is due to non-uniformity of the retina: non-uniform pigmentation of the macula lutea, and different orientation and size of the photosensitive cells. In the standardization two particular regions are considered, which are circular with different radii, centred on the centre of the fovea, and subtending an angular field of 2° and 10°, respectively (Sections 9.2 and 9.3).

The colour-matching functions obtained with these two methods (maximum and minimum saturation) are not exactly equal and the reason for this discrepancy is not yet understood.

The tristimulus vectors in the RGB reference frame are specified by sets of three numbers (R, G, B), that substitute the cone activation (L, M, S) of the fundamental reference frame, defined by the integrals

$$R = \int_{380}^{780} L_{e,\lambda}\, r(\lambda)\, d\lambda, \ G = \int_{380}^{780} L_{e,\lambda}\, g(\lambda)\, d\lambda, \ B = \int_{380}^{780} L_{e,\lambda}\, b(\lambda)\, d\lambda \tag{6.37}$$

Generally, the colour-matching functions in the different regions of the spectrum have different signs and their units are such that the equal-energy stimulus $\mathbf{E}_E$, with radiance $E_{E,\lambda} = 1$, has three tristimulus values equal to 1 (or conventionally to 100):

$$\int_{380}^{780} r(\lambda)\,d\lambda = \int_{380}^{780} g(\lambda)\,d\lambda = \int_{380}^{780} b(\lambda)\,d\lambda = 1 \tag{6.38}$$

The tristimulus vectors representing the monochromatic stimuli with unit radiance have the tristimulus values equal to the colour-matching functions (Figure 6.27):

$$\mathbf{U}_{E,\lambda} = \big(r(\lambda), g(\lambda), b(\lambda)\big) \tag{6.39}$$

Also in this reference frame the chromaticity of a colour stimulus $\mathbf{Q} = (R, G, B)$ is defined by the intersection point between the tristimulus vector and the unit plane $R + G + B = 1$ (Figure 6.28a) and the chromaticity coordinates are

$$r = \frac{R}{R+G+B}, \; g = \frac{G}{R+G+B}, \; b = \frac{B}{R+G+B}, \; r+g+b = 1 \tag{6.40}$$

Since $r + g + b = 1$, usually only two components are considered, (r, g), which are plotted as two orthogonal Cartesian coordinates. In this case the chromaticity diagram is obtained as a projection of the equilateral diagram from infinity on the plane $B = 0$ (Figure 6.28b). This representation of the chromaticity diagram is more frequent than the equilateral one because the Cartesian coordinates are easier to use.

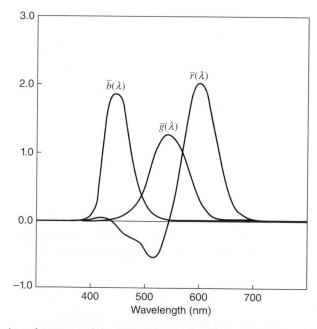

Figure 6.27   *Colour-matching functions of the CIE 1931 standard colorimetric observer in the RGB reference frame, denoted by $(\overline{r}(\lambda), \overline{g}(\lambda), \overline{b}(\lambda))$.*

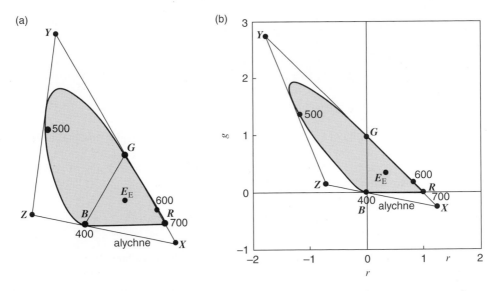

Figure 6.28 (a) Plane $R + G + B = 1$ in tristimulus space with a chromaticity diagram, spectrum locus, chromaticity of the equal-energy stimulus $\mathbf{E}_E = (r = 1/3, g = 1/3)$, alychne and triangle $\mathbf{XYZ}$, with vertices representing the chromaticity of the vectors $\mathbf{X}$, $\mathbf{Y}$ and $\mathbf{Z}$ assumed below as a reference basis. (b) Chromaticity diagram in the (r, g) coordinate system with alychne and triangle $\mathbf{XYZ}$ obtained by projecting the figure (a) on the plane $B = 0$ from infinity.

Any trichromatic device (monitor, scanner, video camera, etc., TV system such as NTSC, PAL, SECAM and HDTV) has its own RGB reference frame and colour-matching functions. The passage between different RGB reference frames is made by linear transformations. Confusing different RGB reference frames is a mistake. Moreover, an RGB system exists, whose (R, G, B) components are obtained as powers of tristimulus values, destroying the original linearity of the space (e.g. sRGB used for images on computers and the WEB).

The colour-matching functions developed here are obtained for foveal vision with a 2° degree visual field – anyway lower than 4°. Since the retina is not uniform, another observer related to a 10° visual field (Section 9.3) was developed and standardized. In this case all the quantities are distinguished by the subscript '10', e.g. (R_{10}, G_{10}, B_{10}).

6.13.6 Luminance and Exner-Schrödinger's 'Helligkeit' Equation

In Section 6.5.1 the luminance associated with heterochromatic radiance $L_{e,\lambda}$ is defined by equation (6.5), which is based on Abney's laws and is similar to the equations that define the tristimulus values (6.18). In particular, the relative luminous efficiency function $V(\lambda)$ has a similar role to that of the colour-matching functions $l(\lambda)$, $m(\lambda)$ and $s(\lambda)$. The search for a physiological justification for the light sensation suggests the existence of a fourth type of photoreceptor used for this sensation. Since there is no fourth kind of photoreceptor with this feature, a link must be sought between equation (6.6), which defines the luminance, and equations (6.18), which define the tristimulus vector $\mathbf{Q} = (L, M, S)$ and represent the transduction. Experiments suggested that the three types of photoreceptors contribute to the luminance in an additive way, proportionally to their activation.[78] This is stated by

Exner-Schrödinger's 'Helligkeit' equation

$$L_v = K_m(L_L L + L_M M + L_S S) = K_m(\begin{array}{ccc} L_L & L_M & L_S \end{array})\begin{pmatrix} L \\ M \\ S \end{pmatrix} = K_m \tilde{\Lambda}\mathbf{Q} \qquad (6.41)$$

This equation is written by applying the matrix multiplication rules: the scalar product is written between a row vector, marked by the tilde, and a column vector. The components L_L, L_M and L_S of the line vector are termed *Exner's brightness weights*, and are the weights with which the three kinds of cones contribute to luminance. From a comparison of (6.6), (6.18) and (6.41) it follows that

$$V(\lambda) = L_L l(\lambda) + L_M m(\lambda) + L_S s(\lambda) = \tilde{\Lambda}\mathbf{U}_{\mathrm{LMS},\lambda} \qquad (6.42)$$

In the experimental visual situation used for measuring $V(\lambda)$, the S-cones have shown almost no contribution to luminance sensation; therefore, $L_S \approx 0$.

The tristimulus vectors $\mathbf{Q}$ such that $\tilde{\Lambda}\mathbf{Q} = constant$ define in the tristimulus space a plane at constant luminance, or an *equiluminant plane*. Moreover, the zero-luminance plane $\tilde{\Lambda}\mathbf{Q} = 0$ intersects the plane of the chromaticity diagram in a line that Schrödinger called *alychne*[78], whose points define the chromaticity of (non-physical) stimuli having zero luminance.

The luminance L_v and the relative luminous efficiency function $V(\lambda)$, defined by the scalar products (6.41) and (6.42), respectively, are scalars, and this requires that they are invariant for each linear transformation in the tristimulus space. It follows that the vectors (L, M, S) and $(l(\lambda), m(\lambda), s(\lambda))$ are transformed in a contravariant way and the vector $\tilde{\Lambda} = (L_L, L_M, L_S)$ in a covariant way.

Equations (6.41) and (6.42), transformed in the RGB reference frame, are

$$L_v = K_m(L_{v,\mathrm{red}}R + L_{v,\mathrm{green}}G + L_{v,\mathrm{blue}}B) = K_m(\begin{array}{ccc} L_{v,\mathrm{red}} & L_{v,\mathrm{green}} & L_{v,\mathrm{blue}} \end{array})\begin{pmatrix} R \\ G \\ B \end{pmatrix} = K_m \tilde{\Lambda}\mathbf{Q} \qquad (6.43)$$

$$V(\lambda) = \frac{1}{K_m}\left[L_{v,\mathrm{red}}r(\lambda) + L_{v,\mathrm{green}}g(\lambda) + L_{v,\mathrm{blue}}b(\lambda)\right] = \tilde{\Lambda}\mathbf{U}_{\mathrm{RGB},\lambda} \qquad (6.44)$$

The photopic photometric observer and the photopic colorimetric observer are constrained by equation (6.44). This result is very important for a unified representation of the whole processing in the human visual system, but practically implies that the systematic errors in the definition of first observer are transferred into the second, and vice versa. Even today, this point is discussed.

Since the luminance is obtained by projecting the tristimulus vector on the direction defined by Exner's coefficients, this direction, termed Y axis, was chosen by the CIE as one of the reference axes in tristimulus space; therefore, the Y component represents the luminance or a quantity proportional to it. This is shown in detail in Section 9.2.2. In the CIE reference frame, the tristimulus vector has coordinates (X, Y, Z) and the chromaticity coordinates (x, y). These coordinates are presented here because they are useful in the presentation of some very general properties of the tristimulus space, but are presented with discussion in Section 9.2.2. Figure 6.29 shows the tristimulus space in the XYZ reference frame; Figure 6.30a shows the chromaticity diagram of the fundamental reference frame and the chromaticities of the RGB and XYZ reference frames; Figures 6.30b and 6.30c show the chromaticity diagram in the plane $X + Y + Z = 1$ and its projection on the plane $Z = 0$, where the coordinates (x, y) are also Cartesian.

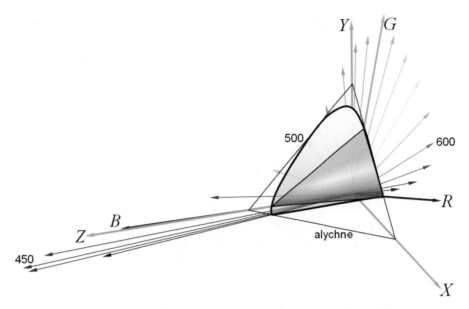

Figure 6.29 *XYZ reference frame in the tristimulus space in which the vectors **X**, **Y** and **Z** constitute a set of three orthogonal vectors. This reference frame is obtained from the fundamental one of Figure 6.22 or from the RGB one of Figure 6.26 by linear transformations. A one-to-one correspondence exists between the elements of the three reference frames as exists between this figure and Figures 6.22 and 6.26. In the figure are plotted the coloured vectors* $\mathbf{U}_{XYZ,\lambda}$ *that represent the monochromatic stimuli with a unity of power and whose components as a function of wavelength are the colour-matching functions in the XYZ reference frame. Here the chromaticity diagram is on the plane* $X + Y + Z = 1$, *where (X, Y, Z) are the coordinates or tristimulus values in this reference frame.*

6.13.7 Dichromats and Fundamental Reference Frame

The normal trichromatic individual observers make trichromatic colour matches, that is, any colour stimulus is matched by a mixing of three independent stimuli. Some colour-deficient observers make different colour matches mainly for two possible different reasons: lack of one or two kinds of cones and different spectral cone sensitivities with respect to the normal observers.

The colour-defective observers are considered more carefully in Section 16.3 dedicated to the colour-vision tests. Here only the dichromats are considered because they are useful to define the fundamental reference frame of the tristimulus space for normal trichromatic observers.

The *König reduction principle*[79] regards the eye physiology of the dichromats related to that of trichromats and postulates that dichromatic vision is a reduced form of trichromatic vision where one cone system is missing and the two others are left unchanged in spectral sensitivity.

This hypothesis has been confirmed by retinal densitometry measurements[80,81] and on the basis of psychophysical arguments.[82]

The dichromats are distinguished into:

1. *protanopes*, if no L-cones,
2. *deuteranopes*, if no M-cones,
3. *tritanopes*, if no S-cones.

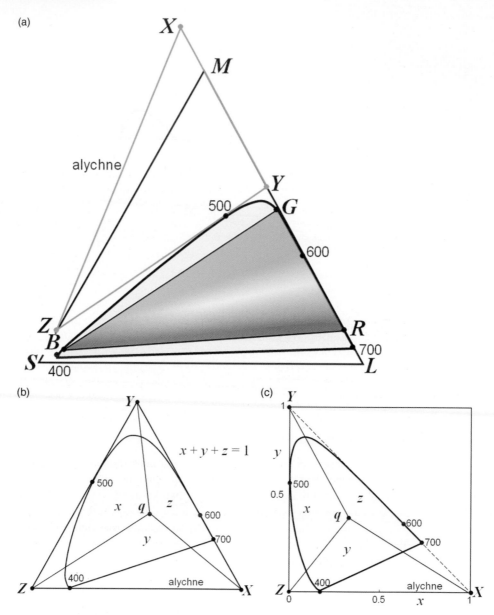

Figure 6.30   *(a) Plane L + M + S = 1 with the chromaticity diagram and the spectrum locus, the chromaticities of two sets of vectors, R, G and B, and X, Y and Z that constitute two other reference frames (Sections 6.13.4, 6.13.6 and 9.2). Moreover, the alychne defined here and in Section 9.2.2 is drawn. (b) Chromaticity diagram on the unit plane X + Y + Z = 1 and (c) its projection on the plane Z = 0.*

The lack of one kind of cones in dichromats implies that different sensations for trichromats can be equal for one kind of dichromats: in fact different stimuli are identical for the dichromats if their difference is a stimulus capable of activating the missing kind of cones. In the fundamental reference frame of the tristimulus space (Figure 6.31), these stimuli lie on a plane containing the axis representative of the activation of the missing kind of cones. This plane intersects the unitary plane $L + M + S = 1$ of the chromaticity diagram

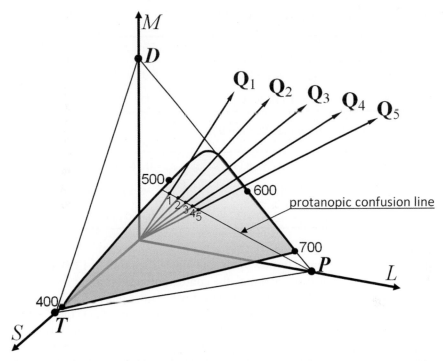

Figure 6.31 *LMS reference frame of the tristimulus space with the unit plane L + M + S = 1. The stimuli Q_1, Q_2, ..., Q_5 differ by a vector parallel to the **L** axis and therefore are coplanar. This plane intersects the unit plane along a straight line – confusion line, which crosses the axis L at the point **P** – protanopic confusion point. This representation can be repeated, mutatis mutandis, for the other dichromats, deuteranopes and tritanopes.*

in a straight line, whose points represent the chromaticity of these stimuli. In this way a bundle of straight lines is generated on the chromaticity diagram, termed *confusion lines*, which have a common point, termed *confusion point*, belonging to the axis that represents the activation of the missing kind of cones. The points of a confusion line represent the chromaticity of stimuli that are chromatically equal for the considered kind of dichromats and different for the trichromats.

The confusion point of the protanopes is denoted by **P**, of the deuteranopes by **D** and tritanopes by **T**.

In the XYZ reference frame the different observers have the following confusion points:

CIE 1931 Observer (Pitt and Wright[83,84]), (see Section 9.2.2)

Protanopic confusion point	$x_p = 0.747$	$y_p = 0.253$
Deuteranopic confusion point	$x_d = 1.080$	$y_d = -0.080$
Tritanopic confusion point	$x_t = 0.171$	$y_t = 0.000$

Vos's Observer (Smith-Pocorny's confusion points[85–86]) (see Section 9.5)

Protanopic confusion point	$x'_p = 0.7465$	$y'_p = 0.2535$
Deuteranopic confusion point	$x'_d = 1.4000$	$y'_d = -0.4000$
Tritanopic confusion point	$x'_t = 0.1748$	$y'_t = 0.0000$

Stockman and Sharpe's 2° Observer (see Section 9.6)

Protanopic confusion point	$x_{Fp} = 0.738\ 4015$	$y_{Fp} = 0.261\ 5986$
Deuteranopic confusion point	$x_{Fd} = 1.326\ 7160$	$y_{Fd} = -0.326\ 7164$
Tritanopic confusion point	$x_{Ft} = 0.158\ 6192$	$y_{Ft} = 0.000\ 0000$

Stockman and Sharpe's 10° Observer (see Section 9.6)

Protanopic confusion point	$x_{F10p} = 0.736\ 8335$	$y_{F10p} = 0.263\ 1665$
Deuteranopic confusion point	$x_{F10d} = 1.350\ 739$	$y_{F10d} = -0.350\ 7391$
Tritanopic confusion point	$x_{F10t} = 0.167\ 0137$	$y_{F10t} = 0.000\ 0000$

The transition between the reference frames XYZ and LMS (Sections 6.13.1, 6.13.4 and 6.13.6) is implemented by a linear transformation, and then the confusion lines (straight lines), given in the LMS reference frame, are transformed into as many straight lines on the plane $X + Y + Z = 1$. To build the LMS reference frame, we must start from the XYZ reference frame, once the confusion points are known, because the latter are located on the reference axes of the LMS reference frame. The confusion points can be experimentally evaluated starting from the confusion lines in the XYZ reference frame. These lines are defined with the technique of *dichromatic colour-matching*. Since only one straight line passes through two points, only one dichromatic colour-matching is sufficient to define a confusion line. Protanopes and deuteranopes require the colour-matching of monochromatic stimuli in the region of medium λ with a mixture of monochromatic stimuli of short and long wavelengths, for example, 450 and 600 nm, while, for the tritanopes, the monochromatic stimuli in the spectral region of short wavelengths with a mixture of monochromatic stimuli of medium and long wavelengths, for example, 540 and 650 nm. Figures 6.32a–c show the chromaticity diagram (x, y) with the confusion lines and confusion points related to the three kinds of dichromates.

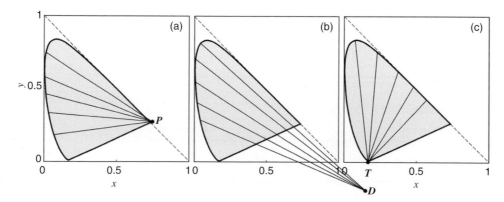

Figure 6.32　*In three reproductions of the CIE 1931 chromaticity diagram, (a), (b) and (c), three sets of confusion lines and three confusion points, **P**, **D** and **T**, are shown separately. These points are the convergence points of the three sets of confusion lines and are the confusion points for the three types of dichromates.*

Once the confusion points are known in the XYZ reference frame or in the RGB one, the transformation to the fundamental reference frame is straightforward by using transformation (6.28).

Section 9.6 presents the CIE fundamental colour-matching functions or simply cone fundamentals.

6.13.8   Newton's Centre-of-Gravity Rule and Chromaticity-Diagram Properties

In practice often the colour stimuli are specified by the luminance Y and the chromaticity (x, y), rather than by the tristimulus values, that is, by $(Y; x, y)$. This is justified by the fact that the chromaticity diagram allows a graphical visual assessment of the colour that the tristimulus values do not give. Moreover, the chromaticity diagrams have remarkable properties, which are presented below.

Colour-Stimulus Mixing and the Centre-of-Gravity Rule

Given two colour stimuli and the corresponding chromaticities (Figure 6.33a)

$$\mathbf{Q}_1 = (X_1, Y_1, Z_1),\ \boldsymbol{q}_1 = (x_1 = X_1 / w_1, Y_1 / w_1, Z_1 / w_1),\ \text{with } w_1 = X_1 + Y_1 + Z_1$$
$$\mathbf{Q}_2 = (X_2, Y_2, Z_2),\ \boldsymbol{q}_2 = (x_2 = X_2 / w_2, Y_2 / w_2, Z_2 / w_2),\ \text{with } w_2 = X_2 + Y_2 + Z_2$$

$$(6.45)$$

the sum of these two stimuli is

$$\mathbf{Q} = \mathbf{Q}_1 + \mathbf{Q}_2 = (X = X_1 + X_2, Y = Y_1 + Y_2, Z = Z_1 + Z_2)$$

$$(6.46)$$

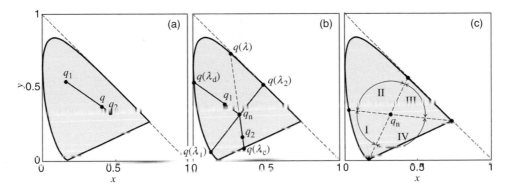

Figure 6.33 *(a) Centre-of-gravity law represented on the diagram (x, y) CIE 1931: the colour stimulus with chromaticity **q** is equated by the sum of two properly weighted stimuli with chromaticity **q**₁ and **q**₂. This represents the additive colour mixing. (b) Any colour stimulus with chromaticity **q** can be obtained by mixing an achromatic stimulus **q**ₙ and a spectral stimulus **q**(λ_d), where **q**, **q**ₙ and **q**(λ_d) are aligned and belong to a segment, and λ_d is the dominant wavelength of the chromaticity **q**. Analogously two spectral stimuli with chromaticities **q**(λ₁) and **q**(λ₂) aligned with **q**ₙ can be mixed in a suitable ratio for obtaining an achromatic stimulus with chromaticity **q**ₙ. The wavelengths λ₁ and λ₂ are termed complementary. Stimuli with a purple hue cannot have a dominant wavelength and the stimulus with chromaticity belonging to the straight line of the purple hues entering the mixture is characterized by the complementary wavelength λ_c. (c) Chromaticity diagram with spectrum locus subdivided into three parts, of which two parts, **I** and **III**, at the end of the spectrum locus, are mutually complementary, while part **II** is complementary to **IV** related to the non-spectral purple stimuli*

which is the result of the process of additive mixing of two stimuli, and has a chromaticity

$$q = (x, y) \text{ with } x = \frac{x_1 w_1 + x_2 w_2}{w_1 + w_2} \text{ and } y = \frac{y_1 w_1 + y_2 w_2}{w_1 + w_2} \tag{6.47}$$

that is, the chromaticity coordinates of the stimulus sum Q are a weighted average of the chromaticity coordinates of the addend stimuli, where the weights w_1 and w_2 are equal to the sum of the corresponding tristimulus values. From this observation, with substitutions and simplifications, a rule, known as *Newton's centre-of-gravity law*, is derivable (Section 10.2):

- point q is inside the segment $q_1 q_2$;
- the lengths of the segments $|qq_1|$ and $|qq_2|$ are linked by the relation

$$|qq_1| \, w_1 = |qq_2| \, w_2 \tag{6.48}$$

that is, by analogy with the scale, the point q is the equilibrium point of the yoke of a scale with the extremes $q_1 q_2$ two plates loaded by the weights w_1 and w_2;

- the property expressed in the previous point applies to all components of the chromaticity, that is,

$$|x - x_1| \, w_1 = |x - x_2| \, w_2, |y - y_1| \, w_1 = |y - y_2| \, w_2, |z - z_1| \, w_1 = |z - z_2| \, w_2, \tag{6.49}$$

where the magnitude between two vertical lines is considered as an absolute value.

This rule can be extended to the sum of any number of addends:

$$x = \frac{\sum_i x_i w_i}{\sum_i w_i}, \, y = \frac{\sum_i y_i w_i}{\sum_i w_i} \tag{6.50}$$

If the colour is specified by $(Y; x, y)$, the weights are also writable as

$$w_i = \frac{Y_i}{y_i} \tag{6.51}$$

which is immediately checked by writing $y_i = Y_i / (X_i + Y_i + Z_i)$.

The practical importance of this rule is very high.

Achromatic Colours and Stimuli

The stimuli with chromaticity belonging to a suitable central region of the chromaticity diagram (Figure 6.33b) are perceived as achromatic:

"*Achromatic colour ...*

(1) in the perceptual sense: perceived colour devoid of hue
  NOTE – The colour names white, grey and black are commonly used or, for transmitting objects, colourless and neutral.
(2) in the psychophysical sense: see 'achromatic stimulus'."[4]

"*Achromatic stimulus* – stimulus that, under the prevailing conditions of adaptation, gives rise to an achromatic perceived colour.

NOTE – In the colorimetry of object colours, the colour stimulus produced by the perfect reflecting or transmitting diffuser is usually considered to be an achromatic stimulus for all illuminants, except for those whose light sources appear to be highly chromatic."[4]

This implies that a unique white stimulus does not exist and the achromatic point $q_n = (x_n, y_n)$ on the chromaticity diagram is defined depending on the visual situation considered. The subscript 'n' means 'neutral'.

Dominant Wavelength

"*Dominant wavelength* λ_d (of a colour stimulus) – wavelength of the monochromatic stimulus that, when additively mixed in suitable proportions with the specified achromatic stimulus, matches the colour stimulus considered.

NOTE – In the case of purple stimuli, the dominant wavelength is replaced by the complementary wavelength."[4]

Each colour stimulus $\mathbf{Q} = (X, Y, Z)$ can be represented as the sum of two stimuli, one achromatic $\mathbf{W} = (X_n, Y_n, Z_n)$ and the other monochromatic $\mathbf{Q}_{\lambda d} = (X_{\lambda d}, Y_{\lambda d}, Z_{\lambda d})$ with wavelength λ_d. As seen above, the chromaticity q, q_n and $q(\lambda_d)$ are points that belong to a segment and therefore, given q and q_n, to know the wavelength of the monochromatic stimulus that enters the mix, just to prolong the segment $q_n q$ to meet the spectrum locus from the point q. The intersection point between this straight line and the spectrum locus represents the chromaticity of the monochromatic stimulus, and then the dominant wavelength of the stimulus $\mathbf{Q}$ with respect to the chosen white with q_n.

If the straight line meets the purple line, the chromaticity of the intersection point is represented by the complementary wavelength λ_c, corresponding to the intersection point between the spectrum locus and the line through q and q_n (Figure 6.33b).

Complementary Wavelength

"*Complementary wavelength* – the wavelength of the monochromatic stimulus that, when additively mixed in suitable proportions with the colour stimulus considered, matches the specified achromatic stimulus."[4]

Complementary colours, already defined in Section 4.3, are reconsidered here by making use of the tristimulus space and of the chromaticity diagram. An achromatic stimulus $\mathbf{Q}_n$ is expressible as a sum of two stimuli of complementary wavelengths λ_1 and λ_2: $\mathbf{Q}_n = \mathbf{Q}_{\lambda 1} + \mathbf{Q}_{\lambda 2}$. These three vectors are coplanar and thus their chromaticities q_n, $q(\lambda_1)$ and $q(\lambda_2)$ are three points of a segment on the chromaticity diagram.

The spectrum locus is thus subdivided into three parts, of which the two parts I and III at the extremities are complementary (Figures 6.33b and 6.33c). The wavelengths of the intermediate part II, relative to the radiation of the green hues, do not have monochromatic stimuli as complementary, but stimuli of purple hue IV, which are obtained as the sum of red and blue stimuli, with chromaticities located at the ends of the spectrum locus. With the stimuli having chromaticity belonging to the purple line, the number of the complementary wavelength with the subscript 'c' is associated.

Excitation Purity

The perceived quantity *saturation* has the purity as the corrispective coordinate on the chromaticity diagram. Here the excitation purity is considered. This coordinate is given by the ratio p_e of the lengths of segments defined by the chromaticity $\boldsymbol{q}_n$, $\boldsymbol{q}$ and $\boldsymbol{q}(\lambda_d)$,

$$p_e = \frac{|\boldsymbol{q}_n\boldsymbol{q}|}{|\boldsymbol{q}_n\boldsymbol{q}(\lambda_d)|} \tag{6.52}$$

and, in the case of intersection with the purple line, it is necessary to consider a point $\boldsymbol{q}(\lambda_c)$ belonging to the purple line:

$$p_e = \frac{|\boldsymbol{q}_n\boldsymbol{q}|}{|\boldsymbol{q}_n\boldsymbol{q}(\lambda_c)|} \tag{6.53}$$

The excitation purity written explicitly as a function of the chromaticity coordinates is

$$p_e = \sqrt{\frac{(x-x_n)^2+(y-y_n)^2}{(x_d-x_n)^2+(y_d-y_n)^2}} = \frac{|y-y_n|}{|y_d-y_n|} = \frac{|x-x_n|}{|x_d-x_n|} \tag{6.54}$$

The excitation purity is a positive number lower than or equal to 1: the value 1 is for monochromatic stimuli or for stimuli with chromaticity belonging to the purple line. The excitation purity is zero for the achromatic stimulus $\boldsymbol{q}_n$.

Figure 6.34a shows the loci of constant excitation purity on the (x, y) chromaticity diagram CIE 1931.

This specification of the excitation purity is obtainable from the chromaticity coordinates (x, y) by using the software presented in Section 16.8.5.

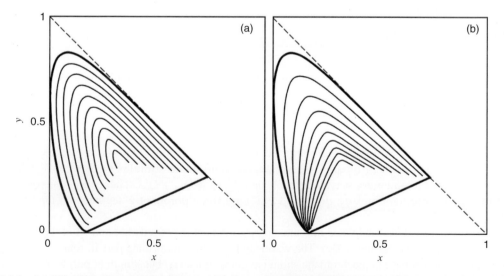

Figure 6.34   *(a) CIE 1931 chromaticity diagram with lines of equal excitation purity with respect to the equal-energy stimulus starting from the centre $p_e = 0.1, 0.2, 0.3, \ldots, 0.9$. (b) CIE 1931 chromaticity diagram with lines of equal colorimetric purity with respect to the equal-energy stimulus starting from the centre $p_c = 0.1, 0.2, 0.3, \ldots, 0.9$.*

Colorimetric Purity

The excitation purity introduced above is proper to the considered chromaticity diagram (x, y). Colorimetric purity p_c is defined to make this concept independent of the chromaticity diagram (Section 6.11):

$$p_c = \frac{L_{v,\lambda_d}}{L_{v,\lambda_d} + L_{v,n}}$$

(6.55)

where $L_{v,n} = K_m Y_n$ and $L_{v,\lambda_d} = K_m Y_{\lambda_d}$ are the luminances of the addend stimuli, the achromatic and monochromatic stimuli, respectively. In the case of purple hue, the complementary wavelength λ_c is considered instead of the dominant wavelength λ_d. The colorimetric purity can be written as a function of the excitation purity:

$$p_c = p_e \frac{y_{\lambda_d}}{y}$$

(6.56)

where y_{λ_d} and y are the y-components of the chromaticity of the monochromatic stimulus $\mathbf{Q}_{\lambda_d}$ and of the stimulus $\mathbf{Q}$, respectively.

Figure 6.34b shows the loci of constant colorimetric purity on the (x, y) chromaticity diagram CIE 1931.

Helmholtz's Coordinates

Chromaticity is specifiable with a more concrete perceptual meaning by other coordinates than the chromaticity coordinates defined on the chromaticity diagram: the *dominant wavelength* and *excitation purity*, defined above and called *Helmholtz's coordinates*. Their concreteness lies in the correspondence of these coordinates with the perceptual variables *hue* and *saturation*, respectively (Sections 4.3 and 4.6).

6.14   Lightness Scales

Fechner's and Stevens's theories differ in the definition of a visual magnitude and its measurement, which in this case could be considered as lightness:

> "*Lightness* (of a related colour) – brightness of an area judged relative to the brightness of a similarly illuminated area that appears to be white or highly transmitting.
> NOTE – Only related colours exhibit lightness."[4]

The measuring instrument of the colour-vision attributes is inevitably the observer, and the derivation of this measurement by psychophysical observations is part of the definition of the visual attribute itself.

Here the achromatic scales are considered to limit the judgement to the lightness without the presence of the chromatic content, which involves other phenomena such as the Helmholtz-Kohrausch effect (Section 6.15).

The various atlases have their own grey scales constructed with different, but similar criteria. Their comparison is meaningful for an evaluation of these criteria.

For a comparison, Figure 6.35 shows grey scales of different atlases (Munsell system in Section 14.4, DIN system in Section 14.5, NCS system in Section 14.7 and OSA-UCS system in Section 14.6). Figure 6.36 shows the plots of the lightness of these grey scales against the luminance factor, including the CIE 1976 L^* formula (Section 11.2), which is very close to Munsell's value curve.

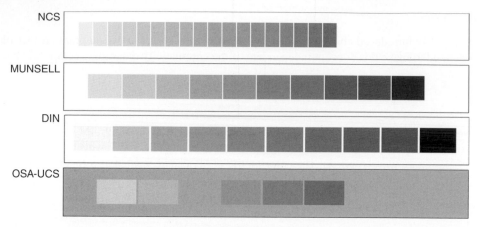

Figure 6.35   *Visual comparison of the grey scales of different atlases: the renotated Munsell atlas (Section 14.4), the 'Dunkelstufe' of the DIN atlas (Section 14.5), the lightness of the OSA-UCS atlas (Section 14.6) and the achromatic scale of the NCS atlas (Section 14.7).*

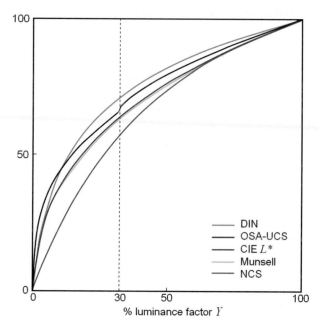

Figure 6.36   *Relationship between lightness scale values and luminance factor Y plotted according to different formulae and empirical data sets: the L* CIE 1976 formula (Section 11.2), fitting of the Munsell renotated data (Section 14.4), the 'Dunkelstufe' of the DIN atlas (Section 14.5), Semmelroth's formula used in the OSA-UCS atlas (Section 14.6) and the formula based on the NCS achromatic samples (Section 14.7).*

6.15   Helmholtz-Kohlrausch Effect

"*Helmholtz-Kohlrausch phenomenon* (or *effect*) – change in brightness of perceived colour produced by increasing the purity of a colour stimulus while keeping its luminance constant within the range of photopic vision.

NOTE – For related perceived colours, a change in lightness can also occur when the purity is increased while keeping the luminance factor of the colour stimulus constant."[4]

This phenomenon concerns the definitions of equi-luminance and brightness sensation. Equi-luminant colour stimuli juxtaposed should appear equally bright and at the same time, since they are equi-luminant, the separation border should be minimally distinct (Section 6.5). Experience says that for stimuli with high colour purity in contact with achromatic stimuli, if the border is with minimal distinction, the purest stimulus looks brighter. The phenomenon is very significant for the red, purple, violet, blue and blue–green hues, more moderate for green and very small for the yellow and orange hues. The Helmholtz-Kohlrausch effect is also dependent on the viewing environment, which includes the background, the surroundings of the object and the lighting. The Helmholtz-Kohlrausch effect is more evident in isolation in darker environments.

An example of this phenomenon is represented by a colour wheel compared with its achromatic transformation made by a customary imaging program, which preserves the luminances (Figure 6.37). The brightnesses of the two figures are obviously different.

The Helmholtz-Kohlrausch effect involves many points: the definition of luminance, then the definition of $V(\lambda)$, Abney's laws and the rule of minimum distinct border between equi-luminant stimuli (Section 6.5).

Helmholtz-Kohlrausch's effect in relation with the technique used to measure the luminous efficiency function $V(\lambda)$ has been examined by Kraft and Werner in 1994.[87] On this effect Kaiser et al.[88] say:

"The border between precisely juxtaposed chromatic (600 nm) and achromatic fields is as distinct when the fields are equally bright as when the chromatic field is brighter. The relative brightnesses of precisely juxtaposed chromatic and achromatic fields can be adjusted so that the border between them is minimally distinct. When the border is least distinct, the chromatic field will appear slightly brighter than the achromatic one. When the brightness of the achromatic field is increased to match that of the chromatic field, the distinctness of the border increases. When this added white light is removed from the achromatic field and added to the already brighter chromatic field, the distinctness of the border remains approximately constant. These results make questionable the usefulness of the concept of brightness contrast."

Kaiser et al. concluded that

"at least for the conditions of their experiment, there was no convincing evidence that brightness differences *per se* contribute to distinctness of a border. At least in conditions where chromatic differences are involved, doubt is cast on the usefulness of the concept of brightness contrast."

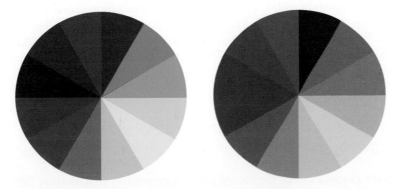

Figure 6.37 Comparison between a colour wheel and its translation into a grey scale that preserves the luminance factor. The differences of brightness between the two figures are obvious.

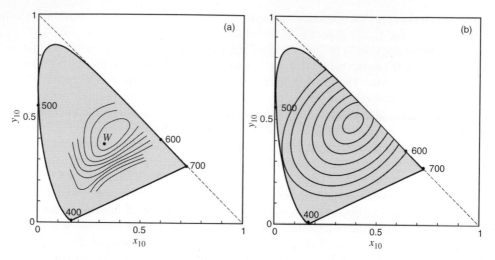

Figure 6.38 *(a) CIE 1964 chromaticity diagram with the constant value lines of the ratio B/Y_{10} produced by Wyszecki in the project OSA-UCS system.*[89] *(b) CIE 1964 chromaticity diagram with the constant value lines of the factor F that relates the equivalent luminance Y_0 of the OSA-UCS system to the luminance factor $Y_0 = F\,Y_{10}$ (Section 14.6.1).*

This phenomenon is considered only in the OSA-UCS system (Section 14.6) and in colour-appearance models, while it is ignored by the standard colorimetry.

Anyway, it is interesting to have a phenomenological knowledge of this phenomenon in order to understand its origin. A quantity useful to represent this phenomenon is the ratio *brightness over luminance B/Y* as a function of the chromaticity, or the analogous *equivalent luminance* $Y_{eq} \equiv KY$, where Y is the luminance and K a suitable function of the chromaticity.

In the OSA-UCS project (Section 14.6), Wyszecki carried out an experiment in which chromatic ceramic tiles were compared with achromatic ceramic tiles under the standard illuminant C (Section 8.7).[89] "The observer was asked to find each chromatic tile, the grey tile that he or she perceived as being of the same lightness as the given chromatic tile". The results are collected in Figure 6.38a, where the lines inside the CIE 1964 chromaticity diagram are the loci with equal ratio B/Y_{10}.

The committee of the Optical Society of America for the OSA-UCS system developed a lightness formula for the 10° CIE 1964 observer adapted to the D65 illuminant, which is dependent on the quantity $Y_0 \equiv FY_{10}$, considered as equivalent luminance with F substituting K in the previous equation $Y_{eq} \equiv KY$. The OSA-UCS formula is represented by equations (14.5) and (14.6) and the function F plotted in Figure 6.38b.

R. Sève in his book[90] gives a modification to the equation of the OSA-UCS system taking into account the experimental data collected by Ware and Cowan[91,92] and considers his proposal a reasonable approximation related to the experimental uncertainties:

$$Y_{eq} = KY_{10}$$
$$K = 1 + 0.04P - 1.55Q + 6.57P^2 - 6.25PQ + 6.29Q^2 - 17.5PQ(P - Q) \qquad (6.57)$$
$$P = x_{10} - x_{10,n} \;,\; Q = y_{10} - y_{10,n}$$

where $(x_{10,n}, y_{10,n})$ is the chromaticity of the neutral point corresponding to illuminant D65.

A review on the Helmholtz-Kohlrausch effect was published by Fairchild and Pirrotta.[93]

6.16   Colour Opponencies and Chromatic Valence

Hering's *opponent-colours theory* has been presented in Section 4.4. This theory is based on the following two chromatic hypotheses:

1.   four unique hues exist: red, green, yellow and blue;
2.   two pairs of opponent hues exist: red-green and yellow-blue.

According to this theory, every hue is defined by the resemblance of only one unique hue or of two contiguous unique hues (red-yellow, yellow-green, green-blue, blue-red), that is, the colours are classified by their hue resemblance of unique red (redness), of unique yellow (yellowness), of unique green (greenness), of unique blue (blueness), and achromatic.

In 1955, Jameson and Hurvich conducted the *hue-cancellation experiment*,[94–97] in which the hue resemblance of a colour is neutralized by mixing such a colour with a suitable amount of the opponent hue, for example, a reddish-yellowish colour mixed with unique green in a suitable amount becomes a unique yellow colour and the redness is completely cancelled.

The content of redness, yellowness, greenness and blueness of monochromatic colours is called *chromatic valence* or "*chromatic response* of the visual system for a given hue", that "is assumed to be proportional to the amount of the opponent cancellation stimulus necessary to extinguish that hue".[94–97]

In the experimental apparatus used by Hurvich-Jameson, "stimuli were seen fused in a $1° \times 0.8°$ slightly elliptical field, which was fixated foveally. The binary mixture field was surrounded by a $37°$ adapting field provided by monochromator ... and this field was adjusted to be a chromatically neutral white for each observer." This experiment was made with achromatic adaptation. In this experiment the observer had to judge the neutrality of the colour stimulus in relation to the opponencies redness-greenness and yellowness-blueness. Before the experiment, the observer detected the wavelength of his/her individual yellow (λ_Y), green (λ_G) and blue (λ_B) unique hues, to be used as cancellation hues. Since the unique red is normally extraspectral, the monochromatic stimulus of $\lambda_R = 700$ nm was used as red cancellation hue. Any monochromatic band that contains red can serve as a cancellation of the greenness. The use of unique hues was preferred only because it simplified analysis.

"All chromatic response measurements were for an equal brightness spectrum of $100/\pi$ cd/m^2. The luminance of the $37°$ white surround was also fixed at $100/\pi$ cd/m^2 throughout the entire experimental series."[94–97]

"At the beginning of each experimental session a 10 minute period of light exclusion was followed by a 5 minute period of bright adaptation to the white surround. The surround remained present and adaptation was maintained throughout each experimental session."

The observer had to adjust the mixing of the radiance $L_{e,\lambda}$ of the monochromatic stimulus with the radiance $L_{e,\lambda C}$ of one of the cancelling stimuli (λ_R, λ_Y, λ_G and λ_B) until the colour stimulus satisfied one of the following two criteria:

- *(YB)* – neutral appearance as between blueness and yellowness; the chromatic valence of the monochromatic stimulus of wavelength λ is defined as

$$\eta_Y(\lambda) = \sigma_Y \frac{L_{e,\lambda_B}}{L_{e,\lambda}}, \ \eta_B(\lambda) = -\sigma_B \frac{L_{e,\lambda_Y}}{L_{e,\lambda}} \ \text{with} \ \frac{\sigma_Y}{\sigma_B} = \frac{L_{e,\lambda_Y}}{L_{e,\lambda_B}} \tag{6.58}$$

for ($L_{e,\lambda Y} + L_{e,\lambda B}$) appearing neutral;

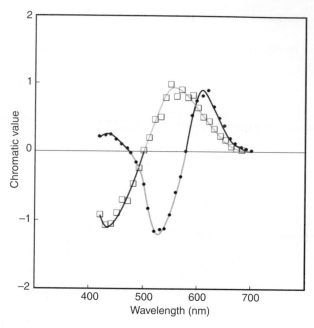

Figure 6.39 *Psychophysical assessments of red-green and blue-yellow chromatic opponencies obtained with the hue cancellation technique.*[94–97]

- (*RG*) – neutral appearance as between redness and greenness; the chromatic valence of the monochromatic stimulus of wavelength λ is defined as

$$\eta_R(\lambda) = \sigma_R \frac{L_{e,\lambda_G}}{L_{e,\lambda}}, \ \eta_G(\lambda) = -\sigma_G \frac{L_{e,\lambda_R}}{L_{e,\lambda}} \ \text{with} \ \frac{\sigma_G}{\sigma_R} = \frac{L_{e,\lambda_G}}{L_{e,\lambda_R}} \tag{6.59}$$

for $(L_{e,\lambda_G} + L_{e,\lambda_R})$ appearing neutral.

The signs of the chromatic valences are assigned by convention to represent the opponent characteristics of the colours: redness has positive values and greenness has negative values, and yellowness has positive values and blueness has negative values. The chromatic responses are measured in radiometric units, that is, are radiances.

The results of one observer of this experiment are shown in Figure 6.39. In ordinates the chromatic valence is plotted as a function of the wavelength. Unique hues, with the exclusion of the unique red, which is extra-spectral, are at the zero crossing of the two chromatic response functions: unique yellow at about 575, unique green at 500 nm and unique blue at 475 nm. Redness is present at long wavelengths, but reappears at short wavelengths, below 475 nm. This result confirms the reports of Helmholtz and others that even the extreme spectral reds are slightly yellowish in hue.[98–100]

This psychophysical experiment made by Jameson and Hurvich was an important confirmation of the Hering theory. In addition, it is compatible with the existence of three types of cones and, in its essentials, is confirmed by the properties of retinal ganglion cells and other cells in the visual system stations subsequent to the retina (Section 5.5). The opponent-process theory became a cornerstone of colour-vision research.

6.17   MacAdam's Chromatic Discrimination Ellipses

MacAdam ellipses – ellipses drawn in the CIE (x, y) chromaticity diagram as a representation of one standard deviation of colour-matching and related to the distribution of equiluminous colours which are perceived equal to a fixed colour.

The perceived colour difference and perceived colour discrimination are important quantities for knowledge and for colorimetric practice. The study of these quantities leads us back to Helmholtz and Shrödinger. In the 1940s, W. D. Wright and M. L. MacAdam provided important empirical data according to different methods:

- the measurement of the chromatic discrimination thresholds at constant luminance – termed *just noticeability method* –.[101];
- the measurement of the standard deviation in colour-matching at constant luminance.[102]

Wright considered the second method a too time-consuming activity, although recognizing this method gives results with greater accuracy and stability. MacAdam prepared an interesting apparatus for colour-matching, with which it is possible to accumulate a large number of colour-matching data and from these obtain very meaningful standard deviation data. MacAdam's work is considered here because great importance has been accorded to it.

MacAdam's colour-matching data are obtained with a 2° bipartite field at a luminance of 48 cd/m² in an adaptation surround of 21° and with the white C (Section 8.7) at a luminance of 24 cd/m² (Figure 6.1). Repeated colour matches were made with 25 colour tests. The distribution of the chromaticities of the matched colours was considered on the CIE 1931 diagram and therefore the distribution is bi-dimensional. In such a way, around any average point, a line was computed which defines the border of one standard deviation. The shape of the chromaticity distribution is important from the statistical point of view. "In order to avoid implications not justified by the observations, the results" were reported in the original paper of MacAdam "in terms of standard deviation rather than in terms of probable error." Anyway, in the hypothesis that the distribution is normal, one standard deviation defines the range containing 68.25% of the observations, two standard deviations 95.44% and three 99.74%. The closeness of 99.74 % to 100% induces us to consider the line at 3 standard deviations as the discrimination threshold line and therefore consider

3 standard deviations – 1 *just noticeable difference* (jnd)

then

MacAdam's unit of colour difference – just perceptible difference in surface colour or chromaticity according to MacAdam's experiment.

The border lines at one standard deviation are known as MacAdam's ellipses, although they cannot be exactly ellipses. The empirical data are represented as mathematical ellipses, specified by the lengths of the semi-axes and the angle of the inclination, but this is an approximation. These 25 ellipses measured by MacAdam were obtained for only one individual observer (Figure 6.40); anyway they are considered well representative of an average observer.

The size and orientation of the ellipses on the diagram vary widely depending on the test colour and this means that the scales of the chromaticity diagram are not uniform, that is, equal differences of perceived colours are not represented by equal distances on the chromaticity diagram. It would be desirable to obtain ellipses identical to circles of equal radius, because in this way the coordinates used to define the chromaticity would have a more direct relationship with the perception, in addition to an obvious practical value (the significant digits of the colorimetric specification would be independent of the colour stimulus). Over time,

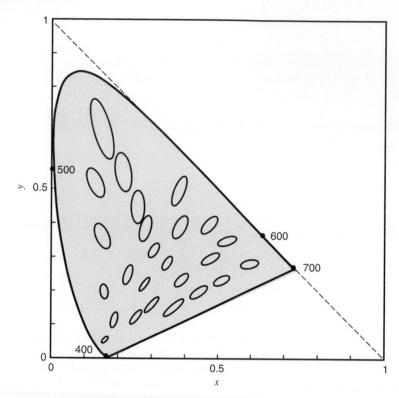

Figure 6.40   *MacAdam's ellipses magnified by a factor 10 on the CIE 1931 chromaticity diagram. The ratio between the minor axis and the greatest one among the axes of all ellipses is 1:30 and this shows how perceptually non-homogeneous are the chromaticity scales of the CIE 1931 diagram.*[102]

the scientific community studied the scales of the colour space with the aim of obtaining a uniform scale and the debate is continuing also today (Sections 6.18, 11.4, 14.4, 14.6 and 16.8.3).

The colour differences evaluated by the values of jnd from colour-matching are considered *small colour differences*, since the colour stimuli in comparison are mutually touching. Conversely, the appearance of separated colour stimuli has a lower dependence on the mutual contrast between stimuli. In this case the colour differences are termed large colour differences.[103]

With the aim of transforming the MacAdam ellipses into circles of equal radius, many of chromaticity diagrams have been made over time using linear transformations, the most significant of which are presented in the software described in Section 16.8.3.

6.18   Perceived Colour Difference

"Perceived colour difference – perceived dissimilarity between 2 colour elements."[4]

This CIE ILV[4] completes this definition of perceived colour difference as of a quantity related to the CIE formulae for colour-difference specification (Section 11.4). Colour-difference formulae and colour-difference judgement have very practical importance.

Section 6.17 presents the experiment of the MacAdam ellipses, which describe the statistical uncertainty in the colour-matching at constant luminance. This experiment is very important because it derives the threshold of chromatic discrimination indirectly from colour-matching data. This is not the measurement of a colour difference but the definition of the jnd. Small colour differences and jnd are certainly related quantities.

The colour-difference measurement is a very important practical task and has to be made in a simple and standardized way. In practice, the perceived colour differences are judged in a different way with respect to the MacAdam apparatus. The colourists visually judge and quantify the *perceived colour difference* as the magnitude of the attributes of the *dissimilarity* between two object colours, one standard and one test, under specified visual conditions and described by such terms as redder, bluer, lighter, darker, greyer or cleaner.

Generally the required visual conditions for judging and quantifying the perceived colour difference are:

- the view is in a light box with grey surfaces (Section 13.2);
- the illumination is made with defined standard light sources and defined geometry of illumination (directional or diffused illumination);
- the colour specimens are observed on a defined grey surround and at defined distance (Figure 6.41):
  1. in *direct edge* contact; or
  2. *large gap* separated – few millimetres –; or
  3. *hairline* separated;
- the geometry of observer vision (direction of view on the specimens, generally orthogonal) is defined.

But, in practice, the colourist's task is to give a response: *pass or fail*. In this case no quantification of colour difference is made. Anyway an approximate quantification is necessary if a colour tolerance is admitted. Consider a few definitions:

- *Colour tolerance* – barely *acceptable / permissible* colour difference between test and specified colour standard.
- *Commercial colour tolerance* – the range of colour difference that is acceptable to satisfy a *commercial colour match*. Most usually an agreement is made between buyer and seller concerning *colour acceptability*.
- *Colour tolerance set* – a set of coloured standards, usually seven painted patches, one exhibiting the desired colour, and two each exhibiting the limits of the barely *acceptable/permissible* range of colour variation in each of the colour attributes, typically the acceptable limits in lightness to darkness, redness to greenness, and yellowness to blueness.

The use of a colour tolerance set is not sufficient to quantify the colour difference. The quantification is obtained in a very approximate way by comparing standard and test colour samples to pairs of grey samples belonging to a suitable grey scale.

- *Grey scale for colour difference* – a system for evaluating the colour difference between two specimens by relating it to that pair in a series of differences between two greys that is visually the closest to the colour difference. This method is analogous to the standard grey scales used in fastness testing for assessing colour change in the textile industry.[104]
- *Grey scale* – an achromatic scale ranging from black through a series of successively lighter greys to white. Such a series may be made up of steps that are perceived to be equally distant from one another – typically the Munsell Value Scale with the unit of lightness difference equal to 1.5 unit of Munsell Value (Section 14.4.3) – or may be arranged according to some other criteria such as a geometric progression based on lightness. Such scales may be used to describe the relative amount of difference between two similar colours.

It follows that the colour differences are classified as

- *threshold colour difference*, that is, threshold of imperceptibility;
- *supra-threshold colour difference*, that is, test and standard colours are different and the difference is classified as (N.B. Section 6.17)
  1. *small colour differences*, if the differences are a few jnds above the threshold;
  2. *moderate colour differences*, if the differences are between 20 and 40 jnds, approximately equivalent to 0.5–1.5 CIELAB units (Sections 11.3.5, 11.4);
  3. *large colour difference,* if the differences are between 100 and 200 jnds.

The large differences are evaluated visually with multiple samples presented spatially separated on a neutral background and the task of the observer is to scale the relative difference between each sample and its neighbours. In contrast, the small colour differences are evaluated between juxtaposed colour samples

Figure 6.41 *A couple of samples of two different colours shown on different grey backgrounds and different mutual distances for visually evaluating their colour difference in different situations. The appearance of colour also depends on the distance between observer and screen.*

and with minimal separation and the observer scales the differences relative to a standard pair or absolutely.[103,105–108]

Two examples of using the grey-scale method are described synthetically in Figure 6.41.

Standard Grey-Scale Test Method

The standardized and most used method for a visual evaluation of the colour differences is the grey-scale method,[104] conceived to determine the colour degradation in textiles.

Here again, the observer has to compare the colour difference between reference and test to a set of grey differences. Only the luminance factor is common to both differences, and in this case the border distinctness could help in the judgement. The comparison of a difference of chroma and particularly of hue with a difference of luminance factor is hard and uncertain.

The visual judgement regards colour differences between reference and test compared with the scale of grey differences. Two grey scales exist (Figure 6.42):

1. The basic scale is composed of five pairs of grey strips, which represent the perceived colour differences corresponding to the *solidity* indices 5, 4, 3, 2 and 1.
2. The complete scale is composed of nine pairs of grey strips, which represent the perceived colour differences corresponding to the *solidity* indices 5, (4-5), 4, (3-4), 3, (2-3), 2, (1-2) and 1.

(The word 'solidity' is typical of textile and represents the resistance to colour fading under the action of various agents.)

The strips are non-glossy.

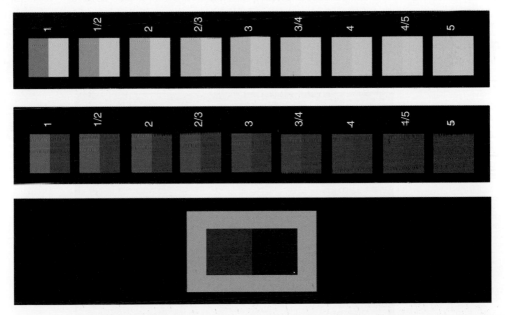

Figure 6.42 *Set of different pairs of juxtaposed grey rectangles, similar to the commercial ones used in fastness testing for assessing change of colour. A pair of colour samples are shown as an example of colour difference to be compared with the grey differences. The difference between the grey pairs in the sequence is in approximate geometrical progression and the scale of differences is not perceived as uniform. N.B. The reference conditions for small colour differences (Section 11.4.5) require that the samples are placed in contact on an achromatic background with a CIE 1976 lightness L* ≈ 50.*

The first part of each pair is a neutral grey, the same for all the pairs, and the second parts of the pairs are grey gradually more clear so that each pair represents increasing perceived colour differences and increasing lightness contrast. For index 5 the two parts of the pair are equal, that is, without any colour difference. For index 1 the colour difference is the highest. In addition, there are two different versions of these two grey scales: one is a light grey and the other dark grey. Figure 6.42 shows both scales.

The illumination should be of 600 lx directional with 45° incidence angle and of a source simulating the illuminant D65 (Section 8.4).

The direction of observation should be approximately perpendicular to the plane of the colour samples.

Bradford Ellipses (BFD)

An important colour-difference data set, called a COM (combined) data set, has been collected for the last standard colour-difference formula, known as CIEDE2000 (Section 11.4.5). An important contribution to the colour-difference data set comes from four laboratories: BFD-P, RIT-DuPont, Leeds and Witt. These colour differences are classified as small-medium.

In the definition of the colour-difference formula of the last decades, an important role had the set of 132 chromaticity discrimination ellipses collected and published by M. R. Luo and B. Rigg.[109] These data came from a collection of colour-difference data of different factories (textile and non-textile) and laboratories, and the basic aim was to produce a self-consistent set of ellipses. This aim was achieved by new experiments conducted following the grey-scale assessment method. The judgement was made in a viewing cabinet with matt-paint background, with a luminance factor approximately equal to 30% and illuminated by a source simulating the standard illuminant D65 (Section 8.6). The scene was constituted by five grey colour samples, two sample pairs (test and reference) and a grey-scale pair. (Figure 6.43) "The observer was asked to pick a sample from the grey scale, to put it alongside the standard, and compare the difference with that of the sample pair. Different grey-scale samples were tried until the one giving a different from the standard closest in magnitude to that for the sample pair was found."[109] The observer was asked to provide the visual results in terms of grade. For example, suppose that a sample pair has a colour difference between two pairs formed by the 'standard' and grade 3, and the 'standard' and grade 4, and is close to the former pair. Hence, the visual result should be in the range of 3.1–3.4. The non-linearity

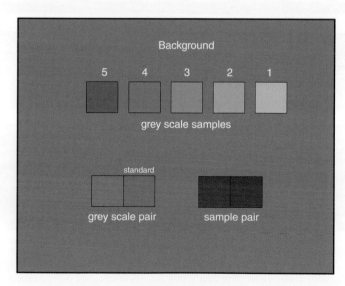

Figure 6.43   *Visual disposition of the colour samples for grading the colour-difference evaluations as described in the text.*

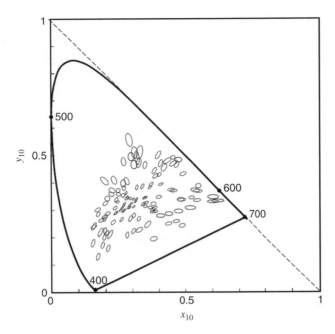

Figure 6.44   *Chromaticity-discrimination ellipses, known as BFD ellipses, plotted in the CIE 1964 (x_{10}, y_{10}) diagram.*[109]

of the grading required an equation (or a correspondence table) to relate the grades to the perceived colour-difference scale.

It must be stressed that colour differences between test and reference were compared to differences on a grey scale and the visual effect could be very different if the colour difference regards the lightness, the colourfulness, the hue or altogether. Therefore, this judgement has high uncertainty. Anyway the obtained ellipses show shapes with meaningful regularities (Figure 6.44). This set of colour-difference ellipses in known as BFD and is used in the definition of almost all colour-difference formulae (Section 11.4).

6.19   Abney's and Bezold-Brücke's Phenomena

A colour stimulus constituted by a mixture of two stimuli, one spectral with a defined wavelength and the other achromatic, is specified by hue, purity and luminance. These three colour attributes are perceived as non-independent, since, at variations of luminance or of purity, the hue changes, with the exclusion of particular hues.

> "*Abney phenomenon,* also *Abney's hue shift* – change of hue produced by decreasing the purity of a colour stimulus while keeping its dominant wavelength and luminance constant."[4]

> "*Bezold-Brücke phenomenon* – change of hue produced by changing the luminance (within the range of photopic vision) of a colour stimulus while keeping its chromaticity constant.
>
> NOTE – With certain monochromatic stimuli, hue remains constant over a wide range of luminances (for a given condition of adaptation). The wavelengths of these stimuli are sometimes referred to as 'invariant wavelengths'."[4]

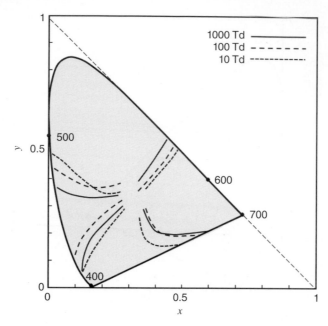

Figure 6.45   *Unique hue loci at 10, 100 and 1000 Td for an individual observer.[112] The constant hue lines are bent according to Abney's phenomenon and the constant hue lines are shifted by changing the illuminance according to Bezold-Brücke's phenomenon.*

Figure 6.45 represents both phenomena. The constant unique hue loci are measured as a function of the purity and for three different retinal illuminations (10, 100 and 1000 Td). Particularly the hue constancy requires that the mixing at changing purity is made while also changing the dominant wavelength. Along a straight line on the chromaticity diagram linking a monochromatic stimulus with the achromatic stimulus, which represents the chromaticities of variable mixtures of these two stimuli, the hue changes. Figure 6.46 gives a general view of the constant hue lines at constant luminance where it is shown that for four hues the hue shift is almost absent.

Consider as an example of Abney's hue shift the measurement of the dominant wavelength associated with a unique hue with two different purities made by Kurtenbach et al.[110] After the usual adaptation process, a test field constituted by a disk of 0.5° diameter on a dark background was proposed to the observer. The stimulus was flashed for 1 second every 19 seconds. First, the white stimulus that each observer judged perceptually neutral was defined. The perceptually neutral white determined by each observer was in the colour-temperature range 6200–6980 K. The stimuli used in the experiment were constituted by a spectral light mixed with neutral white light with controlled purity. These stimuli were presented to the observer with a wavelength ranging on the spectrum locus. Then, for example, in determining the wavelength that produces the perception of unique blue, the observer was asked to respond to each stimulus with either 'reddish' or 'greenish'. The wavelength for unique hue blue was the wavelength that received the same number of both responses. The same procedure was used for compound hues, for example, red-blue and blue-green. Figure 6.47 shows the frequency-of-seeing curve for red–blue.

The classical measurements of the Bezold-Brücke phenomenon are those of Purdy[111] (Figure 6.48). He found that a decrease of intensity in the long-, middle- and short-wavelength regions of the spectrum produced a shift toward red, green and violet, respectively. Conversely, an increase of intensity produced a trend toward yellow in the middle- and long-wavelength regions and toward blue in the short-wavelength region. Purdy found three invariant hues to any luminance variation in correspondence with the spectral lights with a wavelength of 478 nm (blue), 503 nm (green) and 572 nm (yellow). Moreover, he found a fourth invariable hue in the purple region of the chromaticity diagram.

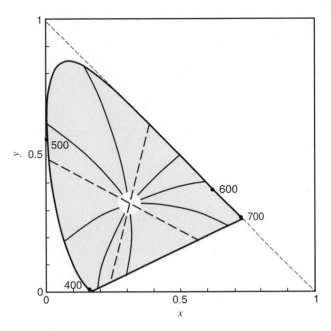

Figure 6.46 *Lines of constant hue measured in connection with the Munsell atlas (Section 14.4.10).*[113]

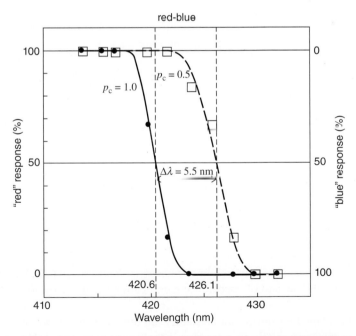

Figure 6.47 *Frequency-of-seeing curve for red-blue. Responses for 'red' (left ordinate) and 'blue' (right ordinate) are plotted as a function of the wavelength. The left curve was obtained with a colorimetric purity $p_c = 1.0$ and; the right one with $p_c = 0.5$. The drawing shows that the hue shift $\Delta\lambda = 5.5$ nm was ascertained.*[110]

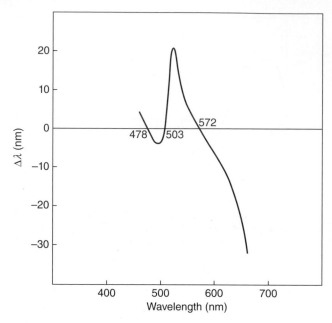

Figure 6.48   *Bezold-Brücke's hue shift corresponding to a reduction in retinal illuminance from 1000 to 100 Td.*[111]

6.20   Chromatic Adaptation and Colour Constancy

"Chromatic adaptation – visual process whereby approximate compensation is made for changes in the colours of stimuli, especially in the case of changes in illuminants."[4]

"Chromatic adaptation – changes in the visual system sensitivities due to changes in the spectral quality of illuminating and viewing conditions."[114]

"Chromatic adaptation – The self-adjustment of the visual system to the colour of the prevailing illumination in such a way that object surfaces appear to have the same colour for all daylight phases and for most artificial lights. The visual system works towards neutralizing (in an unknown way) the effect of the colour of the illumination. For example, a white surface appears white even if the illumination changes from bluish daylight to yellowish incandescent light."[115]

"Color constancy – the general tendency of the colors of an object to remain constant when the color of the illumination is changed."[114]

"Color inconstancy – the violation of the colour constancy in the presence of colour modification produced by a change of illuminant and of appearance modification produced by adaptation."

In literature chromatic adaptation and colour constancy are often considered as synonymous.

Chromatic adaptation refers especially to those transient changes in sensitivity that are ascribable to photopic chromatic stimulation and are reflected in changes in chromatic perception.

Visually, chromatic adaptation is experienced when a sudden change in illuminant (e.g. in a room the passage from daylight to a tungsten light) causes a global change in perceived colour of the object that is recognized by the observer as a change in illuminant. In spite of this evident change in the perceptual appearance of the non-self-luminous objects caused by significant changes in the wavelength composition of the light reflected from different objects under the new illuminant, the perceived colour of the objects remains largely unchanged. This adaptation phenomenon is termed *instantaneous colour constancy*.

The chromatic, achromatic and white sensations are a results of chromatic adaptation, and particularly:

> "*Adapted white* – colour stimulus that an observer who is adapted to the viewing environment would judge to be perfectly achromatic."[4]

6.20.1 Asymmetric Colour Matching

Chromatic adaptation phenomenon has been intensively studied in many laboratories over recent decades. The studied experiment is called the *corresponding colour* experiment.

> "*Corresponding colour stimuli* – pairs of colour stimuli that look alike when one is seen in one set of adaptation conditions, and the other is seen in a different set."[4]

The corresponding colour stimuli, when seen by the observer fully adapted to the reference illuminant, should look the same as the test specimen when seen by the observer adapted to the test illuminant. Generally this kind of colour-matching, termed *asymmetric colour-matching*, is performed in four different ways:

1. *Haploscopic matching* – also called *differential ocular conditioning and comparison*
   The observer sees through two eye pieces and the viewing apparatus presents different adapting stimuli to each of the two eyes: for example, one eye is adapted to illuminant D65 and the other to illuminant A (Section 8.6) (Figure 6.49). The observer is then asked to adjust the stimulus seen by one eye to match that seen by the other eye by adjusting the amounts of red, green and blue lights. In general, the matching results have a good precision. The accuracy depends on the existence of cross talk between the two eyes differently adapted and on the dominancy of one eye with respect to the other. Haploscopic matching has been used with different apparatuses.[116,117]

2. *Local-adaptation matching*[118]
   This technique is based on the matching of stimuli that activate two differently adapted areas of the retina in the same eye. The two different areas are associated with the two halves of a classical bipartite field of 8° for colour-matching. "Two halves of a colorimeter are filled with different adapting colours. Every ten seconds, for one second only, a test colour replaces the adapting colour in one half, and an adjustable combination of three primaries replaces the adapting colour in the other. The observer sees the complete field with both eyes (natural pupils) at all times. He looks directly at the centre of the dividing line, and adjusts the primaries during the adaptation periods, so as to make the two halves of the field appear to match during the brief exhibition of the test and comparison colours."[118] It is assumed that the two areas of the retina do not affect each other.

3. *Memory matching*
   Memory matching is made in binocular viewing conditions using both eyes and without optical devices. In such experiments, the observer produces a match to a previously memorized colour.[119–121] Particular memory matches are made to mental stimuli such as an ideal achromatic colour (grey) and unique hues.

4. *Magnitude estimation method.* (Section 6.1.7)
   The observer is asked to scale the visual sensations of lightness, colourfulness and hue under fully adapted viewing conditions. The results of magnitude estimation are colour appearance grids plotted on a standard CIE diagram in correspondence with different adaptation illuminants. The pairs of corresponding points belonging to two different grids represent pairs of corresponding colours. Unfortunately, this technique has a low precision but has the advantage of normal binocular vision and a short training period.

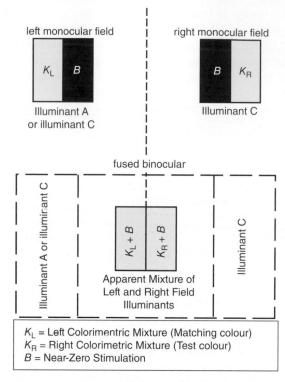

Figure 6.49　*Central region of the viewing fields proposed in monocular vision as separated fields (upper figure) and as overlapped fields in binocular vision (lower figure).*

Corresponding colour data are used for defining the chromatic adaptation transforms that constitute the ground of any colour-appearance model. In Figure 6.50 are plotted as an example the chromaticities of corresponding colour pairs measured by Breneman and related to the adaptations to the D65 and A illuminants (Section 8.6) with the white at 1500 cd/m^2.[116]

6.20.2　Empirical Data

The technical committee CIE TC 1-52 published "A Review of Chromatic Adaptation Transforms", which selected nine corresponding colour data sets to be used to define a CAT.[122]

6.20.3　Von Kries's Coefficient Law

Chromatic adaptation is a very important property of the human visual system for understanding colour appearance. Historically, the oldest and most widely used law to quantify the chromatic adaptation is the *von Kries coefficient law*, which describes the relationship between the illuminant and the sensitivity of the human visual system. The law accounts for the approximate colour constancy in the human visual system. This law compensates for the illumination change using a pure scaling of the cone absorptions supposing that the three kinds of cones operate in a mutually independent way. Frequently applications of this law reported systematic discrepancies between prediction and experiment. Anyway, this law was used for the definition

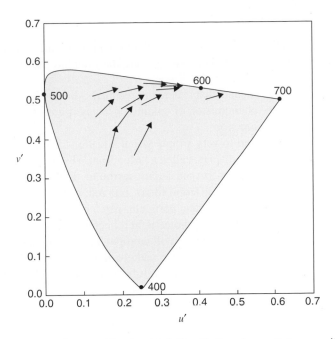

Figure 6.50 *(u′, v′) chromaticity diagram (Section 11.3.3) with the chromaticity coordinates of corresponding colour pairs represented by arrows measured related to the adaptations to the D65 and A illuminants with the white at 1500 cd/m² in complex visual fields.*[116] *The tip of the arrow corresponds to illuminant A (Section 8.6) and the tail to illuminant D65.*

of the chromatic adaptation transform for colour-appearance models[123] (Section 11.3.5); therefore, this law is here recalled. The law assumes that in the chromatic adaptation process the adapted responces of each of the three cone mechanisms remain unchanged and the responses of the three adapted cone types (L_a, M_a, S_a) are obtained by a modulation of the activations (L, M, S): the modulation factors for the three cone types (k_L, k_M, k_S) are inversely related to their activations (L_{white}, M_{white}, S_{white}) by the spectral power distribution of the particular illuminant used. The modern interpretation of the von Kries hypothesis to be applied to a chromatic adaptation model is

$$L_a = k_L L, \; M_a = k_M M, \; S_a = k_S S \tag{6.60}$$

where the coefficients are

$$\left(k_L = \frac{1}{L_{white}}, \; k_M = \frac{1}{M_{white}}, \; k_S = \frac{1}{S_{white}} \right) \text{ or } \left(k_L = \frac{1}{L_{max}}, \; k_M = \frac{1}{M_{max}}, \; k_S = \frac{1}{S_{max}} \right) \tag{6.61}$$

and (L_{max}, M_{max}, S_{max}) is used if (L_{white}, M_{white}, S_{white}) is not the maximum in the scene.

This law states that different illuminants seen in adaptation have the same set ($L_a = 1$, $M_a = 1$, $S_a = 1$) and produce equal colour sensations, which is not exactly true. Therefore, von Kries' law is certainly approximate. Over time, many experiments carried on to test the von Kries hypothesis have indicated the necessity to search for alternative ways to describe chromatic adaptation and colour constancy.

6.20.4   Retinex

In 1971 E. H. Land formulated the *retinex theory* to explain colour constancy experiments. The word is a fusion of *retina* and *cortex*, suggesting that both the eye and the brain are involved in the visual processing.[124–126]

Land with his hypotheses opened a debate that is here not recalled. Almost all that belongs to history. Here only the experiment that showed the phenomenon of colour constancy is considered and the assumptions of normalization rules to meet this phenomenon.

The experiments are no longer related to very simple visual situations, with few different colour stimuli. The vision is binocular and the scene is a display, called a "Mondrian" (Figure 6.51), consisting of numerous coloured patches, which is shown to an observer in a dark surround. The Mondrian display is illuminated by three white-light projectors through three different filters, one red, one green and one blue. The observer is asked to adjust the intensity of the lights so that a particular patch on the Mondrian appears white and the experimenter measures the fluxes of red, green, and blue light reflected from this patch, which appears white. Then the observer is asked to select a neighbouring coloured patch, for example, green, and then to adjust the illuminating lights so that the intensities of red, blue and green light reflected from the selected patch are equal to those originally measured from the white patch. It happens that the observer continues to perceive the original colours of all the patches of the Mondrian, although differently illuminated, and particularly the selected green patch continues to appear green, although reflecting the same light originally reflected by the white patch. This is the colour constancy phenomenon. Land's experiment recalls and confirms a previous experiment of Judd.[127]

With this empirical demonstration, Land proposed the retinex algorithm, which, over time, has been investigated and improved. Land proposed three colour mechanisms with the spectral responsivities of the cone

Figure 6.51   *Example of the 'colour Mondrian' similar to that used by Land and McCann.*[124–126]

photoreceptors, which he called retinexes. Then he hypothesized a three-dimensional colour appearance space, whose elements are the cone responses normalized in different ways: first, by normalizing the maximum activation of the three kinds of *retinexes*, and more recently by taking the ratio of the signal at any given point in the scene and normalizing it with an average of the signals throughout the scene – called *grey hypothesis* –.

Applications and improvements to this theory have been made by McCann[128] and later by Jobson et al.[129] Although retinex models are still used, they do not describe the human colour perception accurately.[130] This retinex normalization corresponds to the von Kries colour adaptation principle, which certainly provides a general approximate foundation, but needs enhancement to address adaptation phenomena.

The works of Land have stimulated other experiments, among which the experiment of the Maximov shoebox made by McCann is here considered. The apparatus has the size of a shoe box. The interior lighting is through a filter placed on the opening of the upper longer side. On the smaller side is placed an eyepiece for viewing of the opposite internal side. On this side there is a square figure, like a simplified Mondrian, consisting of four coloured rectangles around a central coloured square (Figure 6.52). There are two embodiments of this figure, the colours of which are chosen in correspondence to two different filters used for the illumination. The colours of the figures and the filters are matched so that the light reflected from the figures

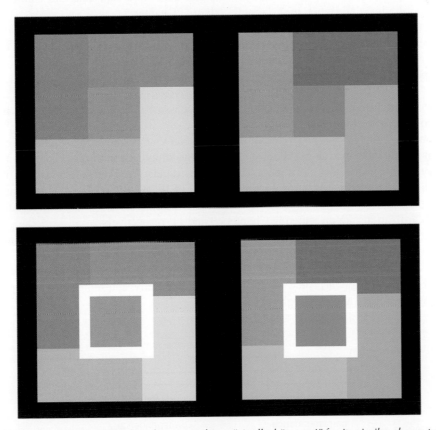

Figure 6.52 *Example of the simplified "colour Mondrians" (called "tatami" for its similar shape to the Japanese tatami) similar to that used by McCann in Maximov's shoebox experiment. These "colour Mondrians" appear here as seen equally illuminated. If the vision is made in the shoebox apparatus, the appearance can be equated by a suitable filtering of the illumination for the upper pair of Mondrians, while the appearance of the lower Mondrians, which have a white strip, is equal and independent of the filtering.*[128]

is the same for the two figure-filter combinations. The observer sees through the eyepiece in different configurations: without filters and the two figures have different appearances, and with filters and the two figures have the same appearance, denying the constancy of the colour.

The addition of a white strip to isolate the central square in the two figures leads the observer to see the same colours that he/she sees without the use of filters, restoring colour constancy. This experiment confirms that the normalization has to be made with respect to the maximum activation of the retinexes, produced by white colour. Anyway the colour constancy is not exact, confirming the limitations of the von Kries rule. Moreover, McCann's conclusion is that this normalization has to be performed in a different way in the different regions of a complex scene. It means that colour constancy is a very complex phenomenon not yet understood, on which A. Valberg wrote that "colour constancy seems to be a composite phenomenon that incorporates adaptation and simultaneous contrast, with contributions from several processing stages in the visual pathway." (p. 174 of [115])

6.21   Colour-Vision Psychophysics and Colorimetry

The colour-vision psychophysics presented in this chapter adds important data and confirmations to data given in the physiological description of the visual system of Chapter 5.

The construction of the tristimulus space is essential for associating the physical colour stimuli to the tristimulus vectors, which are in a many-to-one correspondence. This correspondence is known as metamerism and represents a characteristic of the transduction phenomenon.

Hering's hypothesis of two chromatic opponency mechanisms is well confirmed by Jameson-Hurvich's experiments.

Both Fechner's and Stevens's laws state that in visual processing the cone activations, that is, the tristimulus vales, are compressed, although in small different ways.

The visual phenomenon can be subdivided into a set of stages according to the chain of Figure 6.53.

All the phenomena, described in psychophysical and physiological experiments, have to be present in a global model of colour vision for colorimetry. The CIE has standardized these two steps of visual processing by proposing psychophysical systems for the first step – CIE 1931 and CIE 1964 standard colorimetric observers/systems – and psychometric systems for the second step – CIELUV and CIELAB colour systems –. These CIE standardizations are presented in Chapters 9 and 11, respectively.

The CIE has also proposed more advanced colour-appearance models, termed CIECAM97 and CIECAM02, which are not considered in this book as they are subjects of debate.

Over time psycho-physiologists have proposed other models of colour-vision processing,[131–134] which are in progress and not completely accepted by the scientific community, and represent visual processing with high attention to physiology and psychophysics.

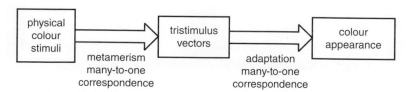

Figure 6.53 *Set of stages in the visual process from the physical colour stimulus to the colour appearance. The correspondence between physical colour stimuli and tristimulus vectors is many-to-one (metamerism), as many-to-one is the correspondence between combinations of tristimulus vectors with adapting stimulus vectors and colour sensations.*

References

1. Fechner GT, *Element Der Psychophysik*, Breitkopf & Härterl Leipzig (1860).
2. Maxwell JC, Theory of the perception of colours, in *Transactions of the Royal Scottish Society of Arts*, **4**, 394–400 (1872). Reprinted in MacAdam DL, Ed., *Sources of Colour Science*. MIT Press: Cambridge, MA, pp. 75–83 (1970)
3. Marks LE, *The Unity of the Senses: Interrelations Among the Modalities*. Academic Press, New York (1978).
4. CIE Publication S 017/E:2011. *ILV: International Lighting Vocabulary*. CIE Central Bureau, Kegelgasse 27, A-1030 Vienna, Austria (2011). Available at: http://eilv.cie.co.at/
5. Tangkijviwat U, Rattanakasamsuk K and Shinoda H, Color preference affected by mode of color appearance, *Color Res. Appl.*, **35**, 50–61 (2010).
6. Evans RM, Variables of perceived color, *J. Opt. Soc. Amer.*, **54**, 1467–1474 (1964).
7. Katz D, *The World of Colour*, Keagan Paul, London (1935).
8. Hunt RWG and Pointer MR, *Measuring Colour*, 4th edn, Wiley–IS&T Series in Imaging Science and Technology, pp. 294 (2011).
9. Wyszecki G and Stiles WS, *ColorScience, Concepts and Methods, Quantitative Data and Formulae*, John Wiley & Sons, New York (1982).
10. Westheimer G, The Maxwellian view, *Vision Res.*, **6**, 669–682 (1966).
11. Ives HE, Studies in photometry of lights of different colors. Spectral luminosity curves obtained by equality of brightness photometer and flicker photometer under similar conditions, *Phil. Mag.* **24**, 149–188, 352–370, 744, 845 (1912).
12. Marks LE and Bornstein MH, Spectral sensitivity by constant CFF: effect of chromatic adaptation, *J. Opt. Soc. Amer.*, **63**(2), 220–226 (1973).
13. Brindley GA, *Physiology of the Retina and the Visual Pathways*. Edward Arnold Ltd, London (1960).
14. Wright WD, The measurement and the analysis of colour adaptation phenomena, *Proc. Roy. Soc. (London)*, 115B, 49–87 (1934).
15. Hunt RWG, The perception of color in 1° fields for different states of adaptation, *J. Opt. Soc. Amer.*, **43**, 479–484 (1953).
16. Gescheider GA, *Psychophysics: The fundamentals*, 3rd edn, Lawrence Erlbaum Associates, Publishers, London (1997).
17. Wundt W, *Grundzüge der physiologischen Psychologie*, Wilhelm Engelmann, Leipzig (1902–3); *Principles of Physiological Psychology*. Translated from the Fifth German Edition (1902) by Edward Bradford Titchener. Swan Sonnenschein, London (1904).
18. Miller G, Wordnet on line http://wordnet.princeton.edu, Princepton University
19. Hunt RGW, Colour terminology, *Color Res. Appl.*, **3**, 79–87 (1978).
20. Hunt RGW, The specification of colour appearance. I. Concepts and terms, *Color Res. Appl.*, **2**, 55–68 (1977).
21. CIE Supplement No. 2 to CIE 15-1971. *Recommendations on Uniform Colour Spaces, Colour-Difference Equations, Psychometric Colour Terms*, Commission Internationale de l'Éclairage, Vienna, Austria (1978). Available at: http://www.cie.co.at/
22. Burns B and Sheep BE, Dimensional interactions and the structure of psychological space: The representation of hue, saturation and brightness. *Percept. Psychophys.* **43**, 494–507 (1988).
23. Hard A, Sivik L and Tonnquist G, NCS, Natural Color System – from concepts to research and applications. Part I, *Color Res. Appl.*, **21**, 180–205 (1996).
24. Kolbe H, How the retina works, *American Scientist*, **91**, 28–35 (2003).
25. Pugh EN, Vision: physics and retinal physiology, in *Steven's Handbook of Experimental Psychology*, 2nd ed., AtkinsonRC, ed., John Wiley and Sons, Inc., New York (1988).

26. SpillmanL and WernerJS, eds, *Visual perception: the neurophysiological foundations*, Academic Press, San Diego (1990).

27. Crawford BH, Visual adaptation in relation to brief conditioning stimuli. *Proceedings of the Royal Society of London*, Series B, **128**, 283–302 (1947).

28. Adelson EH, Saturation and adaptation in the rod system. *Vision Res.*, **22**, 1299–1312 (1982).

29. Hecht S, Vision II: The nature of the photoreceptor process, in MurchisonC, ed., *A Handbook of General Experimental Psychology*. Clark University Press, Worchester, MA (1934).

30. Hecht S, Haig C and Chase AM, The dark adaptation of retinal fields of different size and location, *J. Gen. Physiol.* **19**(2), 321–337 (1935).

31. Hecht S, Haig C and Chase AM, The influence of light adaptation on subsequent dark adaptation of the eye. *J. Gen. Physiol.* **20**, 831–850 (1937).

32. Riggs LA, Vision, in *Woodworth and Schlosberg's Experimental Psychology*, 3rd edn, Kling JW and Riggs LA, eds., Holt, Rinehart, and Winston, New York (1971).

33. Bartle JNR, Dark and light adaptation, in *Vision and Visual Perception*, Graham CH, ed., Chap. 8, John Wiley and Sons, Inc., New York (1965).

34. Davson H, *Physiology of the Eye*, 5th edn, Macmillan Academic and Professional Ltd, London (1990).

35. Hecht S, Shlaer S and Pirenne MH, Energy, quanta and vision, *J. General Physiol.*, **25**, 819–841 (1942).

36. Ludwig E and MacCarthy EF, Absorption of visible light by refractive media of the human eye, *Archives Ophtalmol.*, **20**, 37–51 (1938).

37. Wald G, Human vision and the spectrum, *Science*, **101**, 653–658 (1945).

38. Hsia Y and Graham CH, Spectral sensitivity of the cones in the dark adapted human eye, *Proc. National Acad. Sci.*, **38**, 80–85 (1952).

39. Stiles WS, Color vision: The approach through increment-threshold sensitivity. *Proc. National Acad. Sci.*, **45**, 100–114 (1959).

40. Wald G, The receptors of human color vision, *Science*, **145**, 1007–1017 (1964).

41. Eisner A and MacLeod DIA, Flicker photometric study of chromatic adaptation: selective suppression of cone inputs by colored backgrounds, *J. Opt. Soc. Am.*, **71**, 705–718 (1981).

42. Stockman A, MacLeod DM and Vivien JA, Isolation of the middle- and long-wavelength sensitive cones in normal trichromats, *J. Opt. Soc. Am.*, **A 10**, 2471–2490 (1993).

43. Estévez O and Spekreijse H, The "silent substitution" method in visual research. *Vision Res.* **22**(6), 681–691 (1982).

44. Yeh T, Smith VC, and Pokorny J, Colorimetric purity discrimination: data and theory. *Vision Res.*, **33**, 1847–1857 (1993).

45. Wyszecki G and Stiles WS, *ColorScience, Concepts and Methods, Quantitative Data and Formulae*, John Wiley & Sons, New York, pp. 410 (1982).

46. Abney W, *Researches in Colour Vision*, Longmans, Green, London (1913).

47. Ikeda M, Yaguchi H and Sagawa K, Brightness luminous-efficiency functions for 2° and 10° fields, *J. Opt. Soc. Am.*, **72**, 1660–1665 (1982).

48. Ikeda M and Nakano Y, Spectral luminous-efficiency functions obtained by direct heterochromatic brightness matching for point sources and for 2° and 10° fields, *J. Opt. Soc. Am.*, **A 3**, 2105–2108 (1986).

49. Wagner G and Boynton RM, Comparison of four methods of heterochromatic photometry, *J. Opt. Soc. Am.*, **62**, 1508–1515 (1972).

50. Barlow HB, Increment thresholds at low intensities considered as signal noise discriminations, *J Physiol*, **136**, 469–488 (1957).

51. Davson H, *Physiology of the Eye*, 5th edn, Macmillan Academic and Professional Ltd, London (1990).

52. Valberg A, *Light Vision Color*, Wiley pp. 151 (2005).

53. Hood D and Finkelstein M, Sensitivity to light. *Handbook of Perception and Human Performance*, **1**(5), 1–66 (1986).

54. Weber EH, *De Pulsu Resorpitione, Auditu Et Tactu: Annotationes Anatomicae et Phylosoficae*. Köhlor, Leipzig (1834).

55. Blackwell R, Studies of psychophysical methods for measuring visual thresholds, *J. Opt. Soc. Am.*, **42**(9), 606–614 (1952).

56. Stevens SS, On the psychophysical law, *Psychol. Rev.*, **64**, 153–181 (1957).

57. Stevens SS and Stevens JC, Brightness function: effects on adaptation, *J. Opt. Soc. Am.*, **53**(3), 375–385 (1963).

58. Falmagne JC, *Elements of Psychophysical Theory*, Oxford University Press, Oxford (1985).

59. Wright WD and Pitt FGH, Hue discrimination in normal color vision, *Proc. Phys. Soc.* (London) **46**, 459–473 (1934).

60. Bouman MA and Walraven PL, Color discrimination data, in *Handbook of Sensory Physiology*, Jameson D, and HurvichLM, eds., Vol. VII/4, *Visual psychophysics* Chap. 19, Springer-Verlag, New York, pp. 484–516 (1972).

61. Wright WD and Pitt FHG, The saturation discrimination of two trichromats, *Proc. Phys. Soc. (London)*, **49**, 329–331 (1937).

62. Jones LA and Lowry EM, Retinal sensibility to saturation differences, *J. Opt. Soc. Am.*, **13**, 25–34 (1926).

63. Martin LC, Warburton FL and Morgan WJ, The determination of the sensitiveness of the eye to differences in the saturation of colours. *Great Britain Med. Res. Council*, **188**, 5–42. (1933).

64. Kaiser PK, Comerford JP and Bodinger DM, Saturation of spectral lights, *J. Opt. Soc. Am.*, **66**, 818–826 (1976).

65. Yeh T, Smith VC and Pokorny J, Colorimetric purity discrimination: data and theory, *Vision Res.*, **33**, 1847–1857 (1993).

66. Rushton WAH, Visual pigments in man, in *Handbook of Sensory Physiology*, Dartnall, ed., Vol. VII/1, Springer-Verlag, New York, pp. 364 369 (1972).

67. Wyszecki G and Stiles WS, *ColorScience, Concepts and Methods, Quantitative Data and Formulae*, John Wiley & Sons, New York, pp. 591–593 (1982).

68. Rushton WAH, Color vision: An approach through the cone pigments, *Investigative Ophthalmol.*, **10**, 311–322 (1971).

69. CIE 15.3:2004, *Colorimetry*, 3rd edn, CIE CIE Central Bureau, Kegelgasse **27**, A-1030 Vienna, Austria

70. Wyszecki G and Stiles WS, *Color Science*, John Wiley & Sons, New York (1982), Chap. 3 pp. 117–118, 130–173, Chap. 5 pp. 346, Chap. 8 pp. 587.

71. Grassmann H, Zur Theorie der Farbenmishung, *Poggendorf', Ann. Phys.* **89**, 69–84 (1853).

72. Krantz DH, Color measurement and color theory: I. Representation theorem for Grassman structures, *J. Mathematical Psychol.*, **12**, 283–303 (1975).

73. Wyszecky G, and Judd DB, *Color in Business, Science and Industry*, John Wiley & Sons, New York (1975).

74. König A, Dieterici C, Die Grundempfindungen und ihre Intensitäts-Vertheilung im Spectrum, *Sitzungsberichte der Akademie der Wissenschaften in Berlin*, 805–829 (29 July 1886).

75. Ostwald W, *Die Farbenlehre*, Vol. 2 Physikalische Farbenlehre, Unesma Leipzig, pp. 237 (1918).

76. Maxwell JC, On the theory of compounds colours, and the relations of the colours of the spectrum, *Phil. Trans. R. Soc.* (London) **150**, 57–84 (1860).

77. International Commission on Illumination, Proc 8th Session, Cambridge, England 1931, Bureau Central de la CIE, Paris, (1931)

78. Schrödinger E. Grundlinien einer Theorie der Farbenmetrik im Tagessehen, *Annalen der Physik* IV, I part 63, 397-426, II part **63**, 427-456, III part **63**, 481-520 (1920); Part I and II have been translated and published in *Sources of Color Science*, The MIT Press, David L MacAdam Ed. (1970).

79. König A and Dieterici C, Die Grundempfindungen und ihre Intensitäts, Vertheilung im Spectrum (Fundamental sensations and their intensity distribution in the spectrum) *Sitzungsberichte der Akademie der Wissenschaften in Berlin*, 805–829 (29 July 1886).

80. Rushton WAH, Colour blindness and cone pigments, *Am. J. Optom.*, **41**, 265–282 (1964).

81. Rushton WAH, Powel SD, White KD, Exchange thresholds in dichromats, *Vision Res.*, **13** (11), 1993–2002 (1973).

82. Vos JJ and Walraven PL, On the derivation of the foveal receptor primaries, *Vision Res.*, **11**(8), 799–818 (1971).

83. Pitt FHG, Characteristics of dichromatic vision, *Med. Res. Council, rep. of Comm. on Physiology of Vision* No. XIV, Her Majesty's Stationery Office, London (1935).

84. Wright WD, Characteristics of tritanopia, *J. Opt. Soc. Am.*, **42**, 509–521 (1952).

85. Smith VC and Pokorny J, Spectral sensitivity of colorblind observers and the cone pigments, *Vision Res.*, **12**, 2059–2071 (1972).

86. Smith VC and Pokorny J, Spectral sensitivity of the foveal cone photopigments between 400 and 50 nm, *Vision Res.*, **15**, 161–171 (1972).

87. Kraft JM and Werner JS, Spectral efficiency across the life span: flicker photometry and brightness matching, *J. Opt. Soc. Am.* **A 11**, 1213–1221 (1994).

88. Kaiser PK and Greenspon TS, Brightness difference and its relation to the distinctness of a border, *J. Opt. Soc. Am.*, **61**, 962–965 (1971).

89. Wyszecki G, Correlate for brightness in term of CIE chromaticity coordinates and luminous reflectance, *J. Opt. Soc. Am.*, **57**, 254–257 (1967).

90. Sève R, *Physique de la couleur*, Masson, Paris (1996).

91. Ware C and Cowan WB, Specification of heterochromatic brightness matches: A conversion factor for calculating luminances of stimuli that are equal in brightness. *National research Council of Canada*. 42 pages, Publication N° 26055 (1983).

92. Kaiser PK and CIE 1.03, Models of heterochromatic brightness matching. *CIE J.,* **5**, 57–59 (1986).

93. Fairchild MD and Pirrotta E, Predicting the lightness of chromatic object colors using CIELAB, *Color Res. Appl.*, **16**, 385–393 (1991).

94. Hurvich LM and Jameson D, Spectral sensitivity of the fovea. I. Neutral adaptation. *J. Opt. Soc. Am.*, 43, 485–494 (1953).

95. Jameson D and Hurvich LM, Some quantitative aspects of the opponent-color stheory. I. Chromatic responses and spectral saturation, *J. Opt. Soc. Am.*, **45**, 546–552 (1955).

96. Hurvich LM and Jameson D, Some quantitative aspects of the opponent-color theory. II. Brightness, saturation, and hue in normal and dichromatic vision, *J. Opt. Soc. Am,* **45**, 602–616 (1955).

97. Jameson D and Hurvich LM, Some Quantitative Aspects of an Opponent-Colors Theory. III. Changes in Brightness, Saturation, and Hue with Chromatic Adaptation, *J. Opt. Soc. Am.*, **46**, 405-415 (1956).

98. Krantz DH, Color measurement and color theory: II. Opponent-colors theory. *J. Mathematical Psychol.*, **12**, 304–327 (1975).

99. Nagy AL and Zacks JL, Effects of psychophysical procedure and stimulus duration in the measurement of Bezold-Brücke hue shifts, *Vision Res.*, **17**, 193–200 (1977).

100. von Helmholtz, H, *Handbuch der Physiologischen Optik*, 2nd ed., VossL, ed., Hamburg and Leipzig, pp. 456 (1896).

101. Wright WD, The sensitivity of the eye to small colour differences, *Proc. Phys. Soc.*, **53**, 93–112 (1941).

102. MacAdam DL, Visual sensitivities to color differences in daylight, *J. Opt. Soc. Am.*, **32**, 247–274 (1942).

103. Kuehni RG, Industrial color difference: progress and problems. *Color Res. Appl.*, **15**, 261–265 (1990).

104. ISO 105-A02:1993, Textiles-Tests for colour fastness, International Organization for Standardization, Genevra, Switzerland (1993)

105. Robertson AR, CIE guidelines for coordinated research on color-difference evaluation, *Color Res. Appl.*, **3**, 149–151 (1978).

106. Danny C, Rich Fred W and Billmeyer Jr, Small and moderate color differences. IV. Color-difference-perceptibility ellipses in surface-color, *Color Res. Appl.*, **8**, 31–39, (1983).

107. Strocka D, Brockes A and Paffhausen W, Influence of experimental parameters on the evaluation of color-difference ellipsoids, *Color Res. Appl.*, **8**(3), 169–175 (1983).

108. Fuchida T, Cowan WB and Wyszecki G, Matching large color differences with achromatic and chromatic surrounds, *Color Res. Appl.*, **10**, 92–97 (1985).

109. Luo MR and Rigg B, Chromaticity-discrimination ellipses for surface colours. *Color Res. App.* **11**, 25–42 (1986).

110. Kurtenbach W, Sternheim CE and Spillmann L, Change in hue of spectral colors by dilution with white light (Abney effect), *J. Opt. Soc. Am.*, **A 1**, 365–372 (1984).

111. Purdy MD, Spectral hue as a function of intensity, *Am. J. Psychol.*, 43, 541–559 (1931).

112. Ayama M, Nakatsue T and Kaiser PK, Constant hue loci of unique and binary balanced hues at 10, 100, and 1000 Td. *J. Opt. Soc. Am.*, **A 4**(6), 1136–1144 (1987).

113. Newhall SM, Nickerson D and Judd DB, Final report of the O.S.A. Subcommittee on the Spacing of the munsell colors, *J. Opt. Soc. Am.*, **33**, 385–418 (1943).

114. ASTM E284 - 13b Standard Terminology of Appearance, ASTM International, West Conshohocken, PA, USA (2013)

115. Valberg A, *Light Vision Color*, John Wiley & Sons, New York, pp. 174 (2005).

116. Breneman EJ, Corresponding chromaticities for different states of adaptation lo complex visual fields, *J. Opt Soc. Am.*, **A 4**, 1115–1129 (1987).

117. Fairchild MD, Pirrotta E and Kim T, Successive-Ganzfeld haploscopic viewing technique for color appearance research, *Color Res. Appl.*, **19**, 214–221 (1994).

118. MacAdam DL, Chromatic adaptation, *J. Opt Soc. Am.*, **46**, 500–513 (1956).

119. Helson H, Judd DB and Warren MH, Object-color changes from daylight to incandescent filament illumination, *Illum. Ellg.* **47**, 221–233 (1952).

120. Braun KM and Fairchild MD, Psychophysical generation of malching images for cross-media colour reproduction, *Proc. 4th Color Imaging Conference*, 214–220, IS&T, Springfield, VA (1996).

121. Lam KM, *Metamerism and Colour Constancy*, Ph.D. Thesis, University of Bradford (1985).

122. Luo MR, A review of chromatic adaptation transforms. *Rev. Prog. Coloration*, **30**, 77–91 (2000).

123. von Kries J, Die Gesichtempfindungen, in *Handbuch der Physiologie der Menschen*, NagelW, ed., Vol. **3**, Braunschweig, pp. 109–282 (1905).

124. Land EH, Recent advances in retinex theory and some implications for cortical computation: color vision and the natural image. *Proc. Natl. Acad. Sci. USA*, **80**, 5163–5169 (1983).

125. Land EH, Recent advances in retinex theory. *Vision Res.*, **26**, 7–21 (1986).

126. Land EH and McCann JJ, Lightness and retinex theory, *J. Opt. Soc. Am.*, **61**, 1–11 (1971).

127. Judd DB, Hue, saturation and lightness of surface colours with chromatic illumination, *J. Opt. Soc. Am.*, **30**, 2–32 (1940).

128. McCann J, Color sensations in complex images, *IS&T/SID 1st Color Imaging Conference*, Scottsdale, 16–23 (1993).

129. Jobson DJ, Rahman Z and Woodell GA, A multi-scale retinex for bridging the gap between color images and the human observation of scenes, *IEEE Trans. Im. Proc.* **6**, 956–976 (1997).

130. Hurlbert AC and Wolf K, The contribution of local and global cone-contrasts to colour appearance: a Retinex-like model, In *Proc. SPIE* 4662, *Human Vision and Electronic Imaging* VII, San Jose, CA, 298 (June 3, 2002).

131. Müller GE, Über die Farbenempfindungen, *Z. Psychol. Ergänzungsb.* **17** and **18** (1930).

132. Judd DB, Response functions for types of vision according to the Müller theory, *J. Res. Nat. Bur. Standards* (Washington, DC) **42**, (1959).

133. Guth SL, Model for color vision and light adaptation, *J. Opt. Soc. Am. A*, **8**, 976–993 (1991).

134. DeValois RL and DeValois KK, A multistage color model, *Vision Res.*, **33**, 1053–1065 (1993).

7

CIE Standard Photometry

7.1 Introduction

Radiometry is presented in Section 2.4. Photometry is presented in a very short form in Section 2.6 with the sole task of providing a tool to read the subsequent chapters. Photometry is taken up in Section 6.5 in relation to the psychophysical measurements of relative spectral luminous efficiency functions $V(\lambda)$ and $V'(\lambda)$. This chapter considers photometry as a separate discipline and the presentation is given in its entirety, in relation to the origin of *photometric units*, and to *radiometry* and *lighting*.

> "*Photometry* – measurement of quantities referring to radiation as evaluated according to a given spectral luminous efficiency function, e.g., $V(\lambda)$ or $V'(\lambda)$."[1]

> "*Radiometry* – measurement of the quantities associated with optical radiation."[1]

Photometry is a chapter of optics, which deals with the formal measurement of visible light, by quantifying *luminous efficiency* with units of the human visual system, unlike radiometry, which deals with the measurement of light in absolute units of power. The photometric measurement is formal because it is not a direct measure of luminous efficacy, but a measure implemented in accordance with the conventional definition of a theoretical photometric observer assumed as reference, which is based on brightness heterochromatic matching (Section 6.1.4). This definition of photometry has its basis in psychophysics (Section 6.5).

Luminous sensation is one of the attributes of colour sensation and therefore comes into all colorimetric systems. The role of photometry in colorimetry is illustrated in Section 6.13.6, 9.2.1 and 9.2.2.

All the relative luminous efficiency functions considered as standard are plotted together in two figures for a comparison. Figure 7.1 has a linear scale and Figure 7.2 has a logarithmic scale. The standard luminous efficiency functions considered are as follows:

- CIE 1924 photopic photometric observer $V(\lambda)$ (Sections 7.3, 7.4 and 16.1.2)
- CIE 1951 scotopic photometric observer $V'(\lambda)$ (Section 7.6 and 16.1.2)

Standard Colorimetry: Definitions, Algorithms and Software, First Edition. Claudio Oleari.
© 2016 John Wiley & Sons, Ltd. Published 2016 by John Wiley & Sons, Ltd.

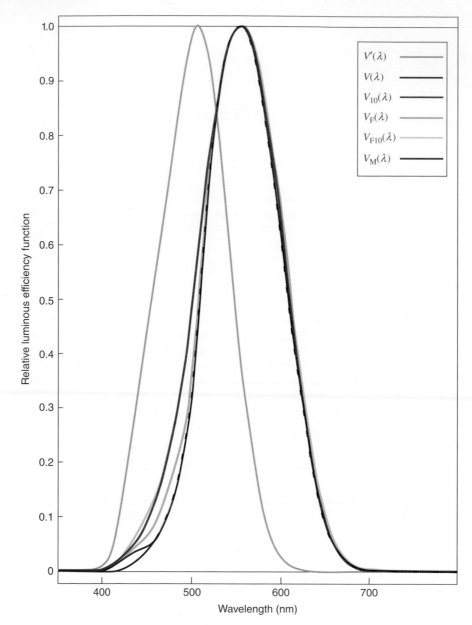

Figure 7.1 *All the relative luminous efficiency functions considered as standard: CIE 1924 photometric photopic observer $V(\lambda)$, CIE 1988 photopic photometric observer $V_M(\lambda)$, CIE 2005 photopic photometric observer with 10° visual field $V_{10}(\lambda)$, CIE fundamental photopic photometric observer with 2° visual field $V_F(\lambda)$ and 10° $V_{F,10}(\lambda)$, and CIE 1951 scotopic photometric observer $V'(\lambda)$. The various curves overlap and, where it is necessary, are represented by dotted lines to identify them. (Numerical values are in Section 16.1.2 data files.)*

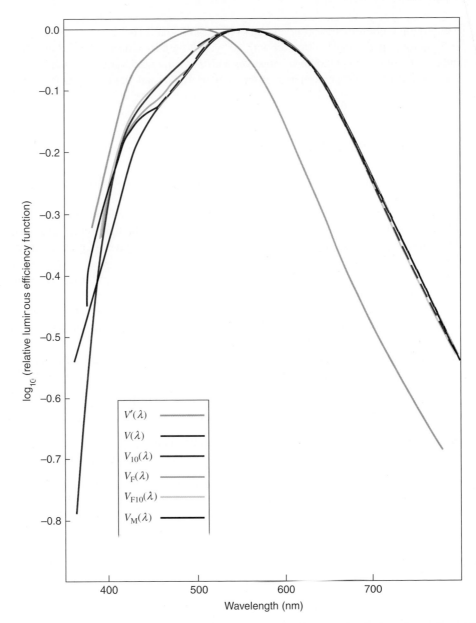

Figure 7.2 *All the relative luminous efficiency functions considered as standard plotted on a logarithmic scale: CIE 1924 photometric photopic observer V(λ), CIE 1988 photopic photometric observer $V_M(λ)$, CIE 2005 photopic photometric observer with 10° visual field $V_{10}(λ)$, CIE fundamental photopic photometric observer with 2° visual field $V_F(λ)$ and 10° $V_{F,10}(λ)$, and CIE 1951 scotopic photometric observer V′(λ). The various curves overlap and, where it is necessary, are represented by dotted lines to identify them. (Numerical values are in Section 16.1.2 data files.)*

- CIE 1988 photopic photometric observer $V_M(\lambda)$ (Section 7.4 and 16.1.2)
- CIE 2005 photopic photometric observer with $10°$ visual field $V_{10}(\lambda)$ (Section 7.7 and 16.1.2)
- CIE fundamental photopic photometric observer with $2°$ visual field $V_F(\lambda)$ (Section 7.8 and 16.1.2)
- CIE fundamental photopic photometric observer with $10°$ visual field $V_{F,10}(\lambda)$ (Section 7.8 and 16.1.2)

7.2   History of the Basic Photometric Unit

Photometry ignores colour and considers only the brightness appearance of visible light and therefore has a central role in lighting.

Science requires the definition of a light-unit source, useful to measure the level of lighting in an environment. Originally, the *candle* was considered and this name specified both the light source and the photometric unit. Over the twentieth century, scientific and technological development gave six definitions of the candle:

1. In 1881 it was the first standard candle as a standard light source, which was made of sperm-whale fat, hence the name *spermacetti candle*.
2. In 1898, with the advent of gas lamps, came the problem of reproducing the candle with this lamp. Only in 1909 the standard *international candle* was defined by a pentane-gas lamp, which was immediately replaced by an electric lamp with a coal filament.
3. In 1901, the importance of the Max Planck theory of *blackbody* radiation (Sections 3.3 and 8.3) required new terms for defining the candle and in 1940

   the *new candle* was defined as the light *intensity* emitted by a blackbody radiator of platinum at a temperature of 2042 K, corresponding to incipient solidification of the molten platinum.

4. In 1948 the new candle was accepted internationally with the name *candela*.
5. In 1979 the *Conférence Générale des Poids et Mesures* gave the current definition of candela:

   "The candela is the luminous intensity, in a given direction, of a source that emits monochromatic radiation of frequency 540×10^{12} hertz – corresponding to a wavelength of 555.016 nm conventionally approximate with 555 nm – and that has a radiant intensity in that direction of $(1/683)$ watt per steradian."

This definition of 'candela' is an approximation of the existing definition of 'new candle'. An exact correspondence requires that the number 683 corresponds to 683.002. The number 683, which defines the translation of a radiometric quantity in a photometric one, has the dimension of [lm/W] and is denoted by K_m. This conceptually new definition is part of the International System of measurement and all photometry as well as colorimetry is based on it.

The definitions of candela given here are physical and independent of the human visual system; then, the definition of the candle is not sufficient to define photometry completely.

7.3   CIE 1924 Spectral Luminous Efficiency Function

Before 1979 all the defined candles emitted light consisting of a wide spectral band that covers the whole range of visible radiation and beyond. Otherwise, the final definition of the International System does not refer to a particular light source and considers the monochromatic radiation of wavelength $\lambda = 555$ nm. The highest luminous efficiency of the human visual system corresponds to this wavelength. To get the same sensation for monochromatic radiations with increasing wavelength over 555 nm, or decreasing below, an increasing power is necessary. Luminous efficiency, already defined at 555 nm, must be extended to the entire range of visible radiation, which leads to the definition of the *relative photopic spectral luminous efficiency function* $V(\lambda)$, already introduced in Section 6.5, where 'relative' means that the maximum value of the function is chosen conventionally equal to 1 and that the function is dimensionless.

As known, the retina is not a uniform membrane and foveal vision is different from extra-foveal and peripheral vision. Historically, photometry is based on foveal vision, which excludes the other part of the retina, and thus the function $V(\lambda)$ is defined for this kind of vision.

The foveal vision field is subtended by a solid angle with the vertex in the pupil of the observer whose section is a plane angle of 2°, or, in any case, lower than 4°, while the extra-macular region by convention subtends a solid angle with a section of 10°. The measurement of $V(\lambda)$ is not simple and it is still under discussion. It is made by visually matching the brightness of lights of different wavelengths. Since these lights have different colours, *brightness matching* is necessary and this requires appropriate techniques already described in Section 6.5.

The CIE 1924 standard photopic photometric observer was defined using *minimum flicker* and *minimally distinct border* techniques.[2] The laboratories involved in these measurements were those of Coblentz-Emerson[3] and Gibson-Tyndall[4] at US NBS.

According to the definition of $V(\lambda)$, two monochromatic radiations with wavelength λ_1 and λ_2 and power $P_{e,1}$ and $P_{e,2}$ are equally bright if $V(\lambda_1)P_{e,1} = V(\lambda_2)P_{e,2}$.

7.4   CIE 1924 and CIE 1988 Standard Photometric Photopic Observers

"*CIE standard photometric observer* – an ideal observer having a relative spectral responsivity curve that conforms to the spectral luminous efficiency function for photopic vision $V(\lambda)$ or to the function for scotopic vision $V'(\lambda)$, and that complies with the summation law implied in the definition of luminous flux."[1]

"CIE, considering the discrepancies between the average human spectral luminous efficiency and the $V(\lambda)$ function, adopted in 1990[5] the *CIE 1988 Modified 2° Spectral Luminous Efficiency Function for Photopic Vision*, $V_M(\lambda)$, and recommended it for applications in visual sciences."[1]

Heterochromatic brightness matching is defined in Section 6.1.4. *Basic laws of heterochromatic brightness matching* are the Abney laws regarding the additivity of the colour stimuli (Section 6.5.1). This property of additivity is used to implement a spectral weighing of different colour stimuli and to quantify the luminous efficiency of the resulting stimulus. Particularly, the luminous flux Φ_v, measured in *lumens* [lm], obtained as the sum of the monochromatic components of the spectral radiant flux $\Phi_{e,\lambda}$ (Section 6.5.1) is

$$\Phi_v = K_m \int_{380}^{780} \Phi_{e,\lambda} V(\lambda)\,d\lambda \text{ lm with } K_m = 683\,\text{lm/W} \tag{7.1}$$

where the constant K_m implements the connection between radiometric and photometric units and therefore has the dimension [lm/W]. In practice, the spectral decomposition of a luminous sensation is never considered.

In 1924, the CIE defined the *CIE (1924) standard photopic photometric observer*, based on the photopic spectral luminous efficiency function $V(\lambda)$ and on the basic laws of heterochromatic brightness matching.

The CIE 1924 photometric standard observer is a fundamental part of photometry, since the *Bureau Inter-national des Poids et Mesures* (CIPM) has approved the use of these functions with the effect that the corresponding photometric quantities are defined in purely physical terms as quantities proportional to the integral of a spectral power distribution, weighted according to a specified function of wavelength.[6]

Moreover, the CIE 1924 photometric standard observer is a fundamental part of 2° colorimetry (Sections 9.2.1 and 9.2.2 show how the function $V(\lambda)$ comes into colorimetry).

Not everything is fully satisfactory. The CIE standard photometric observer shows two limitations:

1. The Helmholtz-Kohlrausch effect (Section 6.15) states that the brightness sensation increases with increasing purity at constant luminance, then the minimum distinct border does not correspond to equal brightness sensations. [The *Optical Society of America* (OSA) takes account of this effect in defining its uniform scale colorimetric system, known as OSA-UCS (Section 14.6.1).] The non-additivity of the exact brightness distribution is a debated point in photometry and attention is focused on the technique of heterochromatic brightness matching (Section 6.15).

2. The function $V(\lambda)$, obtained as the average of the heterochromatic matching data of several observers, shows at wavelengths below 460 nm the presence of a systematic deviation with respect to the value today considered more significant. In 1951, it was corrected by Judd[7] – *Judd correction* – and refined by Vos[8] and this, denoted with $V_M(\lambda)$, was taken as a function of *CIE 1988 Modified 2° Spectral Luminous Efficiency Function for Photopic Vision*[5] (Figures 7.1 and 7.2). In photometric and colorimetric practice, the CIE 1924 standard photometric observer is used because the deviation from the CIE 1988 observer has often negligible relevance. However, it has to be noted that the gap between these two observers is relevant for blue lights: for example, for the blue light emitted by the phosphor of a cathode ray tube, the difference between the two observers is 6.7%; for LEDs with a maximum bandwidth emission at 470 nm, it is ~1%; for a maximum at 450 nm, it is ~4 %; for a maximum at 430 nm, it is ~12%; and for a maximum at 410 nm, it is ~25%. At the moment, the function $V_M(\lambda)$ does not enter the definition of the photometric units of the International System.

7.5   Photometric and Radiometric Quantities

Radiometric and photometric quantities constitute two parallel systems of corresponding quantities and the second quantities are derivable from the first ones by using the function $V(\lambda)$ that, representing the spectral luminous efficiency, weighs in an appropriate way the radiant power distributed on the various wavelengths translating it into a quantity that represents the global luminous efficiency. The role of this function is given by equation (7.1) and is repeated for all the photometric quantities, as shown in Table 7.1.

As written above, the photometric quantities are obtained by spectral weighing and the integrals are defined on the range of the visible wavelengths 380–780 nm. The factor K_m is a constant that depends on the system of units chosen and in *The International System of Units* (SI) it is 683 lm/W.

The quantity that in radiometry is involved in defining all the other quantities is power [W], whose corresponding photometric quantity is *luminous flux* with unit *lumen* [lm]. All other photometric quantities consider luminous flux in their geometric conditions, the choice of which is typical of the different practical and instrumental situations.

Luminous exitance, a photometric quantity corresponding to radiant exitance [W/m²], is the surface density of the luminous flux emerging from a surface and is measured in lumens per m² [lm/m²]. This quantity is suitable for characterizing an extended light source.

Illuminance of a surface is represented by the surface density of the illuminating flux on the surface itself, a photometric quantity corresponding to irradiance [W/m²], and is measured in *lux* [lx], equivalent to lumens per m². This quantity is suitable to characterize, for example, the illuminance on a table in a place of work.

Table 7.1   *Photometric and radiometric quantities.*

Radiometric quantity $\Leftrightarrow$	Photometric quantity	Symbol and definition	SI unit
Radiant flux $\Phi_{e,\lambda}$	Luminous flux	$\Phi_v = K_m \int\limits_{380}^{780} \Phi_{e,\lambda} V(\lambda)\,d\lambda$	lumen [lm]
Radiant exitance $M_{e,\lambda}$	Luminous exitance	$M_v = K_m \int\limits_{380}^{780} M_{e,\lambda} V(\lambda)\,d\lambda$ $= \dfrac{d\Phi_v}{dA} = \int\limits_{2\pi} L_v \cos\vartheta\,d\Omega$	lumen/m^2 [lm/m^2]
Irradiance $E_{e,\lambda}$	Illuminance	$E_v = K_m \int\limits_{380}^{780} E_{e,\lambda} V(\lambda)\,d\lambda$ $= \dfrac{d\Phi_v}{dA} = \int\limits_{2\pi} L_v \cos\vartheta\,d\Omega$	lux [lx], [lm/m^2]
Radiant intensity $I_{e,\lambda}$	Luminous intensity	$I_v = K_m \int\limits_{380}^{780} I_{e,\lambda} V(\lambda)\,d\lambda = \dfrac{d\Phi_v}{d\Omega}$	candela [cd], [lm/strad]
Radiance $L_{e,\lambda}$	Luminance	$L_v = K_m \int\limits_{380}^{780} L_{e,\lambda} V(\lambda)\,d\lambda =$ $= \dfrac{d^2\Phi_v}{dA \cos\vartheta\,d\Omega}$	candelas per metre^2 [cd/m^2], [lm/(strad m^2)]

Luminous intensity is the illuminating flux density per unit solid angle, photometrically corresponding to radiant intensity [W/strad], and is measured in *candelas* [cd], equivalent to lumens per unit solid angle. This quantity is suitable for characterizing point-like light sources or ones considered as point like.

Luminance, a photometric quantity corresponding to radiance [W/(m^2 strad)], is particularly important because of its direct correlation with visual perception and is therefore the quantity most used in the study of human colour vision and colorimetry. It is measured in *candelas per m^2* [cd/m^2]. In vision, there is a correspondence between the observed object and the retinal image (Figures 5.1, 2.3 and 2.4). The light that forms the generic point of the retinal image is emitted from a point of the observed body and is collected within the solid angle having its vertex at the object point and subtended by the opening of the eye pupil. The definition of luminance considers the power density coming out of the observed body calculated with respect to the apparent emitting surface, that is, perpendicular to the optical axis of the eye, and with respect to the solid angle that collects the light. Luminance is calculated considering the retinal surface illuminated and the solid angle between the pupil and the retina. These concepts are considered in Sections 2.4 and 2.6. The definition of luminance is obtained from that of radiance by replacing the words 'flux of radiant power' with 'luminous flux'.

Tables 7.2 and 7.3 report the data concerning the most frequent luminance and illuminance in everyday life in order to give a practical order of magnitude to the most common photometric quantities.

Table 7.2   *Luminance scale.*

Object emitting or reflecting light	cd/m^2	Vision
Surface of the sun at noon	$>10^7$	Dangerous vision
Tungsten filament	10^6	Photopic vision
	10^5	
White paper illuminated by the sunlight	10^4	
	10^3	
White light emitted by CRT television	10^2	
White paper lit for comfortable reading	10	
	1	Mesopic vision
	10^{-1}	Scotopic vision
White paper illuminated by the moonlight	10^{-2}	
	10^{-3}	
	10^{-4}	
	10^{-5}	
Absolute threshold of the visul system	10^{-6}	

Table 7.3   *Illuminance scale.*

lux	Environmental illumination
~100 000	Daylight hours 12 clear sky
~65 000	Daylight hours 10 clear sky
~35 000	Daylight hours 15 overcast sky
~32 000	Daylight hours 12 overcast sky
~25 000	Daylight hours 10 overcast sky
~2000	Natural light an hour after dawn, overcast sky
~1000	Natural light an hour before dawn
~1000	Internal, fluorescent lamps and window light
700-500	Department stores
500-400	Interior with fluorescent lamps
300-200	Interior
200-100	Internal stairs

7.6   CIE 1951 Standard Scotopic Photometric Observer

In 1951, the CIE defined the standard scotopic photometric observer[9] with the relative spectral luminous efficiency function $V'(\lambda)$ (Section 6.4). This photometric observer should be used only in situations where the light level is low enough to exclude the activation of the cones in the visual process. The luminous flux Φ'_v associated with the scotopic spectral radiant flux $\Phi_{e,\lambda}$ is defined by

$$\Phi'_v = K'_m \int_{380}^{780} \Phi_{e,\lambda} V'(\lambda) \, d\lambda \; \text{lm} \quad \text{with } K'_m = 1700 \, \text{lm/W} \tag{7.2}$$

where the definition of K'_m is made in agreement with the definition of candela given for the CIE 1924 photometric observer, imposing $[K'_m \times V'(\lambda = 555.016)] = 683$ lm/W. The value $K'_m = 1700$ is obtained by approximation from 1700.06.

The photometric scotopic observer is mentioned here only for the sake of completeness, but it does not have an effective role in colorimetry, which concerns only photopic and mesopic vision.

7.7   CIE 2005 Photopic Photometric Observer with 10° Visual Field

In 2005, the CIE defined the *photometric photopic observer for 10° visual field*[10] with the definition of the relative spectral luminous efficiency function $V_{10}(\lambda)$ (Figures 7.1 and 7.2). This photometric observer should be used only in situations in which the visual field presents a section of more than 4°, or objects to be observed with the extra-macular region of the retina, and with a light level sufficiently high to exclude rod intrusion in the visual process. The luminous flux for this observer is defined by

$$\Phi_{10,v} = K_{10,m} \int_{380}^{780} \Phi_{e,\lambda} V_{10}(\lambda) \, d\lambda \; \text{lm} \quad \text{with } K_{10,m} = 683.6 \, \text{lm/W} \tag{7.3}$$

where the definition of $K_{10,m}$ is made in agreement with the definition of candela given for the CIE 1924 photometric observer, imposing $[K_{10,m} \times V_{10}(\lambda = 555.016)] = 683.002$ lm/W. Since by interpolation $V_{10}(\lambda = 555.016) = 0.9991236$, it follows that $K_{10,m} = 683.002/0.9991236 = 683.6011$, and in practical photometry, it is approximated with $K_{10,m} = 683.6$ lm/W.

Section 9.3 discusses how the function $V_{10}(\lambda)$ comes into colorimetry.

7.8   CIE Fundamental Photopic Photometric Observer with 2°/10° Visual Field

In 2006, the CIE (TC 1-36)[11] defined two fundamental observers for 2° and 10° visual fields and then the corresponding two photometric observers[12] (Section 9.6).

Afterwards, the CIE (TC 1-36) defined the transformations from the fundamental reference frame to the *XYZ* reference frame, in which the *Y* component is proportional to the luminance. This meant that the relative spectral sensitivity of light was equal to the *Y* component of the colour matching functions $V_F(\lambda) = \bar{y}_F(\lambda)$ and $V_{F,10}(\lambda) = \bar{y}_{F,10}(\lambda)$ for observers at 2° and 10° visual fields, respectively (Section 9.6). The functions $V_F(\lambda)$ and $V_{F,10}(\lambda)$ define two new photometric observers, which should be compared with the previous ones (Figures 7.1 and 7.2).

These luminous efficiency functions are determined for a background adaptation of daylight D65 (Section 8.6) in agreement with the recommendation of the CIE.

In line with the basic assumption of the CIE (TC 1-36), the S-cone system does not notably contribute to luminance for both observers.

The photometric analysis of the two observers must be done separately.

7.8.1   Photopic Spectral Luminous Efficiency Functions for the 2° Fundamental Observer

After the original definition of the CIE 1924 standard photometric observer, corrections and refinements were proposed by Judd (1951) and Vos (1978). All that suggested for an improved $V(\lambda)$, which resulted in the $V_M(\lambda)$ function as described in a CIE report.[5] Today, $V_F(\lambda)$ has to be compared with $V_M(\lambda)$. The CIE (TC 1-36) said that the discrepancies "are rather small, and actually insignificant in comparison with individual variability, so that switching from $V_M(\lambda)$ to $V_F(\lambda)$ does not mean a denouncement of $V_M(\lambda)$, but rather one step further to define an integrated Photometric and Colorimetric Observer".

7.8.2   Photopic Spectral Luminous Efficiency Functions for the 10° Fundamental Observer

$V_{F,10}(\lambda)$ has to be compared with $V_{10}(\lambda)$. The 10° fundamental observer is based only on the Stiles-Burch[13] data, while the colour-matching functions of the CIE 1964 observer were defined by using data of the Stiles-Burch and Speranskaya[14] laboratories, with several other adjustments by the CIE. The consequence is that between $V_{F,10}(\lambda)$ and $V_{10}(\lambda)$ a discrepancy exists. The CIE (TC 1-36) said that "a straightforward comparison with an experimentally measured $V_{10}(\lambda)$ function would be preferred. It should be further stressed that this provisional solution is offered mainly as a means of defining an $(x_{F,10}, y_{F,10})$ chromaticity diagram similar to the CIE 1964 (x_{10}, y_{10}) diagram". Again, the CIE (TC 1-36) noted "that further study is required and that this is by no means a proposal to replace the CIE 1964 $\bar{y}_{10}(\lambda)$" or the CIE 2005 $V_{10}(\lambda)$.

References

1. CIE Publication S 017/E:2011, *ILV: International Lighting Vocabulary*, Commission Internationale de l'Éclairage, Vienna (2011). Available at: www.eilv.cie.co.at/
2. CIE Publication, Principales decisions (6e Session, 1924), CIE Sixième Session, Genève, Juillet, 1924. Recueil des Travaux et Compte Rendu de Séances, Cambridge, Cambridge University Press, 67–69 (1926).
3. Coblenz WW and Emerson WB, Relative sensibility of the average eye to the light of different colors and some practical application of radiation problem, *US Bureau Standards Bull.* **14**, 167–236 (1918).
4. Gibson KS and Tyndall EPT, Visibility of radiant energy, *Bulletin Bureau Standards* **19**, 131–191 (1923).
5. CIE publication N°. 086, *CIE 1988 2° Spectral Luminous Efficiency Function for Photopic Vision*, Commission Internationale de l'Éclairage, Vienna (1990).
6. Bureau International des Pois et Mesures. *Monographie. Principles Governing Photometry* (1983) Pavilloon de Breteuil, F-92310 Sèvres, Imprimerie Durand, 28600 Luisant (France).
7. Judd DB, Report of U.S. Secretariat Committee on colorimetry and artificial daylight, in *CIE Proceedings*, Vol. 1. Part 7, p. 11 (Stockholm 1951), Bureau Central de la CIE (1951).
8. Vos JJ, Colorimetric and photometric properties of a 2° fundamental observer, *Color Res. Appl.*, **3**, 125–128 (1978).
9. *CIE Proceedings 1951*, Vol.1,Sec.4; Vol.3, p.37, Bureau Central de la CIE, Paris, (1951).
10. CIE publication 165:2005, *CIE 10 Degree Photopic Photometric Observer*, Commission Internationale de l'Éclairage, Vienna, Austria (2005).
11. CIE 170-1:2006. *Fundamental Chromaticity Diagram with Physiological Axes-Part 1*. Commission Internationale de l'Éclairage, Vienna (2006).

12. CIE TC 1-36 *Fundamental Chromaticity Diagram with Physiological Axes-Part II*. Draft, Commission Internationale de l'Éclairage, Vienna (2015).
13. Stiles WS and Burch JM, N.P.L. colour-matching investigation: final report (1958). *Opt. Acta*, **6**, 1–26 (1959).
14. Speranskaya NI, Determination of spectrum color co-ordinates for twenty-seven normal observers, *Optics Spectrosc.* **7**, 424–442 (1959).

8

Light Sources and Illuminants for Colorimetry

8.1 Introduction

The colour of a non-self-luminous body depends on the radiation that illuminates it, and then colorimetric practice poses the problem of defining the illuminating radiation that is important for colorimetry. The problem is not simple, because in nature there are different radiations, which are changing and then unidentifiable in a single light reproducible in the laboratory. For example, there is no single light of day to refer to, this being dependent on the hour of day, latitude, season, the atmospheric haze, etc. Practical needs require the definition of some radiations with spectral power distributions considered to be significant. This leads us to distinguish between the radiations emitted by light sources and the illuminants:

"*Source* – an object that produces light or other radiant flux."[1]

"*Illuminant* – radiation with a relative spectral power distribution defined over the wavelength range that influences object colour perception."[1]

Then it follows[1]:

"*CIE standard illuminants* – illuminants A and D65 defined by the CIE in terms of relative spectral power distributions (Section 8.3, 8.4 and 8.6).
   NOTE 1 – These illuminants are intended to represent:

- A: Planckian radiation at an absolute temperature of about 2848 K (Sections 8.3 and 11.3.1).
- D65: The relative spectral power distribution representing a phase of daylight with a correlated colour temperature of approximately 6500 K (called also 'nominal correlated colour temperature of the daylight illuminant') (Sections 8.4 and 8.6).

Standard Colorimetry: Definitions, Algorithms and Software, First Edition. Claudio Oleari.
© 2016 John Wiley & Sons, Ltd. Published 2016 by John Wiley & Sons, Ltd.

> NOTE 2 – Illuminants B, C and other D illuminants, previously denoted as standard illuminants, should now be termed CIE illuminants. (Section 8.9)"

> "*CIE standard sources* – artificial sources specified by the CIE whose radiation approximate CIE standard illuminants."

The illuminants are mathematical tables of values (relative power versus wavelength) used for colorimetric computations. The CIE has standardized a few spectral power distributions and recommended that they should be used whenever possible. The spectral power distributions of all the considered illuminants are obtainable in the form of ASCII files from the directory 'Colour files' enclosed with the software described in Section 16.1.2.

> *Simulators* of the CIE illuminants are sources with a spectral power distribution as close as possible to that of the illuminants.

The light sources and illuminants are specified by their spectral distribution of the emitted power, *absolute* S_λ [W/nm] or *relative* $S(\lambda)$ [dimensionless]. If possible, the relative magnitude is related to the value assumed at the wavelength of 560 nm, set conventionally equal to 100. If not possible, as in the case of narrow band spectra, other wavelengths are chosen to set the relative value equal to 100.

The spectral range considered in defining the lighting is from 300 to 830 nm. The extension in the ultraviolet region is required to assess the fluorescence when the fluorescent bodies are illuminated.

> Conventionally, light sources and illuminants for illumination
>
> 1.   have colorimetric specification related to the CIE 1931 observer, unless specified otherwise;
> 2.   have a luminance reflected by the *perfect reflecting diffuser* (Section 3.5.1) which is set equal to 100 and defines the relative luminance scale called the *percentage of the luminance factor*; and
> 3.   have all the components of the tristimulus vector rescaled by an equal factor to have $Y = 100$, but generally the chromaticity (x, y) CIE 1931, and particularly (u', v') CIE 1976, is preferred to the tristimulus values.

The spectral power distribution of the electromagnetic radiation emitted by each body depends on the nature of the body and on the temperature. The description of the physical principles on which the various sources are based, although very interesting, is not made here. A list of the possible light sources is given in Sections 3.2 and 3.3. Here just a description of the light sources and illuminants for colorimetry is given.

Radiometry is presented in Section 2.4. Photometry is presented in Chapter 7, after a very short introduction in Section 2.6. The colorimetric specification of the light illuminants is defined in Section 9.7.

8.2   Equal-Energy Illuminant

The *equal-energy* radiator, or *equi-energy* radiator, is an ideal light source with constant spectral power distribution inside the visible spectrum. In practice, only the corresponding illuminant E_E defined by the spectral power distribution $E_E(\lambda) \equiv 1$ is used as a theoretical reference in the definition of the units in the tristimulus space, therefore it is specified by the tristimulus values $X = Y = Z = 100$ and the chromaticity ($x = 1/3$, $y = 1/3$) (the same for the other CIE standard observers, Tables 8.2 and 8.3). This illuminant has a correlated colour temperature (*CCT*, or T_{cp}) of 5455 K close to that of the daylight illuminant D55 (Sections 8.4 and 11.3.1).

8.3   Blackbody Illuminant

The most important light source is the *blackbody radiator* – or *full radiator*, or *Planckian radiator* – whose spectral power distribution is given by *Planck's formula*. This light source and the Planck formula are presented in Section 3.3. Here the spectral power distribution of the Planckian radiator is considered, that renders this kind of light source important in colorimetry and

1. is defined by Planck's formula, which has a strong theoretical basis, and
2. depends only on the absolute temperature, which leads to define the *colour temperature* and the *correlated colour temperature*, important quantities useful for a classification of light sources and illuminants.

   Colorimetry and photometry consider generally the relative spectrum of the Planckian radiator, normalized to 100 at 560 nm, and the relative spectral power distribution at absolute temperature T defined by the CIE is

$$S_T(\lambda) = 100\left(\frac{560}{\lambda}\right)^5 \frac{\exp\left(\frac{1.4388 \times 10^7}{560 \times T}\right) - 1}{\exp\left(\frac{1.4388 \times 10^7}{\lambda T}\right) - 1} \quad \text{with } \lambda \text{ in nm and } T \text{ in K} \tag{8.1}$$

as written in the definition of the illuminant A given in CIE 15:2005 *Colorimetry 3rd edition*[2] (N.B. In the previous edition CIE 15.2:1986 *Colorimetry 2nd edition*[3] the number 1.435 of Planck's formula instead of 1.4388, and the T_c 2848 K of the illuminant A instead of 2856 K are used.) (Figure 8.1)

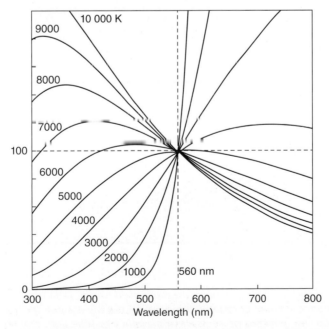

Figure 8.1   *Relative spectral power distribution of the blackbody emission at different temperatures according to Planck's formula.*

Table 8.1   *CIE 1931 chromaticity coordinates of blackbody radiation at various temperatures.*

T [K]	1000	2000	3000	4000	5000	6000	7000	8000	10 000
x	0.6526	0.5266	0.4368	0.3804	0.3450	0.3220	0.3063	0.2952	0.2806
y	0.3446	0.4133	0.4041	0.3767	0.3516	0.3318	0.3165	0.3048	0.2883

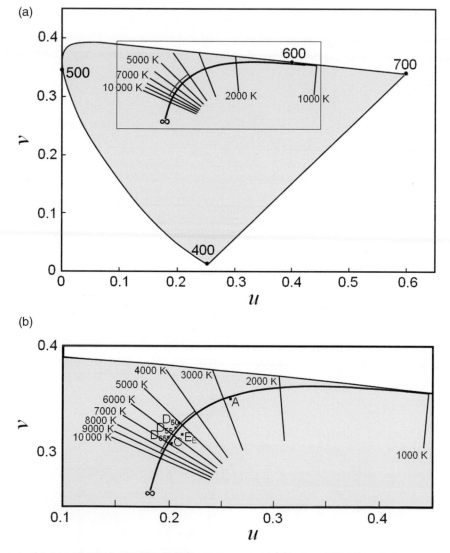

Figure 8.2   *(a), (b) CIE 1960 (u, v) chromaticity diagram, used for the definition of the correlated colour temperature. The loci of equal correlated temperature are on straight lines orthogonal to the Planckian locus (black curve), which represents the chromaticity of the light emitted by a blackbody at the same temperature. Daylight locus (red curve) and chromaticity of the illuminants A, D50, D55, D65, C and E_E.*

Planck emitters are not sources of practical use, but incandescent filament lamps (usually tungsten lamps) crossed by an electric current have an exitance that can be well approximated with that of the blackbody at a slightly lower temperature. This temperature, expressed in kelvin and denoted by T_c, is called the *colour temperature*. This concept is extended to any kind of light source and illuminant by defining the *correlated colour temperature* T_{cp}, or *CCT*[4] (Section 11.3.1).

The radiations of a blackbody at different colour temperatures T_c are specified colorimetrically by their chromaticities, shown in Table 8.1 and Figure 8.2, where the set of points representing their chromaticities is termed *Planckian locus*.

The colour of light emitted by the blackbody radiator ranges from red to yellow and to blue with increasing temperature and the best representative temperature for a Planckian radiator, in order that its radiant energy shall evoke the achromatic neutral sensation in observers adapted to the same radiation is approximately 6200–6900 K.

The graphical comparison between the spectral power distribution of the blackbody and the day light at the same colour temperature is made by the software described in Section 16.9.1.

The most widely used lamps based on the Planckian radiator are:

- *Tungsten filament lamp* – incandescent lamp whose luminous element is a filament of tungsten.[1]
- *Tungsten halogen lamp* – gas-filled lamp containing halogens or halogen compounds, the filament being of tungsten. Iodine lamps belong to this category.[1]

8.4   CIE Daylights

Daylight has a very important role in everyday life and is considered the most important light in colorimetry. Naturally daylight fluctuates depending on various factors such as the time of day, latitude, season, weather and the scene. An almost systematic set of measurements of the spectral power distribution of the daylight in the wavelength range 330–700 nm has been made over time.[5,6] The analysis of these spectra is distinguished in colorimetric and radiometric analysis.

The chromaticity coordinates of the collected spectra plotted on the CIE 1931 chromaticity diagram define a region well approximated by a line, termed *daylight locus* and fitted by an equation which is approximately parallel to the Planckian locus, where each point is specified by a correlated colour temperature T_{cp} (Section 8.3 and 11.3.1) (Figure 8.2). The equation of the daylight locus is

$$y_D = -3.000 x_D{}^2 + 2.870 x_D - 0.275 \text{ for } 0.250 < x_D < 0.380 \tag{8.2}$$

where the x_D component of the chromaticity depends on the correlated colour temperature and is defined by the formulas:
for $4000 < T_{cp} < 7000$ K,

$$x_D = \frac{-4.6070 \times 10^9}{T_{cp}{}^3} + \frac{2.9678 \times 10^6}{T_{cp}{}^2} + \frac{0.09911 \times 10^3}{T_{cp}} + 0.244063 \tag{8.3}$$

for $7000 < T_{cp} < 25\,000$ K,

$$x_D = \frac{-2.0064 \times 10^9}{T_{cp}{}^3} + \frac{1.9018 \times 10^6}{T_{cp}{}^2} + \frac{0.24748 \times 10^3}{T_{cp}} + 0.237040 \tag{8.4}$$

These equations give a chromaticity specification of a daylight illuminant whose correlated colour temperature is approximately equal to the nominal value T_{cp}.

The spectral distributions were analysed by the method of principal component analysis. The result was that the various spectral power distributions of daylight are well represented by a linear combination of three basic functions $S_0(\lambda)$, $S_1(\lambda)$ and $S_2(\lambda)$ (Figure 8.3):

$$S_D(\lambda) = S_0(\lambda) + M_1 S_1(\lambda) + M_2 S_2(\lambda) \tag{8.5}$$

where

$$M_1 = \frac{-1.3515 - 1.7703 x_D + 5.9114 y_D}{0.0241 + 0.2562 x_D - 0.7341 y_D}, \quad M_2 = \frac{0.0300 - 31.4424 x_D + 30.0717 y_D}{0.0241 + 0.2562 x_D - 0.7341 y_D} \tag{8.6}$$

The CIE says that the spectral power distributions of these recommended daylight illuminants, defined in the wavelength range 330–700 nm, can be extrapolated in the wavelength ranges 300–330 nm and 700–830 nm and the extrapolated values are believed to be accurate enough only for colorimetric purposes.

The spectrum of these illuminants overlaps that of the blackbody of the same correlated temperature but with overlapped 'irregularities' mainly due to the absorption that occurs in the earth's atmosphere (Section 16.9).

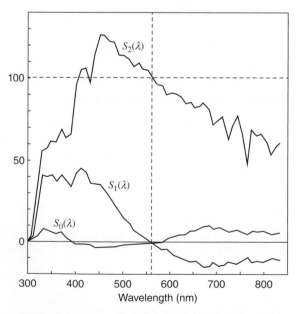

Figure 8.3 *Plot of three spectral functions $S_0(\lambda)$, $S_1(\lambda)$, $S_2(\lambda)$, by a linear combination of which the daylight spectral power distribution is obtained. The coefficients of the linear combination are a function of the chromaticity of the daylight illuminant considered. (Numerical values are in Section 16.1.2 data files.)*

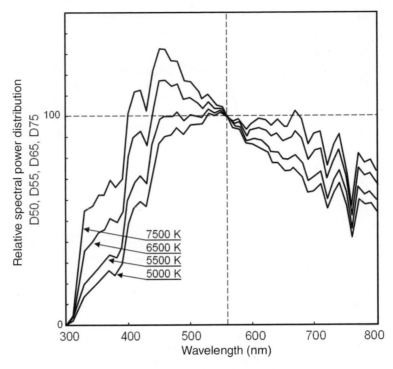

Figure 8.4 *Relative spectral power distributions of the illuminants D50, D55, D65, and D75. (Numerical values are in Section 16.1.2 data files.)*

The daylight illuminants are denoted by the 'D', that stands for daylight, and a number specifying the correlated colour temperature expressed in kelvins and divided by 100, for example, D50, D55, D60, …, D75 (Figure 8.4).

> The CIE recommends "that, in the interest of standardization, D65 be used whenever possible. When D65 cannot be used, it is recommended that one of the daylight illuminants D50, D55 or D75 be used. When none of these daylight illuminants can be used, a daylight illuminant at a nominal correlated colour temperature T_{cp} can be calculated using the Equations (8.2)–(8.6)."[2] (Section 11.3.1)

8.5 CIE Indoor Daylights

The CIE daylight illuminants at different correlated colour temperature have a content of UV and IR that is representative of natural outdoor daylight. Indoor daylight enters through the glass of the windows which filters and reduces the content of the power below 400 nm and over 600 nm. This filtering particularly effects the fluorescent phenomena. For practical reasons, in 2009 the CIE defined two indoor daylight illuminants, termed ID50 and ID65 with approximately 5000 K and 6500 K correlated colour temperature, respectively[7]. Figures 8.5a and 8.5b show a comparison of the spectral power distribution of these two daylights D50 and D65 with the corresponding indoor ID50 and ID65.

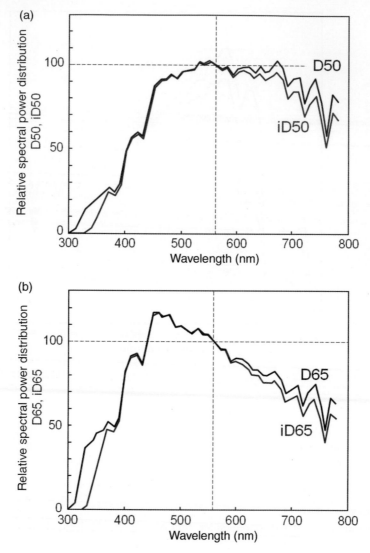

Figure 8.5   *Relative spectral power distributions of the illuminants D50 and D65 compared to the indoor illuminants iD50 (a) and iD55 (b), respectively. (Numerical values are in Section 16.1.2 data files.)*

8.6   CIE Standard Illuminants

Over time, with the evolution of light source technology, the CIE modified the set of Standard illuminants[2,8]. Today the CIE standard illuminants are:

1.   *Standard illuminant A*
     Planckian radiation at a temperature of about 2848 K (previously 2856 K) with relative spectral power distribution (8.1)

$$S_A(\lambda) = S_{T=2848}(\lambda)$$

(8.7)

2.   *Standard illuminant D65*
     This intended to represent the relative spectral power distribution related to a phase of the daylight with a correlated colour temperature of approximately 6500 K – also called *nominal correlated colour temperature of the daylight illuminant –*.

The CIE also proposes other D illuminants to be used in case the D65 or the other are inappropriate: D50, D55, D60 and D75 (Figure 8.4). These illuminants are no longer CIE standard (although some times in CIE 15.3:2004, *Colorimetry 3rd ed.*[2] are still called standard).

8.7   CIE Light Sources: A, B and C

Over time, the CIE defined some standard sources, denoted by the letters A, B and C, as representative of the homonymous illuminants (Figure 8.6):

- The *light source A* has a spectral exitance equal to that of a blackbody at a temperature of 2848 K (previously 2856 K) and in practice is well approximated by a bulb with a tungsten filament of approximately equal temperature (Tables 8.2 and 8.3).
- The *light source B*, whose practical use is rare and today deprecated by CIE, is close to the average light of day at noon and corresponds to a correlated colour temperature of approximately 4900 K (Tables 8.2 and 8.3).

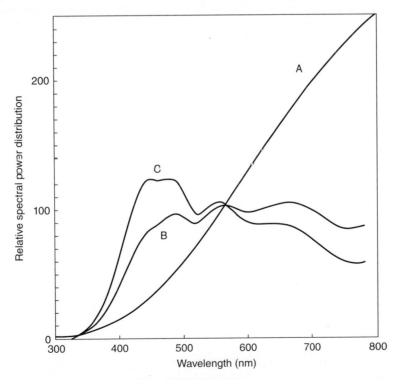

Figure 8.6   *Relative spectral power distributions of the illuminants* A, B *and* C. *(Numerical values are in Section 16.1.2 data files.)*

- The *light source C* is approximately the average light of day with overcast skies and corresponds to a colour temperature of approximately 6800 K (Tables 8.2 and 8.3). Its use was widespread in the past and today is replaced by D65. The CIE declared the illuminant C obsolete in the publication CIE 15:2005 *Colorimetry 3rd edition*[2].

The light sources B and C are obtainable in the laboratory by the source A combined with suitable filters defined by CIE recipes.[9–11]

The chromaticity of illuminants A and C is compared with the Planckian locus in Figure 8.2b.

8.8   CIE Sources for Colorimetry

The CIE recommends that the following artificial light sources for laboratory inspection be used if it is desired to realise an illuminant among those previously defined.

1. *Source A.*

   CIE standard illuminant A is constituted by a gas-filled tungsten filament lamp operating at a corre-lated colour temperature of 2848 K (previously 2856 K). If the source is also to be used in the UV region, a lamp having an envelope or window made of fused-quartz or silica must be used, because glass absorbs the UV component of the radiation emitted from the filament.

2. *Source D65 – Daylight at 6500 K.*

   At present, no artificial source is recommended to create the CIE standard illuminant D65 or any other illuminant D of different correlated colour temperature. The fluorescent lamp FL3.15 was conceived as a D65 simulator (Figure 8.7).

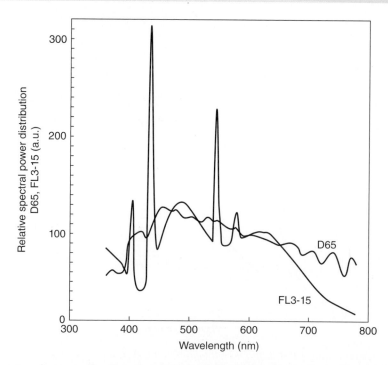

Figure 8.7   *Comparison between the illuminant D65 and the FL3.15, that was conceived for simulating it. (Numerical values are in Section 16.1.2 data files.)*

The CIE suggests that "whenever the highest accuracy of the spectral power distribution of a standard is required, it is advisable to make a spectroradiometric calibration of the actual source used, because the relative spectral power distribution of the source may not exactly coincide at all wavelengths with that defining the corresponding illuminant."[2]

The CIE considered the assessment of the quality of a simulator for colorimetry of the daylight illuminants D50, D55, D65 and D75. The purpose of this assessment is to quantify the suitability of a test source as a practical reproduction of the CIE standard illuminant D50, D55, D65 or D75 for colorimetric tasks. The assessment is based on the 'special metamerism index: change in illuminant' (Section 11.4.8), employing specified colour samples that are metameric matches for the CIE illuminants D50, D55, D65 and D75, respectively, and the CIE 1964 standard colorimetric observer.

8.9   CIE Illuminants: B, C and D

Illuminants B, C and other D illuminants, previously denoted as CIE standard illuminants, should now be termed *CIE illuminants*.

With regard to daylight, the CIE defined some standard illuminants, which are only illuminants and not sources, denoted by D55, D65, D75 (the notation is as for the standard D illuminants)[2]. The results given by the use of these illuminants are not exactly related to situations of general actual observation, but with acceptable deviation attributable to many real situations, and this property makes them the most widely used in colorimetry, particularly D65, the only daylight illuminant that today is a CIE standard. The chromaticity of these illuminants is compared with the Planckian locus in Figure 8.2.

At the moment, D illuminants do not correspond to light sources, the production of which is still an open problem.[12,13]

8.10   Fluorescent Lamps

One type of source used for its very high yield is the fluorescent lamp, whose luminous efficacy (lumen/watt) exceeds the efficacy of a tungsten incandescent source with comparable visible light output by several times.

A fluorescent lamp, or fluorescent tube, is a low-pressure mercury-vapour gas-discharge lamp that uses fluorescence to produce visible light. Common fluorescent lamps operate at a pressure of about 0.3% of atmospheric pressure. An electric current in the gas, obtained by a thermo-ionic effect, excites the mercury vapour by inelastic collision. The excited atoms of mercury de-energize with the emission of ultraviolet light – mainly at 253.7 nm – and visible light – mainly at 404.7 (violet), 435.8 (blue), 546.1 (green) and 578.2 nm (yellow–orange) –. The ultraviolet light excites a phosphor coating on the inside of the bulb of the lamp. The excited phosphors de-energize with the emission of visible light. Suitable mixtures of phosphors are used in the different light sources.

Similar fluorescent lamps are based on the phenomenon of induction.

8.10.1   Typical Fluorescent Lamps

The typical fluorescent lamps, denoted with the abbreviations F1, F2, ..., F12, are divided into three groups according to their spectrum[2]:

1. *normal*, where *cool white* F2 is the most representative;
2. *broadband*, where *daylight fluorescent* F7 is the most representative; and
3. *three narrow bands*, where *white fluorescent* F11 is the most representative.

(Tables 8.2 and 8.3; Figures 8.8a–c)

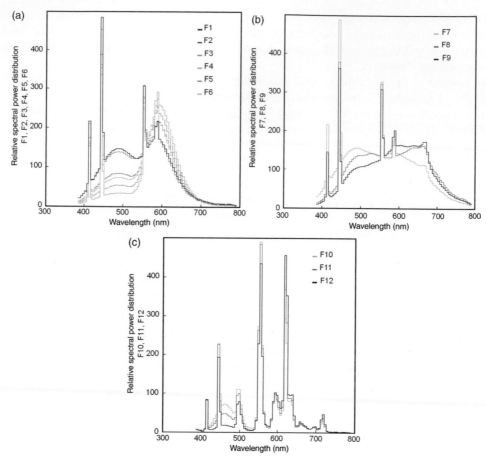

Figure 8.8 *Relative spectral power distributions of the different typical fluorescent lamps subdivided into three groups (a), (b) and (c), according to their spectral characteristics. (The spectrum is represented by a wavelength histogram with 5 nm steps.) (Numerical values are in Section 16.1.2 data files.)*

The illuminants F have line spectra and their spectral power distribution cannot be related to the value of 560 nm, however, we assume by convention a relative tabulation $S(\lambda)$. In these lamps the ultraviolet emission is negligible because the glass tube coated with phosphors is not transparent to such a radiation, while inside the tube the ultraviolet radiation – 254, 313 and 365 nm – produced by an electric discharge in low-pressure mercury vapours feeds the phosphor fluorescence.

The wide use of these lamps has also brought up their importance from the colorimetric point of view (Tables 8.2 and 8.3).

8.10.2   New Set of Fluorescent Lamps

The latest edition of the publication CIE 15:2005 *Colorimetry 3rd edition*, are published other illuminants of type F, denoted by FL3.k, $k = 1$-14, whose spectral power distributions can be downloaded as ASCII files from a folder of the software described in Section 16.1.2.

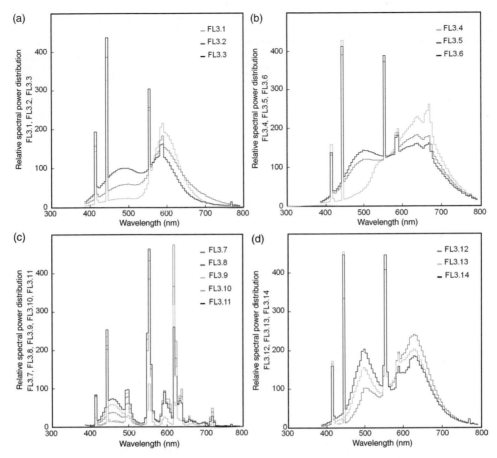

Figure 8.9   *Relative spectral power distributions of the different new set fluorescent lamps subdivided into four groups (a), (b), (c) and (d), according to their spectral characteristics. (The spectrum is represented by a wavelength histogram with 5 nm steps.) (Numerical values are in Section 16.1.2 data files.)*

These illuminants FL3.k are divided into four groups[2]:

1. The illuminants FL3.1–FL3.3 of the first group, known as the *Standard-Halo*, have spectral power distribution consisting of two wide bands due to the emission of a halogen-phosphorus of calcium phosphate with antimony and manganese as activators.
2. The illuminants FL3.4–FL3.6 of the second group, known as the *DeLuxe*, have almost flat spectral power distribution, due to the presence of more phosphors, to get a better colour rendering.
3. The illuminants FL3.7–FL3.11 of the third group with *three-band*s have spectral power distribution constituted by three narrow bands due to the emission of rare earth phosphors of ternary composition in the regions of the wavelengths blue, green and red.
4. The illuminants FL3.12–FL3.14 of the fourth group are related to multi-band fluorescent lamps.
5. The illuminant FL3.15 of a fluorescent lamp designed to simulate D65 by the American National Standards Institute (ANSI) and known by the acronym JIS Z 8716:1991. (Figure 8.7).

(Tables 8.2 and 8.3; Figures 8.9a–d)

Also these fluorescent lamps have importance from the colorimetric point of view (Tables 8.2 and 8.3).

Table 8.2 *Illuminant colorimetric specification related to the CIE 1931 standard observer. (The colour rendering index Ra is defined in Section 16.8.8.)*

Illuminant	X	Y	Z	x	y	$T_{cp}[K]$	R_a
E_E	100	100	100	1/3	1/3	5455	96
A	109.85	100.00	35.58	0.4476	0.4074	2856	100
B	99.09	100.00	85.31	0.3401	0.3562	4874	80
C	98.07	100.00	118.23	0.3101	0.3162	6774	99
D50	96.41	100.00	82.50	0.3457	0.3585	5003	100
D55	95.68	100.00	92.14	0.3324	0.3474	5503	100
D65	95.04	100.00	108.88	0.3127	0.3290	6504	100
D75	94.97	100.00	122.57	0.2991	0.3149	7504	100
iD50	95.28	100.00	82.53	0.3432	0.3602	5096	99
iD65	93.95	100.00	108.46	0.3107	0.3309	6596	99
F1	92.87	100.00	103.78	0.3131	0.3371	6425	76
F2	99.19	100.00	67.39	0.3721	0.3751	4225	64
F3	103.80	100.00	49.93	0.4091	0.3941	3447	57
F4	109.20	100.00	38.88	0.4402	0.4031	2943	51
F5	90.90	100.00	98.82	0.3138	0.3452	6301	72
F6	97.34	100.00	60.26	0.3779	0.3882	4149	59
F7	95.04	100.00	108.75	0.3129	0.3292	6498	90
F8	96.43	100.00	82.42	0.3458	0.3586	4998	95
F9	100.38	100.00	67.94	0.3741	0.3727	4149	90
F10	96.38	100.00	82.35	0.3458	0.3588	5000	81
F11	100.95	100.00	64.35	0.3805	0.3769	3999	83
F12	108.12	100.00	39.27	0.4370	0.4042	3000	83
FL 3.1	109.27	100.00	38.69	0.4407	0.4033	2932	51
FL 3.2	101.99	100.00	65.85	0.3808	0.3734	3966	71
FL 3.3	91.69	100.00	99.13	0.3153	0.3439	6280	72
FL 3.4	109.54	100.00	37.78	0.4429	0.4043	2904	87
FL 3.5	102.11	100.00	70.25	0.3749	0.3672	4086	97
FL 3.6	96.89	100.00	80.89	0.3488	0.3600	4894	97
FL 3.7	108.38	100.00	38.82	0.4384	0.4045	2979	82
FL 3.8	99.69	100.00	61.29	0.3820	0.3832	4006	79
FL 3.9	97.43	100.00	81.05	0.3499	0.3591	4853	79
FL 3.10	97.07	100.00	83.87	0.3455	0.3560	5000	88
FL 3.11	94.51	100.00	96.72	0.3245	0.3434	5854	78
FL 3.12	108.43	100.00	39.30	0.4377	0.4037	2984	93
FL 3.13	102.85	100.00	65.65	0.3831	0.3724	3896	96
FL 3.14	95.51	100.00	81.55	0.3447	0.3609	5045	95
FL 3.15	95.11	100.00	109.07	0.3127	0.3288	6509	98

Illuminant	X	Y	Z	x	y	T_{cp}[K]	R_a
HP 1_NaHP	128.45	100.00	12.54	0.5330	0.4150	1959	8
HP 2_NaHPe	114.90	100.00	25.58	0.4778	0.4158	2506	83
MB (G3)	86.97	100.00	75.95	0.3308	0.3803	5592	15
MBF (G4)	104.94	100.00	57.00	0.4006	0.3818	3538	48
MBTF (G5)	103.16	100.00	56.98	0.3965	0.3844	3652	49
HMI (G6)	103.75	100.00	116.80	0.3237	0.3120	5988	88
Xenon (G7)	99.78	100.00	110.21	0.3219	0.3226	6044	94
Na LP (G1)	135.24	100.00	0.24	0.5743	0.4246	1726	–
Na HP (G2)	128.74	100.00	14.59	0.5291	0.4110	1967	28

Table 8.3 *Illuminant colorimetric specification related to the CIE 1964 standard observer. (The colour rendering index Ra is defined in Section 16.8.8.)*

Illuminant	X_{10}	Y_{10}	Z_{10}	x_{10}	y_{10}
E_E	100	100	100	1/3	1/3
A	111.14	100.00	35.20	0.4512	0.4059
B	99.19	100.00	84.35	0.3498	0.3527
C	97.29	100.00	116.15	0.3104	0.3190
D50	96.71	100.00	81.41	0.3477	0.3596
D55	95.80	100.00	90.93	0.3341	0.3488
D65	94.81	100.00	107.33	0.3138	0.3310
D75	94.42	100.00	120.60	0.2997	0.3174
iD50	95.68	100.00	81.16	0.3456	0.3612
iD65	93.81	100.00	106-80	0.3121	0.3327
F1	94.82	100.00	103.20	0.3101	0.3355
F2	103.28	100.00	69.03	0.3793	0.3672
F3	109.01	100.00	52.00	0.4176	0.3831
F4	115.01	100.00	41.00	0.4492	0.3906
F5	93.39	100.00	98.70	0.3197	0.3424
F6	102.18	100.00	62.11	0.3866	0.3784
F7	95.79	100.00	107.69	0.3156	0.3295
F8	97.12	100.00	81.19	0.3490	0.3593
F9	102.13	100.00	67.87	0.3783	0.3704
F10	98.96	100.00	83.29	0.3506	0.3543
F11	103.86	100.00	65.61	0.3854	0.3711
F12	111.48	100.00	40.37	0.4427	0.3971
FL 3.1	115.27	100.00	40.99	0.4498	0.3902
FL 3.2	105.79	100.00	67.62	0.3869	0.3658

(continued)

Table 8.3   *(Continued)*

Illuminant	X_{10}	Y_{10}	Z_{10}	x_{10}	y_{10}
FL 3.3	94.32	100.00	99.36	0.3212	0.3405
FL 3.4	112.85	100.00	38.99	0.4481	0.3971
FL 3.5	103.05	100.00	67.71	0.3778	0.3666
FL 3.6	97.47	100.00	79.46	0.3520	0.3611
FL 3.7	111.98	100.00	40.05	0.4443	0.3968
FL 3.8	103.00	100.00	62.75	0.3876	0.3763
FL 3.9	100.35	100.00	82.58	0.3547	0.3534
FL 3.10	98.41	100.00	83.29	0.3493	0.3550
FL 3.11	97.17	100.00	97.94	0.3293	0.3389
FL 3.12	110.23	100.00	39.01	0.4422	0.4012
FL 3.13	103.20	100.00	63.98	0.3863	0.3743
FL 3.14	94.67	100.00	77.96	0.3473	0.3668
FL 3.15	94.37	100.00	105.59	0.3146	0.3334
HP 1_NaHP	134.06	100.00	12.68	0.5433	0.4053
HP 2_NaHPe	117.49	100.00	25.93	0.4826	0.4108
MB (G3)	96.95	100.00	84.63	0.3443	0.3551
MBF (G4)	112.44	100.00	63.12	0.4080	0.3629
MBTF (G5)	110.44	100.00	62.78	0.4042	0.3660
HMI (G6)	105.45	100.00	120.71	0.3233	0.3066
Xenon (G7)	98.97	100.00	108.74	0.3216	0.3250
Na LP (G1)	139.39	100.00	0.09	0.5890	0.4106
Na HP (G2)	131.11	100.00	14.65	0.5335	0.4069

8.11   Gas-Discharge Lamps

Gaseous discharge lamp – lamp in which the light is produced, directly or indirectly, by an electric discharge through a gas, a metal vapour or a mixture of several gases and vapours.

   NOTE – According to whether the light is mainly produced in a gas or in a metal vapour, one distinguishes between

1.   *gaseous discharge lamps*, for example, xenon, neon, helium, nitrogen, carbon dioxide lamps and
2.   *metal vapour lamps*, such as mercury vapour and sodium vapour lamps.[1]

The discharge lamps considered by the CIE[2] are high pressure and are subdivided into three sets:

HP1: *Standard high pressure sodium lamp.*
HP2: *Colour enhanced high pressure sodium lamp.*
HP3-5: *Three types of high pressure metal halide lamps.*

High-pressure lamps have a discharge in gas at a pressure slightly less to greater than atmospheric pressure.

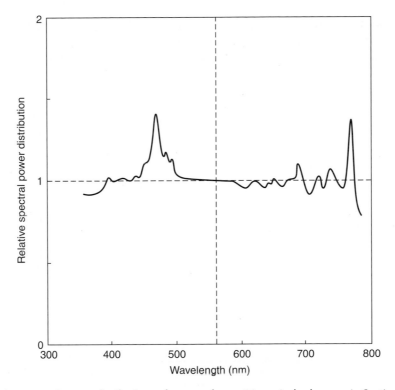

Figure 8.10   *Relative spectral power distributions of a xenon lamp. (Numerical values are in Section 16.1.2 data files.)*

Today, the metal halide lamp and the high pressure sodium lamp replace in most applications the older *high-pressure mercury-vapour lamps*, denoted by MB, MBF, MBTF, HMI. Anyway these old kinds of lamps are considered in the software attached to this book.

A *xenon-arc lamp* is a lamp with discharge in ionized xenon gas at high pressure. Xenon arc lamps are very important in colorimetry because they are used in movie projectors, in theatres and, as *Xenon-flash lamps*, in spectrophotometric instrumentation and in photographic shooting. Their most important characteristic is the spectral power distribution that is almost constant in all the visible range and is an approximate imitation of sunlight (Figure 8.10).

8.12   Light-Emitting Diodes

Electroluminescence is the form of light emission by inorganic semiconductors, termed *light-emitting diode* (LED), and by organic semiconductors, termed OLED.[13] Here only LEDs are concisely considered. The light colour emitted by LEDs has band spectra dependent on the composition of the semi-conductors used and belonging to a range from infrared, visible, to the near ultraviolet region. Here our interest is in LEDs for illumination, which generate high-intensity white light and are termed WLEDs. There are two primary ways of producing WLEDs:

1. *Multi-colour WLEDs* – individual LEDs, that emit three or more colours – red, amber, green, blue – are combined to mix all their coloured lights to form white light. This method has many uses because of the flexibility of mixing different colours and, in principle, also has higher quantum efficiency in producing white light.
    * *Dichromatic* WLEDs have the best luminous efficacy (120 lm/W), but the lowest colour rendering capability (Section 16.8.8).

- *Trichromatic* WLEDs are in between, having both good luminous efficacy (>70 lm/W) and fair colour rendering capability (e.g. RGB WLED) (Figure 8.11a).
- *Tetrachromatic* WLEDs have excellent colour rendering capability (Figure 8.11b).

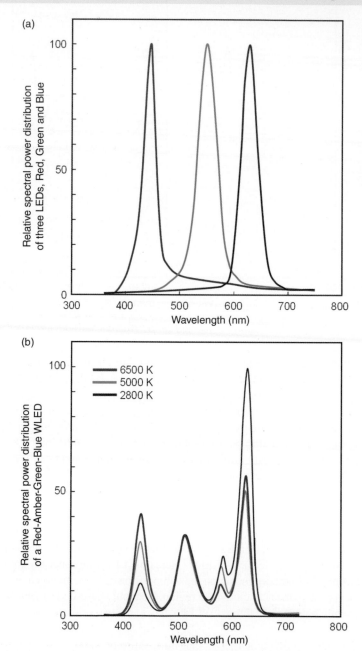

Figure 8.11   *(a) Relative spectral power distributions of a white LED composed of three LEDs, one red, one green and one blue. (b) Relative spectral power distributions of a white LED composed of four LEDs, one red, one green, one amber and one blue. The correlated colour temperature can be set by modulating the intensities of the four LEDs.*

2. *Phosphor-based LEDs* – a suitable phosphor material is used to convert a fraction of the almost monochromatic light emitted from a blue or UV LED to broad-spectrum white light (this fluorescence is produced in the same way as in the fluorescent light sources). If several phosphor layers of distinct colours are applied, the emitted spectrum is broadened [Figure 8.12a and b), and the

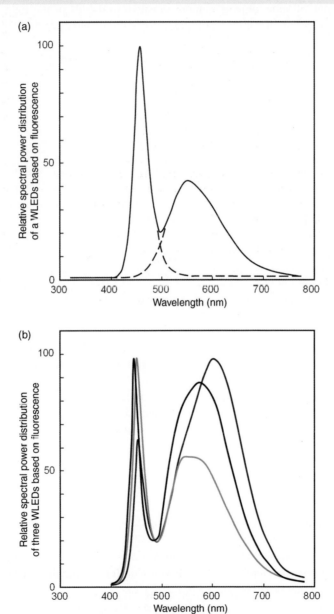

Figure 8.12   (a) Relative spectral power distributions of a white LED constituted by a blue LED and a phosphor, that partially absorbs the blue light and emits by fluorescence a broad band light over 500 nm. (b) Relative spectral power distributions of a white LED constituted by a blue LED and a different phosphor that partially absorbs the blue light and emits by fluorescence a broad band light over 500 nm. The correlated colour temperature depends on the phosphors used.

colour rendering increases. These phosphor-based WLEDs have efficiency losses due to the heat loss from the Stokes shift. The manufacturing of these WLEDs is quite simple.

Because of metamerism, it is possible to have WLEDs with different spectra that appear white. It follows that the colour appearance of some objects is more sensitive to details of the light spectrum illuminating them. It renders the classification of colour rendering of these light sources difficult and the old rules have to be updated.

The technology of WLEDs is in progress and other methods for producing white LEDs are investigated. At the moment the market offers mainly phosphor-based LEDs, but RGB WLEDs are used where a higher luminous efficacy is required.

References

1. CIE Publication S 017/E:2011. *ILV: International Lighting Vocabulary*. Commission Internationale de l'Éclairage, Vienna (2011). Available at: www.eilv.cie.co.at/ (accessed on 3 June 2015).
2. CIE Publication 15.3:2004, *Colorimetry* 3rd ed., Commission Internationale de l'Éclairage, Vienna (2004). Available at: www.cie.co.at (accessed on 3 June 2015).
3. CIE Publication 15.2 (1986), *Colorimetry* 2nd ed. Commission Internationale de l'Éclairage, Vienna (1986). Available at: www.cie.co.at (accessed on 3 June 2015).
4. McCamy CS, Correlated color temperature as an explicit function of chromaticity coordinates, *Color Res. Appl.*, **17**, 142–144 (1992).
5. Judd DB, MacAdam DL and Wyszecki G, (with the collaboration of Budde HW, Condit HR, Henderson ST, Simonds JL), Spectral distribution of typical daylight as function of correlated color temperature, *J. Opt. Soc. Amer.*, **54**, 1031–1040 (1964).
6. CIE Publication 14, Recommendations on standard illuminants for colorimetry, in *Proceedings of the CIE Washington Session*, Vol. A, Commission Internationale de l'Éclairage, Vienna, pp. 95–97 (1967). Available at: www.cie.co.at (accessed on 3 June 2015).
7. CIE publication 184:2009, *Indoor Daylight Illuminants*, Commission Internationale de l'Éclairage, Vienna (2009). Available at: www.cie.co.at (accessed on 3 June 2015).
8. CIE Standard S 014-2E, *Colorimetry—Part 2: CIE Standard Illuminants*, Commission Internationale de l'Éclairage, Vienna (2006). Available at: www.cie.co.at (accessed on 3 June 2015).
9. *CIE Proceedings 1931*, Cambridge University Press, Cambridge (1932).
10. *CIE proceedings*, Vol. 1, Sec. 4; Vol. 3, Bureau Central de la CIE, Paris (1951).
11. Judd DB, The 1931 ICI standard observer and the coordinate system for colorimetry, *J. Opt. Soc. Amer.*, **23**, 359–374 (1933).
12. CIE Technical Report, *A Method for Assessing the Quality of Daylight Simulators for Colorimetry* (with supplement 1-1999), CIE 51.2–1999 (1999).
13. CIE Standard S 012/E, *Standard Method of Assessing the Spectral Quality of Daylight Simulators for Visual Appraisal and Measurement of Color*, Commission Internationale de l'Éclairage, Vienna (2004). Available at: www.cie.co.at (accessed on 3 June 2015).

9

CIE Standard Psychophysical Observers and Systems

9.1 Introduction

Colour-vision psychophysics has shown that the operation of 'colour matching' is the easiest to characterize the human observer only intended as an image-colour sensor. Colour matching asks the viewer for a simple judgment of equality in the comparison between spectrally different colour stimuli. From the radiometric measurement of stimuli matching in colour are built the spectral sensitivities of the image-colour sensor, called colour-matching functions. This characterization of the human visual system defines the activation, the action that initiates the transduction, that is, the conversion of the colour stimulus into a visual nerve signal, and does not affect the subsequent processing that take place in the visual process within the visual system of the human being. Psychophysical colorimetry considers only the phenomenon of activation of the visual system and finds complete and accurate confirmation in the physiology of the visual system. The CIE has for obvious practical uses standardized psychophysical colorimetry, of which there are multiple versions, dependent on the non-uniformity of the retina, and multiple versions related to a refinement of the knowledge of colour-matching functions over time. These versions are:

- *CIE 1931 standard colorimetric observer and system (X, Y, Z) for 2° visual field[1]*
- *CIE 1964 standard colorimetric observer and system (X_{10}, Y_{10}, Z_{10}) for 10° visual field[1]*
- *CIE standard deviate observer[2]*
- *CIE standard 2° physiologically relevant fundamentals[3,4]*
- *CIE standard 10° physiologically relevant fundamentals[3,4]*

(Numerical values are in Section 16.1.2 data files.) This chapter presents all of these versions of the CIE psychophysical colorimetry and the Vos colorimetric observer,[5] which was produced as a correction of the observer CIE 1931 and has become de facto the standard in physiological studies of colour vision.

Standard Colorimetry: Definitions, Algorithms and Software, First Edition. Claudio Oleari.
© 2016 John Wiley & Sons, Ltd. Published 2016 by John Wiley & Sons, Ltd.

9.2   CIE 1931 Standard Colorimetric System and Observer

The CIE makes a distinction between colorimetric system and colorimetric observer:

"*CIE 1931 standard colorimetric observer* — ideal observer whose colour-matching properties correspond to the CIE colour-matching functions $\bar{x}(\lambda)$, $\bar{y}(\lambda)$ and $\bar{z}(\lambda)$ adopted by the CIE in 1931".[1]

"*CIE 1931 standard colorimetric system* (X, Y, Z) — system for determining the tristimulus values of any spectral power distribution using the set of reference colour stimuli $\hat{\mathbf{X}}, \hat{\mathbf{Y}}, \hat{\mathbf{Z}}$ and the three CIE colour-matching functions $\bar{x}(\lambda)$, $\bar{y}(\lambda)$ and $\bar{z}(\lambda)$ adopted by the CIE in 1931.

NOTE 1. $\bar{y}(\lambda)$ is identical to $V(\lambda)$ and hence the tristimulus values Y are proportional to values of luminance.

NOTE 2. This standard colorimetric system is applicable to centrally viewed fields of angular subtense between about 1° and about 4° (0.017 rad and 0.07 rad).

NOTE 3. The CIE 1931 standard colorimetric system can be derived from the CIE 1931 RGB colorimetric system using a transformation based on a set of three linear equations. The CIE 1931 RGB system is based on three real monochromatic reference stimuli."[6]

The CIE wrote[1]:

"The colour-matching functions $\bar{x}(\lambda)$, $\bar{y}(\lambda)$ and $\bar{z}(\lambda)$ are the tristimulus values of monochromatic stimuli of equal radiant power related to a set of reference stimuli $\hat{\mathbf{X}}, \hat{\mathbf{Y}}$ and $\hat{\mathbf{Z}}$, whose level is determined by convention, thus it is more appropriate to use capital letters to denote the colour-matching functions: $X(\lambda)$, $Y(\lambda)$, $Z(\lambda)$. For an interim period both forms of writing the symbols of colour-matching functions are permitted, but it is recommended that the new, capital letter version be used in all new publications. In the case of the CIE 1931 standard colorimetric observer no subscript is used, in the case of the CIE 1964 standard colorimetric observer the subscript '10' is used."

In this book, the old and highly consolidated notation is used.

The colorimetric system is based on the 'reference frame' of the tristimulus space while the observer is based on the 'colour-matching functions'. Anyway, since the definition of the colour-matching functions without the definition of a reference frame is impossible, here system and observer are presented together. The psychophysical basis of the colorimetric system/observer is given in section 6.13.

A psychophysical photopic observer is defined by the colour-matching functions. Individual differences in crystalline-lens-pigment density, macular pigment density, photopigment optical density and in the cone photopigments themselves (genetic reason) influence colour matches. The first stage of the visual processing is based on colour-matching functions, which are dependent on these individual differences. Colorimetric practice requires a definition of a standard psychophysical photopic observer, whose colour-matching functions are good representatives of the colour-matching functions of normal trichromatic individual observers. This has been achieved by an average of a population of individual colour-matching functions measured in the laboratories of Wright[7] (1928–1929) and Guild[8] (1931). The definition of the average colour-matching functions was not easy because the individual colour-matching functions were different. Nevertheless the chromaticity coordinates associated with these colour-matching functions were very similar, so it was meaningful to consider an averaged chromaticity diagram. The fact that the individual colour-matching functions were different while the individual chromaticity diagrams were similar to a good approximation emphasized that the differences between individuals were predominantly in the optical density of the macula lutea, which had a variable pigmentation between individuals. It remained to build the colour-matching functions from the chromaticity diagram. To do this, the luminous efficiency function $V(\lambda)$ (Sections 6.5 and 7.3) of the standard CIE 1924 photopic photometric observer and the

Exner-Schrödinger 'Hellighkeit' equation (Section 6.13.6) were used. The path followed in defining the colour-matching functions CIE 1931 is summarised as follows:

(i) Representation of the empirical individual colour matching function normalised on the WDW chromaticity diagram.
(ii) Rescaling of the colour-matching functions with the introduction of the CIE 1924 luminous efficiency function $V(\lambda)$.
(iii) Empirical definition of the luminances of the three reference stimuli to match the equal energy stimulus.

9.2.1 CIE 1931 RGB Reference Frame and WDW Chromaticity-Coordinates Normalization

The procedure described by G. Wyszecki and W Stiles on pages 133–140 of reference[9] is here reinterpreted as follows. The individual colour-matching functions are assumed to be the product of two functions:

$$\left(r(\lambda) = t_i(\lambda)r_0(\lambda),\ g(\lambda) = t_i(\lambda)g_0(\lambda),\ b(\lambda) = t_i(\lambda)b_0(\lambda)\right) \tag{9.1}$$

where the function $t_i(\lambda)$ is the individual transmittance of the ocular media – macula lutea and crystalline lens –, which is very dependent on the individual observers. The function $t_i(\lambda)$ takes into account also systematic radiometric errors made in the different laboratories. This hypothesis states that these individual observers have equal chromaticity coordinates and the same chromaticity diagram. These colour-matching functions are given in an RGB reference frame, whose reference stimuli are monochromatic of wavelengths $\lambda_b = 460$ nm, $\lambda_g = 530$ nm and $\lambda_r = 650$ nm, and are normalized according to the following rule, known as *WDW normalization* rule,

$$r*(\lambda) = \frac{t_i(\lambda)r_0(\lambda)}{W_Y},\ g*(\lambda) = t_i(\lambda)g_0(\lambda),\ b*(\lambda) = \frac{t_i(\lambda)b_0(\lambda)}{W_{BG}} \tag{9.2}$$

with

$$W_Y = \frac{r_0(\lambda_Y)}{g_0(\lambda_Y)},\ W_{BG} = \frac{b_0(\lambda_{BG})}{g_0(\lambda_{BG})} \tag{9.3}$$

In this way, the three colour-matching functions are scaled assuming equal values at two wavelengths $\lambda_Y = 582.5$ nm and $\lambda_{BG} = 494$ nm. The corresponding chromaticity coordinates are independent of individual absorption of the ocular media (and of systematic radiometric errors)

$$\bar{r}*(\lambda) = \frac{r*(\lambda)}{r*(\lambda) + g*(\lambda) + b*(\lambda)} = \frac{r_0(\lambda)}{W_Y\left[\dfrac{r_0(\lambda)}{W_Y} + g_0(\lambda) + \dfrac{b_0(\lambda)}{W_{BG}}\right]}$$

$$\bar{g}*(\lambda) = \frac{g*(\lambda)}{r*(\lambda) + g*(\lambda) + b*(\lambda)} = \frac{g_0(\lambda)}{\left[\dfrac{r_0(\lambda)}{W_Y} + g_0(\lambda) + \dfrac{b_0(\lambda)}{W_{BG}}\right]} \tag{9.4}$$

$$\bar{b}*(\lambda) = \frac{b*(\lambda)}{r*(\lambda) + g*(\lambda) + b*(\lambda)} = \frac{b_0(\lambda)}{W_{BG}\left[\dfrac{r_0(\lambda)}{W_Y} + g_0(\lambda) + \dfrac{b_0(\lambda)}{W_{BG}}\right]}$$

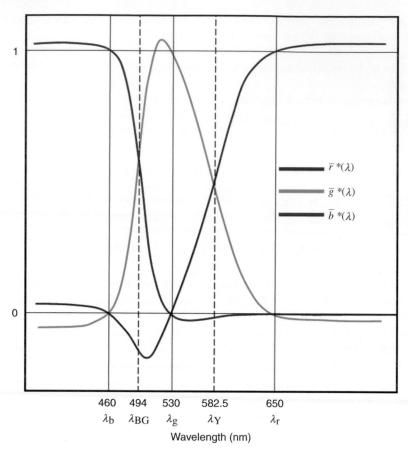

Figure 9.1   *Mean chromaticity coordinates plotted as functions of wavelength with WDW normalization.*

and the resulting averaged WDW normalised chromaticity coordinates (Figure 9.1) are assumed representative for all the normal individual observers (the bar means average among individual observers).

   Once the WDW chromaticity coordinates are obtained, the colour-matching functions have to be built. This is done with Schrödinger-Exner's 'Hellighkeit' equation (Section 6.13.6), in which the function $V(\lambda)$ contains an averaged effect of the ocular media absorption and is supposed a good representative of the average photometric observer (Sections 7.3 and 7.4)

$$V(\lambda) = L_R \overline{r}(\lambda) + L_G \overline{g}(\lambda) + L_B \overline{b}(\lambda) =$$
$$\equiv s(\lambda) \left[L_R \overline{r} * (\lambda) + L_G \overline{g} * (\lambda) + L_B \overline{b} * (\lambda) \right]$$

(9.5)

where $\left(\overline{r}(\lambda), \overline{g}(\lambda), \overline{b}(\lambda) \right)$ are the searched colour-matching function and

$$s(\lambda) = \overline{r}(\lambda) + \overline{g}(\lambda) + \overline{b}(\lambda)$$

(9.6)

The colour-matching functions are obtained by a substitution of $s(\lambda)$, defined by Equation (9.5), in the WDW normalization

$$\bar{r}(\lambda) = \bar{r}*(\lambda)s(\lambda) = \frac{\bar{r}*(\lambda)V(\lambda)}{L_R\bar{r}*(\lambda) + L_G\bar{g}*(\lambda) + L_B\bar{b}*(\lambda)}$$

$$\bar{g}(\lambda) = \bar{g}*(\lambda)s(\lambda) = \frac{\bar{g}*(\lambda)V(\lambda)}{L_R\bar{r}*(\lambda) + L_G\bar{g}*(\lambda) + L_B\bar{b}*(\lambda)} \qquad (9.7)$$

$$\bar{b}(\lambda) = \bar{b}*(\lambda)s(\lambda) = \frac{\bar{b}*(\lambda)V(\lambda)}{L_R\bar{r}*(\lambda) + L_G\bar{g}*(\lambda) + L_B\bar{b}*(\lambda)}$$

In this way photometry enters colorimetry and the definitive colour-matching functions are bound to the relative luminous efficiency function $V(\lambda)$.

It remains to define the Exner coefficient L_R, L_G and L_B. These coefficients have been defined by CIE in a few ways (page 136 of reference[9]). The reference stimuli were specified as monochromatic stimuli of wavelengths $\lambda_R = 700.0$ nm, $\lambda_G = 546.1$ nm and $\lambda_B = 435.8$ nm, respectively. Their units were chosen to make a mixture of equal quantities of the three reference stimuli match the equal-energy spectrum. The luminances of the units of the three spectral stimuli were in the ratios 1.0000: 4.5907: 0.0601. This was the final choice.

The reference colour stimuli used in the laboratories, in the definition of the averaged chromaticities and the final standardization were different. The CIE 1931 standard colorimetric observer have been defined in the RGB reference frame with reference stimuli at λ_R, λ_G and λ_B. (Figure 6.26).

The tristimulus values of a radiance $L_{e,\lambda}$ in the standard RGB reference frame are defined as

$$R = \sum_{\lambda=360,\Delta\lambda=1}^{830} L_{e,\lambda}\bar{r}(\lambda)\Delta\lambda, \ G = \sum_{\lambda=360,\Delta\lambda=1}^{830} L_{e,\lambda}\bar{g}(\lambda)\Delta\lambda, \ B = \sum_{\lambda=360,\Delta\lambda=1}^{830} L_{e,\lambda}\bar{b}(\lambda)\Delta\lambda \qquad (9.8)$$

where $\bar{r}(\lambda)$, $\bar{g}(\lambda)$ and $\bar{b}(\lambda)$ are the standard colour-matching functions in this reference frame and the summation is made with a step $\Delta\lambda = 1$ nm (Figure 6.27). The integrals (6.37) are substituted by summations because the colour-matching functions are defined by a table of numbers. For a proper computation, the summation has to be made with a step equal to 1 nm because in a wavelength interval $\Delta\lambda = 1$ the human visual system cannot discriminate between different hues (Section 6.10).

In this RGB reference frame of tristimulus space, the vectors $\mathbf{Q} = (R, G, B)$ are still called *tristimulus vectors* and their three components, R, G and B, are called the *tristimulus values*.

The colour-matching functions are partly positive and partly negative and their units are chosen so that the components of the equal energy stimulus $E_{E,\lambda} = 1$ are equal to each other, namely

$$\sum_{\lambda=360,\Delta\lambda=1}^{830} \bar{r}(\lambda)\Delta\lambda = \sum_{\lambda=360,\Delta\lambda=1}^{830} \bar{g}(\lambda)\Delta\lambda = \sum_{\lambda=360,\Delta\lambda=1}^{830} \bar{b}(\lambda)\Delta\lambda \qquad (9.9)$$

The tristimulus vectors associated with the monochromatic stimuli at wavelength λ and unit power are

$$\bar{\mathbf{U}}_{RGB}(\lambda) = \left(\bar{r}(\lambda), \bar{g}(\lambda), \bar{b}(\lambda)\right) \qquad (9.10)$$

(monochromatic means narrow band centered in the wavelength λ and a bandwidth within which the cone sensitivity functions are constant.)

Also in this reference frame, it is possible to define the chromaticity (r, g, b) of the tristimulus vector $\mathbf{Q} = (R, G, B)$ as the intersection point between the tristimulus vectors and the plane $R + G + B = 1$ (Figures 6.26 and 6.28)

$$r = \frac{R}{R+G+B}, \; g = \frac{G}{R+G+B}, \; b = \frac{B}{R+G+B}, \; r + g + b = 1 \tag{9.11}$$

As mentioned, the three chromaticity coordinates are mutually dependent, because they are bound by the equation $r + g + b = 1$, then only two of them are sufficient to define the chromaticity. It is customary to consider the (r, g) and these, if represented as two orthogonal Cartesian coordinates, are coordinates of a diagram obtained by a projection of the chromaticity diagram (r, g, b) on the plane $B = 0$ from infinity (Figure 6.28). This representation of the chromaticity diagram is much more frequent than that on the plane $R + G + B = 1$ because it is easier to use.

The CIE 1924 photopic photometric observer and 1931 CIE colorimetric observer are constrained by Exner-Schrödinger's Equation (6.43), which implies that the errors/uncertaninties in the definition of first observer are transferred onto the second one. Even today, this point is discussed.

After the definition of the CIE 1931 standard colorimetric observer in the reference frame RGB, the passage to the reference frame XYZ has to be defined.

9.2.2   CIE 1931 XYZ Reference Frame

Luminance has great practical importance, moreover, luminance is expressible as a linear combination of the cone activations (6.41). This leads to implement a linear transformation in tristimulus space so that one of the components of the tristimulus vector precisely represents the luminance of the colour stimulus. This is what was implemented by D.B. Judd and accepted by the CIE in 1931 with the reference frame XYZ, in which the Y component is proportional to the luminance associated with the colour stimulus[9] (Figure 6.29). The transformation from RGB coordinates to XYZ is straightforward and is accomplished by requiring that:

- The X, Y and Z are a set of three orthogonal axes.
- The Y component of the tristimulus vector is proportional to the luminance L_v (Section 6.13.6): that means that the corresponding component of the colour-matching functions is equal to the relative spectral luminous efficiency function $V(\lambda)$ and that the Exner coefficients are all equal to zero except for L_y component, which is equal to 1, that is,

$$L_v(\lambda) = K_m Y \quad \text{cd/m}^2$$
$$\overline{y}(\lambda) = V(\lambda) \tag{9.12}$$
$$\tilde{\Lambda} = \left(L_x = 0 \; L_y = 1 \; L_z = 0 \right)$$

- The X and Z axes lie on the zero-luminance plane $Y = 0$.
- The planes $X = 0$ and $Z = 0$ are tangential to the spectrum locus at the short-medium and long wavelengths, respectively.
- The *standard colour-matching function* $\overline{x}(\lambda)$, $\overline{y}(\lambda)$ and $\overline{z}(\lambda)$ in the space *XYZ* are defined between 360 nm and 830 nm and with 1 nm steps and represent the monochromatic stimuli with unit power (Figure 9.2).

$$\overline{\mathbf{U}}_{\text{XYZ}}(\lambda) = \left(\overline{x}(\lambda), \overline{y}(\lambda), \overline{z}(\lambda) \right) \tag{9.13}$$

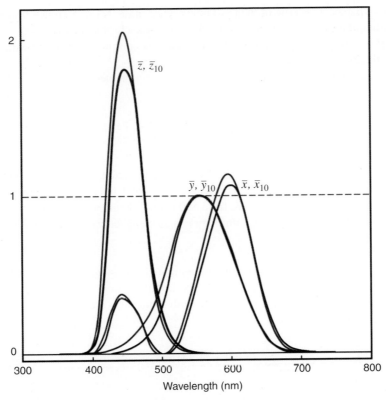

Figure 9.2 *Comparison of the CIE 1931 (black line) and CIE 1964 (red line) colour matching functions in the XYZ reference frame. (Numerical values are in Section 16.1.2 data files.)*

(monochromatic means narrow band centered in the wavelength λ and a bandwidth inside which the cone sensitivity functions are constant).

- The tristimulus vector $\mathbf{Q} = (X, Y, Z)$ in the standard XYZ reference frame produced by a radiance $L_{e,\lambda}$ is defined as

$$X = \sum_{\substack{\lambda=360 \\ \Delta\lambda=1}}^{830} L_{e,\lambda} \bar{x}(\lambda)\Delta\lambda$$

$$Y = \sum_{\substack{\lambda=380 \\ \Delta\lambda=1}}^{780} L_{e,\lambda} \bar{y}(\lambda)\Delta\lambda = \sum_{\substack{\lambda=380 \\ \Delta\lambda=1}}^{780} L_{e,\lambda} V(\lambda)\Delta\lambda = \frac{L_{v}}{K_{m}} \qquad (9.14)$$

$$Z = \sum_{\substack{\lambda=360 \\ \Delta\lambda=1}}^{830} L_{e,\lambda} \bar{z}(\lambda)\Delta\lambda$$

and the summation is made with a step $\Delta\lambda = 1$ nm. L_{v} is the luminance measured in cd/m^2 and $K_{m} = 683$ lm/W.

Figure 6.29 gives a perspective view of the tristimulus space in the XYZ reference frame, the spectral tristimulus vectors with unitary power and the unitary plane with the chromaticity diagram.

Figure 6.30 (a) shows the plane $L + M + S = 1$ in the fundamental reference frame with the chromaticities of the RGB and of the XYZ reference stimuli.

In the XYZ reference frame, the chromaticity coordinates are obtained as usual

$$x = \frac{X}{X+Y+Z}, \; y = \frac{Y}{X+Y+Z}, \; z = \frac{Z}{X+Y+Z}, \; x+y+z = 1 \tag{9.15}$$

(These numbers in the standard data file of the CD-ROM associated to the publication CIE 15 *Colorimetry* range from wavelength $\lambda = 360$ nm to 830 nm with a step of 1 nm and the numbers are rounded to 12 decimals, while in the publication the tables are from 380 nm to 780 nm with a step of 5 nm and numbers are rounded to 6 decimals. The values given at 5-nm intervals are selected values from the standard, rounded to six decimal places)

- The tristimulus values of the *equi-energy stimulus* computed with standard colour-matching functions are equal to each other, that is, $X_E = Y_E = Z_E$; particularly the summations of the colour-matching function of the printed table between 380 nm and 780 nm and with 5-nm steps are[1]

$$\sum_{380, \Delta\lambda=5}^{780} \overline{x}(\lambda) = 21.371524,$$

$$\sum_{380, \Delta\lambda=5}^{780} \overline{y}(\lambda) = 21.371327, \tag{9.16}$$

$$\sum_{380, \Delta\lambda=5}^{780} \overline{z}(\lambda) = 21.371540$$

The tristimulus values (X, Y, Z) are differently normalized depending on the context. In photometry, the Y value multiplied by K_m represents the luminance measured in cd/m^2. Tristimulus values of object colours are considered in Section 9.7.

Figures 6.30b and 6.30c give a view of the (x, y) chromaticity diagram with barycentric coordinates in the equilateral triangle and with Cartesian coordinates in the projection on the plane $Z = 0$. Figure 9.3 shows the chromaticity diagram (x, y) CIE 1931 with an approximate coloration in order to give an idea of the perceived colours associated to the different regions of the chromaticity diagram.

Today, the CIE 1931 standard colorimetric observer data deviate from the original data in several ways.

The latest edition of the 'CIE publication 15:2004, *Colorimetry*, 3rd ed.' [1] says:

"The colour-matching functions $\overline{x}(\lambda)$, $\overline{y}(\lambda)$ and $\overline{z}(\lambda)$ given in Table T.4 agree closely with those defined originally in 1931. Three minor changes have been introduced: at $\lambda = 775$ nm the new value of $\overline{x}(\lambda)$ is 0.000 059 instead of 0,0000; at $\lambda = 555$ nm $Y(\lambda)$ is 1.0000 instead of 1.0002; and at $\lambda = 740$ nm $\overline{y}(\lambda)$ is 0.000 249 instead of 0.0003. These changes are considered insignificant in most colorimetric computations. From these corrected tables, the CIE standard colorimetric observer (CIE, 1986) was determined."

The CIE 1986 tables published in 'CIE No. 15.2 (1986) Colorimetry 2nd ed.' [11] have values with 4 decimals from 380 nm to 780 nm at 5-nm intervals. The tables published in 'CIE 15 Colorimetry' [1] have values rounded to 6 decimals.

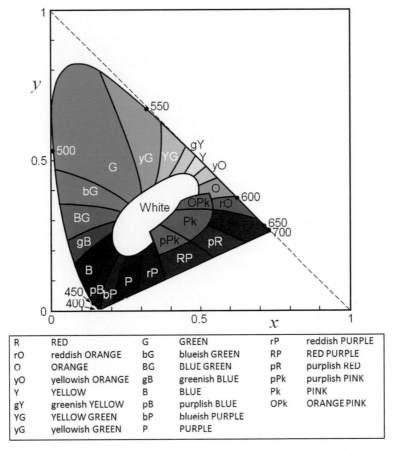

R	RED	G	GREEN	rP	reddish PURPLE
rO	reddish ORANGE	bG	blueish GREEN	RP	RED PURPLE
O	ORANGE	BG	BLUE GREEN	pR	purplish RED
yO	yellowish ORANGE	gB	greenish BLUE	pPk	purplish PINK
Y	YELLOW	B	BLUE	Pk	PINK
gY	greenish YELLOW	pB	purplish BLUE	OPk	ORANGE PINK
YG	YELLOW GREEN	bP	blueish PURPLE		
yG	yellowish GREEN	P	PURPLE		

Figure 9.3 *CIE 1931 (x, y) chromaticity diagram with a very approximate coloration for having an idea of the perceived colours associated with the various regions of the diagram.*[10]

A large portion of the original work for the definition of the CIE 1931 standard observer was performed by W. D. Wright. In a CIE committee meeting Wright was asked to give a talk on the origin of the 1931 CIE system. This talk is reported on page 541–542 of the book 'Human colour vision' by P. K. Kaiser-R. M. Boynton.[12] The last sentences are reproduced here: "I would make one final comment before I stop. The CIE Colorimetry Committee recently in their wisdom have been looking at the old 1931 observer and have been smoothing the data to obtain more consistent calculations with computers. This has also involved some extrapolation and, in smoothing, they have added some additional decimal places. When I look at the revised table of the $\overline{x}(\lambda)$, $\overline{y}(\lambda)$ and $\overline{z}(\lambda)$ functions, I am rather surprised to say the least. You see, I know how inaccurate the actual measurements really were. Guild did not take any observations below 400 nm and neither did I, and neither did Gibson and Tyndall on the $V(\lambda)$ curve, and yet at a wavelength of 362 nm, for example, we find a value $\overline{y}(\lambda)$ of 0.000004929604! This, in spite of the fact that at 400 nm the value of $\overline{y}(\lambda)$ may be in error by a factor of 10. I cannot help wondering what Mr Guild thinks if he happens to see these tables. I know we can put the blame on the computer but we must not abdicate our common sense altogether. I think on that note I had to better stop!"

It must be said that a number of figures strictly limited by the measurement uncertainty has the consequence that, when the colour-matching functions approach to zero at the ends of the visible spectrum, the spectrum locus on the chromaticity diagram, obtained by Equations (9.15), becomes a scribble. This can be avoided by adding extrapolated non-meaningful figures to the colour-matching functions, but in this way the spectrum locus represents the extrapolation rule and not the meaning of the empirical data.

The transformations between the reference frames RGB and XYZ are

$$
\begin{pmatrix} X \\ Y \\ Z \end{pmatrix} = \begin{pmatrix} 0.4900 & 0.3100 & 0.2000 \\ 0.1770 & 0.8124 & 0.0106 \\ 0 & 0.0100 & 0.9900 \end{pmatrix} \begin{pmatrix} R \\ G \\ B \end{pmatrix}
$$

$$
\begin{pmatrix} R \\ G \\ B \end{pmatrix} = \begin{pmatrix} 2.3646 & -0.8965 & -0.4681 \\ -0.5152 & 1.4264 & 0.0888 \\ 0.0052 & -0.0144 & 1.0092 \end{pmatrix} \begin{pmatrix} X \\ Y \\ Z \end{pmatrix}
$$

(9.17)

These matrices are computed by using the algorithms of Section 6.13.4 and the software of Section 16.8.6, with the constraint that the equal energy stimulus is specified as $(X = 1, Y = 1, Z = 1)$ and $(R = 1, G = 1, B = 1)$ and therefore are equal to those defined in the CIE 15.3 Colorimetry[1] up to the factor 0.177.

In practice often the colour stimuli are specified using the tristimulus value Y, proportional to the luminance, and the chromaticity (x, y), rather than via the tristimulus values, and written as $(Y; x, y)$. This choice is made because the chromaticity diagram allows a visual graphical assessment of the colour that is not immediately given by tristimulus values. Moreover the chromaticity diagrams have remarkable properties, which are presented in Section 6.13.8.

9.3   CIE 1964 (Supplementary) Standard Colorimetric Observer/System (10°-Standard Colorimetric Observer)

"CIE 1964 standard colorimetric system … is applicable to centrally viewed fields of angular subtense greater than about 4° (0.07 rad).

When this system is used, all symbols that represent colorimetric measures are distinguished by use of the subscript 10."[1]

The CIE 1931 standard observer was defined for foveal vision, that is restricted to a field of view subtending a solid angle with the maximum section less than 4° and generally approximately 2°. In this case, vision is accomplished with that part of the retina with high acuity, devoid of rods and filtered by inert pigments in the macula lutea. Outside of this region, the retinal tissue changes and colour vision changes as well, although, in normal visual situations of daily life, individual observers are not aware of it. In particular visual situations colour matching and colour discrimination are phenomena that show the differences between foveal vision (2°) and extramacular vision (10°). The inert pigments of the macula lutea appear to be responsible for most of these differences. These differences induced the CIE to define the CIE 1964 (*supplementary*) *standard colorimetric observer* for vision with an extra-foveal visual field of 10°. A 10° field of view has a diameter of about 9 cm at a viewing distance of 50 cm.

In 10° visual fields, there is also the presence of rods, dedicated to low-light vision, producing *rod intrusion* in the cone vision, which is present mainly at low light levels and manifests itself in the comparison between stimuli of different scotopic luminance. However, this phenomenon is negligible for the purposes of the definition of the CIE 1964 standard colorimetric observer, which was obtained taking care to be free of rod intrusion. For daylight, the possible intrusion of rod vision in colour matches is likely to diminish progressively above about 10 cd/m^2 and be completely absent at approximately 200 cd/m^2.[13]

Formally, this tristimulus space appears in shape as the CIE 1931.

The process to define this observer was similar to that for the 1931 CIE standard observer, although it was defined not starting from the chromaticity diagram, but directly from averaged colour matching data. The experimental basis consisted of the colour-matching functions of individual observers obtained by requiring that the colour matching hold true with a 10° visual field. The reference stimuli used for colour-matching were monochromatic, one red with a wavelength $\lambda_r = 645.2$ nm, a green with $\lambda_g = 526.3$ nm and a blue with $\lambda_b = 444.4$ nm. The CIE 1964 observer in the reference $R_{10}G_{10}B_{10}$ is defined by the colour-matching functions $\bar{r}_{10}(\lambda)$, $\bar{g}_{10}(\lambda)$ and $\bar{b}_{10}(\lambda)$, which were obtained by the weighted average of data from two laboratories, Stiles-Burch[13] and Speranskaya[14] and for multiple observers. The reference $X_{10}Y_{10}Z_{10}$ was derived from $R_{10}G_{10}B_{10}$ by a linear transformation defined with requirements analogous to those used for the CIE 1931 observer, that are:

- In the space $X_{10}\,Y_{10}\,Z_{10}$ the *standard colour-matching function* $\bar{x}_{10}(\lambda)$, $\bar{y}_{10}(\lambda)$ and $\bar{z}_{10}(\lambda)$, are defined between 360 and 830 nm and with 1 nm steps. Figure 9.2 shows the colour-matching functions CIE 1964 in comparison with the CIE 1931.
- In the space $X_{10}Y_{10}Z_{10}$ the colour-matching function $\bar{y}_{10}(\lambda)$ is the spectral luminous efficiency function $V_{10}(\lambda)$ for 10° visual field photopic vision, and consequently the Y_{10} is proportional to the 10° luminance, that is $L_{10,v} = K_{10,m}\,Y_{10}$ with $K_{10,m} = 683.6$ lm/W (Section 7.7). [Figures 7.1 and 7.2 show a comparison of the spectral relative photopic luminous efficiency functions in linear and logarithmic scale, including $V(\lambda) = \bar{y}(\lambda)$ and $V_{10}(\lambda) = \bar{y}_{10}(\lambda)$.]
- The tristimulus vector in the standard $X_{10}Y_{10}Z_{10}$ reference system produced by a radiance $L_{e,\lambda}$ is defined as

$$X_{10} = \sum_{\lambda=360,\Delta\lambda=1}^{830} L_{e,\lambda}\bar{x}_{10}(\lambda)\Delta\lambda$$

$$Y_{10} = \sum_{\lambda=360,\Delta\lambda=1}^{830} L_{e,\lambda}\bar{y}_{10}(\lambda)\Delta\lambda = \sum_{\lambda=360,\Delta\lambda=1}^{830} L_{e,\lambda}V_{10}(\lambda)\Delta\lambda = \frac{L_{10,v}}{K_{10,m}} \qquad (9.18)$$

$$Z_{10} = \sum_{\lambda=360,\Delta\lambda=1}^{830} L_{e,\lambda}\bar{z}_{10}(\lambda)\Delta\lambda$$

and the summation is made with a step $\Delta\lambda = 1$ nm. L_v is the luminance measured in cd/m^2 and $K_{m,10} = 683.6$ lm/W.
- the tristimulus values of the equal-energy stimulus are equal

$$\sum_{360,\Delta\lambda=1}^{830}\bar{x}_{10}(\lambda) = \sum_{360,\Delta\lambda=1}^{830}\bar{y}_{10}(\lambda) = \sum_{360,\Delta\lambda=1}^{830}\bar{z}_{10}(\lambda) \qquad (9.19)$$

for the standard colour-matching. For the tables between 380 nm and 780 nm and with 5-nm steps (printed file[1]) the summations are

$$\sum_{\lambda=380,\Delta\lambda=5}^{780} \overline{x}_{10}(\lambda) = 23.329353$$

$$\sum_{\lambda=380,\Delta\lambda=5}^{780} \overline{y}_{10}(\lambda) = 23.332036 \tag{9.20}$$

$$\sum_{\lambda=380,\Delta\lambda=5}^{780} \overline{z}_{10}(\lambda) = 23.334153$$

(The small differences are due to the limited number of decimal digits given in the printed table[1]).

It follows that the chromaticity diagrams of the two observers, (x_{10}, y_{10}) and (x, y) have very similar shape.

Figure 9.4 shows a comparison of the chromaticity diagrams of the two observers CIE 1931 and CIE 1964. The difference between the spectrum loci of the two observers is really small in shape, but the most significant difference is in the locations of the chromaticities of monochromatic stimuli that are almost everywhere mutually shifted, except for the wavelengths at the extremities of the visible spectrum. A set of differences exist between the standard observers CIE 1931 and CIE 1964, all seemingly small, but such that the two observers are substantially distinguishable.

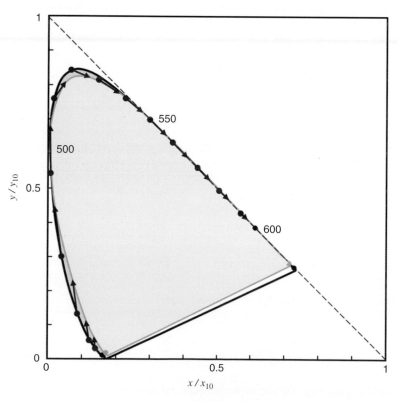

Figure 9.4 *Comparison of the CIE 1931 and CIE 1964 chromaticity diagrams with highlighting of the wavelength shift. The tip of the arrow refers to the observer CIE 1964 and its tail to CIE 1931.*

In daily practice, the CIE 1964 colorimetric standard observer is essential in all those situations where the colour covers large areas, while the standard observer CIE 1931 is used in cases where the colours regard visual fields subtended by small angles, as is the view of colour in prints, photographs and television monitors, although the luminance of the reference blue of the CRT monitors shows a discrepancy with respect to the perceived brightness.

9.4   CIE 1989 Standard Deviate Observer/System

"*CIE standard deviate observer* — standard observer whose colour matching functions deviate from those of the CIE standard colorimetric observer in a defined manner.

NOTE 1 — The defined deviations can be applied to the CIE 1931 standard colorimetric observer or the CIE 1964 standard colorimetric observer.

NOTE 2 — The use of the standard deviate observer is intended to generate differences that are typical of those that occur when colour matches are made by different real observers whose colour vision is classified as normal."[2,6]

Over time it appeared evident that the CIE standard observers were not representative of all the normal observers.[15] This is certainly a consequence of the variability between individual observers, that is especially great because it involves differences in pigmentation in the macula lutea and differences in the degree of yellowing of the crystalline lens dependent on age.

In 1989 the CIE developed the *standard deviate observer*[2] that should represent the variability limits among individual observers and standard observers. This deviate observer is based on an analysis of the Stiles and Burch data,[13] used to define the CIE 1964 standard observer.[1] These observers are reported here because of their role in the definition of the CIE special metamerism index for changes in observer (Section 11.4.7).

The standard deviate observer is defined by the following colour-matching functions

$$
\begin{aligned}
x_{\mathrm{d}}(\lambda) &= \overline{x}(\lambda) + \Delta\overline{x}(\lambda) \\
\overline{y}_{\mathrm{d}}(\lambda) &= \overline{y}(\lambda) + \Lambda\overline{y}(\lambda) \\
\overline{z}_{\mathrm{d}}(\lambda) &= \overline{z}(\lambda) + \Delta\overline{z}(\lambda)
\end{aligned}
\tag{9.21}
$$

where $\left(\overline{x}(\lambda), \overline{y}(\lambda), \overline{z}(\lambda)\right)$ are the standard colour-matching function CIE 31 or CIE 64 (in this case the subscript '10' is required), and $\left(\Delta\overline{x}(\lambda), \Delta\overline{y}(\lambda), \Delta\overline{z}(\lambda)\right)$ are the so-called *first deviation functions.*[2]

9.5   Vos' 1978 Modified Observer for 2° Visual Field

The colour-matching functions, the tristimulus values and chromaticity coordinates in the Vos' modified tristimulus space for foveal vision are identified by the prime index:

$$
\begin{aligned}
&\left(\overline{x}'(\lambda), \overline{y}'(\lambda), \overline{z}'(\lambda)\right),\ (X', Y', Z'),\ (x', y', z') \\
&\left(\overline{l}'(\lambda), \overline{m}'(\lambda), \overline{s}'(\lambda)\right),\ (L', M', S'),\ (l', m', s')
\end{aligned}
\tag{9.22}
$$

(In literature, the Vos quantities have their apex only in the XYZ reference frame. In this book the apex is used also for the quantities in the fundamental reference frame in order to avoid any confusion with the corresponding quantities of the CIE fundamental observer introduced in the next Section 9.6).

The definition of the CIE 1931 standard observer showed over time the presence of a systematic error in the short wavelengths. In 1951 D. B. Judd[16] proposed a revision of the 1931 colour-matching functions, which in 1978 was perfected by J. J. Vos.[5] The colour-matching functions corrected by Judd were given from 370 nm to 770 nm with 10-nm steps and given with four decimals, while those perfected by Vos were given from 380 nm to 825 nm with 5-nm steps and with the number of five significant figures. Figure 9.5 gives a comparison of the Vos colour matching functions and the CIE 1931 ones. The differences between the corresponding chromaticity diagrams are very small and regard the region of the short wavelengths. The modification of the colorimetric functions implies a difference considered significant only in colour-vision science but not for colorimetric practice, which continues to use the CIE 1931 colour-matching functions:

Particularly, the relative spectral luminous efficiency $\bar{y}'(\lambda)$ of the Vos observer, denoted by $V_M(\lambda)$, has become a CIE standard in 1988 (Section 7.4).

Let us recall the final sentence written by Vos in his original publication[5]: We have the opinion that the values tabulated here reflect the present state of knowledge. "We can be sure, though, that new experimental data will sooner or later lead to better tables. Therefore, and to underline its temporary character by definition, it may be suggested to refer to them as the '1978 2° fundamental observer data'."

In recent years, Vos was a member of the CIE TC 1-36 for the definition of the new CIE standard fundamentals proposed by Stockman and Sharpe (Section 9.6).

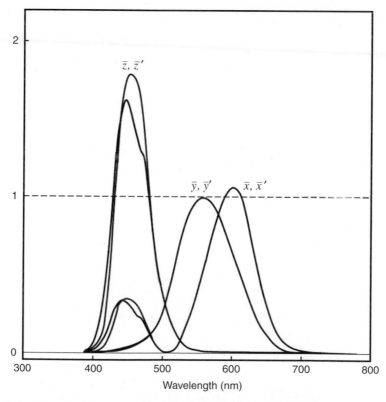

Figure 9.5 *Comparison of the CIE 1931 (red line) and Vos (black line) colour matching functions in the XYZ reference frame.*

9.5.1   Smith-Pokorny's Cone Fundamentals

Research into colour vision considers the observer defined by the colour-matching functions of Judd-Vos, but the L′M′S′ cone-activation reference frame is used, instead of the X′Y′Z′ reference frame. The L′M′S′ reference frame is much more significant from the physiological point of view. The tristimulus values L', M' and S', that is, the components of the tristimulus vector $(L'M'S')$, are the activations of the L, M and S cones, respectively. To get this reference frame from X′Y′Z′ it is necessary to implement a linear transformation based on knowledge of the three directions of the colour stimuli capable of stimulating individually the three kinds of cones. The definition of these directions is indirect, since the corresponding colour stimuli are physically impossible, and is based on the study of confusion points of dichromates (Section 6.13.7). The confusion points measured by V.C. Smith and J. Pokorny[17] were used

$$\begin{cases} \text{for protanopes} \quad \left(x'_p = 0.7465, y'_p = +0.2535\right) \\ \text{for deuteranopes} \left(x'_d = 1.4000, y'_d = -0.4000\right) \\ \text{for tritanopes} \quad \left(x'_t = 0.1748, y'_t = +0.0000\right) \end{cases} \tag{9.23}$$

Vos' colour-matching functions represented in the fundamental reference frame defined by the confusion points given by Smith-Pokorny became a *de facto* standard.[17] The colour-matching functions in the fundamental reference frame are also termed *cone fundamentals* or simply *fundamentals*.

9.5.2   Vos' 1978 2° Fundamental Observer Data and MacLeod-Boynton's Chromaticity Diagram

Generally, the chromaticity diagram is defined on the unit plane, but other choices are possible with the intent of a better description of visual phenomena. MacLeod-Boynton proposed an equiluminant chromaticity diagram based on physiology, emphasizing the separation between the luminous and the chromatic components of the colour stimuli[18]. The reference frame is with L′M′S′ Cartesian coordinates representing the excitations of the three cone types involved in colour vision, where the $(L'$, M', $S')$ tristimulus values are weighted taking into account the equal contribution of L- and M-cones to the luminance (the S-cones do not contribute to luminance), that is, the units of the reference stimuli are set so that $(L' + M') = Y'$.

The linear transformation between the X′Y′Z′ and L′M′S′ reference frames with the MacLeod–Boynton units is the following

$$\begin{pmatrix} L' \\ M' \\ S' \end{pmatrix} = \begin{pmatrix} 0.15516 & 0.54308 & -0.03287 \\ -0.15516 & 0.45692 & 0.03287 \\ 0.00000 & 0.00000 & 0.01608 \end{pmatrix} \begin{pmatrix} X' \\ Y' \\ Z' \end{pmatrix}$$

$$\begin{pmatrix} X' \\ Y' \\ Z' \end{pmatrix} = \begin{pmatrix} 2.94483 & -3.50013 & 13.17449 \\ 1.00000 & 1.00000 & 0.00000 \\ 0.00000 & 0.00000 & 62.18906 \end{pmatrix} \begin{pmatrix} L' \\ M' \\ S' \end{pmatrix} \tag{9.24}$$

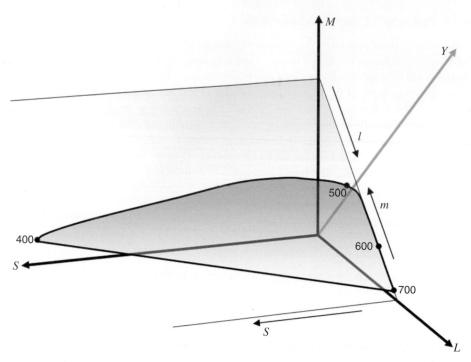

Figure 9.6   *Perspective view of the tristimulus space in the LMS reference frame as defined by MacLeod and Boynton,[18] with the equal luminance plane L + M = 1 and the chromaticity diagram.*

(The very small difference between this matrix given in the Kaiser-Boynton book,[12] denoted by (A.3.14), and that given by MacLeod-Boynton in their original paper is due to the observer, that was the Judd observer and not that of Vos.)

The linear transformations considered here are used in the software presented in Section 16.8.4.

The chromaticity diagram on the constant luminance plane $L' + M' = 1$ (Figures 9.6 and 9.7), known as *MacLeod-Boynton chromaticity diagram*,[18] is spanned by the chromaticity coordinates

$$l' = \frac{L'}{L' + M'}, \; m' = \frac{M'}{L' + M'}, \; s' = \frac{S'}{L' + M'}, \; l' + m' = 1 \tag{9.25}$$

where *m'* and *l'* are mutually dependent.

The Vos observer and the MacLeod-Boynton diagram are considered in the software presented in Section 16.8.4.

9.6   CIE Standard Stockman-Sharpe's 'Physiologically Relevant' Fundamentals and XYZ Reference Frame

All the quantities related to Stockman-Sharpe's fundamental observers are distinguished by use of the subscript 'F', which means 'fundamental', and the additional subscript '10' distinguishes the 10° observer from the 2°.

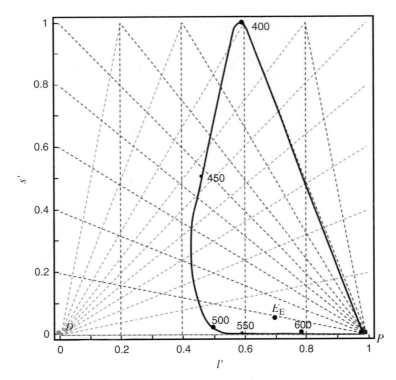

Figure 9.7 *Vos' observer and MacLeod-Boynton's chromaticity diagram with the confusion lines of the three kinds of dichromats (red for protanopes, green for deuteranopes and blue for tritanopes), confusion points **P** and **D** and chromaticity of the equal-energy stimulus **E**ₑ.*

Cone fundamentals or *fundamental colour-matching functions* are the *spectral sensitivity functions* of the L-, M- and S-cones, measured on the corneal plane.[3,4]

In 2005, after an impressive work (CIE TC 1-36), the CIE published the standard cone fundamentals for 2° and 10° observers, which are representative of the cone spectral sensitivities of an average observer. These fundamentals are not intended as a substitute for the CIE standard colorimetric observers, but serve as a link between colorimetry and physiology with the hope that it will improve the understanding of colour, "will be useful for education and will offer novel opportunities to solve problems of colour measurement and colour perception in everyday life and industry. [a,1]

Both fundamentals are based on the 10° colour-matching functions of Stiles and Burch.[13] These data are probably the safest and most comprehensive set of existing colour-matching functions for a centrally viewed 10° visual field. These colour matching data are relative to a set of monochromatic $\hat{\mathbf{R}}$, $\hat{\mathbf{G}}$ and $\hat{\mathbf{B}}$ reference stimuli. Following the ideas of Stockman and Sharpe,[19] by applying König's reduction hypothesis[20] (Section 6.13.7) and using the most modern data on the spectral sensitivity functions of dichromats, the spectral sensitivity functions of the L-, M- and S-cones, measured in the corneal plane for a 10° viewing field were derived: $\left(\bar{l}_{10}(\lambda), \bar{m}_{10}(\lambda), \bar{s}_{10}(\lambda)\right)$. These spectral sensitivity functions are called 'cone fundamentals'.

The optical absorbances of cone visual pigments for a 10° visual field were obtained by taking into account the absorption of the ocular media and the macular pigment. Analogously the optical densities of the visual pigments for a 2° visual field were derived. The cone fundamentals for 2° visual field $\left(\bar{l}(\lambda), \bar{m}(\lambda), \bar{s}(\lambda)\right)$ were reconstructed starting from the absorbances of cone visual pigments for a 10° visual field and introducing the absorption of the ocular media and of the macula.

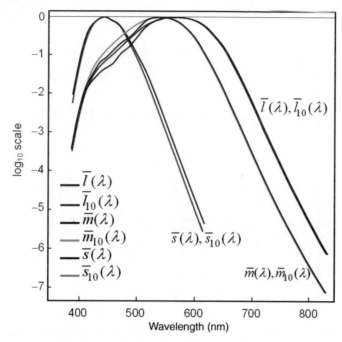

Figure 9.8   *Comparison of the fundamental colour matching functions of the 2° and 10° fundamental observers on a logarithmic scale. (Numerical values are in Section 16.1.2 data files)*

Both colour-matching functions $\left(\bar{l}_{10}(\lambda), \bar{m}_{10}(\lambda), \bar{s}_{10}(\lambda)\right)$ and $\left(\bar{l}(\lambda), \bar{m}(\lambda), \bar{s}(\lambda)\right)$ are normalized such that the maximum value of each function is set equal to 1.

The 10° cone fundamentals are defined in the wavelength range 390–830 nm with a step of 5 nm and six significant figures, while the 2° cone fundamentals are defined in the same wavelength range with an equal step and seven significant figures.

Figure 9.8 shows the colour matching functions of the two fundamental observers on a logarithmic scale for a comparison.

Using an analogous procedure, the cone fundamentals for every viewing angle between 1° and 10° and for any age can be constructed. In this last case, the optical density of the crystalline lens is given as a function of age. Here only the 2° and 10° observers are considered.

Figure 9.9 shows the Vos colour matching functions and the 2° fundamental ones on a logarithmic scale for a comparison.

9.6.1 $X_F Y_F Z_F$ and $X_{F,10} Y_{F,10} Z_{F,10}$ Reference Frames

These CIE standard cone fundamentals, proposed by Stockman and Sharpe,[19] are used by the physiologists. The CIE TC 1-36 continued its work and today the $X_F Y_F Z_F$ and $X_{F,10} Y_{F,10} Z_{F,10}$ reference frames are also defined for these observers ('10' denotes the visual field of 10°). As known, the $X_F Y_F Z_F$ and $X_{F,10} Y_{F,10} Z_{F,10}$ reference frames are defined with the requirement that

- the Y_F and $Y_{F,10}$ axes represent the luminance, that is, the components of the colour-matching functions $\bar{y}_F(\lambda)$ and $\bar{y}_{F,10}(\lambda)$ are equal to the relative spectral luminous sensitivity functions, $\bar{y}_F(\lambda) = V_F(\lambda)$ and $\bar{y}_{F,10}(\lambda) = V_{F,10}(\lambda)$;

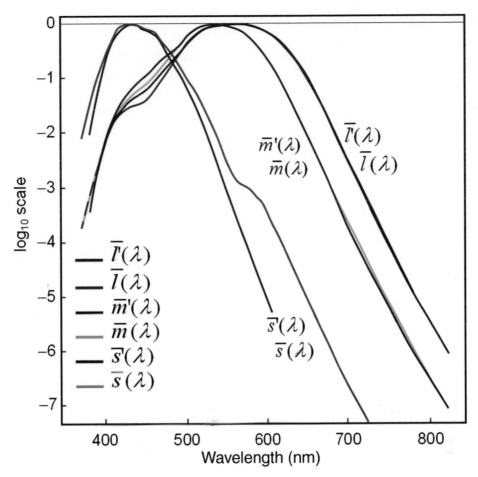

Figure 9.9 *Comparison of the fundamental colour matching functions of the 2° fundamental observers and of Vos on a logarithmic scale. (Numerical values are in Section 16.1.2 data files)*

- the equal energy stimulus E_E is specified by a vector with three equal components;
- the planes $X_F = 0$ and $Z_F = 0$, and similarly $X_{F,10} = 0$ and $Z_{F,10} = 0$, are tangent to the spectrum locus in the medium-short wavelengths and in the long wavelengths, respectively, with the consequence that the tristimulus values are positive.

These requirements define the dichromatic confusion points for 2° and 10° viewing fields.
The 2° *observer* is defined by
- confusion points:

$$
\begin{cases}
\text{protanopes} \left(x_{F,P} = 0.73840, y_{F,P} = 0.26160, \ z_{F,P} = 0.00000 \right) \\
\text{deuteranopes} \left(x_{F,D} = 1.32672, y_{F,D} = -0.32672, z_{F,D} = 0.00000 \right) \\
\text{tritanopes} \left(x_{F,T} = 0.15862, y_{F,T} = 0.00000, \ z_{F,T} = 0.84138 \right)
\end{cases}
\tag{9.26}
$$

- luminous efficiency function

$$\bar{y}_F(\lambda) = V_F(\lambda) = 0.68990272\,\bar{l}(\lambda) + 0.34832189\,\bar{m}(\lambda) \tag{9.27}$$

- reference frame transformation

$$\begin{pmatrix} \bar{x}_F(\lambda) \\ \bar{y}_F(\lambda) \\ \bar{z}_F(\lambda) \end{pmatrix} = \begin{pmatrix} 1.94735469 & -1.41445123 & 0.36476327 \\ 0.68990272 & 0.34832189 & 0 \\ 0 & 0 & 1.93485343 \end{pmatrix} \begin{pmatrix} \bar{l}(\lambda) \\ \bar{m}(\lambda) \\ \bar{s}(\lambda) \end{pmatrix}$$

$$\begin{pmatrix} \bar{l}(\lambda) \\ \bar{m}(\lambda) \\ \bar{s}(\lambda) \end{pmatrix} = \begin{pmatrix} 0.21057582 & 0.85509764 & -0.03969827 \\ -0.41707637 & 1.17726110 & 0.07862825 \\ 0 & 0 & 0.51683501 \end{pmatrix} \begin{pmatrix} \bar{x}_F(\lambda) \\ \bar{y}_F(\lambda) \\ \bar{z}_F(\lambda) \end{pmatrix} \tag{9.28}$$

The 10° *observer* is defined by
- confusion points:

$$\begin{cases} \text{protanopes} & \left(x_{F,10,P} = 0.73683, y_{F,10,P} = 0.26317, z_{F,10,P} = 0.00000\right) \\ \text{deuteranopes} & \left(x_{F,10,D} = 1.35074, y_{F,10,D} = -0.35074, z_{F,10,D} = 0.00000\right) \\ \text{tritanopes} & \left(x_{F,10,T} = 0.16701, y_{F,10,T} = 0.00000, z_{F,10,T} = 0.83299\right) \end{cases} \tag{9.29}$$

- luminous efficiency function

$$\bar{y}_{F,10}(\lambda) = V_{F,10}(\lambda) = 0.69283932\,\bar{l}_{10}(\lambda) + 0.34967567\,\bar{m}_{10}(\lambda) \tag{9.30}$$

- reference frame transformation

$$\begin{pmatrix} \bar{x}_{F10}(\lambda) \\ \bar{y}_{F10}(\lambda) \\ \bar{z}_{F10}(\lambda) \end{pmatrix} = \begin{pmatrix} 1.93986443 & -1.34664359 & 0.43044935 \\ 0.69283932 & 0.34967567 & 0 \\ 0 & 0 & 2.14687945 \end{pmatrix} \begin{pmatrix} \bar{l}_{10}(\lambda) \\ \bar{m}_{10}(\lambda) \\ \bar{s}_{10}(\lambda) \end{pmatrix}$$

$$\begin{pmatrix} \bar{l}_{10}(\lambda) \\ \bar{m}_{10}(\lambda) \\ \bar{s}_{10}(\lambda) \end{pmatrix} = \begin{pmatrix} 0.21701045 & 0.83573367 & -0.04351060 \\ -0.42997951 & 1.20388946 & 0.08621090 \\ 0 & 0 & 0.46579234 \end{pmatrix} \begin{pmatrix} \bar{x}_{F10}(\lambda) \\ \bar{y}_{F10}(\lambda) \\ \bar{z}_{F10}(\lambda) \end{pmatrix} \tag{9.31}$$

The 2° and 10° colour-matching functions, $\left(\bar{x}_F(\lambda), \bar{y}_F(\lambda), \bar{z}_F(\lambda)\right)$ and $\left(\bar{x}_{F,10}(\lambda), \bar{y}_{F,10}(\lambda), \bar{z}_{F,10}(\lambda)\right)$, are defined in the wavelength range 390–830 nm with step of 1 nm and seven significant figures.

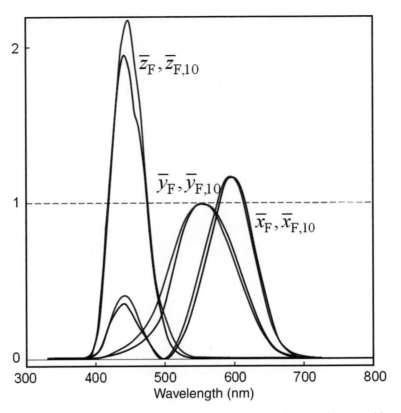

Figure 9.10 *Comparison of the colour matching functions of the 2° (black line) and 10° (red line) fundamental observers in the XYZ reference frame. (Numerical values are in Section 16.1.2 data files)*

Figures 9.10 and 9.11 propose a comparison of the 2° and 10° observers related to the cone fundamentals, the $(\bar{x}_F(\lambda), \bar{y}_F(\lambda), \bar{z}_F(\lambda))$ and $(\bar{x}_{F,10}(\lambda), \bar{y}_{F,10}(\lambda), \bar{z}_{F,10}(\lambda))$ colour-matching functions and the (x_F, y_F) and $(x_{F,10}, y_{F,10})$ chromaticity diagrams.

Particularly, Figure 9.12 represents the ratios of the two fundamentals, which should be compared with the optical density of the macula lutea. This figure shows that the difference between the two observers is mainly due to the macula lutea.

9.6.2   MacLeod-Boynton's Tristimulus Space and Chromaticity Diagram

As written in the previous Section 9.5.2, the MacLeod-Boynton chromaticity diagram is defined on the equi-luminant plane and the units of the (L, M, S) tristimulus value take into account the contribution of L- and M-cones to the luminance with equal weights.

For the 2° observer, the *MacLeod-Boynton tristimulus values* due to a radiance $L_{e,\lambda}$ are

$$L = \alpha_L \int_{390}^{830} L_{e,\lambda} \bar{l}(\lambda)\,d\lambda, \; M = \alpha_M \int_{390}^{830} L_{e,\lambda} \bar{m}(\lambda)\,d\lambda, \; S = \alpha_S \int_{390}^{830} L_{e,\lambda} \bar{s}(\lambda)\,d\lambda \qquad (9.32)$$

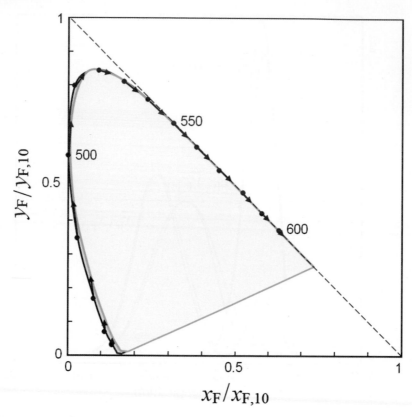

Figure 9.11 *Comparison of the chromaticity diagrams (x, y) of the 2° and 10° fundamental observers with highlighting of the wavelength shift. The tip of the arrow refers to the 10° observer and its tail to 2° observer.*

where $(\bar{l}(\lambda), \bar{m}(\lambda), \bar{s}(\lambda))$ are the unit-peak normalized 2° cone fundamentals, $\alpha_L = 0.68990272$, $\alpha_M = 0.34832189$, $\alpha_S = 0.03710971$ are the weights such that $L + M = Y_F$ and the factor α_S normalizes $\bar{s}(\lambda)/V_F(\lambda)$ to a maximum value of one. This reference frame is named *MacLeod-Boynton tristimulus space* and the (L, M, S) are named tristimulus values. The chromaticity plane is the equiluminant plane defined by the equation $L + M = 1$, which is parallel to the S axis (Figure 9.6). The chromaticity diagram (Figure 9.13) is spanned by two orthogonal coordinates named *MacLeod-Boynton chromaticity coordinates* and defined as

$$l = \frac{L}{L+M}, \; m = \frac{M}{L+M} = 1-l, \; s = \frac{S}{L+M}, \; l+m = 1 \tag{9.33}$$

where l and m are mutually dependent.

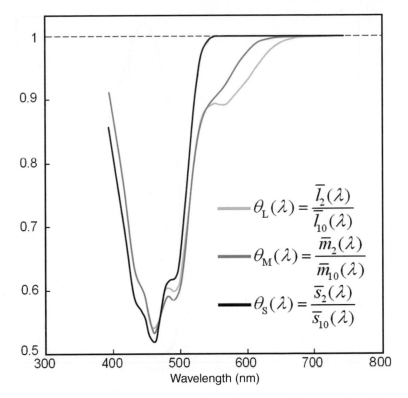

Figure 9.12 *Ratios of the fundamental colour matching functions of the 2° and 10° fundamental observers.*

For the 10° observer, the *MacLeod-Boynton tristimulus values* due to a radiance $L_{e,\lambda}$ are

$$
\begin{cases}
L_{10} = \alpha_{L,10} \displaystyle\int\limits_{390}^{830} L_{e,\lambda} \bar{l}_{10}(\lambda)\,\mathrm{d}\lambda \\[2ex]
M_{10} = \alpha_{M,10} \displaystyle\int\limits_{390}^{830} L_{e,\lambda} \bar{m}_{10}(\lambda)\,\mathrm{d}\lambda \\[2ex]
S_{10} = \alpha_{S,10} \displaystyle\int\limits_{390}^{830} L_{e,\lambda} \bar{s}_{10}(\lambda)\,\mathrm{d}\lambda
\end{cases}
\tag{9.34}
$$

where $\left(\bar{l}_{10}(\lambda), \bar{m}_{10}(\lambda), \bar{s}_{10}(\lambda)\right)$ are the unit-peak normalized 10° cone fundamentals, $\alpha_{L,10} = 0.69283932$, $\alpha_{M,10} = 0.34967567$, $\alpha_{S,10} = 0.05547860$ are the weights such that $L_{10} + M_{10} = Y_{F,10}$ and the factor $\alpha_{S,10}$ normalizes $\bar{s}_{10}(\lambda)/V_{F,10}(\lambda)$ to a maximum value of one. The chromaticity plane is the equiluminant plane

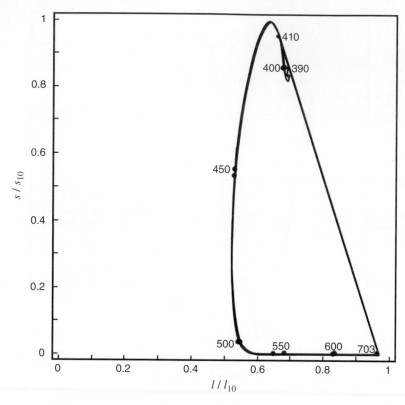

Figure 9.13 *Comparison of the MacLeod-Boynton chromaticity diagrams of the 2° (black line) and 10° (red line) fundamental observers.*

defined by the equation $L_{10} + M_{10} = 1$ (Figure 9.13). The chromaticity diagram is spanned by two orthogonal coordinates named *MacLeod-Boynton chromaticity coordinates* and defined as

$$l_{10} = \frac{L_{10}}{L_{10} + M_{10}}, \; m_{10} = \frac{M_{10}}{L_{10} + M_{10}} = 1 - l_{10}, \; s_{10} = \frac{S_{10}}{L_{10} + M_{10}}, \; l_{10} + m_{10} = 1 \qquad (9.35)$$

where l_{10} and m_{10} are mutually dependent because $l_{10} + m_{10} = 1$.

9.7 CIE Colorimetric Specification of Primary and Secondary Light Sources

Conventionally, in the case of primary and secondary light sources, the radiance entering the eye is suitably normalized producing a relative colour specification, independent of the absolute intensity of the colour stimulus. This leads to a relative colour specification, in fact

1.   the light sent to the observer by the *perfect reflecting diffuser* lit by the light source used is measured instead of the light emitted directly by the light source itself, and the reflected luminance is placed conventionally equal to 100 (or 1) for all the illuminants;

2.   the scale of the reflected or transmitted luminance of a non-luminous body viewed in reflection or transmission is referred to the scale of the reflected luminance of a perfect reflecting diffuser equally illuminated and is termed *percentage luminance factor* denoted by Y. The percentage luminance factor divided by 100 is simply *luminance factor* and is denoted by $\beta = Y/100$.

In this way the

- *primary light sources* and illuminants are represented by their relative spectral power distribution $S(\lambda)$, that is proportional to radiance entering the eye, and the colorimetric specification is given by

$$
\begin{cases}
X = K \displaystyle\sum_{\lambda=360,\Delta\lambda=1}^{830} S(\lambda)\bar{x}(\lambda)\Delta\lambda \\[2ex]
Y = K \displaystyle\sum_{\lambda=380,\Delta\lambda=1}^{780} S(\lambda)\bar{y}(\lambda)\Delta\lambda \quad \text{with } K = 100 \Bigg/ \displaystyle\sum_{\lambda=380,\Delta\lambda=1}^{780} S(\lambda)\bar{y}(\lambda)\Delta\lambda \\[2ex]
Z = K \displaystyle\sum_{\lambda=360,\Delta\lambda=1}^{830} S(\lambda)\bar{z}(\lambda)\Delta\lambda
\end{cases}
\tag{9.36}
$$

and Y is the percentage luminance factor;

- illuminated surfaces — *secondary light sources* — are represented by the spectral reflectance factor $R(\lambda)$ (Section 12.2.2) and the radiance entering the eye is proportional to $S(\lambda)\,R(\lambda)$; the colorimetric specification is given by

$$
\begin{cases}
X = K \displaystyle\sum_{\lambda=360,\Delta\lambda=1}^{830} S(\lambda)R(\lambda)\bar{x}(\lambda)\Delta\lambda \\[2ex]
Y = K \displaystyle\sum_{\lambda=380,\Delta\lambda=1}^{780} S(\lambda)R(\lambda)\bar{y}(\lambda)\Delta\lambda \quad \text{with } K = 100 \Bigg/ \displaystyle\sum_{\lambda=380,\Delta\lambda=1}^{780} S(\lambda)\bar{y}(\lambda)\Delta\lambda \\[2ex]
Z = K \displaystyle\sum_{\lambda=360,\Delta\lambda=1}^{830} S(\lambda)R(\lambda)\bar{z}(\lambda)\Delta\lambda
\end{cases}
\tag{9.37}
$$

and Y is the percentage luminance factor;

- illuminated transmitting objects — secondary light sources — are represented by the spectral transmittance $\tau(\lambda)$ (Section 12.2.2) and the radiance entering the eye is proportional to $S(\lambda)\,\tau(\lambda)$; the colorimetric specification is given by

$$
\begin{cases}
X = K \sum_{\lambda=360,\Delta\lambda=1}^{830} S(\lambda)\tau(\lambda)\bar{x}(\lambda)\Delta\lambda \\[2ex]
Y = K \sum_{\lambda=380,\Delta\lambda=1}^{780} S(\lambda)\tau(\lambda)\bar{y}(\lambda)\Delta\lambda \quad \text{with } K = 100 \Bigg/ \sum_{\lambda=380,\Delta\lambda=1}^{780} S(\lambda)\bar{y}(\lambda)\Delta\lambda \\[2ex]
Z = K \sum_{\lambda=360,\Delta\lambda=1}^{830} S(\lambda)\tau(\lambda)\bar{z}(\lambda)\Delta\lambda
\end{cases} \tag{9.38}
$$

and *Y* is the percentage luminance factor.

Equations (9.36), (9.37) and (9.38) regard the CIE 1931 observer, but the generalization to any observer is made by changing the colour-matching functions. The summations obtained by steps of $\Delta\lambda = 5$ nm are considered practical approximations.

References

1. CIE Publication 15.3:2004, *Colorimetry* 3rd ed., Commission Internationale de l'Éclairage, Vienna (2004). Available at: www.cie.co.at/ (accessed on 3 June 2015)
2. CIE Publication 80, *Special Metamerism Index: Change in Observer*, Commission Internationale de l'Éclairage, Vienna (1989). Available at: www.cie.co.at/ (accessed on 3 June 2015)
3. CIE Publication 170-1:2006, *Fundamental Chromaticity Diagram with Physiological Axes-Part I*, Commission Internationale de l'Éclairage, Vienna (2006). Available at: www.cie.co.at/ (accessed on 3 June 2015)
4. CIE TC 1-36, *Fundamental Chromaticity Diagram with Physiological Axes- Part II*, Draft, Commission Internationale de l'Éclairage, Vienna (2015). Available at: www.cie.co.at/ (accessed on 3 June 2015).
5. Vos JJ, Colorimetric and photometric properties of a 2' fundamental observer. *Color Res. Appl.*, **3**, 125–128 (1978).
6. CIE Publication S 017/E:2011, *ILV: International Lighting Vocabulary*, Commission Internationale de l'Éclairage, Vienna (2011). Available at: www.cie.co.at/ (accessed on 3 June 2015).
7. Wright WDA, Re-determination of the mixture curves of the spectrum, *Trans. Opt. Soc. Lond.*, **30**, 141–164 (1928–1929).
8. Guild J, The colorimetric properties of the spectrum, *Philos. Trans. R Soc. Lond.*, **230**A, 149–187 (1931).
9. Wyszecki G and Stiles WS, *Color Science, Concepts and Methods, Quantitative Data and Formulae*, 2nd ed., John Wiley & Sons, New York (1982).
10. Kelly K L, Color designations for lights, *J. Opt. Soc. Amer.*, **33**, 627–632 (1943).
11. CIE Publication 15.2:1986, *Colorimetry*, 2nd ed., Commission Internationale de l'Éclairage, Vienna (1986). Available at: www.cie.co.at/ (accessed on 3 June 2015).
12. Kaiser PK and Boynton RB, *Human Color Vision*, 2nd ed., Optical Society of America (1996).
13. Stiles WS and Burch JM, N.P.L.colour-matching investigation: final report (1958). *Opt. Acta*, **6**, 1–26 (1959).
14. Speranskaya NI, Determination of spectrum color co-ordinates for twenty-seven normal observers, *Optics Spectrosc.*, **7**, 424–442 (1959).
15. Nimeroff I, Rosenblatt JR and Dannemiller MC, Variability of spectral tristimulus values. *J. Res. Natl. Bureau Stand.* A; **65** (6):475–483 (1961).

16. Judd DB, Report of U.S. secretariat committee on colorimetry and artificial daylight, in *Proceedings of the Twelfth Session of the CIE, Stockholm*, Vol. 1, Bureau Central de la CIE, Paris, pp. 11 (1951).
17. Smith VC and Pokorny J, Spectral sensitivity of the foveal cone photopigments between 400 and 500 nm. *Vision Res.*, **15**, 161–171 (1975).
18. MacLeod DIA and Boynton RM, A chromaticity diagram showing cone excitation by stimuli of equal luminance, *J. Opt. Soc. Amer.*, **69**, 1183–1186 (1979).
19. Stockman A and Sharpe LT, The spectral sensitivities of the middle- and long-wavelength-sensitive cones derived from measurements in observers of known genotype. *Vision Res.*, **40**, 1711–1737 (2000).
20. König A and Dieterici C, Die Grundenempfindungen und ihre Intensitäts, Vertheilung im Spectrum (Fundamental sensations and their intensity distribution in the spectrum) *Sitzungsberichte der Akademie der Wissenschaften in Berlin* 29 July 1886, pp. 805–829 (1886).

10

Chromaticity Diagram from Newton to the CIE 1931 Standard System

10.1 Introduction

This section considers the historical steps necessary for the definition of the chromaticity diagram used today, that represents not only the visual process of additive colour synthesis, but is also a part of the tristimulus space which has profound physiological and psychophysical justifications. The scientists who contributed to this history are all gigantic: Newton, Young, Helmholtz, Grassmann, Maxwell, Schrödinger. This chapter is not meant to replace the many chapters and books written about these scientists and the history of colour vision.[1-3] This chapter aims to highlight the intuitions of these scientists, their difficulties in proceeding and especially their great intellectual honesty. In addition, their various scientific achievements are often represented in manner that is a historically not rigorous way but makes reference to the knowledge of today.

10.2 Newton and the Centre of Gravity Rule

Isaac Newton wrote the 'New Theory About Light and Colour'[4] in 1671, but this theory was presented in a more complete way in his 1704 book 'Opticks, or a treatise on the reflections, refractions inflections & Colours of Light'.[f] From that year to 1931, the year of the definition of the 'Colorimetric Standard Observer' by the 'Commission Internationale de l'Éclairage' (CIE), 260 years elapsed. A long time necessary for an understanding of Newton's intuition, but in addition it was necessary to invent a new mathematics for formalizing the centre of Newton's gravity rule, regarding the definition of the linear vector space. Today it is possible to say that Newton was also the founder of this important field of mathematics. Meanwhile a deep intuition of the physiology of the human visual system was realized. This long path is covered here in a very concise way.

Before Newton, the general convincement was that white light is a pure light, while any coloured light is the product of a contamination of white light. In 1671–1672, Newton[4] demonstrated by a well-known experiment, called

> "Experimentum Crucis", that white "light is a Heterogeneous mixture of differently refrangible Rays" and "no one sort of Rays alone … can exhibit" whiteness. "As the Rays of light differ in degrees of

---

Standard Colorimetry: Definitions, Algorithms and Software, First Edition. Claudio Oleari.
© 2016 John Wiley & Sons, Ltd. Published 2016 by John Wiley & Sons, Ltd.

Refrangibility, so they also differ in their disposition to exhibit this or that particular colour. Colours are not Qualifications of Light, …, but original and connate properties, which in divers Rays are divers."

Moreover, Newton[5] had a new and different concept of colour:

"For the rays to speak properly are not coloured. In them there is nothing else than a certain power and disposition to stir up a sensation of this or that colour."

Then, Newton produced an optical apparatus for combining lights of all the possible colours, already described in Section 4.3.1 and Figure 4.2, and proposed a geometrical correspondence law between light spectra and colour sensations.[4,5] As shown in Figure 4.2, the light source is the sun, of which a beam enters into a room through a hole in the window. This beam is first dispersed by a prism into seven coloured beams and then, by a lens and a second prism, recombined into a beam. Newton called 'Primary' the colours of these seven beams, which are Red, Orange, Yellow, Green, Blue, Indigo, Violet. For Newton, the primary colours are directly related to the refrangibility degree of the corresponding rays, as he defined:

"DEFIN. VII. The Light whose Rays are all alike Refrangible, I call Simple, Homogeneal and Similar; and that whose Rays are some more Refrangible than others, I call compound, Heterogeneal and Dissimilar."

"DEFIN. VIII. The Colours of Homogeneal Lights, I call Primary, Homogeneal and Simple and those of Heterogeneal Lights, Heterogeneal and Compound."

The amounts of primary colours of the initial beam can be modulated by introducing a comb with teeth of different length/size on a plane close to the lens (Section 4.3.1, Figure 4.2). After this modulation, the recombined beam appears coloured. The dispersion by a third prism is used to check the preservation of the spectral modulation, after the recombination.

"In a mixture of Primary Colours, the Quantity and Quality of each beam given, to know the colour of the Compound" is defined by a very original geometrical construction, termed the "Center of Gravity" rule.

This rule was first given by Newton and its understanding was difficult and almost impossible for more than 150 years. This rule is synthesized into a circle represented in Figure 10.1 and its original description is:

"With the Center O and Radius OD describe a Circle ADF, and distinguish its Circumference into seven parts DE, EF, FG, GA, AB, BC, CD, …. Let the first part DE represent a red Colour, the second EF orange, the third FG yellow, the fourth CA green, the fifth AB blue, the sixth BC indigo and the seventh CD violet."

The colour names are thus given without definition, here and throughout Newton's book. The primary colour names thus define the Newton colour categorization and each name corresponds to an arc of the spectrum.

"And conceive that these are all the Colours of uncompounded Light gradually passing into one another, as they do when made by Prisms; the circumference DEFGABCD, representing the whole series of Colours from one end of the Sun's colour'd Image to the other, so that from D to E be all degrees of red, at E the mean Colour between red and orange, from E to F all degrees of orange, at F the mean between orange and yellow, from F to G all degrees of yellow and so on."

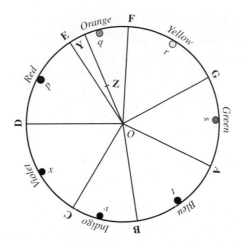

Figure 10.1 *Chromaticity diagram sketched by Newton*[5] *(Newton's original figure).*

Newton defined also the border between two contiguous spectral regions, that is, between two contiguous colour categories.

"Let *p* be the center of gravity of the Arch DE, and *q, r, s, t, v, x*, the centers of gravity of the Arches EF, FG, GA, AB, BC and CD, respectively, and about those centers of gravity let Circles proportional to the number of Rays of each Colour in the given Mixture be describ'd; that is, the Circle *p* proportional to the Number of the red-making Rays in the Mixture, the Circle *q* proportional to the number of the orange-making Rays in the Mixture, and so of the rest."

Then more detailed words were used by Newton to describe the Centre of Gravity rule:

"To give an instance of this Rule; suppose a Colour is compounded of these homogeneal Colours, of violet one part, of indigo one part, of blue two parts, of green three parts, of yellows five parts, of orange six parts and of red ten parts. Proportional to these parts describe the Circles *x, v, t, s, r, q, p*, respectively, that is, so that if the Circle *x* be one, the Circle *t* two, the Circle *s* three and the Circle *r, q* and *p*, five, six and ten. Then I find Z the common centre of gravity of these Circles, and through Z drawing the Line OY, the Point Y falls upon the circumference between E and F, something nearer to E than to F, and thence will be an orange, verging a little more to red than to yellow. Also I find that OZ is a little less than one half of OY, and thence I conclude, that this orange hath a little less than half the fullness or intenseness of an uncompounded orange; that is to say, that it is such an orange as may be made by mixing an homogeneal orange with a good white in the proportion of the Line OZ to the Line ZY, this Proportion being not of the quantities of mixed orange and white powders, but of the quantities of the Lights reflected from them."[5]

Newton gave two ways for producing the compound colour Z:

1. by a mixture of the seven primary colours represented by the circles *p, q, r, s, t, v, x*;
2. a mixture of the white colour O and the orange Y.

The use of a planar diagram to represent all the colours, primaries and compounds, has enormous conceptual content (that, probably, was not entirely considered by Newton himself):

1.   The centre of gravity rule regards colour sensations, that is, it is not a physical law.
2.   The multiplication of all the primary colours that enter the mixture by an equal factor does not change the centre of gravity, therefore the centre of gravity rule specifies only the chromatic quality of the colour – *chromaticity*, as termed today –, and does not consider its luminous quality.
3.   A point on the diagram represents a chromatic quality of a colour, that is a sensation.
4.   The specification of a chromatic quality of a colour is given by a point on a planar figure, therefore, it is a numerical specification.
5.   A one-to-one correspondence exists between spectral lights and chromatic sensations.
6.   Any compound colour is producible as infinite mixtures of spectral colours and, for point (5) is producible by infinite mixtures of spectral lights. The compound colour and the lights are in a one-to-many correspondence – this phenomenon was termed *metamerism* by Ostwald (Section 6.13.2) –.
7.   The minimum number of spectral colours necessary to match a compound colour is 3, except the case that the compound colour is a mixture of two spectral colours and the case that the compound colour is a spectral colour.

Points (5) and (6) say that the amounts of spectral colours necessary to match a compound colour are unique if the number of spectral colours is three and the spectral colours are fixed, while the amounts are in an infinite number if the number of spectral colours is greater than 3. In this last case, the amounts of spectral colours are mutually constrained by the centre of gravity rule. This phenomenon represents the *trichromacy* of the impalpable colours, that is, colours produced by lights, and represents *metamerism*. The usual definition of trichromacy given in Section 5.5.2 is strictly related to physiology, but psychophysics has shown that for a trichromat observer the colour appearance of any stimulus is matched by a set of three independent colour stimuli. Therefore the term trichromacy is extended to include colour matching, as in this case.

Newton's words need some didactical comments in order to easily grasp the new ideas contained in the centre of gravity rule:

(i)   Consider the mixture of only two different colours, white and orange associated with the segment OY of Newton's diagram. The segment OY can be considered as the yoke of a balance (Figure 10.2a), where on the yoke are shown all the colours obtainable by mixing the two colours – colour stimuli – considered as weights in continuous variable ratios. At any point Z of the yoke, there corresponds a ratio W_{white}/W_{orange}, where W_{white} and W_{orange} are the amounts of the two colours entering the mixture. The arrangement of the colours on the bar can be such that any colour subdivides the bar into two segments with lengths YZ and ZO, that we fix proportional to W_{white} and W_{orange}, respectively. The bar can be considered as the yoke of a balance, on whose plates are placed two weights equal to the amounts W_{white} and W_{orange} of the two colours entering the mixture. Of course, by definition, $W_{white} / ZY = W_{orange} / ZO$.

(ii) The generalization to a mixture of three independent colours (three colours are independent if none of these is matched by a mixture of the other two) is obtained by a balance with a triangular yoke (Figure 10.2b), where the three amounts of colours are proportional to the areas r, g and b of the three triangles constituting the yoke, having a common vertex, which is the equilibrium point of the balance and representing the compound colour. In this case $W_{red}/r = W_{green}/g = W_{blue}/b$ holds true. The quantities r, g and b are also the coordinates of the equilibrium point and are termed barycentric coordinates.

(iii)   The generalization to a mixture of four independent colours does not exist and the colour of any light is matched by a mixture of three independent colours. The consequence is that the spectral colours are mutually constrained.

Point (iii) states the trichromacy represented by Newton's Centre of Gravity rule. This rule is a generalization to the mixing of an infinite and continuous set of spectral colours, reduced in Figure 10.1 to the seven primary colours. The circle of Figure 10.1 can be considered as a bidimensional yoke of a balance (Figure 10.2c), on whose plates is placed a weight equal to the amount of spectral colours entering the mixture. The equilibrium point represents the compound colour.

The centre of gravity rule renders the Newton diagram not a simple colour wheel, but a *chromaticity diagram*. Particularly, equal amounts of spectral colours at the ends of any circle diameter produce white colour. Today, the pairs of spectral lights associated to these spectral colours are called complementary lights. Anyway the existence of pairs of complementary spectral lights was not securely established until the middle of the nineteenth century:

i.   Christian Huygens (1673) said that "two colours alone (yellow and blue) might be sufficient to yield white."

ii.   Newton wrote: (1671/1672) "There is no one sort of Rays which alone can exhibit" Whiteness.[4] White "is ever compounded, and to its composition are requisite all the aforesaid primary colours."[5] (1704) "Also if only two of the primary colours which in the circle (Figure 10.1) are opposite to one another be mixed in an equal proportion, the point Z shall fall upon the centre O and yet the colour compounded of these two shall not be perfectly white, but some faint anonymous Colour. For I could never yet by mixing only two primary Colours produce a perfect white."

The mixing rule given by the balances of Figures 10.2a and 10.2b for mixing two or three primary colours is only a way of representing the compound colours, but when the centre of gravity rule is applied to the circle of Figure 10.2c, a constraint is introduced on the spectral colours, which is not simply a geometrical trick and can be explained only by the physiology of the vision system.

The content of the Centre of Gravity rule, although today accepted as correct, did not completely convince Newton himself: "This Rule I conceive accurate enough for practice, though not mathematicallyaccurate; ..." (N.B. the mathematics is new) Three open problems were still present in this rule:

1.   Angular positions of the spectral colours (primary colours) on the colour diagram were erroneously placed in relation to the musical notes and not to the exact colour complementarities, that were not yet verified.

2.   Only spectral colours are placed on the external border of the diagram, therefore all the magenta hues, obtainable by variable mixings of Red and Violet primaries, are not considered in the rule, though discussed by Newton (Section 4.3.1).

3.   Circular shape is only an ideal approximation, because any radiometric measurement was impossible for Newton.

All these open problems found solution after the middle of the nineteenth century.

The centre of gravity rule holds true for colour compounds produced by light mixtures, not for pigment mixtures. Although Newton was very clear, this has been misunderstood by many people. Let us recall Newton's words referred to the diagram of Figure 10.1:

"... it is such an orange as may be made by mixing an homogeneal orange with a white in the proportion of the Line OZ to the Line ZY, this Proportion being not of the quantities of mixed orange and white Powders, but the quantities of the Lights reflected from them."

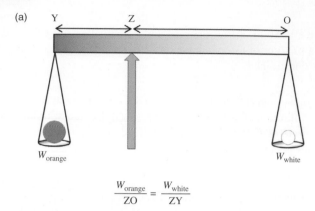

(a)

$$\frac{W_{\text{orange}}}{ZO} = \frac{W_{\text{white}}}{ZY}$$

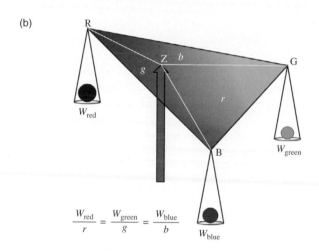

(b)

$$\frac{W_{\text{red}}}{r} = \frac{W_{\text{green}}}{g} = \frac{W_{\text{blue}}}{b}$$

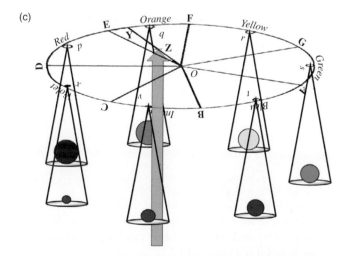

(c)

Figure 10.2   *(a) Barycentric coordinates related to a balance with one-dimensional yoke and two plates; (b) barycentric coordinates related to a balance with bidimesional yoke and three plates; (c) balance with a bidimensional yoke with the shape of Newton's diagram and a set of seven plates, related to the seven primary colours of Newton, loaded so as to have Z as a point of balance. This figure is a primeval sketch of the linear vector space termed tristimulus space and the center of gravity rule is the internal composition law that mathematically defines this space.*

10.3   Material Colours and Impalpable Colours in the Eighteenth Century

This short section does not contribute to the definition, analysis and completion of the chromaticity diagram, initiated by Newton, but serves to clarify the ambiguity of the words trichromacy and three-colour process. The term trichromacy regards the physiology and the colour matching without confusion, but it is also used as synonymous or is confused with the three-colour process with regard to the material colours. This confusion was great mainly in the eighteenth century.

Before Newton the colour mixing rule regarded the material colours, produced by pigments and dyes. Its origin went back to the medieval ages and was well synthesised by François d'Aguilon[6] (1613) with the sketch of Figure 10.3: white and black (i.e., light and dark) were considered as 'primaries' and yellow, red and blue were the basic or 'noble' hues, from which all other colours are derived. Yellow, red and blue are the only colours not obtainable by mixing other material colours. All that agrees with Aristotle.

People, familiar with d'Aguilon's rule for the material colours produced by dyes and pigments, were unable to understand Newton's theory of the impalpable colours produced by lights. Throughout the eighteenth century great confusion existed, that could be solved with the distinction between 'material colours' and 'impalpable colours'. Many people were against Newton. In spite of this fact, the ideas at the basis of the colour reproductions in prints have been empirically defined in this century by Jakob Christoffel Le Blon, who has to be remembered for his booklet 'Coloritto: or the Harmony of Coloring in Painting Reduced to Mechanical Practice'.[7] This work is clearly stressing the distinction between 'material colours' and 'impalpable colours' (Figure 10.4). For Le Blon all was clear and his work was very important. Le Blon is also famous because he understood the important role of black ink in printing and invented the *four-colour process* (Section 15.3).

Based on the four-colour process, the best and most fascinating prints of the eighteenth century were made by Jacques Fabian Gautier d'Agoty. These prints are of didactical sections of the human body. An example is shown in Figure 10.5. A correct understanding of this technique started only two centuries later, in 1924, with M. E. Demichel[8] to conclude in 1937 with Hans E. J. Neugebauer[9], who used the updated Newton chromaticity diagram (Section 15.3).

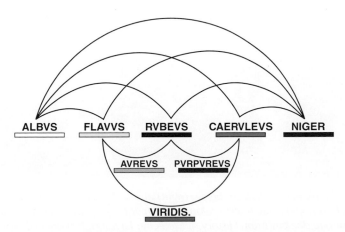

Figure 10.3   *François d'Aguilon's[6] material colour-mixing model (original figure).*

PAINTING can reprefent all *vifible* Objects with three Colours, *Yellow*, *Red*, and *Blue*; for all other Colours can be compos'd of thefe *Three*, which I call *Primitive*; for Example,

Yellow and *Red* } make an *Orange Colour*.

Red and *Blue* } make a *Purple* and *Violet Colour*.

Blue and *Yellow* } make a *Green Colour*.

And a *Mixture* of thofe *Three* Original Colours makes a *Black*, and all *other* Colours whatfoever; as I have demonftrated by my Invention of *Printing* Pictures *and* Figures *with their* natural *Colours*.

I am only fpeaking of *Material* Colours, or thofe ufed by *Painters*; for a *Mixture* of *all* the primitive *impalpable* Colours, that cannot be felt, will not produce *Black*, but the very Contrary, *White*; as the Great Sir ISAAC NEWTON has demonftrated in his Opticks.

White, is a Concentering, or an *Excefs* of Lights.
Black, is a deep Hiding, or *Privation* of Lights.

Figure 10.4   *Selected text from 'Coloritto: or the Harmony of Coloring in Painting Reduced to Mechanical Practice' written by Jakob Christoffel Le Blon[7].*

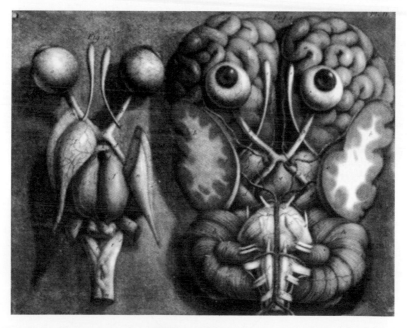

Figure 10.5   *A view of the human brain from below (right) and only a part from above (left) with in evidence the eyes, optic nerves, chiasm and optic tracts (colour prints made by Jacques Fabian Gautier d'Agoty, Paris ~1752). (Image in the public domain from 'History of Medicine Division, National Library of Medicine, National Institutes of Health, Department of Health and Human Services, Rockville Pike, Bethesda, MD').*

10.4   Physiological Intuitions and the Centre of Gravity Rule–Young, Grassmann, Helmholtz, Maxwell and Schrödinger

The physiological way for understanding the colour vision process was opened up by George Palmer[10] (1777), who postulated the existence in the retina of the eye of three kinds of "particles", and "each of these particle is moved by his own ray."

The deep intuition of Palmer was discussed by Thomas Young[11] (1802), who postulated three kinds of fibres as transducers of the visible lights in colour sensations:

"The sensation of different colors depends on the different frequency of vibrations excited by light in the retina." "The human eye is capable of three distinct primitive sensations of colour, which by their composition in various proportions, produce the sensations of actual colours in all their varieties."

This intuition was confirmed physiologically in 1956 by Gunnar Svaetichin.[12]

Only in the mid-1800s was it possible to combine this physiological intuition with the Newton chromaticity diagram. This model with three kinds of transducers confirmed the trichromacy and assumed the same name.

Hermann Günther Grassmann[13] in 1852–1853 mathematically formalised Newton's ideas giving a geometrical representation to the colour mixing. The mathematical tools used by Grassmann were stated by him in 1844 and are the mathematical ideas used to define the modern linear vector spaces. The colour space has three dimensions associated to three perceived quantities, that in German are called *Farbton, Intensität der Farbe* and *Intensität des beigemischten Weiss*. The Grassmann contribution is only mathematical, while the open problems of Newton are almost all present in the Grassmann theory.

Shortly before 1852, Hermann Ludwig Ferdinand von Helmholtz began his study on colour vision analysing the open problem of the complementary colours. He came to confirm the existence of continuous sets of complementary spectral lights according to Newton's centre of gravity rule and particularly "It was found that the colours from the red to green-yellow were complementary to colours ranging from green-blue to violet, and that the colours between green-yellow and green-blue have no homogeneous complementaries, but must be neutralized by mixtures of red and violet" (Figure 10.6a). This induced Helmholtz to transform Newton's circle into a half-moon like shape (Figure 10.6b), where the red light is connected to the violet light by a straight line, whose points represent purple or magenta colours.[14]

Another step had to be made to combine the chromaticity diagram with Young's hypothesis of three transducers. This step was made by two scientists in different ways:

1. Helmholtz produced a sketch of a diagram in the reference frame that, after Arthur König[15] (1886), is called *fundamental*. Helmholtz supposed that any spectral light excites the three kinds of fibres postulated by Young together and in different amounts. These excitation curves with the corresponding chromaticity diagram, obtained by applying the centre of gravity rule, are reproduced in Figures 10.7a and 10.7b (1866). The three barycentric coordinates related to the equilateral triangle containing the chromaticity diagram represent the activations of the three kinds of fibres. Now, these activations may be considered as components of vectors in a three-dimensional space. A perspective view of this space (Figure 10.9a) can be considered to show that the chromaticity diagram is a figure obtained by the intersection of the vectors, representing fibre activations, and a conventional plane.
2. James Clerk Maxwell[16–17] (1855) performed extensive experiments on additive mixture of lights by superposing spinning disks, whose coloured sectors are equivalent to weights in the light mixture (Section 4.3.2). In this way, Maxwell tested the correctness of Newton's centre of gravity rule.

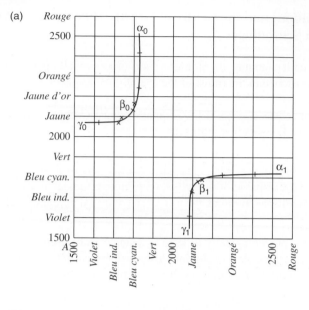

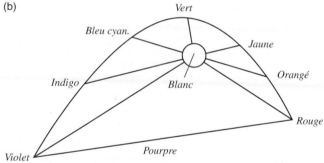

Figure 10.6   *(a) Helmholtz's correspondence between complementary spectral colours. (b) Helmholtz's chromaticity diagram obtained with the centre of gravity rule applied to complementary spectral colours (Helmholtz's original figures[14]).*

In 1860, Maxwell[18] produced a chromaticity diagram supported by measurements in the reference frame of the tristimulus space that today is said to be referred to the laboratory, that is, to a set of three spectral colours – one scarlet, one green and one blue – called 'standard' by Maxwell. An instrument for colour matching was projected and constructed by Maxwell, whose simplicity is striking. The white light of the sun was matched by a mixture of the three standard spectral lights, whose amounts enter a first equation. Moreover the same sun white can be matched by a mixture of any spectral light with two of the standard lights, whose choice depends on the wavelength of the spectral light considered. A second equation is written regarding the amounts of lights in any matching. The combination of these two equations at any wavelength gives the spectral sensitivities of the Young fibres referred to the three 'standard' lights, that constitute the laboratory-reference frame (Section 6.13.5). All that holds true in the hypothesis that the colour sensations are specified by vectors in a linear vector space according to Grassmann's laws. Figure 10.8a reproduces the three spectral sensitivity functions and Figure 10.8b the chromaticity

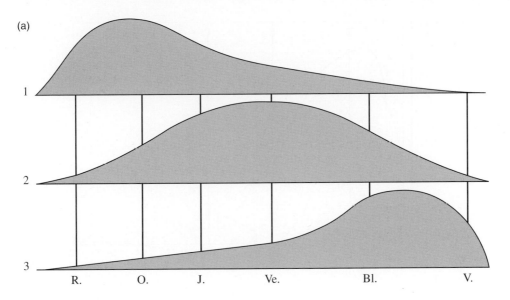

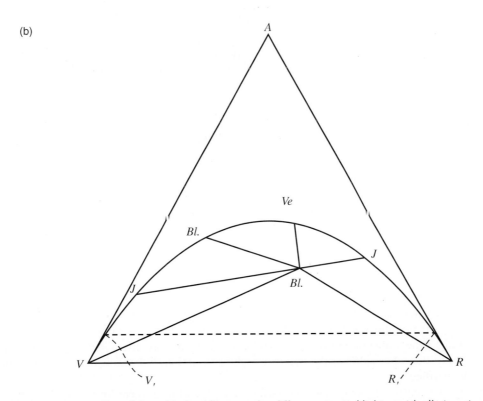

Figure 10.7 *(a) Sensitivities of three kinds of fibres to the different spectral lights as ideally imagined by Helmholtz (Helmholtz's original figure[14]); (b) Chromatic diagram as sketched by Helmholtz in correspondence to the imagined sensitivity curves (a) (Helmholtz's original figure[14]).*

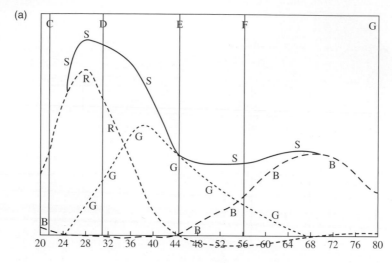

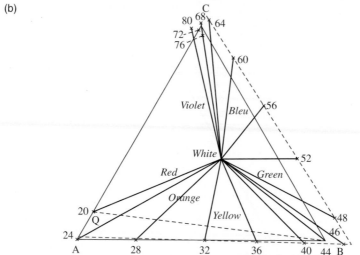

Figure 10.8 *(a) Colour-matching functions and (b) corresponding chromaticity diagram as measured by Maxwell (Maxwell's original figures[16–18]).*

diagram obtained by measurement with Maxwell's wife, Katherine, as observer. All three functions have a zero crossing and where two functions are positive the third is negative (Section 6.13.5). A negative sensitivity or a negative colour sensation – that is a negative light source – seems unconceivable, therefore these functions appear absurd, but Maxwell said that this is not a problem, because "by transposing the negative term to the other side" of the colour matching equation "it becomes positive, and then the equation may be verified."

After Maxwell, the centre of gravity rule was a true mathematical instrument for the colour specification based on measurements of spectral lights reflected or transmitted by coloured bodies or emitted by light sources. Any other step was a refinement.

The passage from the fundamental reference frame of Helmholtz to the laboratory-reference frame of Maxwell is represented in Figures 10.9a and 10.9b. A linear transformation exists between these two reference frames (Section 6.13.4).

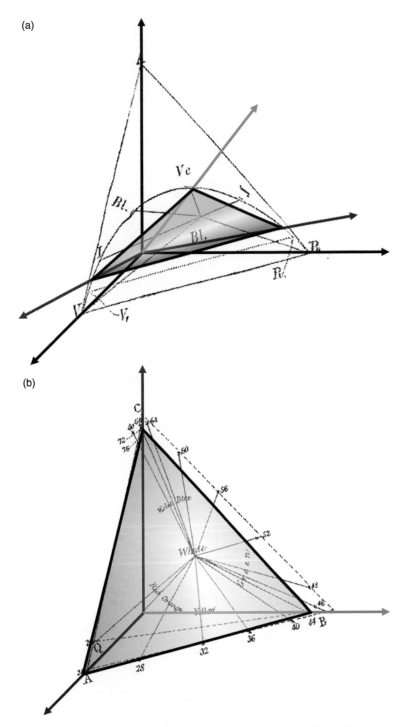

Figure 10.9 *(a) A perspective view of the tristimulus space with the plane of the chromaticity diagram in the fundamental reference frame according to Helmholtz; (b) and in the laboratory-reference frame according to Maxwell. In the two reference frames, the coloured axes, that represent the three 'standard' lights of Maxwell, and the thick black line triangles are in correspondence.*

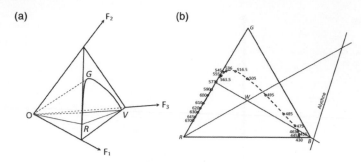

Figure 10.10   *(a) Perspective view of the tristimulus space and chromaticity-diagram plane in the fundamental reference frame and (b) normal view of the corresponding chromaticity diagram, where the spectrum locus is specified by the wavelengths of the spectral radiations and the Alychne line is drawn (Schrödinger's original figures[19] that could be representative also of the actual knowledge of today).*

For completeness, two figures drawn by Ervin Schrödinger are reported, because these figures are the first ones in which the tristimulus space is represented clearly in a perspective view and the chromaticity diagram as a flat figure on a plane in tristimulus space[19] (Figures 10.10a and 10.10b).

Since the luminance is the projection of the tristimulus vector on the direction defined by the Exner brightness weights, the stimuli with equal luminance belong to a plane orthogonal to such a direction, and particularly a plane of unreal colour stimuli with zero projection and zero luminance exists. The 'Alychne' line in Figure 10.10b, first given by Schrödinger, is the intersection line between the plane of the chromaticity diagram and the zero-luminance plane and was chosen as the abscissa line in the CIE 1931 (x, y) chromaticity diagram[20] (Figures 6.29, 6.30b and 6.30c).

The XYZ reference frame of the 1931 CIE standard observer is presented in Section 9.2.2 and the Newton[20] centre-of-gravity rule is explained in Section 6.13.8 and in Figure 10.11.

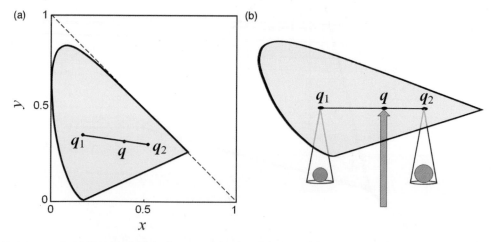

Figure 10.11   *(a) CIE 1931 (x, y) chromaticity diagram with the mixture of two colour stimuli, whose chromaticities are $\boldsymbol{q}_1$ and $\boldsymbol{q}_2$ and (b) chromaticity diagram used as the bidimensional yoke of a balance. The mixture of the two stimuli with chromaticities $\boldsymbol{q}_1$ and $\boldsymbol{q}_2$ follows the centre of gravity rule, represented by a balance.*

10.5   Conclusion

In colour science, the diagrams in which the centre-of-gravity rule holds true are chromaticity diagrams. It is very surprising that Newton's ingenious idea, first proposed in 1671–1672, met with such strong opposition for more than 150 years and such a long time was necessary to arrive at testing its correctness (Maxwell 1855), to define its geometrical shape on the basis of the complementary colours (von Helmholtz 1866), and to define its mathematics (Grassmann 1853). The choice of the reference frame made by Maxwell was absolutely necessary to be able to measure the three spectral sensitivities of the human visual system by colour matching. This view in retrospect renders Newton's, Young's, Helmholtz's and Maxwell's intuitions gigantic and Newton's doubts the honest ground of the scientists. The history of the Chromaticity Diagram is centuries-old and fascinating.

References

1. Mollon JD, The origins of modern color science, in *The Science of Color*, 2nd edn., Shevel Steven K, ed. OSA-Elsevier, Oxford, pp. 512–519 (2003).
2. MacAdam DL, Selected papers on colorimetry-fundamentals, *SPIE Milestones series* **77**, (1993).
3. Barry R and Masters A, History of human color vision from Newton to Maxwell. *OPN Optic. Photon. News*, 43–47 (January, 2011).
4. Newton I, Letter to the publisher containing Newton's new theory about light and colors, *Philos. Trans. R Soc. Lond.*, **80**, 3075–3087 (February 19, 1671–1672).
5. Newton I, *Opticks, or a Treatise on the Reflections, Refractions Inflections & Colours of Light*. Sam. Smith and Benj. Walford, London (1704), reprinted by Dover, New York (1952).
6. d'Aguilon F, *Opticorum Libri Sex*, Antwerp, Plantin (1613).
7. Le Blon JC, *Coloritto: or the Harmony of Coloring in Painting Reduced to Mechanical Practice*. London (1725).
8. Demichel ME, *Procédé*, **26**, 17–21 and 26–27 (1924).
9. Neugebauer HEJ, Die theoretische rundlagen der Mehrfarbendrucks, *Z. Wiss. Photogr.*, **36**, 73–89 (1937).
10. Palmer G, *Theory of Colours and Vision*, S. Leacroft, London (1777).
11. Young T, On the theory of light and colours, *Philos. Trans. R. Soc.*, **92**, 12–48 (1802).
12. Svaetichin G, Spectral response curves from single cones, *Actaphysiol. Scand.*, **39**(134), 17–46 (1956).
13. Grassmann HG, Zur theorie der Farbenmischung, *Poggendorf Ann. Phys.*, **89**, 69–84 (1853).
14. von Helmholtz HLF, *Handbuch der Physiologischen Optik*, 3rd ed., Dritte Auflage, ed., English translation of the third edition, Vol. 3, (1909–1011) by J-P.C, Southhall, Optical Society of America, Washington, DC (1924).
15. König A and mit Dieterici C, Die grundempfindungen und ihre intensitäts vertheilung im spectrum, in *Sitzungsberichte der Akademie der Wissenschaften,* Berlin, 29 July 1886, pp. 805–829 (1886).
16. Maxwell JC, On the theory of colours in relation to colour-blindness, *Transactions of the Royal Society of Arts*. **IV**, Part III, VI, A letter to Dr G. Wilson (1855).
17. Maxwell JC, Experiments on colour, as perceived by Eye, with remarks on colour-blindness, *Transac. R Soc. Edinburg*, **21**, 275–298 (1855).
18. Maxwell JC, On the theory of compounds colours, and the relations of the colours of the spectrum, *Phil. Trans. of the Royal Soc.* (London) **150**, 57–84 (1860).
19. Schrödinger E, Grundlinien einer theorie der farbenmetrik im tagessehen (I. II. III. Mitteilungen), *Ann. Phys.*, IV Folge, **63**, 397–426, 427–456, 481–520 (1920).
20. Judd DB, The 1931 ICI standard observer and the coordinate system for colorimetry, *J. Opt. Soc. Amer.*, **23**, 359–374 (1933).

11

CIE Standard Psychometric Systems

11.1 Introduction to Psychometric Systems in Colour Vision

As the psychophysical colorimetric systems, presented in Chapter 9, are based on the matching in colour of different colour stimuli,

the psychometric colorimetric systems are based[1, 2]

1. on the judgement of the differences between magnitudes of attributes of light and colour "such that equal scale intervals represent approximately equal perceived differences in the attribute considered" and
2. on the radiometric measurement of the colour stimuli present in the scene.

As almost 'unrelated colours' are considered in psychophysical colorimetry, 'related colours' are considered in psychometric colorimetry (Section 4.6).

In colour matching experiments two colour stimuli, matching in colour, are present in the visual field. At least two colour stimuli are needed to have related colours, but the effect is seen if the stimuli are not matching. In psychometric colorimetry, at least three colour stimuli are present: the two stimuli in comparison and the white (or grey) neutral stimulus, produced by the perfect reflecting diffuser lit by the chosen illuminant.

The first goal of psychometric colorimetry is to have the colour specification on uniform perceived scales, that is, equal differences of colour specifications represent equal differences of perceived colour.
    The second goal of psychometric colorimetry is to have the colour specification dependent on the illuminant in accordance with chromatic adaptation phenomena.

The CIE produced different psychometric systems starting from the psychophysical observers CIE 1931 and CIE 1964, and following two different ways:

1. Preserving the centre of gravity rule of the chromaticity diagrams in order to have an easy application of the additive colour mixing (Section 6.13.8);

Standard Colorimetry: Definitions, Algorithms and Software, First Edition. Claudio Oleari.
© 2016 John Wiley & Sons, Ltd. Published 2016 by John Wiley & Sons, Ltd.

2. Having a colour space structure with uniform scales close to that suggested by colour atlases, for example, the Munsell atlas (Section 14.4), and with a coordinate system with an evident correspondence with the Hering opponencies (Sections 4.4, 6.16).

Today, the second way could be considered as starting from psychophysics and physiology, which have shown that there are several steps in visual signal processing, which are summarized in the following sequence:

- Transduction, according to the Rushton univariance principle (Section 6.13.1);
- The adaptation to the illuminant (*chromatic adaptation transform*), according to the von Kries law (Section 6.20.3);
- Linear mixing of signals, analogously to the Exner-Schrödinger 'Helligkeit' equation (Section 6.13.6);
- Signal compression, according to the Weber-Fechner law or the Stevens power law (Section 6.9);
- Transformation of signals in colour opponencies and lightness, as shown by Hurvich and Jameson (Section 6.16).

In 1976, the CIE proposed two psychometric systems:

1. the *CIE 1976 (L*, u*, v*)*, officially termed CIELUV and made according to way 1.;
2. the *CIE 1976 (L*, a*, b*)*, officially termed CIELAB and defined on the Adams-Nickerson colour difference formula with a cube root expression.[3] Today CIELAB could be considered conceived according to way 2.

R. Hunt on this twofold CIE proposal wrote[1]: "The existence of two different CIE 1976 Colour Spaces and associated colour-difference formulae has arisen partly from different industrial requirements, but also partly from the limited success of either formula in predicting the perceptual magnitude of colour differences accurately."

Only in 1990 was an historical development of the CIELAB and CIELUV spaces with related colour difference formula published,[3] and its reading is instructive for understanding the origin of these systems, which have been conceived mainly following a 'technical' way.

The proposal of two CIE colour systems produced several misconceptions about the distinction between the two systems and the circumstances in which one should be used and not the other.[3] After 40 years, it is evident that CIELAB has prevailed over CIELUV: this is due to the development of small colour difference formulae in CIELAB and not in CIELUV, and not to a solving of the misconceptions. The proposal of two systems with unclarified aims has the flavour of a political and technical compromise,[3] that is not a good service to science.

In particular, the psychometric lightness L^* of the spaces CIELUV and CIELAB is the same, and this quantity is defined here separately.

11.2 CIE Lightness L^*

The lightness is defined for related colours (Section 4.6), therefore the test sample (X, Y, Z) is considered in relation to the perfect reflecting diffuser $(X_n, Y_n = 100, Z_n)$ lit by the same illuminant, that is supposed white or neutral [Y and Y_n are percentage luminance factor (Section 9.7)]. The CIE lightness

$L*$ has been conceived for substituting the unwieldy quintic expression of the Munsell value[3] [Section 6.14, Equation (14.2) of Section 14.4.3]. The lightness defined by the CIE is a function of the percentage luminance factor Y of test colour over the percentage luminance factor Y_n of a perfect reflecting diffuser

$$L* = \left[116 \left(\frac{Y}{Y_n} \right)^{1/3} - 16 \right] = \left[(29 \times 4) \left(\frac{Y}{Y_n} \right)^{1/3} - 16 \right] \text{ for } \left(\frac{Y}{Y_n} \right) > \left(\frac{6}{29} \right)^3 \approx 0.008856$$

$$L* = \left[\left(\frac{29}{3} \right)^3 \left(\frac{Y}{Y_n} \right) \right] \approx 903.3 \left(\frac{Y}{Y_n} \right) \qquad\qquad \text{for } \left(\frac{Y}{Y_n} \right) \le \left(\frac{6}{29} \right)^3 \approx 0.008856$$

$$(11.1)$$

This definition applies to the CIE 1931 observer and to the CIE 1964 observer (in this case the subscript '10' should be used).

The scale of the lightness $L*$ is written with the choice of the cubic root compression in accordance with the Stevens power law (Section 6.8).

This function has negative values for $(Y/Y_n) < (16/116)^3$, therefore, in order to avoid these negative values, a linear shape is chosen for $(Y/Y_n) \le (2/29)^3$: this line is radiating from the origin of the coordinates and is tangent to the cubic root function. This is a simple, but very technical solution.[4]

The adjustment of the parameters of equation (11.1) is shown in the Robertson paper.[3]

The Helmholtz-Kohlraush effect (Section 6.15) and the 'crispening' (Section 4.7) are not considered in defining the lightness $L*$.

11.3   Psychometric Chromaticity Diagrams and Related Colour Spaces

The known CIE chromaticity diagrams (x, y) and (x_{10}, y_{10}) have non-uniform perceptual scales, that is, the Euclidean distances measured on them are not equal or proportional to the differences of the corresponding perceived colours. There was a great practical need to have a perceptual uniform scale diagram. This led scientists to propose projective transformations aimed at transforming the MacAdam ellipses (Section 6.17) approximately into equal radius circles (Section 16.8.3). The construction of psychometric colour spaces first began with the search for a chromaticity diagram with uniform scales. It was a mathematical game apparently separated from physiology and psychophysics. On this line stands the construction of the *CIE 1960 UCS* chromaticity diagram (UCS stands for Uniform Colour Space, but also Uniform Colour Scale, Uniform Chromaticity Scale, Uniform Chromaticity Space).

Two psychometric systems were built from this chromaticity diagram:

- the *CIE 1964* ($U*$, $V*$, $W*$) colour space, also known as CIEUVW,
- the *CIE 1976* ($L*$, $u*$, $v*$) colour space, commonly known as CIELUV.

11.3.1   CIE 1960 (u, v) UCS Psychometric Chromaticity Diagram

Here the CIE 1931 observer is considered.

A chromaticity diagram with more uniform chromaticity scales is derived from a projective transformation. For a better understanding of this transformation, we shall do it in two steps. For the chromaticity diagram, the plane $X + 15Y + 3Z = 1$ is considered instead of the unitary plane $X + Y + Z = 1$

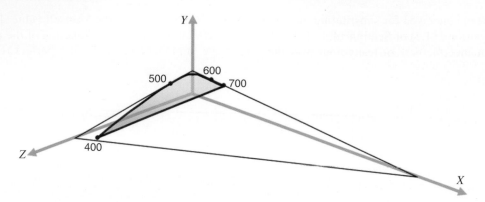

Figure 11.1   *Perspective view of the CIE 1931 tristimulus space with spectrum locus plotted on the plane X + 15Y + 3Z = 1. This figure plotted on this plane and projected from infinity on the plane Z = 0 becomes the chromaticity diagram (u, v).*

(Figure 11.1). The coordinates of the intersection point between a tristimulus vector $\mathbf{Q} = (X, Y, Z)$ and this plane are

$$
\begin{aligned}
x_Q &= \frac{X}{X + 15Y + 3Z} = \frac{x}{x + 15y + 3z} \\
y_Q &= \frac{Y}{X + 15Y + 3Z} = \frac{y}{x + 15y + 3z} \\
z_Q &= \frac{Z}{X + 15Y + 3Z} = \frac{z}{x + 15y + 3z}
\end{aligned}
\tag{11.2}
$$

Then the projection of these intersection points from infinity on the plane $Z = 0$ is made, and finally this diagram is dilated in two directions four and six times, respectively. In this way, the new coordinates of the CIE 1960 UCS chromaticity diagram are obtained (Figure 8.2a)

$$
u = \frac{4x}{x + 15y + 3z}, v = \frac{6y}{x + 15y + 3z}
\tag{11.3}
$$

Colour Temperature and Correlated Colour Temperature

Colour temperature, denoted by T_c, of a given Planckian radiation (Sections 3.3, 8.3) is the temperature measured in kelvin [K]. The colour stimulus of a Planckian radiation is specified by its chromaticity or its colour temperature.

For colour stimuli with chromaticity equal to no one Planckian radiator at whatever temperature, the *correlated colour temperature*, denoted by $T_{c,p}$ or by *CCT* and measured in kelvin [K], is defined as the temperature of the Planckian radiation having the chromaticity nearest to that of the considered colour stimulus, specified on the chromaticity diagram (u, v) CIE 1960 UCS (Figure 8.2). It follows that in the (u, v) CIE 1960 chromaticity diagram the lines orthogonal to the Planckian locus are conventionally assumed as *colour isotherm (isotemperature) lines* (Figure 8.2).

The correlated colour temperature is defined only for colour stimuli with chromaticity (u, v) with a distance from the chromaticity (u_P, v_P) of the nearest Planckian radiator $[(u - u_P)^2 + (v - v_P)^2]^{1/2} \leq 0.05$.

The minimum perceptible difference between two illuminants depends on the difference between the reciprocal of their temperatures, rather than the difference between their colour temperatures, therefore in 1932 Irwin G. Priest[5] introduced the *mired* (contraction of *micro reciprocal degree*).

$$M = \frac{10^6}{T_c} \ \text{MK}^{-1}$$

(11.4)

The corresponding unit of measure in the SI is the *mutual megakelvin*, also termed *micro reciprocal kelvin* and abbreviated with *mirek*, but this term is less used than mired. The just noticeable difference between illuminants is approximately equal to 5.5 mired.

11.3.2 CIE 1964 (*U**, *V**, *W**) Uniform Colour Space – CIEUVW Colour Space

Over time, many proposals for uniform scale spaces have been made. In 1964 the CIE proposed the CIEUVW space defined on the (*u*, *v*) CIE 1960 UCS chromaticity diagram for the CIE 1931 observer

$$W^* = 25Y^{1/3} - 17$$
$$U^* = 13W^*(u - u_n)$$
$$V^* = 13W^*(v - v_n)$$

(11.5)

where *Y* is the percentage luminance factor (Section 9.7) of the test sample and the chromaticity coordinates (u_n, v_n) refer to a defined white or neutral colour stimulus (X_n, $Y_n = 100$, Z_n) specifying the illuminant used. The illuminant adaptation is obtained by applying the *Judd-type white-point shift* (translation), that is, by subtracting the neutral-colour-stimulus chromaticity (u_n, v_n) from the test-stimulus chromaticity (*u*, *v*).

The Euclidean distance was chosen for the colour-difference formula (subscripts '1' and '2' distinguish the two colour samples in comparison)

$$\Delta E_{\text{CIEUVW}} = \sqrt{(U^*_1 - U^*_2)^2 + (V^*_1 - V^*_2)^2 + (W^*_1 - W^*_2)^2}$$

(11.6)

These coordinates are an approximation of the subsequent CIELUV 1976 (*L**, *u**, *v**):

$$W^* \simeq L^*, U^* \simeq u^*, V^* \simeq (2/3)v^*$$

(11.7)

This coordinate system is considered obsolete, but it is reported here because this space is used for the definition of the *colour-rendering index* (Section 16.8.8).

11.3.3 CIE 1976 (*u′*, *v′*) UCS Psychometric Chromaticity Diagram

The diagram CIE 1976 (*u′*, *v′*) holds true when derived from both the CIE 1931 and CIE 1964 observers (in this second case the subscript '10' should be written).

In 1976, CIE improved the uniformity of scale of the previous CIE 1960 (*u*, *v*) chromaticity diagram with a simple rescaling of the coordinate *v* by the factor 1.5, and so defined the *CIE 1976 UCS psychometric chromaticity*

$$u' = u, \ v' = 1.5v$$

(11.8)

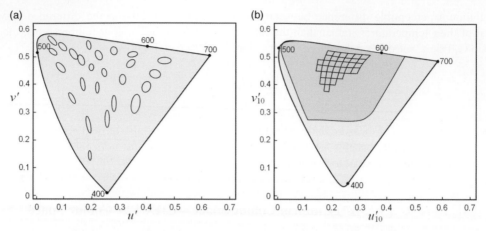

Figure 11.2   *(a) MacAdam's ellipses magnified by a factor of 10 plotted on the (u′, v′) chromaticity diagram and (b) lattice of chromaticities of the samples of the OSA-UCS system and MacAdam's limit at lightness $L_{OSA} = 0$ plotted on the (u′₁₀, v′₁₀) chromaticity diagram. The ellipses are not equal radius circles on (u′, v′) chromaticity diagram, anyway the chromaticity scales in this diagram are much better than those of the (x, y) CIE 1931 diagram (Figure 6.40); the OSA-UCS net on the (u′₁₀, v′₁₀) chromaticity diagram is evidently not square. Therefore these diagrams do not have uniform scales.*

The inverse transformation, from the coordinates (u', v') to (x, y), is

$$x = \frac{9u'}{6u' - 16v' + 12}, \; y = \frac{4v'}{6u' - 16v' + 12} \tag{11.9}$$

This chromaticity diagram was obtained by projections implemented in tristimulus space with the aim of improving the uniformity of the chromaticity scales. To confirm this improvement of scale, Figure 11.2(a) shows the new diagram (u', v') with the MacAdam ellipses and Figure 11.2(b) shows the lattice of chromaticities of the OSA-UCS system at lightness $L_{OSA} = 0$ (Section 14.6).

On the new chromaticity diagram the illuminant chromaticity is considered as that of the achromatic or neutral point (u'_n, v'_n) to which the observer is adapted, according to the *Judd-type white-point shift* proposal.

Moreover, a polar coordinate system is defined for better representing the perceived colour:

- the *hue angle* in (u', v') chromaticity diagram, which replaces the concept of dominant wavelength defined in the psychophysical system,

$$h_{uv} = \arctan\left(\frac{v' - v'_n}{u' - u'_n}\right) \text{deg} \tag{11.10}$$

- the *saturation* in (u', v') chromaticity diagram, written as the Euclidean distance between the chromaticity of the test sample and that of the neutral point

$$s_{uv} = 13\sqrt{(u' - u'_n)^2 + (v' - v'_n)^2} \tag{11.11}$$

The factor 13 has the purpose of making the scale of the saturation more practical and meaningful.

This diagram is obtained from the diagram (x, y) by a sequence of projections and translations; therefore, it is possible to continue to apply the law of the centre of gravity rule.

The transformations between the different CIE colorimetric systems are possible with the software presented in Sections 16.8.1 and 16.8.2.

11.3.4 CIE 1976 (*L**, *u**, *v**) Colour Space – CIELUV Colour Space

The system CIELUV applies to both the CIE 1931 and CIE 1964 observer (in this case the subscript '10' should be used).

CIE defines the CIELUV 1976 colour space as "approximately uniform" and says that CIELUV 1976 "is intended to apply to comparisons of colour differences between object colours of the same size and shape, viewed in identical white to middle-grey surroundings, by an observer photopically adapted to a field of chromaticity not too different from that of average daylight."[6] This sentence induces us to suppose that this system has been conceived for large colour differences with colour samples not in direct contact. Anyway, the chromaticity diagram (*u'*, *v'*), used for constructing CIELUV, has been defined by optimizing the uniformity of scale by transforming the MacAdam ellipses as close as possible to equal radius circles and these ellipses represent one standard deviation in colour matching on C-white background (Figure 11.2a).

The CIELUV system is defined for related colours, therefore the test sample (X, Y, Z) is considered in relation to the perfect reflecting diffuser (X_n, $Y_n = 100$, Z_n) lit by the same illuminant, that is supposed white or neutral. The lightness is L^*, already defined (Section 11.2). This system is based on the (*u'*, *v'*) CIE 1976 psychometric chromaticity diagram and any colour is specified by the coordinates (L^*, u^*, v^*), where u^* and v^* are defined as follows

$$u^* = 13L^*(u' - u'_n)$$
$$v^* = 13L^*(v' - v'_n)$$

(11.12)

The illuminant adaptation is obtained by applying the *Judd-type white-point shift*.

The three coordinates are dimensionally homogeneous.

The definition of the coordinates (*u'*, *v'*) is such that on this chromaticity diagram the Newton centre of gravity rule holds true. Newton's rule is very useful for the additive mixing of colours (Section 6.13.8).

A polar coordinate system is also defined on each plane at constant defined lightness:

- the CIELUV *hue angle*, equal to the corresponding quantity defined in the CIE 1976 UCS chromaticity diagram [Section 11.3.3, Equation (11.10)],

$$h_{uv} = \arctan\left(\frac{v^*}{u^*}\right) \text{ deg}$$

(11.13)

- the CIELUV *chroma*

$$C^*_{uv} = \sqrt{u^{*2} + v^{*2}} = L^* s_{uv}$$

(11.14)

which differs from the saturation *suv* defined on the CIE 1976 UCS chromaticity diagram (Section 11.3.3) by a factor equal to the lightness.

Here Equation (11.14) shows the difference between chroma and saturation.

The space (L^*, u^*, v^*) consists in a continuous set of parallel planes at constant lightness, on which there are diagrams with approximately uniform scales, that, having scale proportional to the lightness, take account of the colour discrimination with increasing lightness.

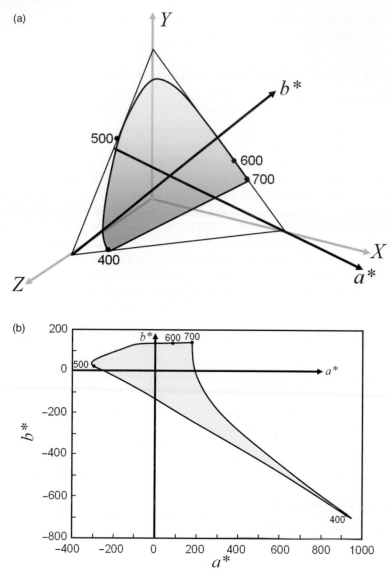

Figure 11.3   *(a) Perspective view of the XYZ CIE 1931 tristimulus space and unit plane X + Y + Z = 1 with the lines representing the a* and b* axes, which subdivide the plane in four quadrants representing approximately the red-green and yellow-blue opponencies. (b) Spectrum locus on the plane at luminance factor Y = 50 in CIELAB space for the CIE 1931 observer. CIELAB transformation maps the points of figure (a) into those of (b). The spectrum locus in CIELAB space shows a very irregular shape: if we do not consider self-luminous colours, the spectrum locus has little importance, in fact it is almost always ignored, while the MacAdam limit is more significant (Figure 11.5 and 11.6).*

In this space the difference between the colours is represented by their Euclidean distance, that is, taking two colours $(L*_1, u*_1, v*_1)$ and $(L*_2, u*_2, v*_2)$, their difference is

$$\Delta E*_{uv} = \sqrt{(L*_1 - L*_2)^2 + (u*_1 - u*_2)^2 + (v*_1 - v*_2)^2}$$
$$= \sqrt{(\Delta L*)^2 + (\Delta u*)^2 + (\Delta v*)^2} \tag{11.15}$$

The chromaticity scales[7] of this space are still not exactly uniform and this can be seen in Figures 11.2a and 11.2b.

The transformations between the different CIE colorimetric systems are possible with the software presented in Sections 16.8.1 and 16.8.2. The CIELAB systems and CIELUV are presented visually with the software described in Section 16.6.

11.3.5   CIE 1976 ($L*$, $a*$, $b*$) Colour Space – CIELAB Colour Space

The CIELAB system applies to both observers CIE 1931 and CIE 1964 (in this second case the subscript '10' should be used).

As for the CIELUV system, the CIE defines the CIELAB 1976 colour space "approximately uniform" and says that CIELAB 1976 "is intended to apply to comparisons of colour differences between object colours of the same size and shape, viewed in identical white to middle-grey surroundings, by an observer photopically adapted to a field of chromaticity not too different from that of average daylight."[6] This induces us to suppose that this system has been conceived for large colour differences with colour samples not in direct contact. Pointer and Attridge[7] wrote:

> "… the origins of CIELAB Uniform Colour Space can be found in the Adams Chromatic Value Diagram,[8] which attempted to regularize the spacing of the samples of the *Munsell Book of Color*. It can thus be argued that this 'space' is not based on just-perceptible difference data at all, but on what are now called 'large' colour-difference data, because the spacing of the Munsell samples in CIELAB space, is usually greater than 4 units."[9–10]

The CIELAB system is defined for related colours, therefore the test sample (X, Y, Z) is considered in relation to the perfect reflecting diffuser $(X_n, Y_n = 100, Z_n)$ lit by the same illuminant, that is supposed white or neutral. The CIELAB 1976 system is represented by a three dimensional space $(L*, a*, b*)$, of which the first coordinate is the CIE 1976 lightness $L*$, already defined (Section 11.2), and the other two are defined by non-linear functions on the tristimulus space (X, Y, Z) (Figures 11.3a and 11.3b) according to the following scheme:

1. The tristimulus values (X, Y, Z) are a linear mixing of the cone activations and are implicitly assumed meaningful in visual processing, although only Y has a psychophysical reason, while X and Z are defined to have positive values for real colour stimuli.
2. The adaptation to the illuminant is made according to the modified von Kries law, where the ratios (X/X_n), (Y/Y_n) and (Z/Z_n) are used instead of the corresponding ratios of the cone activations.
3. A signal compression is made, according to Stevens's power law with power 1/3 (Section 6.8).
4. A signal mixing defines the coordinates $a*$ and $b*$ to reproduce the colour opponencies (Sections 4.4 and 6.16):

$$a* = 500\left[\left(\frac{X}{X_n}\right)^{1/3} - \left(\frac{Y}{Y_n}\right)^{1/3}\right], \quad b* = 200\left[\left(\frac{Y}{Y_n}\right)^{1/3} - \left(\frac{Z}{Z_n}\right)^{1/3}\right] \tag{11.16}$$

The adjustment of the parameters is shown in the Robertson[3] paper.

Similarly to what has been done in defining the lightness L^* for (Y/Y_n) close to zero, here it is done with regard to all three ratios: if one of the ratios (X/X_n), (Y/Y_n) and (Z/Z_n) is lower than or equal to $(6/29)^3 \approx 0.008856$, its cube root in Equations (11.16) is replaced by $[(841/108)\, F + (16/116)]$, where F is the ratio itself and $(841/108) \approx 7.787$.

The coordinate a^* approximately represents the red-green opponency and b^* the yellow-blue opponency, that is:

- if $(X/X_n)^{1/3} > (Y/Y_n)^{1/3}$ the hue is reddish, while if $(X/X_n)^{1/3} < (Y/Y_n)^{1/3}$ the hue is greenish and
- if $(Y/Y_n)^{1/3} > (Z/Z_n)^{1/3}$ the hue is yellowish, while if $(Y/Y_n)^{1/3} < (Z/Z_n)^{1/3}$ the hue is bluish.

It must be said that in 1976 this space was defined for (X/X_n), (Y/Y_n) and (Z/Z_n) greater than 0.01, and the extension to lower values was made on the suggestion of Pauli[4], which was approved by CIE in 1977 and published as an appendix.

On each plane at constant defined lightness also a polar coordinate system is defined (Figure 11.4):

- the *CIELAB hue angle*

$$h_{ab} = \arctan\left(\frac{b^*}{a^*}\right)\ \text{deg} \tag{11.17}$$

- the *CIELAB chroma*,

$$C^*_{ab} = \sqrt{a^{*2} + b^{*2}} \tag{11.18}$$

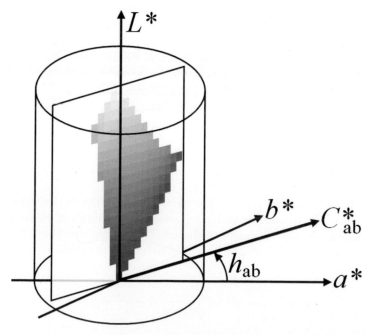

Figure 11.4   *Perspective view of cylindrical coordinates in CIELAB space.*

In this space, the difference between the colours is assumed to be represented by their Euclidean distance, that is, taking two colours $(L^*_1, a^*_1, b^*_1, C^*_{ab,1}, h_{ab,1})$, and $(L^*_2, a^*_2, b^*_2, C^*_{ab,2}, h_{ab,2})$, the *CIELAB colour difference* is

$$\Delta E^*_{ab} = \sqrt{(L^*_1 - L^*_2)^2 + (a^*_1 - a^*_2)^2 + (b^*_1 - b^*_2)^2}$$
$$= \sqrt{(\Delta L^*)^2 + (\Delta a^*)^2 + (\Delta b^*)^2} \tag{11.19}$$

The CIELAB hue difference is

$$\Delta H^*_{ab} = 2\sqrt{C^*_{ab,1} \, C^*_{ba,2}} \, \sin\left(\frac{h_{ab,1} - h_{ab,2}}{2}\right) \tag{11.20}$$

where the argument of the sine function has to be measured in radians.

The chromatic scales of this space are non-uniform. This is evident on considering on planes at constant lightness MacAdam ellipses (Section 6.17) (Figure 11.5a) for the 2°observer and the OSA-UCS grid for the 10°observer (Section 14.6) (Figure 11.5b). Pointer and Attridge wrote[7]: "It is reasonable and of interest to consider the application of these colour difference formulae to the analysis of large colour differences, with magnitudes typical of those achieved by photographic and other colour reproduction systems, typically in the range of 0–20 CIELAB units."

Joy Turner Luke[11] wrote that "Charles Reilly was largely responsible for the CIELAB colour difference formula, but did not introduce Semmelroth's correction (OSA-UCS Section 14.6.1, lightness scale Section 6.14), even though he believed it made an improvement, because he said that the international body was close to agreement and any new element might delay approval for several years. An internationally accepted colour difference formula was urgently needed to bring order into the chaos created by the many different formulas in commercial use."

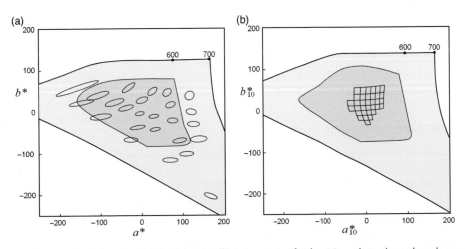

Figure 11.5   *(a) MacAdam's limit and MacAdam's ellipses – magnified × 10 – plotted on the plane at luminance factor Y = 50 in the CIELAB space for CIE 1931 observer. The ellipses show an increased regularity of scale only in the region of achromatic colours, that is in proximity with the coordinate origin a* = 0 and b* = 0. (b) Lattice of chromaticity of the samples of the OSA-UCS system at lightness L_{OSA} = 0 and MacAdam's limit on the plane at equal luminance factor in CIELAB space for the CIE 1964 standard observer.*

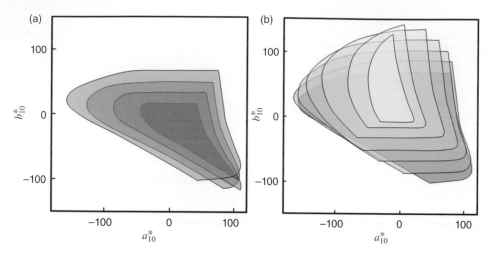

Figure 11.6 *(a) MacAdam's limit related to the illuminant D65 and on the various planes spanned by the* (a^*_{10}, b^*_{10}) *coordinates at percent reflectance factor Y = 10, 20, 30, 40 and (b) Y = 40, 50, 60, 70, 80, 90, 95 (grey level is representative of the Y value).*

Figure 11.6 shows the Macadam limit for different values of the luminance factor percentage in the CIE-LAB space, defined by the illuminant D65 (Section 14.2.1). The white D65 is a point at the top and the black is a point at the bottom. Let us recall that the MacAdam limit defines a region in colour space containing all the non-self-luminous colours.

The transformations between the different CIE colorimetric systems are possible with the software presented in Sections 16.8.1 and 16.8.2. The CIELAB systems and CIELUV are presented visually with the software described in Section 16.6.

11.4   Colour Difference Specification

In practice, colour specifications often involve, not merely the quotation of nominal values, but also the definition of tolerances for colour differences from them (Section 6.18). In this section, the colour difference is considered according to the definition of colour difference formulae in the CIE 1976 systems.

11.4.1   Colour Difference Data

The CIELUV and CIELAB systems are intended to apply to comparisons of colour differences[6]. As already written, their viewing condition given above induces us to suppose that the colour samples are not in direct contact, but are separated, and then the CIELUV and CIELAB have been conceived for the large colour differences (This is not clearly written in the CIE publication.[6]) The representation of the small colour differences in the CIELAB space confirms this supposition (Figure 11.7).

Both CIELUV and CIELAB systems appeared inadequate for a representation of Euclidean colour differences, therefore it was necessary to define colour difference formulae in these spaces. In 1978, Roberson[12] published the CIE guidelines for a coordinated research into colour-difference evaluation. The goal was "to establish a comprehensive set of data describing the perceptibility of small, moderate and large colour differences under a variety of viewing conditions." After a few years, there was a considerable amount of data of empirical colour differences, although not following completely the CIE guidelines. A very important part

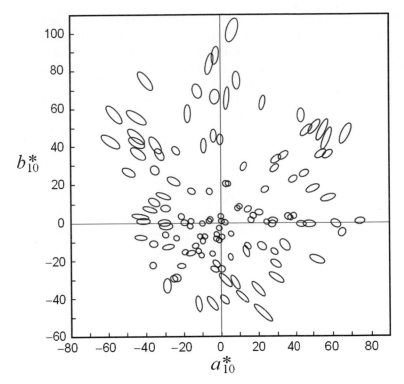

Figure 11.7 *BFD ellipses plotted on a constant lightness plane in CIELAB space. The ellipses have the major axis approximately oriented along straight lines outgoing from the origin of the axes, excluding the ellipses in the region of the blue colours. This regularity, although very approximate, suggests using cylindrical coordinates (Section 6.18).*

of this data is summarized in a set of ellipses at constant lightness, known as BFD ellipses[13] (Section 6.18), that define the distance from a sample, the centre of the ellipse, corresponding to the discrimination threshold of the perception or the acceptance by the market (Figure 11.7). Over time, to these data are added pairs of colours produced by various laboratories, whose colour differences are empirically specified. This amount of data is the empirical basis on which the various colour-difference formulas produced in the last 30 years are based.[14]

Almost all these colour-difference data are associated to the CIE 1964 observer and to the D65 illuminant. Therefore the colour-difference formulae based on these data are expected to be valid only for this illuminant and this observer. Anyway the CIELAB and CIELUV systems take into account the chromatic adaptation to a change in illuminant, although very raw and approximate, and it induces us to use these formulae for other illuminants and both CIE observers, but considering the computed quantities as only approximate and indicative.

11.4.2   CIE 1976 Colour-Difference Formulae

Consider again that CIELUV and CIELAB spaces "are intended to apply to comparisons of differences between object colours of the same size and shape, viewed in identical white to middle-grey surroundings, by an observer photopically adapted to a field of chromaticity not too different from that of average daylight."[6] Anyway, these systems are not with completely uniform scale and therefore the formulas with which to evaluate the colour differences proposed by the CIE (11.15) and (11.19) need corrections. There have been many suggested corrections over time. The formulas for these corrections depend on two or more parametric factors, that have to be defined according to the user needs and to the sector of application (textiles, paints, plastics, …). In fact, two colours can have *perceptible* difference but this cannot prevent the market from considering the difference as *acceptable*. Acceptability and perceptibility relate science and practice: what does not apply to the first can be true for the second, but not vice versa. The value assigned to these parametric factors is now typical of each sector of use.

In industrial colorimetric practice, the CIELAB for the CIE 1964 observer is the most widely used system and it is mainly for this space that colour-difference formulas are given, although these formulae are used also for the CIE 1931 observer.

The BFD colour-difference empirical data (Section 6.18) show a certain regularity in the CIELAB space, particularly the ellipses at constant lightness (Figure 11.7), therefore the (L^*, C^*_{ab}, h_{ab}) cylindrical coordinate system is preferable to represent colour differences and to define colour-difference formulae. In cylindrical coordinates, the differences ΔL^*, ΔC^*_{ab} and ΔH^*_{ab} are defined and the colour difference formula (11.19) becomes

$$\Delta E^*_{ab} = \sqrt{(\Delta L^*)^2 + (\Delta C^*_{ab})^2 + (\Delta H^*_{ab})^2} \tag{11.21}$$

where the chroma difference is

$$\Delta C^*_{ab} = C^*_{ab,1} - C^*_{ab,2} \tag{11.22}$$

and the absolute hue-difference, derived from Equations (11.21) and (11.22), is

$$\left|\Delta H^*_{ab}\right| = \sqrt{(\Delta E^*_{ab})^2 - (\Delta L^*)^2 - (\Delta C^*_{ab})^2} \tag{11.23}$$

with ΔE^*_{ab} defined by equation (11.19). For small colour differences, the difference of hue $\left|\Delta H^*_{ab}\right|$ can be expressed by the difference in the hue angles

$$\left|\Delta H^*_{ab}\right| = \sqrt{C^*_{ab,1} C^*_{ab,2}} \left|\Delta h_{ab}\right| \left(\frac{\pi}{180}\right) \text{ with } \left|\Delta h_{ab}\right| = \left|h_{ab,1} - h_{ab,2}\right| \text{ deg} \tag{11.24}$$

11.4.3   CMC(*l* : *c*) Colour-Difference Formula

The colour-difference formula recommended by the 'Colour-Measurement Committee' of the Society of Dyers and Colourists[15–17] (UK), denoted with CMC(*l* : *c*), describes an ellipsoid in the space $(L^*, C^*_{ab}, H^*_{ab})$ in which the parameters *l* and *c* are involved. For values of the parameters *l* = 1 and *c* = 1, the ellipsoid defines the region of the *imperceptibility* below threshold.[18] On increasing the value of the parameters in the region of the ellipsoid, there are differences in colour that may be perceptible but *acceptable* by the market. In practice for market acceptability, *c* = 1 and *l* varies between 1 and 2.

The CMC equation that defines the colour difference between a test colour $(L^*_2, C^*_{ab,2}, H^*_{ab,2})$ and a standard $(L^*_1, C^*_{ab,1}, H^*_{ab,1})$ is

$$\Delta E_{CMC}(l:c) = \sqrt{\left(\frac{\Delta L^*}{lS_L}\right)^2 + \left(\frac{\Delta C^*_{ab}}{cS_C}\right)^2 + \left(\frac{\Delta H^*_{ab}}{S_H}\right)^2} \tag{11.25}$$

where

$$S_L = \begin{cases} 0.040975\dfrac{L_1^*}{1+0.01765L_1^*} & \text{for } L_1^* \geq 16 \\[2mm] 0.511 & \text{for } L_1^* < 16 \end{cases}$$

$$S_C = \frac{0.0638C_{ab,1}^*}{1+0.0131C_{ab,1}^*} + 0.638$$

$$S_H = S_C\left(Tf + 1 - f\right)$$

with

$$f = \sqrt{\frac{\left(C_{ab,1}^*\right)^4}{\left(C_{ab,1}^*\right)^4 + 1900}},$$

$$T = \begin{cases} 0.36 + |0.4\cos(h_{ab,1}+35)| & \text{for } h_{ab,1} \geq 345° \text{ or } h_{ab,1} \leq 164° \\[2mm] 0.56 + |0.2\cos(h_{ab,1}+168)| & \text{for } 164 < h_{ab,1} < 345° \end{cases}$$

The relative goodness of this formula for small colour differences is confirmed by colorimetric practice.

The intricacy of the parameters of the ellipsoid as a function of the point in space reveals that these parameters are obtained by optimization of the ellipsoid on experimental data and does not reveal any interpretive model.

This formula is now an ISO standard.[19]

11.4.4   CIE 1994 Colour-Difference Formula

In 1994, the CIE[20] proposed a new colour-difference formula for improving the previous 1976 (11.19), which is much simpler than the CMC (Sections 11.4.3 and 11.4.2)

$$\Delta E_{94}^* (k_L : k_C : k_H) = \sqrt{\left(\frac{\Delta L^*}{k_L S_L}\right)^2 + \left(\frac{\Delta C_{ab}^*}{k_C S_C}\right)^2 + \left(\frac{\Delta H_{ab}^*}{k_H S_H}\right)^2} \tag{11.26}$$

with

$$S_L = 1$$
$$S_C = 1 + 0.045C_{ab,s}^*$$
$$S_H = 1 + 0.015C_{ab,s}^*$$

where $C_{ab,s}^*$ is the chroma of the standard. If the standard of a sample pair is not clearly defined, the chroma $C_{ab,s}^*$ in S_C and S_H may be replaced by the geometric average $(C_{ab,s}^*\ C_{ab,b}^*)^{1/2}$, where $C_{ab,b}^*$ is the chroma of the test (batch).

For parametric factors $k_L = 1$, $k_C = 1$ and $k_H = 1$, the ellipsoid should define the region within which the differences in colour with respect to the centre of the ellipsoid are not perceptible. Here, too, the values of the parameters are defined on the basis of acceptable colour differences in the various fields of application, for example $k_L = 2$ is used within the textile industry.

A comparison of the CMC and CIE 1994 formulas reveals that the CIE 1994 tolerances are greater than the CMC ones and the difference seems to be a scaling factor.

11.4.5   CIEDE2000 Total Colour-Difference Formula

In 2000, "for correcting the non-uniformity of the CIELAB colour space for small colour differences under reference conditions", the CIE gave the *total colour-difference formula*, known as CIEDE2000[14,21] and denoted with ΔE_{00}. For the construction of this formula, the CIE

- considered an expanded set of experimental data, termed COM dataset[14];
- required observers with normal trichromatic colour vision;
- recommended a *reference conditions* for assessing the colour differences visually:
  - (i) the lighting source simulates the spectral power distribution of the standard illuminant D65;
  - (ii) the illuminance is approximately 1000 lux;
  - (iii) the vision is in object mode (Section 6.1.3);
  - (iv) the samples are compared on a uniform and achromatic background at lightness $L^* \approx 50$;
  - (v) the samples are in a field of view with a section greater than $4°$;
  - (vi) samples are placed in contact;
  - (vii) the difference in colour between the samples is less than 5 CIELAB units.

The CIEDE2000 formula is defined on the coordinates (L', a', b', C', h'), which are written as function of the known coordinates $(L^*, a^*, b^*, C^*_{ab})$ of the CIELAB system

$$
\begin{cases}
L' = L^* \\[2mm]
a' = (1+G)a^* \quad \text{with} \quad G = 0.5\left(1 - \sqrt{\dfrac{\overline{C}^{*\,7}_{ab}}{\overline{C}^{*\,7}_{ab} + 25^7}}\right) \\[4mm]
b' = b^* \\[2mm]
C' = \sqrt{a'^2 + b'^2} \\[2mm]
h' = \tan^{-1}\left(\dfrac{b'}{a'}\right) \text{ deg}
\end{cases}
\tag{11.27}
$$

where $\overline{C}^*_{ab}$ is defined by the arithmetic average of the values C^*_{ab} related to the sample pair in comparison. The formula is

$$
\Delta E_{00} = \sqrt{\left(\frac{\Delta L'}{k_L S_L}\right)^2 + \left(\frac{\Delta C'}{k_C S_C}\right)^2 + \left(\frac{\Delta H'}{k_H S_H}\right)^2 + R_T \left(\frac{\Delta C'}{k_C S_C}\right)\left(\frac{\Delta H'}{k_H S_H}\right)}
\tag{11.28}
$$

with

$$
S_L = 1 + \frac{0.015(\overline{L}' - 50)^2}{\sqrt{20 + (\overline{L}' - 50)^2}}
$$

$$
S_C = 1 + 0.045\overline{C}'
$$

$$
S_H = 1 + 0.015\overline{C}'T
$$

$$
T = 1 - 0.17\cos(\overline{h}' - 30°) + 0.24\cos(2\overline{h}') + 0.32\cos(3\overline{h}' + 6°) - 0.20\cos(4\overline{h}' - 63°)
$$

$$R_C = 2\sqrt{\frac{\bar{C}'^7}{\bar{C}'^7 + 25^7}}$$

$$R_T = -R_C \sin(2\Delta\theta) \quad \text{with} \quad \Delta\theta = 30\exp\left[-\left(\frac{\bar{h}' - 275°}{25}\right)^2\right]$$

$$\Delta L' = L'_b - L'_s$$

$$\Delta C' = C'_b - C'_s$$

$$\Delta H' = 2\sqrt{C'_b C'_s} \sin\left(\frac{\Delta h'}{2}\right) \quad \text{with} \quad \Delta h' = h'_b - h'_s$$

where

- the suffix 's' denotes the standard sample and 'b' the test (batch),
- $(\bar{L}', \bar{C}', \bar{h}')$ are the geometric averages of (L', C', h') related to standard and test values,
- the parametric factors k_L, k_C, k_H are *correction factors* for variation in experimental conditions (!?). Under reference conditions, they are all set at 1.

    The computation of $\bar{h}'$ must be careful in the case where the hue angles of the two colours belong to different quadrants, for example, if the standard and the test sample have hue angles of 90° and 300°, respectively, and the averaged value 195° is different from the correct value that is 15°. To avoid the error, it is necessary to assess the absolute difference between the two angles of the same colour. If this difference is less than 180°, it is considered the arithmetic average of the hue angles, otherwise the greater hue angle has to be reduced by an angle of 360° and then the arithmetic average is computed.

    An alternative form of the CIEDE2000 colour-difference equation, defined with three terms, is obtainable with the ordered computation of the following quantities[22]

$$\tan(2\phi) = R_T \frac{(k_H S_H)(k_C S_C)}{(k_H S_H)^2 - (k_C S_{HC})^2}$$

$$\Delta C'' = \Delta C' \cos\phi + \Delta H' \sin\phi$$

$$\Delta H'' = \Delta H' \cos\phi - \Delta C' \sin\phi$$

$$S_C'' = (k_C S_C)\sqrt{\frac{2(k_H S_H)}{2(k_H S_H) + R_T(k_C S_C)\tan\phi}}$$

$$S_H'' = (k_H S_H)\sqrt{\frac{2(k_C S_C)}{2(k_C S_C) - R_T(k_H S_H)\tan\phi}}$$

$$\Delta L_{00} = \frac{\Delta L'}{K_L S_L}, \quad \Delta C_{00} = \frac{\Delta C''}{S_C''}, \quad \Delta H_{00} = \frac{\Delta H''}{S_H''},$$

$$\Delta E_{00} = \sqrt{(\Delta L_{00})^2 + (\Delta C_{00})^2 + (\Delta H_{00})^2} \tag{11.29}$$

The publication of the formula CIEDE2000 rendered obsolete all previous formulas.

All the formulas for the colour difference given here are used in the software presented in Section 16.8.7).

11.4.6 Small Colour Differences in OSA-UCS Space

The colour - difference formulas proposed in the last 30 years have been defined mainly in the CIELAB space. The mathematical complexity to correct the irregular shape of empirical discrimination ellipsoids shows that the CIELAB space is not suitable to represent these data. The Committee, which studied the OSA-UCS system (Section 14.6), declared that the Euclidean formula for the perceived colour differences is incompatible with that system and this led to neglect it. In the OSA-UCS space a Euclidean formula represents the large colour differences in an approximate way. Instead, a graphical representation of small colour differences in this space shows a high degree of regularity to lead to propose a *hue-independence hypothesis*, which states that in the OSA-UCS space the discrimination ellipses have a hue independent shape and the discrimination ellipse centred in the coordinate origin is a circle with unitary radius. This hypothesis has led to define a formula for the small colour differences that is very simple and with only six parameters dependent on experimental colour-difference data[23] (Figure 11.8a).

Subsequently, by introducing a suitable logarithmic compression of the chroma, the formula assumed Euclidean shape[24], albeit approximate (of course, the Euclidean formula does not exist) (Figure 11.8b). In this case, the parameters in the formula dependent on the experimental colour-difference data are only 4. The quality of these formulae is equal to that of CIEDE2000 formula.[25]

11.4.7 Metamerism Indices

Metamerism is already defined in Sections 6.13.2. Generalizing, metamerism phenomenon exists if the summations (9.36), (9.37) and (9.38), do not change with the changing of the illuminant or reflectance\transmittance factor or observer (Although this definition clearly regards the colour match of different colour stimuli, the word *metamerism* is often incorrectly used to indicate a metameric failure rather than a match, or an easily degraded match by a slight change in illuminant conditions.):

(i)   light sources with different spectral power distribution $S(\lambda)$ and equal colorimetric specification are *metameric* (9.36) (the lights look the same despite having different spectra);

(ii)   surface colours having different reflectance factors $R(\lambda)$ and equal colorimetric specification (9.37) under a particular illuminant are *metameric* [in this case, the colours appear the same in one illuminant and can appear different in others; this is usually checked in the light box with the use of multiple sources (Section 13.2)];

(iii)   surface colours are *metameric* when they have different reflectance factors $R(\lambda)$ and equal colour specification (9.37), under a particular illuminant, for the colorimetric standard observer CIE 1931 but not for the supplementary standard observer CIE 1964 (in this case, the colours may look the same in foveal vision and different in extrafoveal vision), and vice versa.

The difference in the spectral compositions of two metameric stimuli is often referred to as the *degree of metamerism*. The sensitivity of a metameric match to any changes in the spectral elements that form the colours depends on the degree of metamerism. Two colour stimuli with a high degree of metamerism are likely to be very sensitive to any changes in the illuminant, material composition, observer, field of view and so on.

Metamerism is here reconsidered because the colour-difference formulas can be used for the definition of appropriate metamerism indexes[26]:

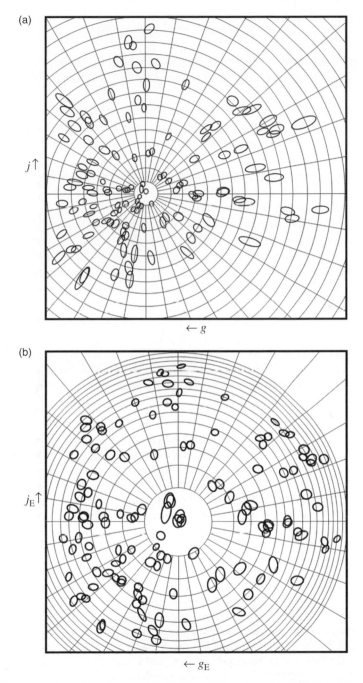

Figure 11.8   *(a) BFD ellipses plotted on the constant lightness plane $L_{OSA} = 0$ in OSA-UCS space. The ellipses have the major axis with good approximation oriented along straight lines outgoing from the origin of the axes for all the hues, suggesting the existence of a hue independence rule for the colour uncertainties in OSA-UCS space and a regular dependence on the distance from the achromatic point (chroma). This regularity suggests the introduction of a logarithmic compression of the chroma for transforming the ellipses into equal radius circles. (b) BFD ellipses plotted on the constant lightness plane $L_{OSA} = 0$ in OSA-UCS space after a suitable logarithmic compression of the chroma. The colour difference formula assumes Euclidean shape in this new coordinate system.*

"*Metamerism index*, degree of colour mismatch caused by substituting

1. a test illuminant of different relative spectral composition for the reference illuminant;
2. a test observer of different relative spectral responsivity for the reference observer."

The CIE recommends "that for two specimens, whose corresponding tristimulus values ($X_1 = X_2$, $Y_1 = Y_2$, $Z_1 = Z_2$) are identical with respect to a reference illuminant and observer, the *metamerism index*, M, be set equal to the colour difference ΔE^*_{ab} between the two specimens computed for the test illuminant or for the test observer."[6]

According to this definition of metamerism index, two special metamerism indexes are distinguished:

Special Metamerism Index: Change in Illuminant M_{ilm}

Case in which the metamerism failure is caused by substituting the illuminant.

The term *illuminant metameric failure* is used to describe situations where two specimens match when viewed under one light source but not another. Most types of fluorescent lights have an irregular or peaky spectral power distribution, so that two specimens lit by a fluorescent lamp might not match, even though they are a metameric match for a Planckian 'white' source, whose spectral power distribution is smooth.

The metamerism index for a change in illuminant regards pairs of surface colour, which are metameric under a particular reference illuminant, that is, $X_{r,1} = X_{r,2}$, $Y_{r,1} = Y_{r,2}$, $Z_{r,1} = Z_{r,2}$ (subscripts 1 and 2 distinguish the two specimens, the subscript 'r' denotes the reference illuminant and the subscript '10' has to be added for the CIE 1964 observer) (Figure 11.9). The CIE suggests the D65 illuminant as reference, otherwise, the reference illuminant has to be specified. At the change of the illuminant, the two metameric colours may have

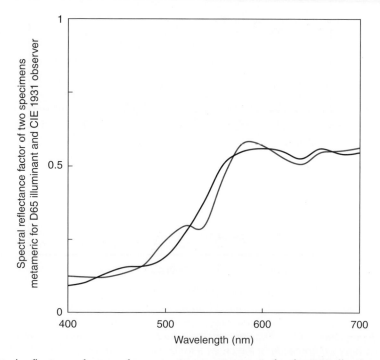

Figure 11.9 *Spectral reflectance factors of two specimens, metameric for the D65 illuminant and CIE 1931 observer.*

different colour specifications, the difference of which can be evaluated with one of the colour difference formulas recommended by CIE, ΔE^*_{ab} (11.19) or with other formulas (Sections 11.4.2–11.4.5). The metamerism index M_S is expressed by this colour difference and depends on the illuminant S, placed as a subscript, and the colour difference formula used.

If the two specimens are not exactly matching under the reference illuminant, that is, ($X_{r,1} \neq X_{r,2}$, $Y_{r,1} \neq Y_{r,2}$, $Z_{r,1} \neq Z_{r,2}$), suitable account should be taken of this failure, the size of which recorded and the tristimulus values ($X_{t,1}$, $Y_{t,1}$, $Z_{t,1}$) and ($X_{t,2}$, $Y_{t,2}$, $Z_{t,2}$), measured under the test illuminant (subscript 't' means test illuminant), have to be adjusted by the multiplicative method as follows ($X'_{t,2} = X_{t,2} (X_{r,1}/X_{r,2})$, $Y'_{t,2} = Y_{t,2} (Y_{r,1}/Y_{r,2})$, $Z'_{t,2} = Z_{t,2} (Z_{r,1}/Z_{r,2})$). Then the metamerism index is computed by a colour difference formula between ($X_{t,1}$, $Y_{t,1}$, $Z_{t,1}$) and ($X'_{t,2}$, $Y'_{t,2}$, $Z'_{t,2}$).

(a) *Special Metamerism Index: Change in Observer M_{obs}*[27].

The CIE 1931 and 1964 standard colorimetric observers should represent the colour vision properties of the average population. Nevertheless colour normal observers have individual deviations in the colour-matching functions and an *observer metameric failure* can occur. In 1989, the CIE defined a *standard deviate observer* (SDO) to classify real trichromatic observers whose colour matching functions deviate from those of the CIE standard colorimetric observer in a defined manner (Section 9.4). If a metamerism failure exists in the comparison of the deviate observers with the standard one, the metamerism index (ΔE^*_{ab}) is useful for the quantification of this deviation.

The *special metamerism index: change in observer* was introduced to describe the average degree of mismatch found among metameric colours if the colour-matching functions of one of the standard colorimetric observers are changed to those of a standard deviate observer of normal colour vision.

For the same pair of metameric object colours specified by ($X_{r,1}$, $Y_{r,1}$, $Z_{r,1}$) of object colour 1, and ($X_{r,2}$, $Y_{r,2}$, $Z_{r,2}$) of object colour 2 (subscript 'r' means reference observer, which is a standard observer; subscript '10' must be added for the CIE 10° observer), their tristimulus values ($X_{t,1}$, $Y_{t,1}$, $Z_{t,1}$) and ($X_{t,2}$, $Y_{t,2}$, $Z_{t,2}$) for the standard deviate observer (test) are computed by using the colour-matching functions of the standard deviate observer (Section 9.4).

The colour difference ΔE^*_{ab} is computed between the tristimulus values ($X_{t,1}$, $Y_{t,1}$, $Z_{t,1}$) of object colour 1, and ($X_{t,2}$, $Y_{t,2}$, $Z_{t,2}$) of object colour 2. Then the metamerism index M_{obs} is defined as $M_{obs} = \Delta E^*_{ab}$.

If a colour difference formula other than CIELAB is used, this should be included in the parenthesis as for example $M_{obs} (u^*v^*)$ for the CIELUV formula.

When the samples are not exactly metameric, that is ($X_{r,1} \neq X_{r,2}$, $Y_{r,1} \neq Y_{r,2}$, $Z_{r,1} \neq Z_{r,2}$), then the tristimulus values ($X_{t,2}$, $Y_{t,2}$, $Z_{t,2}$) are adjusted by the multiplicative method as follows ($X'_{t,2} = X_{t,2} (X_{r,1}/X_{r,2})$, $Y'_{t,2} = Y_{t,2} (Y_{r,1}/Y_{r,2})$, $Z'_{t,2} = Z_{t,2} (Z_{r,1}/Z_{r,2})$) and the colour difference is computed between ($X_{t,1}$, $Y_{t,1}$, $Z_{t,1}$) and ($X'_{t,2}$, $Y'_{t,2}$, $Z'_{t,2}$).

The CIE[6] writes: "Each colour normal observer shows a colour difference ΔE^* for a sample pair metameric with respect to a reference observer and an irradiating illuminant. About 95% of ΔE^*s for colour normal observers are usually found to be within $2M_{obs} (a^*b^*$ or $u^*v^*)$." Anyway, the individual differences, that cause observer metamerism failure, are not only in the colour-matching functions but also: 1. the filtering by optical media; 2. the macular pigment density; 3. the effects of field size.

11.4.8   Daylight-Simulator Evaluation and 'Special Metamerism Index: Change in Illuminant'

Daylight is the most important illuminant and source in colorimetry, particularly D65. D65 exists as natural light in the day and as an illuminant, but a source emitting a light equal to that of the day does not exist. Simulated daylight can be achieved through many different methods: wide band fluorescent lamps, for example, F7 and FL 3.15, filtered tungsten halogen lamp and other technologies. Certainly in the near future other daylight simulators will be produced. All of these simulators can produce very good correlated colour temperature and chromaticity of the simulated daylight, but the reproduction of an equal spectral power distribution is almost impossible. These sources have a very high colour rendering index (Section 16.8.8), but this index is not adequate for an accurate colour judging. A quantitative assessment of the suitability of a practical daylight simulator is made by computing the special metamerism index for a change in illuminant.[28,29] Since daylight has a content of UV radiation with consequent fluorescence, the daylight simulation regards fluorescent or non-fluorescent specimens. The CIE method considers pairs of virtual specimens, specified by their reflecting and fluorescing properties, which are metameric matches under the reference CIE daylight illuminant and for the CIE 1964 observer. Two different indices are used, one to evaluate the visible region of the spectrum, the other to evaluate the ultraviolet region. This method quantifies the mismatch when the pairs of virtual specimens are illuminated by the simulator under test. All this is based on the computation and the necessary numerical data are given by CIE.[28,29] The requirements for the computation and the requirements and the steps are in the following table:

The observer is the CIE 1964
The spectral power distribution of the daylight simulator under test is known $S(\lambda)$
The daylight simulator must have the chromaticity coordinates on the (u'_{10}, v'_{10}) diagram within a circle having a radius of 0.015 centered in the chromaticity coordinates of the standard daylight illuminant.

Evaluation of spectral power distributions in the visible region	*Evaluation of spectral power distributions in the ultraviolet region*
Five pairs of virtual non-fluorescent specimens constituted by five standard and five metameric samples are considered. The standards are the same for all the daylight illuminants (D50, D55, D65, D75). The spectral reflectances $R(\lambda)$ are given by CIE[28,29,]	Three virtual fluorescent samples are considered which are metameric with three non-fluorescent samples illuminated by one of the standard daylight illuminants (D50, D55, D65, D75). The total spectral radiance factor of the non-fluorescent specimen is directly known from the CIE tables, while the total spectral radiance factors of the three fluorescent specimens are computed as follows (Section 12.2.2): 1. the spectral power distribution $S(\lambda)$ is normalized $$S_n(\lambda) = \frac{100\,S(\lambda)}{\sum_{\lambda=400}^{700} S(\lambda)\bar{y}_{10}(\lambda)\Delta\lambda}$$ 2. a factor N taking into account the fluorescent emission is computed $$N = \sum_{\lambda'=300}^{460} S_n(\lambda')Q(\lambda')\Delta\lambda'$$ where $Q(\lambda')$ is the excitation efficiency of the fluorescence; 3. The total spectral radiance factor $\beta_T(\lambda)$ for the fluorescent sample lit by the simulator is $$\beta_T(\lambda) = \beta_R(\lambda) + N\frac{F(\lambda)}{S_n(\lambda)}$$

	where $F(\lambda)$ is the spectral distribution of the fluorescent component normalized to 1 $$\sum_{\lambda=300}^{700} F(\lambda) = 1.$$ The functions $Q(\lambda)$, $\beta_R(\lambda)$ and $F(\lambda)$ are given by CIE[28,29].
The computation of (X_{10}, Y_{10}, Z_{10}) for each specimen lit by the simulator $S(\lambda)$ is made by using Equations (9.37) with summations with 5 nm steps in the wavelength range 300–700 nm.	The computation of (X_{10}, Y_{10}, Z_{10}) for each specimen lit by the simulator $S(\lambda)$ is made by using Equations (9.37) with summations with 5 nm steps in the wavelength range 400–700 nm. The total spectral radiance factor $\beta_T(\lambda)$ substitutes the spectral reflectance factor $R(\lambda)$ in Equations (9.37).
Evaluation of spectral power distributions in the visible region	***Evaluation of spectral power distributions in the ultraviolet region***
The colour difference between the five standards and the corresponding metameric samples is computed by $\Delta E^*_{ab,10}$ (11.19) or $\Delta E^*_{uv,10}$ (11.15) colour difference formulae	The colour difference between the three non-fluorescent samples and the three fluorescent ones is computed by $\Delta E^*_{ab,10}$ (11.19) or $\Delta E^*_{uv,10}$ (11.15) colour difference formulae
The metamerism index in the absence of fluorescence is obtained by an average M_v of the differences of five colour pairs $$M_v = \frac{1}{5}\sum_{i=1}^{5} \Delta E_i$$ where ΔE_i may be $\Delta E^*_{ab,10}$ or $\Delta E^*_{uv,10}$.	The metamerism index in the presence of fluorescence is obtained by an average M_u of the colour differences of three colour pairs $$M_u = \frac{1}{3}\sum_{i=1}^{5} \Delta E_i$$ where ΔE_i may be $\Delta E^*_{ab,10}$ or $\Delta E^*_{uv,10}$.

A grade, A, B, C, D, E, is obtained from the metamerism indices as defined in the following table:

Grade	M_v or M_u based on $\Delta E^*_{ab,10}$	M_v or M_u based on $\Delta E^*_{uv,10}$
A	< 0.25	< 0.32
B	0.25 to 0.50	0.32 to 0.65
C	0.50 to 1.00	0.65 to 1.30
D	1.00 to 2.00	1.30 to 2.60
E	> 2.00	> 2.60

For instance the CIE illuminant C is characterised by $M_v = 0.27$ and $M_u = 3.28$ (CIELAB colour difference formula), and is graded as BE, while the xenon lamp $M_v = 0.40$ and $M_u = 1.18$, and is graded BD.

11.5   Conclusion

Chapter 11 presents the standard psychometric colorimetry systems, which are very useful to technicians involved in the comparisons and measurements of object colours. Now, in surface-colour reproduction in industry, it is an accepted practice to specify the colour by the CIELAB system for the CIE 1964 supplementary standard observer with different illuminants, and among these D65 is standard. In graphic arts, painting, photography, television, generally in imaging, the CIE 1931 observer is used with the D50 illuminant, but, although the CIE proposed the CIELUV psychometric system for this field, today the CIELAB is used. Furthermore, in cases where a comparison is required between a standard colour and a test, it is customary to quantify the colour differences according to the formulas CMC, CIE 1994, CIEDE2000 and even CIELAB 1976 colour-difference formula.

The value of these systems and formulas is especially practical. They are the result of a compromise between scientific knowledge, still unfinished, and the daily need of those who work in factories and laboratories. The phases of colorimetry called 'color difference' and 'colour appearance' are still open and the systems described above, despite their limitations, but with consolidated diffusion and use in personal computers, make the diffusion of new systems and new algorithms difficult.

References

1. Hunt RWG, The specification of colour appearance. I. concepts and terms, *Color Res. Appl.*, **2**, 55–68 (1977).
2. Hunt RWG, Colour terminology, *Color Res. Appl.*, **3**, 79–87 (1978).
3. Robertson AR, Historical development of CIE recommended color difference equations, *Color Res. Appl.*, **15** 167–170 (1990).
4. Pauli H, Proposed extension of the CIE recommendation on "uniform color spaces, color difference equations, and metric color terms, *J.Opt. Soc. Am.*, **66**, 866–867 (1976).
5. Priest IG, A proposed scale for use in specifying the chromaticity of incandescent illumnants and various phase of daylight, *J. Opt. Soc. Am.*, **23**(2), 41–45 (1932).
6. CIE publication N. 15.3:2004, *Colorimetry*, 3rd ed., Commission Internationale de l'éclairage, Vienna (2004). Available at: www.cie.co.at/ (accessed on 18 June 2015).
7. Pointer MR and Attridge GG, Some aspects of the visual scaling of large colour differences, *Color Res. Appl.*, **22**, 298–307 (1997).
8. Adams EQ, X-Z planes in the 1931 ICI System of colorimetry. *J. Opt. Soc. Am.* **32**, 168–173 (1942).
9. Pointer MR, A comparison of the CIE 1976 colour spaces, *Col. Res. Appl.*, **6**, 108–118 (1981).
10. Hunt RWG, *Colour Measurement*, 2nd ed., Ellis Horwood, Chichester (1991).
11. Luke JT, OSA instrumental development of the uniform color scales, *Optics & Photonics News/* September 1999, 29–33 (1999).
12. Roberson AR, CIE guidelines for coordinated research on colour-difference evaluation, *Color Res. Appl.*, **3**, 149–151 (1978).
13. Luo MR and Rigg B, Chromaticity-discrimination ellipses for surface colors, *Color Res. Appl.*, **11**, 25–42 (1986).
14. Luo MR, Cui G and Rigg B, The development of the CIE 2000 colour-difference formula: CIEDE2000, *Color Res. Appl.*,**26**, 340–350 (2001).
15. Clarke FJJ, McDonald R and Rigg B, Modification to the JPC79 color-difference formula, *J. Soc. Dyers Colourists*, **100**, 1218–132 (1980).
16. BS 6923:1988, *Method for Calculation of Small Colour Differences*, British Standard Institution, London (1988).

17. McLaren K, *The Colour Science of Dyes and Pigments*, 2nd ed., Hilger, Bristol, 143pp. (1986).

18. Hunt RWG and Pointer MR, *Measurung Colour*. Wiley and IS&T (2011), Section 3.11.

19. ISO 105-J03:1995, *Textiles – Tests for Colour Fastness*, part J03; *Calculation of colour difference*. ISO 105-J03:1995/Cor.2: 2006(E) (1995).

20. CIE Publication 119:1995, *Industrial Colour-Difference Evaluation*, Commission Internationale de l'Éclairage, Vienna (1995). Available at: www.cie.co.at/ (accessed on 18 June 2015).

21. CIE Pubblication 142:2001, *Improvement to Industrial Colour Difference Evaluation*, Commission Internationale de l'Éclarage, Vienna (2001). Available at: www.cie.co.at/ (accessed on 18 June 2015).

22. Nobbs JH, A lightness, chroma and hue splitting approach to CIEDE2000 colour difference, *Advan. Colour Sci. Technol.*, **5**(2), 46–53 (2002).

23. Huertas R, Melgosa M and Oleari C, Performance of a color-difference formula based on OSA-UCS space using small-medium color differences. *J. Opt. Soc. Am. A.* **23**, 2077–2084 (2006).

24. Oleari C, Mclgosa M, and Huertas R, Euclidean color-difference formula for small-medium color differences in log-compressed OSA-UCS space, *J. Opt. Soc. Am. A* **26**, 121–134 (2009).

25. Raj Pant D and Farup I, Riemannian formulation and comparison of color difference formulas, *Color Res. Appl.*, **37** (6), 429–440 (2012).

26. CIE publication S 017/E:2011, *ILV: International Lighting Vocabulary*, Commission Internationale de l'Éclairage, Vienna (2011). Available at: www.eilv.cie.co.at/ (accessed on 18 June 2015).

27. CIE Publication 80:1989, *Special Metamerism Index: Change in Observer*, Commission Internationale de l'Éclairage, Vienna (1989). Available at: www.cie.co.at/ (accessed on 18 June 2015).

28. CIE Publication 51.2:1999, *A Method for Assessing the Quality of Daylight Simulators for Colorimetry*, Commission Internationale de l'Éclairage, Vienna (1999). Available at: www.cie.co.at/ (accessed on 18 June 2015).

29. CIE Publication CIE S 012/E:2004, *Standard Method of Assessing the Spectral Quality of Daylight Simulators for Visual Appraisal and Measurement of Color*, Commission Internationale de l'Éclairage, Vienna (2004). Available at: www.cie.co.at/ (accessed on 18 June 2015).

12

Instruments and Colorimetric Computation

12.1 Introduction

The basic psychophysical quantities related to human colour vision are the tristimulus values. Luminance is a linear combination of the tristimulus values and similarly all other quantities, that were shown in Chapters 9 and 11, are obtainable as a function of the tristimulus values. The tristimulus values of primary and secondary light sources are defined by summations (9.36), (9.37) and (9.38), which are either calculated mathematically from spectral radiometric measurements, or obtained as output signals of three photo-detectors coupled with suitable filters to simulate the three kinds of cones of the retina. In order to perform the calculation, we need to know the spectral distribution of the radiance that enters the eye and activates the photoreceptors of the retina. We must always consider the radiance, that, as it is defined, requires that the instrument has an objective lens which operates in analogy with the optical system of the eye (cornea, crystalline lens, etc.) focussing the light radiation on a photosensitive surface. This surface can be

1. the surface of a set of filters matched to as many photo-detectors that operate in complete analogy with the photosensitive cells for photopic vision of the retina, that is, the three kinds of cones and
2. a geometric surface that is the opening, through which the light radiation enters a device for spectral analysis – a *spectrometer* –.

Regardless of how the spectral analysis of the light is made, the instrument must be capable to collect a radiance from a surface of which we want to measure the tristimulus values.

The radiance entering the eye or the instrument comes directly from a light source or from an illuminated surface. The spectral measurement of this light is made by a

> "*Spectroradiometer* – instrument for measuring radiometric quantities in narrow wavelength intervals over a given spectral region."[1]

Standard Colorimetry: Definitions, Algorithms and Software, First Edition. Claudio Oleari.
© 2016 John Wiley & Sons, Ltd. Published 2016 by John Wiley & Sons, Ltd.

A spectroradiometer is a device designed to measure the relative/absolute spectral radiance of a source and has inside a spectrometer calibrated on the wavelength scale and the radiometric scale. The spectral radiance is measured in correspondence to a defined set of wavelengths and is proportional to the spectral power distribution. The spectrometer consists of an optical element dispersing the light in its monochromatic components and a photo-detection system. This latter comprises a single photodetector or an array of photodetectors. In the case of a single photo-detector, the contributions at the various wavelengths are measured in succession and switching from one wavelength to the next by rotating the dispersing element – the instrument is a *monochromator* –. Instead in the case of an array of photo-detectors, the measurement is made simultaneously for all wavelengths and there is no mechanical movement – the instrument is called a *polychromator* –.

The photometric and colorimetric quantities are computed from the spectral power distribution.

In everyday experience, the apparent colour of a specimen depends on the observer point of view and on the illumination of the specimen. This appearance phenomenon depends on the optical properties of the specimen and on the spectral composition of the illuminating light, that is, properties to be distinguished into

1. spectral power distribution $S(\lambda)$ of the light emitted by the lamp used to illuminate a specimen – standard/test – (Chapter 8);
2. geometry according to which the light emitted from the lamp illuminates the surface of the specimen (Section 12.2.2); and
3. spectral power distribution of light that exits the surface of the specimen in the direction of the observer's point of view (Section 12.2.2).

Points 2. and 3. say that the flux $S(\lambda)$ of the light source is modified by an interaction process with the colour specimen. This process, that modifies the flow of radiant energy, and its effect is termed *optical modulation.*[2]

The ratio $M(\lambda)$ between the radiant flux crossing the pupil of the eye and the illuminating flux $S(\lambda)$ is the optical modulation, then the spectral flux entering the eye is $S(\lambda)M(\lambda)$.

"*Spectrophotometer* – is the instrument for measuring the ratio $M(\lambda)$ of two values of a radiometric quantity at the same wavelength."[1]

(The suffix *photometer* derives from the time the light transducer used was the human eye. It is now almost always superseded by an optoelectronic sensor. Many instruments are called spectrometers, but with other functions.)

Also spectrophotometer have inside a spectrometer, for generating a signal related to the spectral power distribution of a light entering it.

For a better understanding of the instrumental apparatus, two sketches with typical spectrophotometric structures are proposed in Figures 12.1a and 12.1b. These two structures are different only in the part related to the illumination of the colour sample. In one case the illumination is a flux of parallel rays incident on the colour sample at 45° and in the second case the lighting is diffused by an integrating sphere, which produces a Lambertian illumination (Section 3.5.1) of the colour sample. The geometry of vision is orthogonal or nearly orthogonal to the colour sample. The analysis of these two illumination geometries is made in the following Section 12.2.2.

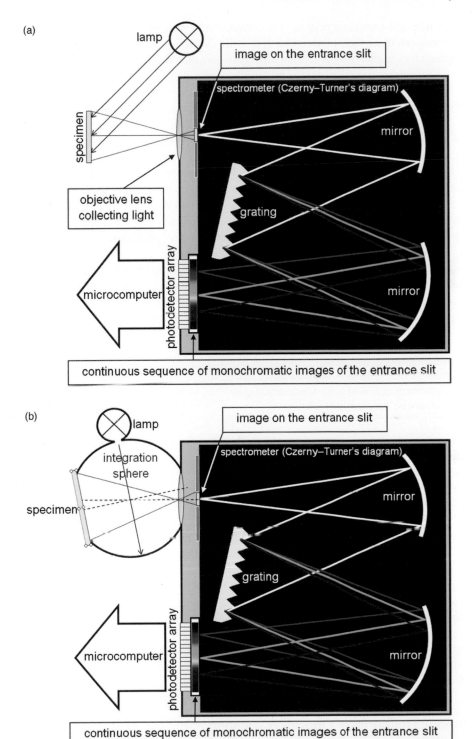

Figure 12.1 *(a) Schematic designs of reflectance spectrophotometers with measurement geometry (45°:0°) and (b) geometry (d:0°) (Section 12.2.2). The spectrophotometric measurement is the ratio of the radiant fluxes reflected by the test sample and the perfect reflecting diffuser (12.4).*

In the case of non-self-luminous bodies, it is convenient to measure the optical modulation $M(\lambda)$ of the colored surface and calculate the radiance that enters the eye $S(\lambda)M(\lambda)$ for each illuminant $S(\lambda)$. The measurements of optical modulation and of radiance are here considered separately.

12.2   Reflection and Transmission Optical-Modulation

The colour of non-luminous and non-fluorescent bodies depends on the spatial distribution of the light that illuminates – *influx geometry* – and from the observer's point of view – *efflux geometry* –. Operationally, this implies choosing the geometry of illumination and the geometry with which the light being measured is collected. The colorimetric characteristics of the bodies depend on their ability to reflect, transmit and absorb light at fixed measurement geometry, and these are summarized in the physical quantities here presented. All the optical modulations defined below are dimensionless ratios of homogeneous radiometric quantities. In addition, they are subdivided into two groups: *absolute* and *relative* quantities.

- The relative quantities are obtained by a comparison with a physical reference standard (reflecting or transmitting) and the word *factor* distinguishes the quantities 'relative' to a reference from the absolute ones. The reference standard is a sample whose optical characterization is retraceable to a metrological laboratory.
- A quantity without the word 'factor' is to be understood in the sense of *absolute*.[3]

The use of the words 'absolute' and 'relative' should not be confused with the metrological terms of 'absolute method' and 'relative method', which refer to methods of measurement. According to the absolute method, measured values are derived from measurement of one or more fundamental quantities, such as mass, length, time or counting a number of objects or events, otherwise the method is relative.

The set of words 'reflectance', 'transmittance', 'reflectance factor' and 'transmittance factor' are grouped together in the single term *optical modulation*.[2–5]

The quantities used in colorimetry are mainly the relative ones. These represent the optical properties of the colour bodies necessary to describe the geometric and spectral characteristics of the light flux entering the observer's eye. The geometries proposed by the CIE are the most significant ones to identify the optical properties of bodies.

Two phenomena are prevalent in the reflection of light with bodies:

1. the *specular reflection* of the light that illuminates and
2. the *diffuse reflection* from the surface, if rough, and from the inside of the body after an interaction with the dyes and pigments contained in it (Sections 3.5–3.9), which involve wavelength selective absorption of light and diffusion.

Analogously, for transparent and translucent bodies, there are two prevailing phenomena: the regular transmission and diffuse transmission produced by the interaction with the pigments and optical inhomogeneities (Sections 3.5–3.9).

12.2.1   Absolute Quantities of Optical-Modulation

The absolute quantities depend on the totality of the incident, reflected, and transmitted radiation. These radiations are contained in a solid angle $\Omega = 2\pi$, equal to the hemispheres defined by planes tangent respectively to the surface of radiation incidence, the surface of reflection and the surface of emission. These are the geometric conditions of the fluxes necessary for the definition of the absolute quantities. These geometries have no direct relation with the solid angle subtended by the pupil of the human eye to collect the light, therefore they are not considered in colorimetry, although important for the optical characterization of the materials.

The absolute quantities are:

- The *spectral reflectance* $\rho(\lambda)$ of a body is the ratio of the total spectral reflected radiant flux $\Phi_{\uparrow,\lambda}$ on the total spectral incident radiant flux $\Phi_{\downarrow,\lambda}$ at fixed wavelength λ

$$\rho(\lambda) = \frac{\Phi_{\uparrow,\lambda}}{\Phi_{\downarrow,\lambda}} \tag{12.1}$$

- The *spectral transmittance* $\tau(\lambda,s)$ of a body of thickness s is the ratio of the total spectral transmitted radiant flux $\Phi_{\downarrow,\lambda}(s)$ and the total spectral incident radiant flux $\Phi_{\downarrow,\lambda}(s=0)$ to fixed wavelength λ and taking into account the reflections on the surface of the body

$$\tau(\lambda,s) = \frac{\Phi_{\downarrow,\lambda}(s)}{\Phi_{\downarrow,\lambda}(s=0)} \tag{12.2}$$

In the case of total reflection, $\rho(\lambda) = 1$ and $\tau(\lambda) = 0$.

In the absence of absorption of light, $\rho(\lambda) + \tau(\lambda) = 1$.

The reflectance and transmittance are called *specular* or *regular* in the absence of light scattering due to diffusion and diffraction.

- The *spectral internal transmittance* $\tau_i(\lambda,s)$ of a thick body at fixed wavelength λ is the ratio of the radiant transmitted internal flux $\Phi_{\downarrow i,\lambda}(s)$ and the radiant incident internal flux $\Phi_{\downarrow i,\lambda}(s=0)$, for which the reflections at the surface of the body are ignored,

$$\tau_i(\lambda,s) = \frac{\Phi_{\downarrow i,\lambda}(s)}{\Phi_{\downarrow i,\lambda}(s=0)} \tag{12.3}$$

12.2.2   Relative Quantities of Optical-Modulation

A colorimetric measurement is made on a flat specimen, placed on a *reference plane*, on which is defined a region termed *sampling aperture*, lit according to an *influx geometry* and observed according to an *efflux geometry*. The optical modulation depends on all these quantities, that have to be defined exactly.

Reference Plane

The measurement geometries in the instrument are referenced to a reference plane which coincides with the plane in the instrument on which is placed the sample surface – which can be a *test* specimen, or a *reference standard*, calibrated by a metrology laboratory, or a copy of it, termed *working standard* –. In the case of reflection measurements, the sample has a single surface and this is coincident with the reference plane, but in the case of transmission measurements, there are two reference planes, one for the illumination geometry and one for the vision. These two planes are parallel and at a mutual distance equal to the thickness of the sample, which is assumed to be negligible. So, conventionally, there is always only one reference plane in the instrument.

Sampling Aperture

Each instrument defines a region on the reference plane to be illuminated and a region from which to collect the light to be sent to the spectrometer. This region, defined as

> "*sampling aperture*, is the area of the reference plane on which measurements are made."[1]

If the illuminated region is greater than the sampling aperture, the region is considered *over filled* and, if smaller, *under filled*. The measurements on translucent samples have to be made carefully because all the illuminating radiation must contribute to the measurement and this is achieved with the illuminated region being greater than the sampling aperture. A good rule is to have the sampling aperture always well below the irradiated region. It is also recommended that the sampling region be uniformly illuminated.

Influx Geometry

> "*Influx geometry* – or *irradiation geometry, incidence geometry* – is the angular distribution of the radiation incident on the sample being measured with respect to the centre of the sampling aperture."[1]

Efflux Geometry

> "*Efflux geometry* is the angular distribution of the receiver responsivity with respect to the centre of the sampling aperture on the sample being measured."[1]

The 'relative' quantities are defined by the ratio of two homogeneous radiometric quantities, the first of which is measured on the test sample and the second, in the case of reflection measurements, is measured on the *perfect reflecting diffuser* (Section 3.5.1), and, in the case of transmission measurements, is measured in the air, that is obtained by removing the specimen from the instrument. More specifically, the 'relative' quantities are defined as follows:

Spectral Reflectance Factor

The spectral reflectance factor $R(\lambda)$ is the ratio of the radiant flux reflected in a given cone Ω, whose vertex is at the centre of the sampling aperture on the surface of the specimen, and the flux reflected in the same direction from the perfect reflecting diffuser likewise illuminated,

$$R(\lambda) = \frac{\Phi_{\text{test},\Omega,\lambda}}{\Phi_{\text{perfect reflecting diffuser},\Omega,\lambda}} \tag{12.4}$$

Spectral Transmittance Factor

The Spectral transmittance factor $T(\lambda)$ is the ratio of the radiant flux transmitted in a given cone Ω, whose apex is located on the specimen, and the flux transmitted in the same direction, after removing the specimen, that is in the air,

$$T(\lambda) = \frac{\Phi_{\text{test},\Omega,\lambda}}{\Phi_{\text{air},\Omega,\lambda}} \tag{12.5}$$

In cases where the measurement in air is equivalent to the measurement of the illuminating flux, the quantity is called *spectral transmittance* and indicated with $\tau(\lambda)$.

Reflectance Standard

The definition of the spectral reflectance factor requires the *perfect reflecting diffuser*, a diffuser exhibiting isotropic diffuse Lambertian reflection with a reflectance factor $R_{\text{perfect reflecting diffuser}, \Omega}(\lambda) = 1$. In practice, it is impossible to have the perfect reflecting diffuser and then a reflectance standard (STD) is used, whose reflectance factor

$$R_{\text{STD},\Omega}(\lambda) = \frac{\Phi_{\text{STD},\Omega,\lambda}}{\Phi_{\text{perfect reflecting diffuser},\Omega,\lambda}} \tag{12.6}$$

specified for the chosen measurement geometry Ω, is traceable to a metrological laboratory.

The use of the reflectance standard implies that the spectral reflectance factor is thus redefined

$$R(\lambda) = \frac{\Phi_{\text{test},\Omega,\lambda}}{\Phi_{\text{perfect reflecting diffuser},\Omega,\lambda}} = \frac{\Phi_{\text{test},\Omega,\lambda}}{\Phi_{\text{STD},\Omega,\lambda}} \frac{\Phi_{\text{STD},\Omega,\lambda}}{\Phi_{\text{perfect reflecting diffuser},\Omega,\lambda}} = \frac{\Phi_{\text{test},\Omega,\lambda}}{\Phi_{\text{STD},\Omega,\lambda}} R_{\text{STD},\Omega}(\lambda) \tag{12.7}$$

Great is the performance of compressed barium sulphate and of polytetrafluoroethylene (PTFE), but in practice also other materials such as ceramics are used with satisfaction.

Given the high cost of a reference standard, often a duplicate, termed *working standard*, is used.

CIE Standard Geometries for Optical Modulation

The efflux solid angle in which the light outgoing from the standard/test specimen is considered corresponding to the solid angle of vision of the eye.

The choice of the measurement geometry needs to be justified and therefore the CIE proposed different geometries for measuring relative quantities with the intention of significantly characterizing the specimens. From what has already been stated, the colour of bodies depends on the irradiation geometry and on the point from which they are observed. The light emerging from an illuminated body has a contribution due to surface reflection and one due to the light scattering on the surface and inside the body. The different influx and efflux geometries intend to distinguish these contributions, as much as possible, and therefore the corresponding measurements have complementary meanings. The lights that illuminate and then reflect off the sample are basically of two kinds,

1. *diffuse irradiation (illumination)* and
2. *directional irradiation (illumination)* with conical convergence.

The influx geometries proposed by the CIE are summarized in two figures: Figure 12.2 for reflection measurements and Figure 12.3 for transmission measurements. For the tolerances on these geometries there is a specific publication of the CIE.[6]

The measuring geometries are defined by the influx solid angle of irradiation and by the efflux solid angle, in which the light outgoing from the standard/test specimen is considered. Therefore the measuring geometries are indicated by the angles with the symbols of degree '°', that define the directions of these two solid angles, separated by ':' and placed in parentheses (it often happens that ':' is replaced by '/', and the symbols '°' and the parentheses are ignored, but that does not create a problem of interpretation).

The hemispherical angle is obtained by an integration sphere (Section 3.10). In this case the light is diffused and the angle is denoted with 'd'. The additions of 'e' or 'i' to 'd' mean that the specular component

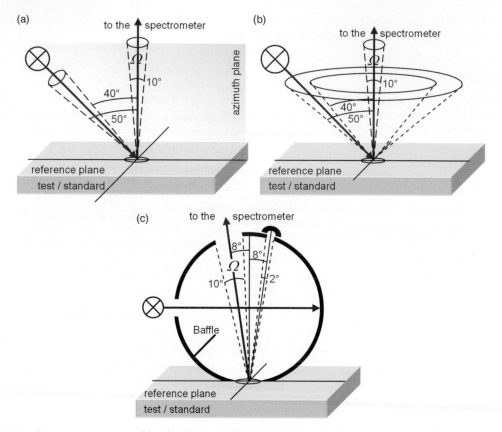

Figure 12.2   *Geometries proposed by the CIE for reflectance-factor measurements: (a) Reflectance factor measured in geometry (45°x:0°). The illuminating light and the light sent to the spectrometer are inside two cones with vertex on the specimen – test or standard – and sections of 10°. The light fluxes inside the cones should be uniform. The axes of these cones belong to an azimuth plane x. Any regular roughness of the surface of the specimen responsible for texture can make the result of the measurement dependent on the azimuth plane x chosen. (b) Reflectance factor measured in geometry (45°a:0°). The light illuminating the specimen – test or standard – is located between two conical surfaces at 40° and 50° from normal and the light sent to the spectrometer is located on the line orthogonal to the specimen. The result of this measurement is equivalent to the average of the measurements made in geometry (45°x:0°) considering all the azimuth planes x and is independent of any texture. (c) Reflectance factor measured in geometry (di:0°) or (de:0°). An integrating sphere renders the light illuminating the specimen diffuse and hemispherical. The light sent to the spectrometer exits from the specimen – test or standard – with an angle of 8° with respect to the orthogonal line and is inside a cone with a section of 10°. This last light is almost equal to that at 0° and the choice of 8° allows two operating modes: one mode including the light specularly reflected by the specimen, denoted by (di:0°), and one excluding, denoted by (de:0°). The right notations should be (di:8°) and (de:8°). The two ways of operating are selected by opening or closing the port that is in a specular position with respect to the port towards the spectrometer. The comparison between the two measurements is an information on the glossiness of the specimen.*

of the light within the hemisphere is excluded or included, respectively. The terms SPIN, for the *specular component included*, and SPEX, for *specular component excluded*, were used past. A cap coated with the same inner coating as the sphere closes a port of the sphere in the measurements with the specular component included, and is removed in measurements with the specular component excluded. With the cap inserted the inner side of the sphere shows a continued and Lambertian surface as all the other parts of the sphere.

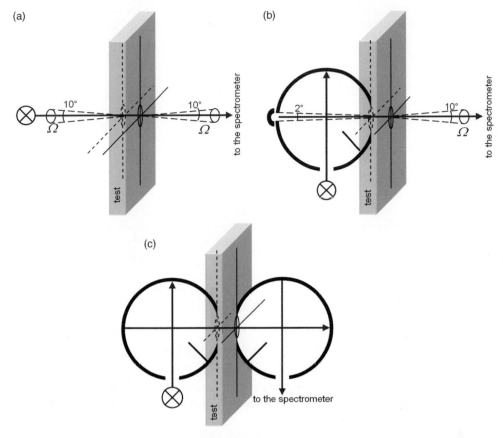

Figure 12.3 *Geometries proposed by the CIE for transmittance measurement: (a) Transmittance measured in geometry (0°:0°). The illuminating light and the light sent to the spectrometer are inside two cones with axes orthogonal to the specimen and sections of 10°. The light fluxes inside the cones should be uniform. (b) Transmittance measured in geometry (di:0°) and transmission factor in geometry (de:0°). The illumination of the test sample is produced by an integration sphere and is diffuse and hemispherical. The light sent to the spectrometer is inside a cone with axes orthogonal to the test sample and section of 10°. A port in the integration sphere on the line orthogonal to the test sample can be close, and the measurement is of the transmittance in geometry (di:0°), or can be open, and the measurement is of the transmittance factor in geometry (de:0°). (c) Transmittance measured in geometry (d:d). The illumination of the test sample is produced by an integration sphere and is diffuse and hemispherical. The light exiting the test sample is completely collected by a second integrating sphere.*

Figure 12.2c shows the position of this port in the sphere, that is specularly symmetrical to the exit port. The section of this port depends on the radius of the sphere and Figure 12.2c shows the section of the solid angle that this port must subtend. Operational rigour requires that, when the cap is removed, this is replaced by a light trap, that is, a device totally absorbing the light.

In the hemispherical geometry with an integrating sphere, the radiation emitted by the specimen reflecting off the sphere wall and re-illuminating the specimen produces an error (*spectral sphere error*), which can be considered by a mathematical correction or in the measuring instrument by using two spectrometers, measuring both the light reflected from the specimen and the light diffused in the sphere that illuminates the specimen.

The computational method for which the CIE proposes a formula for the correction of the spectral reflectance factor is given here without proof,

$$R(\lambda) = R_{uc}(\lambda) \frac{1 - \rho(\lambda) \cdot (1 - f)}{1 - \rho(\lambda) \cdot (1 - f) - f_e \left(R_{STD}(\lambda) - R_{uc}(\lambda) \right)} \tag{12.8}$$

where

$R_{uc}(\lambda)$ is the spectral reflectance factor non-corrected and referred to the ideal diffuser,
$R_{STD}(\lambda)$ is the spectral reflectance factor of the white standard specimen,
$\rho(\lambda), f, f_e$ are defined in Section 3.10.

This correction assumes that the sphere has an inner surface absolutely Lambertian and that all ports, except the exiting port closed by the specimen or by the standard reflectance sample, have zero reflectance. In the case of fluorescent samples, this correction is inadequate and the use of the integrating sphere is to be avoided, unless the diffused light, which is inside the sphere and illuminates the sample, is measured separately through a third port.

The term $(45°x:0°)$ means that the incident light, in the $45°$ direction, and the reflected light, in the $0°$ direction, are belonging to the azimuth plane 'x' (Figure 12.2a).

The term $(45°a:0°)$ means that:

1. the incident light irradiates in an *annular* way, that is, radiates uniformly in all directions in between two coaxial cones with the apex at the centre of the sampling aperture, with the axis orthogonal to the reference surface and forming with this respective angles of $40°$ and $50°$ (Figure 12.2b) and
2. the light being measured is collected in the $0°$ direction, orthogonal to the sampling aperture.

In the particular case that the illumination is obtained through a circular array of sources, which are lighting the sampling aperture with incidence at $45°$, the geometry is denoted by $(45°c:0°)$ and is termed *circumferential geometry*.

It is observed that:

1. for the efflux viewing angle $\Omega \to 2\pi$ the spectral reflectance factor is called *spectral reflectance* and is denoted with $\rho(\lambda)$ and
2. for the efflux viewing angle $\Omega \to 0$, spectral reflectance factor is called *spectral radiance factor* at the various geometries and is denoted by $\beta_{45:0}(\lambda)$, $\beta_{0:45}(\lambda)$ and $\beta_{di:8}(\lambda)$.

In the case of a glossy specimen, arranged orthogonally to the beam of illuminating light, there may be multiple reflections between the specimen and the optics that guide the beam itself. This is an origin of error, which is avoided by placing the specimen not exactly perpendicular to the illuminating beam.

In accordance with *Helmholtz's principle of reciprocity*, whereby the reflectance does not change if the direction of illumination and viewing are exchanged[7], for measurements in reflection geometry $(8°:di)$, $(8°:de)$, $(0°:45°)$ and $(0°:45°x)$ are in addition to those shown in Figure 12.2 by reversing the direction of the rays, and, for measurements in transmission geometry $(0°:di)$ $(0°:de)$ are in addition to those shown in Figure 12.3. For these geometries, the source and the spectrometer positions are exchanged.

It is customary to put the geometry as a subscript of the magnitude considered.

For fluorescent bodies, these definition of optical modulation have to be revised (Section 12.3.2) and the instrumental structure reconsidered.

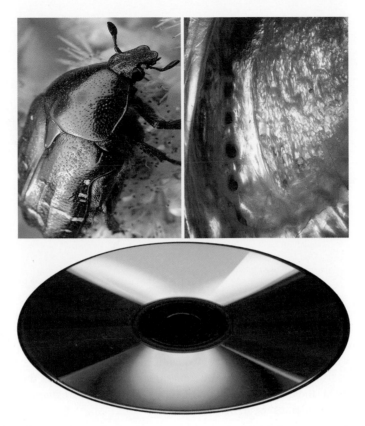

Figure 12.4   *Examples of gonio-dependent colour in nature – a beetle and nacre – and artificial – a CD rom –.*

Optical Modulation of Gonio-Apparent Colour

Before now, consideration has been given to the physical phenomena and the physical quantities related to the colour of non-luminous bodies, whose appearance, in the absence of specularly reflected light, changes little with the geometry of illumination and viewing. However, it is common experience that there are bodies whose colour changes with the direction of illumination and viewing (Figure 12.4). An everyday example is that of metallic paints for cars. In this case, the pigment of the paint is constituted by micrometric aluminium mirrors and the effect is due to the highly non-Lambertian scattering of light. Another example, still in automobiles, is to paint with mica pigments. In this case the pigment is constituted by micrometric mica flakes – ~50 μm – coated with a layer of an oxide, for example, SiO_2, whose thickness is in the order of 20–200 nm. This coating, if lit, generates iridescent effects that change with the influx and efflux angles. This visual phenomenon is generated by interference and therefore the colour is called *interference colour*. These colours are also called *gonio-apparent colours* or, more generally, *special-effect colours*.

Bodies with interference colour are generally coated by a transparent layer with thickness in the order of the wavelength of visible light and a different refraction index. The colours of the optical coatings of optical lenses and in general of glasses are of this type.

In nature we see interference colours in the wings of butterflies, insects, feathers of birds and soap bubbles.

This kind of colour, because it is strongly dependent on the geometry of illumination and vision, cannot be meaningfully specified on the basis of measurements of the spectral reflectance factor in accordance with the

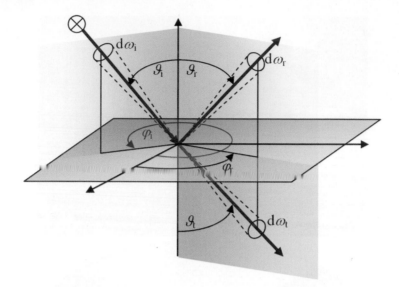

Figure 12.5 *Geometries for the definition of BRDF and BTDF.*

geometries already defined by the CIE, but a set of measurements with different combinations of illumination and viewing geometries must be implemented. New quantities have to be defined for a detailed specification of the radiation outgoing from the specimen in different directions as function of the illumination geometry. In the case of reflected light, the quantity is the *bidirectional reflectance distribution function*, always called BRDF, as proposed by F. Nicodemus[8] et al. in 1970

$$f_r(\vartheta_i, \varphi_i; \vartheta_r, \varphi_r; \lambda) = \frac{dL_r(\vartheta_i, \varphi_i; \vartheta_r, \varphi_r; \lambda; E_i)}{dE_i(\vartheta_i, \varphi_i; \lambda)} \quad \text{sr}^{-1} \tag{12.9}$$

which is the ratio between the spectral radiance $dL_r(\vartheta_i, \varphi_i; \vartheta_r, \varphi_r; \lambda)$ outgoing from the sample in the direction (ϑ_r, φ_r) within an elementary solid angle $d\omega_r$ and the spectral irradiance $dE_i(\vartheta_i, \varphi_i; \lambda)$ crossing the same region with the direction (ϑ_i, φ_i). (Figure 12.5). The word *directional* refers to the fact that in practice the solid angles, in which we consider the radiation beams, are very small and within them the rays are considerable as parallel.

In the case of transmitted light, the *bidirectional transmittance distribution function*, called BTDF, is defined in total symmetry with the BRDF

$$f_t(\vartheta_i, \varphi_i; \vartheta_t, \varphi_t; \lambda) = \frac{dL_t(\vartheta_i, \varphi_i; \vartheta_t, \varphi_t; \lambda; E_i)}{dE_i(\vartheta_i, \varphi_i; \lambda)} \quad \text{sr}^{-1} \tag{12.10}$$

Both these quantities are dependent on the light polarization.

The instrument for measuring these quantities is the gonio-spectrophotometer, which enables illuminating the specimen in each direction and collecting the light emerging in directions ϑ_r belonging to a wide range, often from $-80°$ to $+80°$ with respect to the normal to the specimen. Instruments of this kind are available on the market, but today their use is an exception. The reason is the high cost. These limitations are partly overcome with the introduction on the market of *multi-angle spectrophotometers*, which operate with a fixed azimuth plane, with lighting angle fixed at $45°$ and with four or five angles for the measurement of the

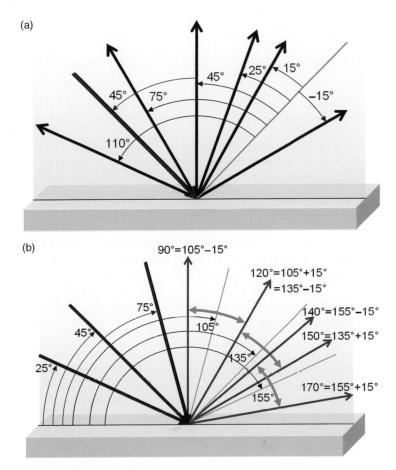

Figure 12.6 *(a) Classic geometry for multiangle spectrophotometry in which the illumination is at 45° from the line orthogonal to the specimen – test or standard – and the directions of the radiation sent to the spectrometers are defined with aspecular angles. The measurement angles are ordered starting from the specular angle, and therefore are termed aspecular angles. (b) Geometry for multiangle spectrophotometry with three lines of interference associated with three angles of incidence of 25°, 45° and 75°. For each of these angles of incidence are considered two directions of observation shifted by ± 15° with respect to the angle of specular reflection (155° ± 15°, 135° ± 15°, 105° ± 15°).*

reflected radiation. As shown in Figure 12.6a, the measurement angles are ordered starting from the specular angle, and therefore are termed *aspecular angles*.[9] Today there are DIN and ASTM standards, which define a method for measuring the spectral reflectance factor with four aspecular directions (25°, 45°, 75° and 110°) in addition to the specular direction.[10,11]

The ASTM has also introduced three lines of interference associated with three angles of incidence of 25°, 45° and 75°. For each of these angles of incidence are considered two directions of observation shifted by ± 15° with respect to the angle of specular reflection (155° ± 15°, 135° ± 15°, 105° ± 15°) (Table 12.1). The two viewing directions are called *cis* and *trans* at −15° and +15°, respectively (Figure 12.6b).

On the market there are also other multiangle geometries, wherein the illumination is circumferential by Xenon Flash according to different angles of incidence (25°, 45°, 75°) and the vision is orthogonal in the 0° direction.

Table 12.1   *Standard multiangle geometries of measurement.*

	ASTM/DIN Aspecular lines				ASTM Interference lines					
Illumination angle	45°	45°	45°	45°	25°	25°	45°	45°	75°	75°
View angle	110°	90°	60°	25°	170°	140°	150°	120°	120°	90°

However for some kinds of surfaces, especially for pearlescent and interference pigments, today used in paint products and in cosmetics, the use of only four aspecular angles is probably insufficient and the use of gonio-spectrophotometers is necessary.

Optical Modulation of Fluorescent Colour

Non-luminous bodies and non-fluorescent bodies illuminated with monochromatic radiation reflect radiation of wavelength equal to the illuminating radiation. This does not happen for most fluorescent bodies (Section 3.7). In this case the emerging radiation must be subdivided into two parts:

1. reflected radiation with wavelength equal to that of the illuminating radiation and
2. radiation produced by fluorescence.

The fluorescent emission of light by fluorescent retro-reflecting specimens is geometrically dependent, therefore angular tolerances on the axes and the angular aperture sizes must be well defined to ensure adequate repeatability and reproducibility. For this reason, the *radiance factor* is used, that is, a reflectance factor with an angle of view close to zero:

"*Radiance factor* – at a surface element of a non-self-radiating medium, in a given direction, under specified conditions of irradiation – is the ratio of the radiance of the surface element in the given direction to that of the perfect reflecting or transmitting diffuser identically irradiated and viewed.

NOTE – For photoluminescent media, the radiance factor contains two components, the *reflected radiance factor*, β_R, and the *luminescent radiance factor*, $\beta_{L,S}$. The sum of the reflected and luminescent radiance factors is the *total radiance factor*, $\beta_{T,S}$:"

$$\beta_{T,S} = \beta_R + \beta_{L,S} \tag{12.11}$$

(The subscript 'S' denotes the light source).[1]

The luminescent radiant factor and total radiance factor depend on the exciting light source 'S', therefore they have to be denoted by $\beta_{L,S}$ and $\beta_{T,S}$, respectively (Figure 12.7).

A complete reflectometric characterization of the test specimen requires the definition of a function that relates the two wavelengths, that is:

"*Bi-spectral luminescent radiance factor* – ratio of the radiance per unit emission bandpass, $\Delta\lambda$ (1 nm), at wavelength λ, due to photoluminescence from the specimen when irradiated at wavelength μ, to the radiance of the perfect reflecting diffuser identically irradiated and viewed,

$$\beta_{L,\lambda}(\mu) = \frac{L_{test,\lambda}}{L_{perfect\ reflecting\ diffuser,\mu}} \quad nm^{-1} \tag{12.12}$$

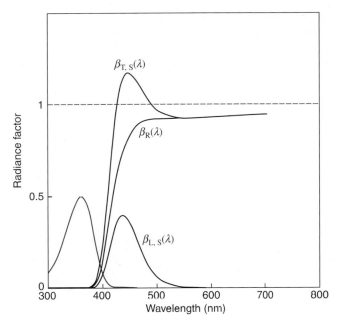

Figure 12.7 *Total spectral radiance factor $\beta_{T,S}(\lambda)$, reflected $\beta_R(\lambda)$ and luminescence $\beta_{L,S}(\lambda)$ of a fluorescent specimen (typically a paper with fluorescent whitening agents) produced by a light source S and corresponding absorption band in the ultraviolet region (red line).*

NOTE 2 – When integrated or summed over the spectral bandpass of the instrument, this quantity becomes the *bandpass-weighted bispectral luminescent radiance factor*, $\beta_{L,\lambda}(\mu)\Delta\lambda$.

NOTE 3 – The tabular form of complete bispectral luminescent radiance factor and reflected radiance factor data is the *Donaldson matrix* representation of *bispectral radiance factor*, $\mathbf{D}(\mu, \lambda)$. The excitation wavelengths μ are tabulated in the vertical direction, the emission wavelengths λ in the horizontal direction, the reflected radiance factors $\beta_R(\lambda)$ appear on the diagonal, where the excitation and emission (detection) wavelengths are equal, and the array of bandpass-weighted bispectral luminescent radiance factor data $\beta_{L,\lambda}(\mu)\Delta\lambda$ appear on the off-diagonals."[1]

$$\left[\mathbf{D}(\mu,\lambda)\right] = \left[\beta_{R,\lambda} + \beta_{L,\lambda}(\mu)\right] =$$

$$\leftarrow \text{exitation } \mu \rightarrow$$

$$= \begin{array}{c} \uparrow \\ \text{emission/} \\ \text{reflection} \\ \lambda \\ \downarrow \end{array} \begin{pmatrix} \beta_R(300) & 0 & \ldots & 0 & 0 \\ \beta_{L,310}(300) & \beta_R(310) & \ldots & 0 & 0 \\ \ldots & \ldots & \ldots & \ldots & \ldots \\ \beta_{L,790}(300) & \beta_{L,790}(310) & \ldots & \beta_R(790) & 0 \\ \beta_{L,800}(300) & \beta_{L,800}(310) & \ldots & \beta_{L,800}(790) & \beta_R(800) \end{pmatrix} \text{nm}^{-1} \qquad (12.13)$$

(for example, matrix written for spectral data at 10 nm steps. The $\Delta\lambda$ is not written in this equation).

The matrix $\mathbf{D}(\mu,\lambda)$ has been written setting the fluorescence emission at wavelengths shorter than that of excitation to zero. In practice this is not strictly true, because the bands of emission and excitation overlap.

Measurements in the overlapping region are very difficult and require a very narrow bandpass. This is another reason for using the radiance factor instead of the reflectance factor.

To ensure inter-instrument agreement tight geometric tolerances are required of the instrument-axis angles and the instrument-aperture angles. Hemispherical geometry with an integrating sphere is not recommended because the radiation emitted by the fluorescent specimen reflecting off the sphere wall and re-illuminating the specimen produces an error (*spectral sphere error*), thereby altering the spectral illuminance distribution on the specimen from that on the standard. Therefore bidirectional geometry, (45°:0°) or (0°:45°), is recommended.

Here the two extreme techniques are considered, the more reduced technique and the most complete:

1. The method of *one-monochromator with known source*, which leads to specify the colour for only the source present in the instrument.
2. The method of *two-monochromators*, which leads to specify the colour for each illuminant.

The method of *one-monochromator*, used to measure the radiance factor, must contain a light source W, which approximates in the best way a standard illuminant (including the ultraviolet part of the spectrum starting from 300 nm), usually D65, obtained from a suitably filtered xenon lamp. The quantity that is obtained directly from the instrument is the spectral total radiance factor $\beta_{T,W}(\lambda)$, typical of the instrument source W

$$\beta_{T,W}(\lambda) = \frac{L_{test,W,\lambda}}{L_{perfect\ reflecting\ diffuser,W,\lambda}} \qquad (12.14)$$

where $L_{test,W,\lambda}$, and $L_{perfect\ reflecting\ diffuser,W,\lambda}$ are the radiances exiting the test and the perfect reflecting diffuser, respectively, illuminated in an equal way by the light source W inside the instrument. The recommended geometry is bidirectional, that is, directional illumination and directional vision. This quantity can also be greater than 1 in the spectral region where there is fluorescence emission (Figure 12.8). Colorimetric spectrometers with one-monochromator are generally designed for the colour measurement of ordinary non-fluorescent specimens and the precision with which they can measure the colour of fluorescent specimens depends on how well the instrument source simulates the standard CIE illuminant. The colorimetric computation uses the spectral total radiance factor $\beta_{T,S}(\lambda)$, that replaces the reflectance factor $R(\lambda)$, with the requirement of using the illuminant simulated by the source $W(\lambda)$ of the instrument. Only the colour specification for this illuminant is meaningful.

The measurement of the *bispectral radiance factor* $\mathbf{D}(\mu, \lambda)$ in the method of *two-monochromators* is made by a bispectrophotometer. A bispectrophotometer is an optical instrument equipped with a source of irradiation, two monochromators and a spectral detection system, such that a specimen can be measured at independently controlled irradiation and viewing wavelengths. The specimen is irradiated with monochromatic radiation of wavelength μ, selected by the first monochromator, and, at this wavelength, the radiance emitted at various wavelengths $\lambda \geq \mu$ is analysed by the second monochromator. The measured quantity is the Donaldson matrix

$$\mathbf{D}(\lambda, \mu) = \beta_{R,\lambda} + \beta_{L,\lambda}(\mu) = \frac{L_{test,\lambda,\mu}}{L_{perfect\ reflecting\ diffuser,\mu}} \qquad (12.15)$$

where

$L_{test,\lambda,\mu}$ = radiance emitted at the wavelength λ by the test specimen lit by a monochromatic light of wavelength μ,

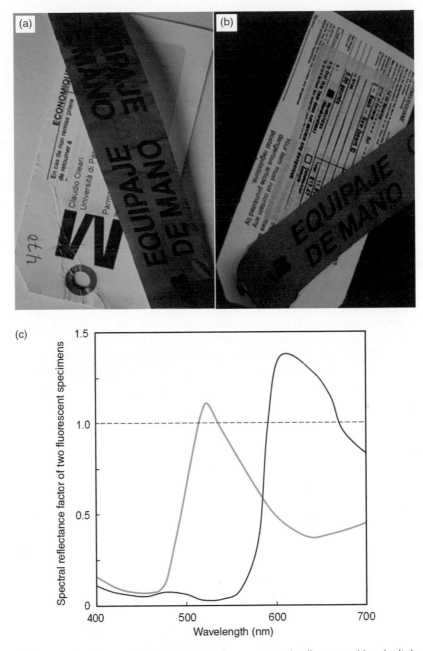

Figure 12.8   *(a) Photograph of paper labels inked with fluorescent inks illuminated by daylight D65 in a light box. (b) Photograph of the same paper labels illuminated by only UV light (a band around 370 nm). (c) Total radiance factor measurement of the same paper labels. The green label is absorbing approximately below 470 nm and the red one below 570 nm. The fluorescent emission renders the radiance factor greater than 1 around 510 nm for the green label, and over 590 nm for the red label.*

$L_{\text{perfect reflecting diffuser},\mu}$ = radiance emitted in the same direction by a perfect reflecting diffuser equally irradiated by the same monochromatic light of wavelength μ.

Knowing $\mathbf{D}(\lambda,\mu)$ for each pair of wavelengths $\lambda \geq \mu$ allows us to evaluate $\beta_{T,S}(\lambda)$ for each illuminant $S(\lambda)$

$$\beta_{T,S}(\lambda) = \frac{\sum_{\mu=300,\Delta\mu=1}^{800} \beta_{T,\mu}(\lambda)S(\mu)\Delta\mu}{S(\lambda)} \tag{12.16}$$

The colorimetric computation uses the spectral total radiance factor $\beta_{T,S}(\lambda)$, that replaces the reflectance factor $R(\lambda)$ of the non-fluorescent colours.

There are other techniques based on a single monochromator and the use of suitable filters that subtract parts of the illuminating radiation exciting the fluorescence and then evaluate the component of the radiation emitted by fluorescence compared to the reflected one.[12–21]

12.3   Spectroradiometric and Spectrophotometric Measurements

12.3.1   Introduction to the Spectrometer

The spectrometer is the essential component for spectrophotometric and spectro-radiometric measurements (Section 12.1). This Section considers the luminous flux from the crossing of the entrance slit of the spectrometer up to the electrical signals generated by the photodetectors, putting these signals in correspondence with spectral radiance and irradiance measurements and with the measurements associated with the various kinds of optical modulation (Section 12.2.2). The electric signal and the quantities measured are connected by a convolution, which is made within the instrument. Here this convolution is analytically described. The optical modulations are obtainable by a deconvolution of the electric signals, but only a complete knowledge of the instrument allows a significant deconvolution. Often the noise associated with the measurements negates the operation of deconvolution.

The Optics of the Spectrometer

Figures 12.1a and 12.1b show diagrams of instruments for measuring light for colorimety. An objective lens focusses the image of the scene on a slit, usually a test/standard specimen, the colour of which is being measured. This is the entrance slit of the spectrometer, in which is made a process of continuous wavelength dispersion, by which the ascending wavelength components are arranged in angularly increasing directions. (We do not enter into the phenomenon of dispersion, albeit interesting and important, because it is not necessary for the definition of the bandpass of the spectrometer we are going to define.)[22]

Lenses and/or mirrors, inside the spectrometer, focus the different wavelength dispersed lights on a plane producing a continuous set of monochromatic images of the entrance slit with increasing wavelengths. The images of different wavelengths are overlapped and the overlap decreases to zero with the wavelength-difference increasing. On this plane there is the exit slit of the monochromator, which is crossed by the part of the various monochromatic images which is located inside the slit (Figures 12.1 and 12.9).

For each wavelength and for that slit (Figure 12.9), a spectral transmittance is defined by the part of the image of that wavelength which crosses the slit, supposing the image area equal to 1. There are images within a wavelength range that pass through the slit entirely, while images with a wavelength outside this range have decreasing transmittance (linearly in the ideal case) as the wavelength difference increases.

The width of the slits selects the wavelength range involved in each spectral measurement. All other conditions being equal, the narrower the slit, the lower the range of wavelengths in each single measurement and

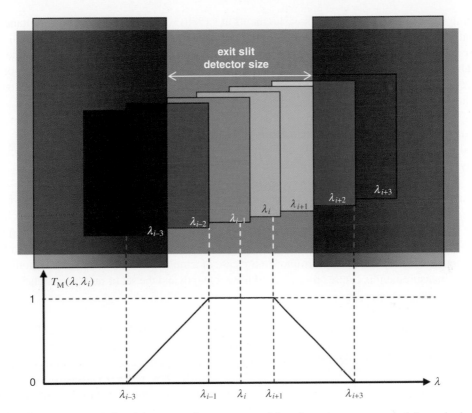

Figure 12.9 *Plane of the exit slit of the monochromator or of the photodetector array of the polychromator. On this plane some monochromatic images of the entrance slit are shown, suitably selected and related to the wavelengths λ_{i-3}, λ_{i-2}, λ_{i-1}, λ_i, λ_{i+1}, λ_{i+2}, λ_{i+3} (in reality these images belong to a continuous distribution in wavelength and are perfectly overlapped; in the figure are mutually shifted in the direction parallel to the exit slit so as to distinguish them). The image corresponding to λ_i is at the centre of the exit slit, and this wavelength is associated with the measurement; λ_{i-1} and λ_{i+1} define the range within which the images pass through the exit slit in their entirety; the images with wavelength between λ_{i+1} and λ_{i+3} pass so gradually decreasing to not go for $\lambda \geq \lambda_{i+3}$. In particular, the image for the λ_{i+2} crosses the exit slit only half. Similarly occurs in the range λ_{i-3} - λ_{i-1}. This analysis associates to the spectrometric system a spectral transmittance with trapezoidal shape $T_M(\lambda, \lambda_i)$, of which the plot below is in correspondence to the upper figure.*

we say that that the *resolving power* is higher. The result of the measurement is given by the signal of the sensor and depends on the sensitivity of the sensor and on the amount of light hitting the sensor, which depends on the area of the slit. The slit width determines the optical response, and slit height is chosen as great as possible in order to obtain an adequate flux on the sensor. Therefore, the choice of the slits is the result of a compromise between sensitivity and resolving power.

Spectrometers with a monochromator have only one exit slit which is located on the photodetector and the various spectral measurements are obtained by mechanically rotating the dispersing element (grating, prism, …). The spectral measurements at various wavelengths follow one another sequentially in time. The monochromator system is also called *single channel*.

Spectrometers with a polychromator have an array of photodetectors on the plane of the monochromatic images of the entrance slit, and the section of each individual detector coincides with the exit slit at the corresponding wavelength. The spectral measurements at various wavelengths are made simultaneously. The polychromator system is termed *multichannel* for the presence of sensors that operate in parallel.

There are other methods of spectral analysis, said to be *abridged*, because they have a simplified structure constituted by a set of filters – generally interference filters – combined with as many photosensors.

The polychromator is a simpler, more robust and less expensive device than the monochromator, but it has two disadvantages:

- array sensors have lower sensitivity than in single-channel systems and
- the bandwidth is not selectable as the exit slit has a fixed width equal to the width of each sensor in the array.

Single channel instruments, the most complex and delicate but most accurate, precise and versatile, are used in laboratories and research. Multi-channel instruments are used for routine measurements in the field, or when the working situation needs it, as in the case of quality control at the end of a production line.

In any case, the spectral measurements are made for points at wavelengths λ_i, freely chosen in the case of a single-channel instrument, and predetermined in the case of a multi-channel instrument. In both cases, each single spectral measurement is centred on a particular wavelength and the set of measurements is discrete and ordered by increasing wavelengths.

The spectral transmittance, associated with the position of the slits and their width, is called the *transmission profile* and is denoted by $T_M(\lambda, \lambda_i)$. This transmittance is a function of the wavelength and has a shape depending on the width of the entrance and exit slits (Figure 12.9). The transmission profile is

- trapezoidal if the exit slit has a width greater than the section of the image of the entrance slit;
- triangular if the exit slit has a width equal to the section of the image of the entrance slit; and
- trapezoidal if the exit slit has a smaller width than the section of the image of the entrance slit, but, in this case, the transmitted power is also decreased for the radiation of wavelength λ_i corresponding to the value on which the exit slit is centred.

The transmission profile represented with accurate geometric figures as a triangle and trapezoid does not represent all the optical phenomena because the images of the entrance slit on the plane of the exit slit are diffraction patterns, even in the case of dispersion by a prism, then the triangular figure is actually close to a Gaussian function. Moreover optical aberrations often make the transmittance shape asymmetrical. The term used for taking into account all the optical phenomena is *optical response bandwidth* that is roughly represented by the difference $\Delta\lambda$ between the upper and lower wavelengths in a continuous set of wavelengths.

The ideal spectral measurements are made by spectrometers with zero-bandwidth transmittance, but in such an ideal situation no power arrives at the photodetector and no measurement is obtainable.

This analysis indicates how important it is for the monochromatic images of the entrance slit on the plane of the exit slit to have a uniform power density. This is only achievable if the entrance slit is uniformly illuminated.

For an optimal performance of the colorimetric calculations, it is a good rule for (Figure 12.10)

1. the optical response bandwidth to have triangular shape and this happens when in the spectrometer the image of the entrance slit on the plane of the exit slit is equal to the exit slit and
2. measurements of the spectrum to be carried out with pitch equal to the bandwidth $\Delta\lambda$, which, in the case of an optical response bandwidth, is equal to half of the base of the triangle.

Rule 2 excludes redundancy in measurements and all parts of the considered spectral range are measured.

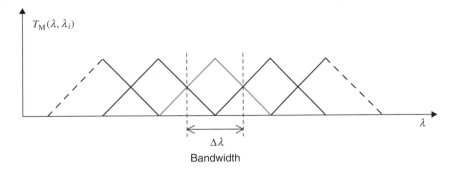

Figure 12.10 *Sequence of wavelength placements for spectrophotometric measurements with a spectrometer having triangular transmission profile $T_M(\lambda, \lambda_i)$. The best situation, represented here, is for bandwidth $\Delta\lambda$ – half-height width – equal to the scanning step.*

Two other definitions of *resolving power*, conceived for different situations, are given:

(a) According to the *Rayleigh resolution* two spectral radiations are considered resolved when the spectral distribution produced by the instrument has two maxima corresponding to two different wavelengths and the function between the two maximum values, supposed equal, drops to a value less than or equal to 81% of the maximum value.

(b) *Digital resolution* or *pixel resolution* is defined in the case where the width of the photosensor of an array corresponds to a defined range of wavelengths and the resolution is expressed in nanometer per pixel.

Spectral Responsivity and Photodetector Signal

A *transducer* is a device used to transform one kind of energy into another.

Sensor is the name given to a transducer that converts a measurable quantity – for example, power of an electromagnetic visible radiation – into an electrical voltage or an electrical current. *Photodetectors,* or *photosensors,* are sensors of light.

The photodetectors used in colorimetric instrumentation are assumed to be linear so that their response is proportional to the stimulus. The proportionality constant is called the *sensitivity* (or the *transducer gain* or, simply, the *gain* of the sensor). The sensitivity is measured with different units and assumes different names.

A photodetector is characterised by the *quantum efficiency*, that is the ratio of the number of electrons output by the photodetector to the number of photons incident on the photodetector. In practice the units of the International System are used, therefore the *responsivity* or spectral response ε_λ, is defined that is conceptually similar to the quantum efficiency and is the ratio of the current generated by the photodetector (ampere) to the electromagnetic power incident on the photodetector (watt). It is measured in A/W.

The responsivity curve of a typical silicon-semiconductor detector is shown in Figure 12.11. The spectral responsivity ε_λ enters the light measurements.

The instruments considered have a spectrometer and measure the light crossing the entrance slit.

In correspondence with a spectral flux $\Phi_{e,\lambda}$ (W/nm) entering the spectrometer, the photodetector centred on the wavelength λ_i is crossed by a flux $[\Phi_{e,\lambda}\, T_M(\lambda, \lambda_i)]$ (W/nm) and converts this flux into an electric signal (ampere or volt)

$$J_{\lambda_i} = \int\limits_0^{+\infty} \Phi_{e,\lambda} T_M(\lambda, \lambda_i)\varepsilon_\lambda \, d\lambda \tag{12.17}$$

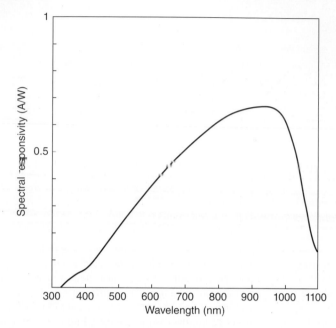

Figure 12.11 *Spectral responsivity of a typical silicon semiconductor photodetector.*

that is a *convolution* on the unknown spectral flux $\Phi_{e,\lambda}$ of the considered light source and the *instrument-spectral-bandpass function* $[T_M(\lambda,\lambda_i)\,\varepsilon_\lambda]$, or simply *bandpass*, of the instrument used. Generally, the term *bandpass* is defined as the *full-width at half maximum* (FWHM) spectral response of a spectrometer channel to a monochromatic light.

Equation (12.17) supposes that the response of the sensor is linear, that is, proportional to the stimulus. This assumption is supposed to hold true hereafter.

Unfortunately, photodetectors generate a signal also with no light flow. This signal is called a *dark signal* or *black signal*. The dark signal must be subtracted from the signal produced by a luminous flux. In the discussion here, every signal is considered decreased by the corresponding dark signal.

Calibration

The formal definition of *calibration* by the International Bureau of Weights and Measures is the following [23]:

"Calibration – Operation that, under specified conditions,

1. in a first step, establishes a relation between the quantity values with measurement uncertainties provided by measurement standards and corresponding indications with associated measurement uncertainties (of the calibrated instrument or secondary standard) and,
2. in a second step, uses this information to establish a relation for obtaining a measurement result from an indication."

Calibration established as a relationship between the value measured by the instrument and the certificate of a *reference* sample, provided that the latter has a guarantee of traceability to recognized standards, through an unbroken chain of comparisons – *traceability chain* –. The calibration certificate includes the value of the quantity in question and its uncertainty. The traceability chain leads to a standard metrology laboratory.

The calibration is the set of operations that *find and eliminate systematic errors of an instrument scale*.

In general use, calibration includes the process of *adjusting* the output or indication on a measurement instrument to agree with the value of the applied standard, within a specified accuracy (Step 2.).

The calibration of spectroradiometers concerns wavelength calibration and radiometric calibration.

Wavelength Calibration

The wavelength calibration is made by a discharge lamp whose spectrum is constituted by spectral lines of wavelength known from the literature and in this way the wavelength is associated with the points of the spectrum in which are the known spectral lines, and for the other points the wavelength is computed with a suitable interpolation.

Consider a polichromator with the array photosensors with ordered numbering $k = 0, 1, 2, …, n$. In the case of a monochromator with only one photosensor, any ordered angular position of the rotating optics is similarly numbered. The wavelength calibration relates the photosensor number of the spectrometer to its corresponding wavelength. The wavelength range considered for the spectrometers used in colorimetry and photometry must be equal to or greater than 300–830 nm. The calibration light entering the spectrometer must have a well-known spectrum constituted by easily identifiable lines. These line spectra can be produced

1. by emission in the case of spectroscopic lamps.

"*Spectroscopic lamp* is a discharge lamp which gives a well-defined line spectrum and which, in combination with filters, may be used to obtain monochromatic radiation."[1]

A lamp that produces several spectral lines, at least 4-6 spectral lines in the wavelength range 300–830 nm, is necessary, for example, a Mercury-Argon lamp (Figure 12.12), of which emission lines have wavelength

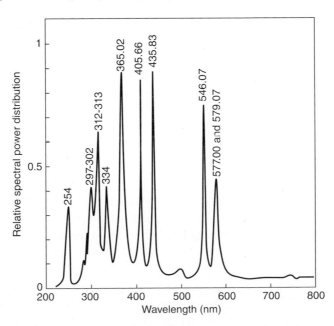

Figure 12.12 *Spectral power distribution of the light emitted by a typical high pressure mercury vapour arc-discharge lamp, used for wavelength calibration.*

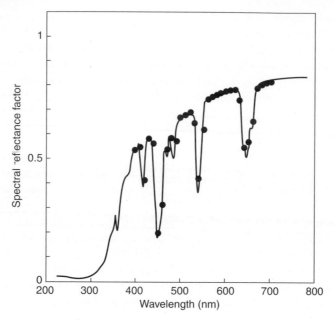

Figure 12.13 *Spectral reflectance factor of a holmium oxide standard tile (continuous line) and corresponding data measured by a spectrophotometer with bandpass and scanning step of 10 nm (black dots).*

determined by quantum mechanics and do not require a certification of a standard metrology laboratory (The atomic emission lines for each source are available in the NIST Atomic Spectra Database Lines Form: http:// physics.nist.gov/cgi-bin/AtData/lines_form).

2. by absorption in the case of reflectance tiles (Figure 12.13): for example, erbium oxide and holmium oxide tiles, retraceable to a standard metrology laboratory.

Once these spectral lines are identified in the spectrum recorded by the spectrometer being calibrated, it is supposed that the correspondence between spectrum wavelengths λ and the number k of the sensor is defined by a *calibration polynomial*

$$\lambda(k) = c_0 + c_1 k + c_2 k^2 + c_3 k^3 + \ldots \tag{12.18}$$

The calibration is made once the coefficients c_i are defined. The comparison of the spectrum of the calibration lamp or tile and the spectrum recorded by the spectrometer shows a correspondence between a set of $i = 1, 2, ..,w$ wavelengths of spectral lines λ_i and sensor numbers k_i. The number of known spectral lines used in the calibration process has to exceed the polynomial order, and the process is improved by considering more lines, although this implies a longer computation. The unknown coefficients c_i are obtained by a polynomial regression based on the least squares method, that is, from the minimization of the following expression

$$\sum_{i=1}^{w} \left[\lambda_i - (c_0 + c_1 k_i + c_2 k_i^2 + c_3 k_i^3 + \ldots) \right]^2 \tag{12.19}$$

The polynomial order could be adjusted in terms of the specific requirements, but typically second and third orders are chosen.

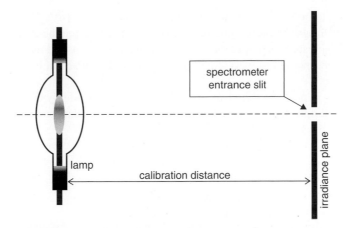

Figure 12.14 *Typical disposition of a standard lamp used for the radiometric calibration of a spectral irradiance meter.*

These are the definitions and assumptions useful for describing the use of a photodetector for light measurement and do not exhaust the characterization of a photodetector.

Radiometric Calibration

Radiometric calibration concerns the radiometric scale and is after the wavelength calibration. This calibration is made by a lamp traceable to a metrology laboratory, usually a lamp with black body emission, of which the metrological certification gives the irradiance in a discrete set of wavelengths and for the other wavelengths provides a mathematical interpolating formula (Section 12.3.1). Figure 12.14 shows a typical disposition of a standard lamp used for the radiometric calibration of a spectroradiometer. The irradiance of the standard lamp at defined calibration distance is defined by the metrology-laboratory certification. The calibration is made associating the certified irradiance data to the signal produced by spectrometer at equal wavelengths.

12.3.2   Instrumental Convolution

The most typical situations of spectral measurements are here considered to define general way for analysing raw empirical data.

Spectral Irradiance Measurement

A spectroradiometer for the spectral irradiance $(W/(m^2 \, nm))$ measurement is considered. The spectral irradiance being measured is on the entrance slit of the spectrometer. The irradiating flux may come

- directly from the lamp, positioned at suitably defined distance,
- from a reflecting or transmitting diffuser,
- from the exit port of an integrating sphere with the light source outside it and
- from the exit port of an integrating sphere with the light source inside the integrating sphere – *total flux mode* –.

The choice of the irradiation mode depends on the lamp considered and is defined by the metrology laboratory that supplies the standard lamp (e.g. Figure 12.14).

The unknown quantity to be measured is the spectral irradiance $E_{e,\lambda} = [\Phi_{e,\lambda} / (area\ of\ the\ entrance\ slit)]$. The instruments generate a signal representing the flux $\Phi_{e,\lambda}$ and not $E_{e,\lambda}$, but these quantities are proportional and therefore the signal represents both quantities well. The measurements are made with a set of number of wavelengths λ_i, $i = integer$, and, at any wavelength λ_i, the photodetector signal is represented by the integral (12.17).

Radiometric calibration is performed by transfer comparison. It is made by illuminating the entrance slit with a retraceable standard light source with an unbroken chain to a standard metrology laboratory (e.g. Figure 12.14), that generates the signal

$$J_{STD,\lambda_i} = \int_0^\infty \Phi_{STD,\lambda} T_M(\lambda,\lambda_i)\varepsilon_\lambda d\lambda \tag{12.20}$$

where $\Phi_{STD,\lambda}$ is the spectral flux crossing the entrance slit with an area A associated to the irradiance $E_{STD,\lambda}$ $= (\Phi_{STD,\lambda}/A)$ of the standard light source. This transfer comparison supposes that

$$\frac{E_{U,\lambda_i}}{E_{STD,\lambda_i}} = \frac{\Phi_{U,\lambda_i}/A}{\Phi_{STD,\lambda_i}/A} = \frac{J_{\lambda_i}}{J_{STD,\lambda_i}} \rightarrow measurement\ equation\ E_{U,\lambda_i} = \frac{J_{\lambda_i}}{J_{STD,\lambda_i}} E_{STD,\lambda_i} \tag{12.21}$$

where the subscript 'U' means 'undeconvoluted quantity'. From Equation (12.21) it follows that

$$E_{U,\lambda_i} = E_{STD,\lambda_i} \int_0^\infty \Phi_{e,\lambda} \frac{T_M(\lambda,\lambda_i)\varepsilon_\lambda}{J_{STD,\lambda_i}} d\lambda \equiv \int_0^\infty \Phi_{e,\lambda}\kappa(\lambda,\lambda_i) d\lambda \tag{12.22}$$

where $\Phi_{e,\lambda}$ is the unknown spectral flux crossing the entrance slit and associated to the irradiance $E_{e,\lambda}$ of the source being measured, that is, the measured undeconvoluted spectral irradiance is a convolution on the unknown true spectral irradiance with kernel

$$\kappa(\lambda,\lambda_i) \equiv \frac{E_{STD,\lambda_i} T_M(\lambda,\lambda_i)\varepsilon_\lambda}{J_{STD,\lambda_i}} \tag{12.23}$$

Spectral Radiance Measurement

In the case of spectral radiance measurement, typical of screens reflecting or emitting light, an objective lens focusses the light exiting the screen on the entrance slit of the spectrometer. This is the difference with respect to the irradiance measurement. The objective lens is linked to the spectrometers constituting a single body instrument.

Spectral Reflectance-Factor Measurement

The situation of the spectrophotometer for the measurement of the spectral-reflectance factor $R(\lambda)$ of a specimen is analogous to the previous one for the spectral irradiance measurement. In this case, the light entering the measurement is produced by a lamp inside the instrument, with spectral power distribution W_λ (W/nm). The illumination (influx) and viewing (efflux) geometry in the instrument has to be one of those standards for optical modulation defined by the CIE (Section 12.2.2). The flux crossing the entrance slit of the spectrometer is proportional to $[W_\lambda R(\lambda)]$ and the corresponding signal at any wavelength λ_i is

$$J_{\lambda_i} = \int_0^\infty W_\lambda R(\lambda) T_M(\lambda, \lambda_i) \varepsilon_\lambda \, d\lambda \qquad (12.24)$$

that is a *convolution* on the unknown spectral reflectance factor $R(\lambda)$ of the considered sample and the instrument-spectral-bandpass function $[W_\lambda T_M(\lambda,\lambda_i) \varepsilon_\lambda]$ of the instrument used.

By definition, the reflectance factor at wavelength λ_i is given by the ratio

$$R_U(\lambda) \equiv \frac{J_{\lambda_i}}{J_{\text{perfect reflecting diffures}, \lambda_i}} \quad \text{with} \quad J_{\text{perfect reflecting diffuser}, \lambda_i} = \int_0^\infty W_\lambda T_M(\lambda, \lambda_i) \varepsilon_\lambda d\lambda \qquad (12.25)$$

But no tile is ideal, therefore a standard reflectance tile $R_{\text{STD}}(\lambda)$, retraceable with an unbroken chain to a standard metrological laboratory, is considered and transfer comparison supposes that

$$\textit{measurement equation } R_U(\lambda_i) \equiv \frac{J_{\lambda_i}}{J_{\text{STD}, \lambda_i}} \frac{J_{\text{STD}, \lambda_i}}{J_{\text{perfect reflecting diffures}, \lambda_i}} = \frac{J_{\lambda_i}}{J_{\text{STD}, \lambda_i}} R_{\text{STD}}(\lambda_i) \qquad (12.26)$$

where

$$J_{\text{STD}, \lambda_i} = \int_0^\infty R_{\text{STD}}(\lambda) T_M(\lambda, \lambda_i) \varepsilon_\lambda d\lambda \ . \qquad (12.27)$$

$R_U(\lambda_i)$ is the usual undeconvoluted spectral-reflectance factor and is the convolution on the unknown function $R(\lambda)$ of the specimen measured at the wavelength λ_i, as it follows from Equation (12.26)

$$R_U(\lambda_i) = \frac{J_{\lambda_i}}{J_{\text{STD}, \lambda_i}} R_{\text{STD}}(\lambda_i) = R_{\text{STD}}(\lambda_i) \int_0^\infty R(\lambda) \frac{W_\lambda T_M(\lambda, \lambda_i) \varepsilon_\lambda}{J_{\text{STD}, \lambda_i}} d\lambda \equiv \int_0^\infty R(\lambda) \kappa(\lambda, \lambda_i) d\lambda \qquad (12.28)$$

and the convolution kernel is

$$\kappa(\lambda, \lambda_i) \equiv \frac{R_{\text{STD}}(\lambda_i) W_\lambda T_M(\lambda, \lambda_i) \varepsilon_\lambda}{J_{\text{STD}, \lambda_i}} \qquad (12.29)$$

The quantity $R_U(\lambda_i)$ is different from the spectral-reflectance factor $R(\lambda_i)$ of the test specimen and can be considered as a measure of $R(\lambda_i)$ only in the particular case that $[R(\lambda) W_\lambda \varepsilon_\lambda]$ is a constant function in the range in which the transmittance $T_M(\lambda, \lambda_i)$ has a meaningful contribution to the integrals, but generally the deconvolution is needed.

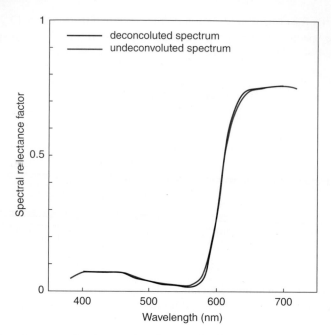

Figure 12.15 *Comparison between the deconvoluted spectral reflectance factor and the undeconvoluted one of a typical orange tile.*

The difference between $R_U(\lambda_i)$ and $R(\lambda_i)$ is termed *bandpass error* and has been investigated in the last few decades by many authors and after the analysis by Sterns & Sterns and Venable, this source of error was considered one of the greatest.[24–32] (Figure 12.15).

Spectral Transmittance and Spectral Transmittance-Factor Measurement

According to the CIE[2], in colorimetry no standard of transmittance exists and, once the test specimen is removed from the instrument, the air is considered as a reference for the transmittance and transmittance factor measurement. The standard transmittance factor is $T_{STD}(\lambda) = 1$. The difference between transmittance and transmittance factor is in the measurement geometry. The illumination and viewing geometry in the instrument has to be one of the standard ones for optical modulation defined by the CIE (Section 12.2.2). The situation of the spectrophotometer for transmittance or transmittance-factor measurement is analogous to the previous one for reflectance-factor. Equation (12.28) becomes

$$T_U(\lambda_i) = \frac{J_{\lambda_i}}{J_{STD,\lambda_i}} T_{STD}(\lambda_i) = \int_0^\infty T(\lambda) \frac{T_M(\lambda,\lambda_i) W_\lambda \varepsilon_\lambda}{J_{STD,\lambda_i}} \, d\lambda \equiv \int_0^\infty T(\lambda) \kappa(\lambda,\lambda_i) d\lambda \qquad (12.30)$$

where
- $T_U(\lambda)$ is the empirical undeconvoluted spectral transmittance or transmittance-factor of the test specimen measured by the instrument,
- $T(\lambda)$ is the unknown spectral transmittance or spectral transmittance-factor of the test specimen,
- $[W_\lambda T_M(\lambda,\lambda_i)\, \varepsilon_\lambda]$ is the same quantity defined for the spectral reflectance factor measurement,

$$- \quad J_{\text{STD},\lambda_i} = \int_0^\infty T_{\text{STD}}(\lambda)\big[W_\lambda T_{\text{M}}(\lambda,\lambda_i)\varepsilon_\lambda\big]\mathrm{d}\lambda \text{ with } T_{\text{STD}}(\lambda)=1\,,$$

$$- \quad \kappa(\lambda,\lambda_i) \equiv \frac{W_\lambda T_{\text{M}}(\lambda,\lambda_i)\varepsilon_\lambda}{J_{\text{STD},\lambda_i}} \text{ is the convolution kernel.}$$

Spectral Radiance Factor Measurement of Fluorescent Objects

The measurement in the case of a one-spectrometer instrument (Section 12.2.2) is equal to that of the spectral reflectance factor measurement.

The measurement in the case of a two-spectrometer instrument is different (Section 12.2.2), where the measured quantity is the *bispectral luminescent radiance factor*

$$\beta_{\text{L},\lambda_i}(\mu) = \frac{L_{\text{test},\lambda_i}}{L_{\text{perfect reflecting diffuser},\mu}} = \frac{J_{\text{test},\lambda_i} C_{\lambda_i}}{J_{\text{perfect reflecting diffuser},\mu} C_\mu} \tag{12.31}$$

where

- $L_{\text{test},\lambda i}$ is the radiance at wavelength λ_i, due to photoluminescence from the specimen irradiated at wavelength μ, which is equal to the signal multiplied by a calibration factor $C_{\lambda i}$ at wavelength λ_i, that is $L_{\text{test},\lambda i} = J_{\text{test},\lambda i} C_{\lambda i}$;
- $L_{\text{perfect reflecting diffuser},\mu}$ is the radiance at wavelength μ of the perfect reflecting diffuser identically irradiated and viewed, which is equal to the signal multiplied by a calibration factor at wavelength μ, that is $L_{\text{perfect reflecting diffuser},\mu} = J_{\text{perfect reflecting diffuser},\mu} C_\mu$;

- $C_{\lambda i}$ (and C_μ) is the calibration factor that takes into account of the spectral responsivity $\varepsilon_{\lambda i}$ as function of the wavelength and is obtainable in a way similar to the spectral irradiance calibration.

The corresponding undeconvoluted quantity is equal to a ratio of the corresponding signals of Equation (12.31)

$$\beta_{\text{U,L},\lambda_i}(\mu) = \frac{J_{\text{test},\lambda_i} C_{\lambda_i}}{J_{\text{perfect reflecting diffuser},\mu} C_\mu} \tag{12.32}$$

but, as before, since no tile is ideal, a standard reflectance tile with certified spectral radiance factor $\beta_{\text{STD}}(\mu)$, retraceable with an unbroken chain to a standard metrology laboratory, is considered and transfer comparison (measurement equation) supposes that

$$\beta_{\text{U,L},\lambda_i}(\mu) = \frac{J_{\text{test},\lambda_i}}{J_{\text{STD},\mu}} \frac{J_{\text{STD},\mu}}{J_{\text{perfect reflecting diffuser},\mu}} \frac{C_{\lambda_i}}{C_\mu} = \frac{J_{\text{test},\lambda_i}}{J_{\text{STD},\mu}} \beta_{\text{STD}}(\mu) \frac{C_{\lambda_i}}{C_\mu} \tag{12.33}$$

Form this equation it follows that

$$\beta_{\text{U,L},\lambda_i}(\mu) = \frac{J_{\text{test},\lambda_i}}{J_{\text{STD},\mu}} \beta_{\text{STD}}(\mu) \frac{C_{\lambda_i}}{C_\mu} = \beta_{\text{STD}}(\mu) \frac{C_{\lambda_i}}{C_\mu} \int_0^\infty \beta_{\text{L},\lambda}(\mu) \frac{W_\lambda T_{\text{M}}(\lambda,\lambda_i)\varepsilon_\lambda}{J_{\text{STD},\mu}} \mathrm{d}\lambda$$

$$\equiv \int_0^\infty \beta_{\text{L},\lambda}(\mu)\kappa(\lambda,\mu,\lambda_i)\mathrm{d}\lambda \tag{12.34}$$

that is, $\beta_{U,L,\lambda}(\mu)$ is the convolution of the unknown function $\beta_{L,\lambda}(\mu)$ with kernel

$$\kappa(\lambda,\mu,\lambda_i) \equiv \frac{\beta_{STD}(\mu)W_\lambda T_M(\lambda,\lambda_i)\varepsilon_\lambda}{J_{STD,\mu}}\frac{C_{\lambda_i}}{C_\mu} \tag{12.35}$$

Colour-Matching-Function Measurement

In the measurement of the colour-matching functions, spectral lights are used, which are constituted by a narrow bandwidth of few manometers. These spectral lights, referred to the wavelengths λ_i, are obtained by interference filters, gratings or prisms with a spectral power distribution $S_{\lambda_i,\lambda}$ different from the ideal Dirac-δ function. Therefore, when the maximum saturation method (Section 6.13.5) is used and in the hypothesis that the Grassmann laws (Section 6.13.1) hold true, the measured undeconvoluted radiances $\left(\bar{r}_U(\lambda_i),\bar{g}_U(\lambda_i),\bar{b}_U(\lambda_i)\right)$ of the three reference lights chosen are the convolution on the unknown colour-matching functions $\left(\bar{r}(\lambda),\bar{g}(\lambda),\bar{b}(\lambda)\right)$ and a kernel $\kappa(\lambda,\lambda_i) = S_{\lambda_i,\lambda}\Big/\int\limits_{360}^{780} S_{\lambda_i,\lambda}\,d\lambda$,

$$\begin{cases} \bar{r}_U(\lambda_i) = \int\limits_{360}^{780} S_{\lambda_i,\lambda}\bar{r}(\lambda)\,d\lambda \Big/ \int\limits_{360}^{780} S_{\lambda_i,\lambda}\,d\lambda \\[2ex] \bar{g}_U(\lambda_i) = \int\limits_{360}^{780} S_{\lambda_i,\lambda}\bar{g}(\lambda)\,d\lambda \Big/ \int\limits_{360}^{780} S_{\lambda_i,\lambda}\,d\lambda \\[2ex] \bar{b}_U(\lambda_i) = \int\limits_{360}^{780} S_{\lambda_i,\lambda}\bar{b}(\lambda)\,d\lambda \Big/ \int\limits_{360}^{780} S_{\lambda_i,\lambda}\,d\lambda \end{cases} \tag{12.36}$$

Very similar equations hold in the case of Maxwell's minimum saturation technique (Section 6.13.5).

12.3.3 Deconvolution

All the considered convolution cases are summarized in the following equation relating the unknown function F_λ, the measured quantities F_{U,λ_i} and the kernel $\kappa(\lambda,\lambda_i)$

$$F_{U,\lambda_i} = \int\limits_{360}^{780} F_\lambda \kappa(\lambda,\lambda_i)\,d\lambda \tag{12.37}$$

The aim of the deconvolution is to write the unknown quantities F_λ as a function of the measured quantities F_{U,λ_i}. In order to perform a deconvolution of the measured data we have to suppose that the kernel $\kappa(\lambda,\lambda_i)$ of the instrumental convolution is known. This is possible if the spectrometer is characterized in all its optical and electro-optical parts.

Some attempts at deconvolution are known in the literature. Here we report an interesting result of reflectance factor deconvolution, known as *Sterns & Sterns formula for bandpass error*.[25] The formula gives the reflectance factor at zero order of the local power expansion of the reflectance factor in the hypothesis that the

1. transmission profile is triangular,
2. spectral responsivity is constant inside the bandpass range,

3. measurements of the spectrum are carried out with pitch equal to the bandpass $\Delta\lambda$,
4. standard reflectance tile has a certified reflectance factor $R_{STD}(\lambda) = constant$

$$R_{i,0} = \frac{1}{12}\left[-R_U(\lambda_{i-1}) + 14R_U(\lambda_i) - R_U(\lambda_{i+1})\right] \qquad (12.38)$$

with $R_U(\lambda)$ defined by Equation (12.26).

With regard to the bandpass error the CIE[2] says:

"Measurement errors arising from the bandwidth of a spectrometer are generally much larger (by an order of magnitude) than the calculation errors associated with data intervals," that is, errors associated to data obtained by interpolation. "Even if the data interval is 1 nm (or interpolated to 1 nm intervals), the colorimetric errors can be significant if bandwidth of the spectrometer is large." How large? "For the highest accuracy, a bandwidth of 1 nm may be used for the measuring instrument, but for most practical purposes a bandwidth and measurement interval of 5 nm may be used. The use of bandwidths of 10 nm, or 20 nm is not recommended; it can lead to a considerable loss of accuracy, and, if applied, should be checked on typical spectra." The standard organization ASTM[33] proposes procedures for colorimetric computation in cases where the bandwidth is 10 and 20 nm. Today also CIE[34] proposes a deconvolution algorithm.

Factories producing instrumentation have proprietary deconvolution algorithms, that are generally unknown to the user and in some cases patented, but almost unusable in different instruments.

In practice, the user of a colorimetric/photometric/radiometric instrument is dependent on the producer as regards to calibration. Anyway the user can check the quality of the calibration by measuring standard lamps, standard reflectance tiles and standard fluorescent tiles and comparing the measured data with the data certified by a standard metrology laboratory (Section 12.6). This check should be required at the time of instrument purchase.

12.4 Colorimetric Calculations

12.4.1 CIE Colour Specification

In the analysis of tristimulus space (Sections 9.2–9.6), the colour stimulus was considered in an absolute way and associated with a spectral radiance, which comes either from a self-luminous body (*primary light source*) or from an illuminated body (*secondary light source*). Moreover the CIE specification of the colour stimulus is made according to a CIE standard colorimetric system, where one of the following sets of three numbers is used, (X, Y, Z), $(Y; x, y)$ or $(Y; \lambda_d, p_e)$. The last two sets are derivable from the first one. In the case of absolute specification, the luminance of the colour stimulus is $L_v = K_m Y$ (Section 9.2).

The colour specification must be associated

- to the chosen observer (CIE 1931, CIE 1964 or other);
- to the chosen illuminant, in the case of surface colour;
- to the geometry of illumination (influx) and vision (efflux) used in the measurement;
- to the type of spectral data (bandpass and step at which the spectral radiance is measured);
- to the technique used in colorimetric computations.

12.4.2   Relative Colour Specification

In colorimetric practice, it is customary to consider the colour of light sources or of illuminated objects regardless of the intensity of the light source, but only considering the relative spectral power distribution. This leads to a *relative* colour specification (Section 9.7), in fact

1. the light sent to the observer by the *perfect reflecting diffuser* lit by the illuminant used is measured instead of the light emitted directly from the illuminant itself, and the reflected luminance is placed conventionally equal to 100 (or 1) for all the illuminants, in accordance with the choice of relative spectral power distribution made for the definition of the illuminants (Section 8.1);
2. the luminance of a non-self-luminous body viewed in reflection refers to the luminance of an equally illuminated perfect reflecting diffuser and is represented by the percentage luminance factor (or to the luminance factor).

A suitable normalization of the tristimulus values renders the colour specification relative. Having chosen an illuminant $S(\lambda)$ and an observer – CIE 1931, or CIE 1964 or other – the computation of the corresponding tristimulus values is given by the following integrals – here the observer CIE 1931 is considered as an example (for the CIE 1964 observer the colour-matching functions $\bar{x}_{10}(\lambda)$, $\bar{y}_{10}(\lambda)$ and $\bar{z}_{10}(\lambda)$ are used) –:

$$
\begin{cases}
X = K \int_{380}^{780} \phi(\lambda_i)\bar{x}(\lambda_i)\,d\lambda_i \\[2mm]
Y = K \int_{380}^{780} \phi(\lambda_i)\bar{y}(\lambda_i)\,d\lambda_i & \text{with } K = 100 \Big/ \int_{380}^{780} S(\lambda_i)\bar{y}(\lambda_i)\,d\lambda_i \\[2mm]
Z = K \int_{380}^{780} \phi(\lambda_i)\bar{z}(\lambda_i)\,d\lambda_i
\end{cases}
\tag{12.39}
$$

where K is the normalization factor and the relative colour stimulus function is

* $\phi(\lambda) = S(\lambda)$ spectral power distribution for a primary light source or an illuminant
* $\phi(\lambda) = S(\lambda)R(\lambda)$ for a reflecting object with reflectance factor $R(\lambda)$ (secondary light source)
* $\phi(\lambda) = S(\lambda)T(\lambda)$ for a transmitting object with transmittance factor $T(\lambda)$ (secondary light source)
* $\phi(\lambda) = S(\lambda)\beta_{T,S}(\lambda)$ for a fluorescent object with total radiance factor $\beta_{T,S}(\lambda)$, which, if it is obtained by one-monochromator spectrophotometer, the illuminant must be that simulated by the lamp inside the instrument, while, if the instrument is a two-monochromator spectrophotometer any illuminant $S(\lambda)$ is possible and Equation (12.16) is used

$$
\beta_{T,S}(\lambda) = \frac{\displaystyle\sum_{\mu=300,\Delta\mu=1}^{800} \beta_{T,\mu}(\lambda)S(\mu)\Delta\mu}{S(\lambda)}
\tag{12.40}
$$

This relative colour specification is independent of the scale of the intensity of the spectral power distribution of the chosen illuminant.

The colour specification is obtained in two steps:

1. the measurement of a spectral radiometric quantity, that is, a spectral radiance or an optical modulation to be made according to a defined geometry (Section 12.2.2);
2. the colorimetric computation of the integrals (12.39).

The quantities X, Y and Z are dimensionless and the Y component is now the *percentage luminance factor* (Section 9.7).

The colour specification is usually given in the form (Y, x, y), where, as known, the chromaticity is defined as $x = X / (X + Y + Z)$ and $y = Y / (X + Y + Z)$.

Any colour specification produced by optical modulation is dependent not only on the chosen observer and the illuminant, but depends also on:

1. the instrument-measuring geometry;
2. the room temperature, because *thermochromism* exists (The optical modulation depends on the temperature. Generally, visually it is imperceptible because variations of temperature are not instantaneous);
3. the room air *humidity* (e.g., the colour of a fresco generally depends on the humidity, because the materials constituting the frescos are hygroscopic).

Colour specifications, which are different for at least one of all the considered points, are comparable only to show the consequences of the different situations. The colour differences have to be understood relating the colour measurement with the physical properties of the colour sample measured.

The Equation (12.39) are integrals, but in practice the colour-stimuli functions are known by points, as well as the colour-matching functions, therefore the integrations are carried out by numerical summations.

The CIE Standard recommends that the CIE tristimulus values of a colour stimulus function be obtained

1. by multiplying at each wavelength λ_i the value of the colour stimulus function $\phi(\lambda_i)$, measured by an instrument built in a *correction to zero bandpass* (i.e., complete deconvolution, but in practice 1 nm bandpass), by the corresponding value of each one of the CIE colour-matching functions;
2. by summation of each set of products over the wavelength range corresponding to the entire visible spectrum – 360-830 nm – at wavelength intervals $\Delta\lambda = 1$ nm

$$\begin{cases} X = K \sum_{380, \Delta\lambda=1}^{830} \phi(\lambda_i)\overline{x}(\lambda_i)\Delta\lambda_i \\[2mm] Y = K \sum_{380, \Delta\lambda=1}^{830} \phi(\lambda_i)\overline{y}(\lambda_i)\Delta\lambda_i \quad \text{with } K = 100 \Big/ \sum_{380, \Delta\lambda=1}^{830} S(\lambda_i)\overline{y}(\lambda_i)\Delta\lambda_i \\[2mm] Z = K \sum_{380, \Delta\lambda=1}^{830} \phi(\lambda_i)\overline{z}(\lambda_i)\Delta\lambda_i \end{cases} \quad (12.41)$$

All rigorous calculations should use summations with 1 nm steps. The CIE recommends "that all numerical calculations be carried out using the full number of significant digits provided by the data in the Tables published in the CIE standards of colorimetry (tables in CD). Final results should be rounded to the number of significant digits indicated by the precision of the measurements."[2]

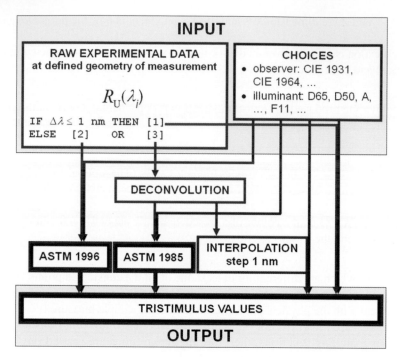

Figure 12.16   *Sequence of operations for the computation of tristimulus values of non-self-luminous colours in relation to the quality of the raw (undeconvoluted) spectral reflectance factor $R_U(\lambda)$ and choices of the observer and illuminant. The ASTM weights are distinguished by observer, illuminant and spectral scanning step. The paths of calculation for $\Delta\lambda > 1$ are equivalent. In this scheme, scanning step and bandpass are considered equal.*

The CIE says that for most practical purposes, the summation may be approximated by using wavelength intervals $\Delta\lambda = 5$ nm over the wavelength range 380 to 780 nm. For summation steps and bandpass of 10 or 20 nm and summation range from 400 to 700 nm, the result is generally a crude approximation. In these cases three phenomena degrade the result: a *band-pass error*, an *under sampling* and a *truncation*. An improvement is possible by *deconvolution*, *interpolation* and *extrapolation* of the spectral data, respectively. Algorithms have been proposed for only the interpolation (ASTM[33] Table 5) and for simultaneous deconvolution and interpolation (ASTM[33] Table 6).

Generally, in practice, the raw spectral data are produced with a bandpass wider than the 1 nm, therefore deconvolution, interpolation and extrapolation are necessary. All the possible situations of colorimetric calculation are summarized in the diagram of Figure 12.16.

Deconvolution, interpolation and extrapolation are considered individually.

12.4.3   Deconvolution

Deconvolution is also considered in Section 12.3.3.

A model to implement deconvolution is possible only if the instrument is well known.

Also the instrumental data must be degraded by noise very little, because practice says that the algorithms for deconvolution could enhance the noise and worsen the outcome.

It is customary to assume that the bandpass is triangular in shape and the step at which the spectrum is measured is equal to the bandpass. The bandpass with a triangular shape is present only in high-class instruments. Generally optical aberrations and diffraction make the bandpass assume a Gaussian and asymmetric shape.[22] There are algorithms based on these assumptions and defined for a bandpass of 10 and 20 nm. The ASTM E 308-1996 document gives a numerical table (known as 'Table 6') of the weights defined for passbands 10 and 20 nm, for the illuminants A, C, D50, D55, D65, D75, F2, F7 and F11, and for the CIE 1931 and CIE 1964 standard observers. These weights take into account the 1 nm interpolation too.

Recently, an explicit algorithm for a triangular bandpass has been given and assumed as CIE proposal that can be used for any observer, any illuminant and any scanning step.[34] This algorithm takes into account the 1 nm interpolation too.

Section 12.3.3 presents the *Stearns & Stearns* deconvolution formula (12.38) for measured reflectance factors with triangular bandpass and scanning step equal to the bandpass. After this deconvolution, the 1 nm interpolation is required.

12.4.4   Interpolation

The radiometric data, deconvoluted to zero bandpass, must be interpolated at a pitch of 1 nm, according to the CIE recommendations. This interpolation can be represented by *adjusted values*, or simply *weights*, replacing the arguments of the summations (12.41) at the measurement wavelengths: for example, in the case of deconvoluted reflectance factor $R(\lambda_j)$, obtained from measurements at a pitch of 10 nm, the summations, that take the interpolation into account, are

$$
\begin{cases}
X = \displaystyle\sum_{380,\Delta\lambda=1\text{nm}}^{830} \left[K\bar{x}(\lambda_j)S(\lambda_j) \right]R(\lambda_j) \equiv \sum_{360,\Delta\lambda=10\text{nm}}^{780} W_{X,O,S,j}R(\lambda_j) \\[2em]
Y = \displaystyle\sum_{380,\Delta\lambda=1\text{nm}}^{830} \left[K\bar{y}(\lambda_j)S(\lambda_j) \right]R(\lambda_j) \equiv \sum_{360,\Delta\lambda=10\text{nm}}^{780} W_{Y,O,S,j}R(\lambda_j) \\[2em]
Z = \displaystyle\sum_{380,\Delta\lambda=1\text{nm}}^{830} \left[K\bar{z}(\lambda_j)S(\lambda_j) \right]R(\lambda_j) \equiv \sum_{360,\Delta\lambda=10\text{nm}}^{780} W_{Z,O,S,j}R(\lambda_j)
\end{cases}
\tag{12.42}
$$

where $W_{X,O,S,j}$, $W_{X,O,S,j}$ and $W_{X,O,S,j}$ are the weights for the observer O (in this case the CIE 1931 standard observer), the illuminant S at the measuring wavelength λ_j. It must be noted that these summations can be considered as the scalar product of two vectors whose components are in correspondence to the wavelengths defined with 1 nm steps: the components of one vector are $[K\bar{x}(\lambda_j)S(\lambda_j)]$, for the X component, and the components of the other vector are the deconvoluted reflectance factors $R(\lambda_j)$.

The ASTM E 308:1985 document gives a numerical table (known as 'Table 5') of the weights defined for passbands 10 and 20 nm, for the illuminating fundamental A, C, D50, D55, D65, D75, F2, F7 and F11, and for the CIE 1931 and CIE 1964 standard observers. This table takes into account only the interpolation. The weights of 'Table 5' are obtainable by applying the Lagrange formula, as shown below.

The CIE recommends implementing the interpolation by one of these four methods[2]:

1. The third-order polynomial interpolation – *Lagrange's formula* – defined on four spectral quantities $F(\lambda_1)$, $F(\lambda_2)$, $F(\lambda_3)$ and $F(\lambda_4)$, at the wavelengths $\lambda_1 < \lambda_2 < \lambda_3 < \lambda_4$, provides an estimate of $F(\lambda)$ at any wavelength λ between λ_2 and λ_3

$$F(\lambda) = \begin{pmatrix} L_1(\lambda) & L_2(\lambda) & L_3(\lambda) & L_4(\lambda) \end{pmatrix} \begin{pmatrix} F(\lambda_1) \\ F(\lambda_2) \\ F(\lambda_3) \\ F(\lambda_4) \end{pmatrix} \tag{12.43}$$

with

$$L_1(\lambda) = \frac{(\lambda - \lambda_2)(\lambda - \lambda_3)(\lambda - \lambda_4)}{(\lambda_1 - \lambda_2)(\lambda_1 - \lambda_3)(\lambda_1 - \lambda_4)}$$

$$L_2(\lambda) = \frac{(\lambda - \lambda_3)(\lambda - \lambda_4)(\lambda - \lambda_1)}{(\lambda_2 - \lambda_3)(\lambda_2 - \lambda_4)(\lambda_2 - \lambda_1)}$$

$$L_3(\lambda) = \frac{(\lambda - \lambda_4)(\lambda - \lambda_1)(\lambda - \lambda_2)}{(\lambda_3 - \lambda_4)(\lambda_3 - \lambda_1)(\lambda_3 - \lambda_2)}$$

$$L_4(\lambda) = \frac{(\lambda - \lambda_1)(\lambda - \lambda_2)(\lambda - \lambda_3)}{(\lambda_4 - \lambda_1)(\lambda_4 - \lambda_2)(\lambda_4 - \lambda_3)}$$

The interpolated function crosses the empirical points exactly.

The practical value of this formula is that the formula can be written in a simple matrix form, that, in the case of spectrum scanning equal to 10 nm, is

$$\begin{pmatrix} F(\lambda_j) \\ F(\lambda_j + 1) \\ F(\lambda_j + 2) \\ \cdot \\ F(\lambda_j + 9) \end{pmatrix} = \begin{pmatrix} 0 & 1 & 0 & 0 \\ L_1(\lambda_j + 1) & L_2(\lambda_j + 1) & L_3(\lambda_j + 1) & L_4(\lambda_j + 1) \\ L_1(\lambda_j + 2) & L_2(\lambda_j + 2) & L_3(\lambda_j + 2) & L_4(\lambda_j + 2) \\ \cdot & \cdot & \cdot & \cdot \\ L_1(\lambda_j + 9) & L_2(\lambda_j + 9) & L_3(\lambda_j + 9) & L_4(\lambda_j + 9) \end{pmatrix} \begin{pmatrix} F(\lambda_j - 10) \\ F(\lambda_j) \\ F(\lambda_j + 10) \\ F(\lambda_j + 20) \end{pmatrix}$$

$$\equiv \mathbf{L}(\lambda_j) \begin{pmatrix} F(\lambda_j - 10) \\ F(\lambda_j) \\ F(\lambda_j + 10) \\ F(\lambda_j + 20) \end{pmatrix} \tag{12.44}$$

and matrix $\mathbf{L}(\lambda_j)$ gives all the values $F(\lambda)$ interpolated at 1 nm in the range $[\lambda_j, \lambda_j + 10]$ starting from values at 10 nm. The substitution of this equation in the Equation (12.42) gives a straightforward computation of the weights $W_{X,O,S,j}$, $W_{X,O,S,j}$ and $W_{X,O,S,j}$, where the generic function $F(\lambda_j)$ is substituted by the deconvoluted reflectance factor $R(\lambda_j)$

$$X = \sum_{\substack{360 \\ \lambda_{j+1}-\lambda_j=10\text{nm}}}^{780} \left(\begin{array}{cccc} w(\lambda_j) & w(\lambda_j+1) & \dots & w(\lambda_j+9) \end{array} \right) \mathbf{L}(\lambda_j) \left(\begin{array}{c} R(\lambda_j-10) \\ R(\lambda_j) \\ R(\lambda_j+10) \\ R(\lambda_j+20) \end{array} \right) \quad (12.45)$$

$$= \sum_{\substack{360 \\ \lambda_{j+1}-\lambda_j=10\text{nm}}}^{780} W_{X,O,S,j} R(\lambda_j)$$

with $w(\lambda_j) = [K\bar{x}(\lambda_j)S(\lambda_j)]$. Equations analogous to (12.45) are for the tristimulus values Y and Z. This algorithm defines the weights of 'Table 5' of the document ASTM.[33] E 308-1985

The optimized weights are calculated for each observer, each illuminant, even non-standard, and for steps of 10 and 20 nm with the software presented in Section 16.12.

2.   The cubic spline interpolation formula.
3.   The fifth order polynomial interpolation formula from the six neighbouring data points around the point to be interpolated.
4.   The Sprague[35] interpolation.

12.4.5   Extrapolation

The CIE says that extrapolation is generally not recommended. However, if necessary, unmeasured values outside the range of measurements of $S(\lambda)$, $R(\lambda)$ or $T(\lambda)$ or $\beta_{T,S}(\lambda)$ may be set as a rough approximation equal to the nearest measured value. This is the typical case of spectrophotometers operating with a pitch of 20 nm, where the measurements are made in the wavelength range 400–700 nm.

The colorimetric results depend on the calculation technique used.

The colorimetric calculations can be performed with the software presented in Section 16.8.1.

12.5   Uncertainty in Colorimetric Measurements

The quantity to be measured is termed *measurand*. No estimate of a measurand is without uncertainty. The measurement uncertainty is an estimate of the range of values within which the true value is with a probabilistic meaning. The measurement uncertainty depends not only on the measurement instruments but also on the measurement conditions. Today, the measurement uncertainty is determined by the rules given in the *Guide to the expression of uncertainty of measurement*, better known as ISO[36] Guide or GUM. Before entering the uncertainty related to colorimetry, a few terms and definitions have to be reconsidered, because the ISO Guide has changed their previous meanings.

Previous meanings:

- *Accuracy* is a quantity that relates to the systematic uncertainty, depends on the calibration and the class of the instrument and its value is greater the smaller the deviation of the estimate made with the considered instrument compared to that of the instrument accepted as reference instrument and located in a standard metrology laboratory.

- *Precision* is a quantity that relates to the repeatability of the observations, depends on random uncertainties and is represented by the standard deviation assessed on the basis of a sufficiently large set of observations.

Meanings according to the ISO Guide:

- Precision is replaced by *uncertainty of type A* (See below).
- The word accuracy has only qualitative value and expresses the closeness of agreement between the result of a measurement and a true value of the measurand.
- The *measurement error* is the difference between the result of a measurement and the true value of the measurand.

The ISO Guide has been synthetized by NIST[37] and here this text is followed. All the colorimetric quantities are defined as functions of the tristimulus values, which are obtained directly from measurements with spectral weighting – (Section 13.4) tristimulus colorimeters, luminance meters, illuminance meters, ... – or by calculation from spectral radiometric measurements – (Section 13.5) spectroradiometers, spectrophotometers, ... –. Almost always the result of a colorimetric measurement is limited to an approximation or *estimate* of the value of the *measurand*. However, a degree of confidence should be given to these estimates. The measurement of the doubt about the quality of the result is represented by the

"*Uncertainty of measurement*, which is a parameter associated with the result of a measurement, that characterizes the dispersion of the values that could reasonably be attributed to the measurand."[37]

The uncertainty of measurement is determined by the rules given in the ISO Guide. Generally,

the uncertainty comprises several components which may be grouped into two categories according to the method used to estimate their numerical values:

(a) *Type A evaluations (of uncertainty)*, those which are evaluated by statistical analysis of a series of observations and result from an estimated standard deviation;
(b) *Type B evaluation (of uncertainty)*, those which are evaluated by means other than the statistical analysis of a series of observations.

Such a distinction is not always simple and commonly the distinction into *random* – approximately type A – and *systematic* – approximately type B – uncertainties is used. The nature of an uncertainty component depends on how the corresponding quantity appears in the mathematical model that describes the measurement process. Each component of uncertainty, that contributes to the uncertainty of a measurement, results from an estimated standard deviation, termed *standard uncertainty* and denoted by u_i.

The random component is calculable on statistical considerations.

The systematic component concerns the difference between an estimate of the measurand and the true value. This discrepancy is an error and should be corrected.

The calculation of the uncertainty in the radiometric, photometric and colorimetric instruments with spectral weighting, although the simplest and crude tools, is more complicated. Here we consider only the quantities based on spectral measurements, that is, spectroradiometers and spectrophotometers, because the instruments with spectral weighting are used in control processes and almost never for a comparison between laboratories.

We need to know the uncertainty of the spectral measurands: spectral power distribution, spectral reflectance factor, spectral transmittance, spectral radiance factor, bispectral radiance factor. The uncertainty of the tristimulus values is calculated as the propagation of the uncertainties of spectral quantities on which they depend. The determination of the uncertainties of spectral magnitudes is complex and is based on the calibration process and the detailed knowledge of the instrument. Although the measuring operation using spectroradiometers or spectrophotometers is a simple pushing a button and instantly obtains an estimate of the radiometric quantity being measured, behind this operation there are several measurements with their corresponding uncertainties. The estimation and the propagation of measurement uncertainties is established by the rules given in the ISO[37] guide.

12.5.1   Laws of Propagation of Uncertainty

Warning: the mathematical symbols used in this section (X_i, x_i, Y, ...) should not be confused with the same ones used in the other sections of the book. The choice of these symbols is the same as the ISO guide.

A.1 In many cases a measurand Y is not measured directly, but is determined from n other quantities $X = X_1, X_2, ..., X_n$ through a functional relation f

$$Y = f(X_1, X_2, ... X_n) \tag{12.46}$$

which is generally known as a *measurement equation*. The convolution Equation (12.37) is only a part of the measurement equation, and the result depends also on the mathematical deconvolution technique used. Included among the quantities X_i are corrections – or correction factors –, as well as quantities that take into account other sources of variability, such as different observers, instruments, samples, laboratories and times at which observations are made for example, different days. Thus the function f of Equation (12.46) should express not simply a physical law but a measurement process, and in particular, it should contain all quantities that can contribute a significant uncertainty to the measurement result.

A.2 An estimate of the measurand or *output quantity Y*, denoted by y, is obtained from Equation (12.46) using *input estimates* $x_1, x_2, ..., x_n$ for the values of the n *input quantities* $X = X_1, X_2, ..., X_n$. Thus the *output estimate y*, which is the result of the measurement, is given by

$$y = f(x_1, x_2, ... x_n) \tag{12.47}$$

A.3 The *combined standard uncertainty* of the measurement result y, designated by $u_c(y)$ and taken to represent the estimated standard deviation of the result, is the positive square root of the estimated *variance* $u_c^2(y)$ obtained from

$$u_c^2(X) = \sum_{i=1}^{n} \left(\frac{\partial f}{\partial x_i} \right)^2 u^2(x_i) + 2 \sum_{i=1}^{n-1} \sum_{j=i+1}^{n} \frac{\partial f}{\partial x_i} \frac{\partial f}{\partial x_j} u(x_i, x_j) \tag{12.48}$$

This equation is based on a first-order Taylor series approximation of $Y = f(X_1, X_2, ..., X_n)$ and is conveniently referred to as the *law of propagation of uncertainty*. The partial derivatives $\partial f / \partial x_i$ – often referred to as *sensitivity coefficients* – are equal to $\partial f / \partial X_i$ evaluated at $X_i = x_i$; $u(x_i)$ is the *standard uncertainty* associated with the input estimate x_i; and $u(x_i, x_j)$ is the *estimated covariance* associated with x_i and x_j.

A.4 As an example of a Type A evaluation, consider an input quantity X_i whose value is estimated from n independent observations $x_{i,k}$ of X_i obtained under the same conditions of measurement. In this case, the input estimate x_i is usually the sample *mean*

$$x_i \equiv \frac{1}{n} \sum_{k=1}^{k=n} x_{i,k} \qquad (12.49)$$

and the standard uncertainty $u(x_i)$ to be associated with x_i is the *estimated standard deviation of the mean*

$$u(x_i) \equiv \sqrt{\frac{1}{n(n-1)} \sum_{k=1}^{k=n} (x_{i,k} - x_i)^2} \qquad (12.50)$$

A.5 As an example of a Type B evaluation, consider the uncertainties related to the spectral irradiance $E_{\mathrm{STD},\lambda}$ of the standard lamp, and the spectral reflectance factor $R_{\mathrm{STD}}(\lambda)$ of the standard reflectance tile, whose values and uncertainties are certified by a Standard Metrological Laboratory.

12.5.2 Uncertainty Computation

In summary,

the key points for the uncertainty evaluation described in the ISO guide are:

1. the indication of the *measurement equation*, that is, the functional relationship between the measurand Y and the n input quantities X_i on which Y depends;
2. the determination of the *best estimate* x_i of the input based on the measurement data or other available information;
3. the evaluation of the *standard uncertainty* $u(x_i)$ of each estimate x_i;
4. the determination of possible correlations;
5. the evaluation of the measurement result, that is, the estimate y of the measurand Y based on the mathematical relationship given at step 1;
6. the estimation of the *combined standard uncertainty* u_c;
7. moreover, if required, the assessment of the *expanded uncertainty* U so that the interval between $(y - U)$ and $(y + U)$ contains the value of the measurand with an estimated value of probability. For industrial measurements, it is now recommended to use an *expanded uncertainty* with a coverage factor $k = 2$, that is, expand the standard uncertainty by a factor of 2, providing a level of confidence of approximately 95%.

There are some sources of uncertainty inherent in the design of the instrument, others in the geometry of measurement (e.g., measurements of the spectral reflectance factor in the geometry (45°x:0°), the polarization of the reflected light can affect the result), others in the choice of the sample to measure and others in ambient conditions (temperature, humidity). Here, as an example, the main sources of uncertainty of Type B given by Clarke-Hanson-Verril[38] are given:

(a) Uncertainty in the level of the absolute scales of diffuse reflectance and radiance factor;
(b) Uncertainty in the spectral slope of the scales (skew uncertainty);
(c) Uncertainty in the transfer of the absolute scales to working scales;

(d) Dark uncertainty (in Section 12.3 the dark current is always subtracted);

(e) Linearity uncertainty standards (in the explanation of Section 12.3 the response of the sensor is considered linear, but, in practice this is an approximation);

(f) Wavelength scale uncertainty (in Section 12.3 the wavelength uncertainty is not considered);

(g) Thermochromism uncertainty (This source of uncertainty depends on the specimen. The temperature is a quantity to be considered combined with the wavelength uncertainty.);

(h) Glossy to matt ratio uncertainty;

(i) Specular beam uncertainty (this depends only on the geometry (di:0°) of the reflectance factor measurement, i.e., only on the geometry with specular component included);

(j) Gloss trap uncertainty (this depends only on the geometry (de:0°) of reflectance factor measurement, i.e., only on the geometry with the specular component excluded).

The explanation of these points is in the original paper.[38] This list is incomplete. Many authors of important metrological laboratories have considered and discussed the most relevant sources of uncertainty.[38–51]

The analysis of these points is based on a thorough knowledge of the instrument and those who carry on this analysis need specific knowledge of optics and material characteristics. The complexity of the measurement equation, especially because the equation is proper of any instrument, makes it difficult to have a general equation. This analysis is not an aim of this book and for the interested reader the main references are given.[38–51] Anyway such an analytical work is intended to identify the sources of uncertainty, to quantify them and the corresponding sensitivity coefficients $\partial f / \partial x_i$. Once all that is done, the uncertainty is computed for the radiometric quantity considered at all the measurement wavelengths and its propagation in the tristimulus values can be computed.

Here the surface colour with reflectance factor $R(\lambda)$ lit by an illuminant $S(\lambda)$ is considered. The tristimulus values are obtained by numerical summation of the spectral data (12.41), here rewritten in a more synthetic form with the index j representing the wavelength λ_j at which the measurements are made,

$$
\begin{cases}
X = K \sum_j S_j R_j \bar{x}_j = K \sum_j E_j \bar{x}_j \\[2mm]
Y = K \sum_j S_j R_j \bar{y}_j = K \sum_j E_j \bar{y}_j \quad \text{with } K = \dfrac{100}{\sum_i S_j \bar{y}_j} \\[2mm]
Z = K \sum_j S_j R_j \bar{z}_j = K \sum_j E_j \bar{z}_j
\end{cases}
\tag{12.51}
$$

(in these equations the summation interval $\Delta \lambda_j$ is not written because it is simplified by the normalization and $E_j \equiv S_j R_j$). Each measured spectral datum is independent of the others with a good approximation, then the uncertainty calculation is relatively simple. For example, the measurement of the spectral reflectance factor at various wavelengths λ_j with relative uncertainty $R_j \pm u(R_j)$, where $u(R_j)$ is the result of the combination of the uncertainty evaluated in the calibration process with the statistical uncertainty. Usually, metrological certificates express the uncertainty $u(R_j)$ in a linear relationship with the reflectance factor R_j

$$
u(R_j) = \alpha_{0,j} + \alpha_j R_j \approx \alpha_j R_j
\tag{12.52}
$$

and the quality of the measurement is represented by the coefficients α_j. It follows that:

• the uncertainties of the tristimulus values are

$$
\begin{cases}
u(X) \approx K \sqrt{\sum_j (\alpha_j S_j R_j \bar{x}_j)^2} \\[2ex]
u(Y) \approx K \sqrt{\sum_j (\alpha_j S_j R_j \bar{y}_j)^2} \\[2ex]
u(Z) \approx K \sqrt{\sum_j (\alpha_j S_j R_j \bar{z}_j)^2}
\end{cases}
\tag{12.53}
$$

where S_j and $(\bar{x}_j, \bar{y}_j, \bar{z}_j)$ have an uncertainty equal to zero;

- the uncertainties of the chromaticity coordinates $x = X/(X+Y+Z)$ and $y = Y/(X+Y+Z)$ are

$$
\begin{cases}
u(x) \approx \dfrac{K}{X+Y+Z} \sqrt{\sum_j (\alpha_j S_j R_j)^2 \left[\bar{x}_j - x \left(\bar{x}_j + \bar{y}_j + \bar{z}_j \right) \right]^2} \\[3ex]
u(y) \approx \dfrac{K}{X+Y+Z} \sqrt{\sum_j (\alpha_j S_j R_j)^2 \left[\bar{y}_j - y \left(\bar{x}_j + \bar{y}_j + \bar{z}_j \right) \right]^2}
\end{cases}
\tag{12.54}
$$

- the uncertainties of the CIELAB coordinates are

$$
\begin{cases}
u(L^*) \approx \dfrac{116}{3} K \sqrt{\sum_j \left\{ \alpha_j S_j R_j F(Y) \bar{y}_j \right\}^2} \\[3ex]
u(a^*) \approx \dfrac{500}{3} K \sqrt{\sum_j \left\{ \alpha_j S_j R_j \left[F(X)\bar{x}_j - F(Y)\bar{y}_j \right] \right\}^2} \\[3ex]
u(b^*) \approx \dfrac{200}{3} K \sqrt{\sum_j \left\{ \alpha_j S_j R_j \left[F(Y)\bar{y}_j - F(Z)\bar{z}_j \right] \right\}^2}
\end{cases}
\tag{12.55}
$$

where

$$F(X) = (X_n X^2)^{-1/3} \text{ for } \frac{X}{X_n} \geq 0.008856 \text{ else } F(X) = 7.787 \frac{1}{X_n}$$

$$F(Y) = (Y_n Y^2)^{-1/3} \text{ for } \frac{Y}{Y_n} \geq 0.008856 \text{ else } F(Y) = 7.787 \frac{1}{Y_n}$$

$$F(Z) = (Z_n Z^2)^{-1/3} \text{ for } \frac{Z}{Z_n} \geq 0.008856 \text{ else } F(Z) = 7.787 \frac{1}{Z_n}$$

12.6   Physical Standards for Colour-Instrument Calibration

The main standard metrology laboratories provide tiles of different materials, whose spectral reflectance factor and their uncertainty are certified. The certificate has to be referred to a measurement geometry identical to that of the instrument in use. These tiles are known as *physical standards* for colour measurement and are of very high practical value. The main uses of these physical standards are the following:

1. Definition of levels of spectral reflectance factor equal to 0 and 100%;
2. Verification of the performance of the instrument;
3. Diagnosis of instrumental errors;
4. Reference of the measurements to a standard metrology laboratory.

Manufacturers equip the instruments for colour measurement of a reference white tile, with which, if it is a standard certified by a metrology laboratory, the spectral reflectance factor is computed. Since this reference standard is white, it allows defining only the maximum value of the spectral reflectance factor corresponding to 100%.

To define the level of reflectance of 0%, a standard black tile is used. Generally, the manufacturers provide a light trap to define 0% of the reflectance factor, but not a standard black tile, traceable to a standard metrological laboratory.

To assess the linearity of the radiometric scale of the instrument, grey tiles are used, of which the spectral reflectance factor between 40 and 60%, is approximately constant to the change in the wavelength and is traceable to a standard metrology laboratory. A measure of the linearity of the radiometric scale of the instrument is obtained by comparing the spectral reflectance factor measured with the certified one. It is essential for the reflectances to have values between 40 and 60%, because the errors due to non-linearity of the response of the sensors go to zero at the extremes of the scale, when the sample has reflectance equal to zero or when the sample is the same white reference tile. Any non-linearity errors are correctable by calculation.

The errors in wavelength scale of a spectrophotometer can be determined with a standard tile of holmium or didymium or a suitable mixture, which has narrow absorption bands in the long wavelength region, of the red, orange and yellow hues (Figure 12.13). The wavelength scale is assessed by comparing the reflectance spectra of tiles of this kind according to the shifts of the measured absorption peaks from their certified values. On the basis of this comparison, one can implement an algorithm for the correction of the wavelength scale.

In the general case, the specimens have a reflectance factor varying with the wavelength. Sets of a dozen coloured tiles, traceable to a standard metrology laboratory, are used to check the behaviour of an instrument when measuring coloured specimens. The number and the choice of colours in these sets of tiles are done in such a way that the measured spectra, compared with the certified ones, are useful for a correction of the deviations.

In this way a computer program can be implemented for the correction of the instrument errors, the use of which, with a good approximation, defines an instrument calibration traceable to a standard metrology laboratory.

However, there are emblematic cases when the linearity assessed with grey tiles looks good, while the non-linearity is confirmed by the tiles with a high spectral reflectance factor in the region of long wavelengths, typical of the yellow, orange and red hues. This is a frequent case. The reasons for this error are probably attributable to stray lights, which concern the design of the instrument. Despite that the measured spectral reflectance factor is affected by an error of about 3%, the calculation of the tristimulus values are not too affected by error because in the region of long wavelengths, the colour matching functions have low values, which mitigate the radiometric error. This is the reason why instruments affected by these errors have a market.

To avoid thermochromism phenomena, the temperature at the time of use should have a difference with respect to the temperature at which they were calibrated within 1°C.

To evaluate the performance of the instrument, metrology laboratories advise carrying out both short- and long-term tests. To examine the short-term, all the standards are measured several times (typically 10 times) and the standard deviation between measured and certified values for each tile is computed. The test is then repeated periodically (e.g., monthly) and then any increase in standard deviations is recorded. The average value of the readings is computed and stored on a monthly basis for a long-term examination.[38–46]

References

1. CIE Publication S 017/E:2011,. *ILV: International Lighting Vocabulary*, Commission Internationale de l'Éclairage, Vienna (2011). Available at: www.eilv.cie.co.at/ (accessed on 18 June 2015).
2. CIE Pulication 15.3:2004, *Colorimetry*, 3rd ed., Commission Internationale de l'Éclairage, Vienna (2004). Available at: www.cie.co.at/ (accessed on 18 June 2015).
3. McCamy CS, Concepts, terminology, and notation for optical modulation, *Photogr. Sci. Eng.*, **10**, 314–325 (1966).
4. Judd DB, Terms, definitions, and symbols in reflectometry, *J. Opt. Soc. Am.*, **57**, 445–452 (1967).
5. CIE Publication No. 44, *Absolute Methods for Reflection Measurement*, Commission Internationale de l'Éclairage, Vienna (1979). Available at: www.cie.co.at/ (accessed on 18 June 2015).
6. CIE Publication No. 176.2006, *Geometrical Tolerances for Colour Measurement*, Commission Internationale de l'Éclairage, Vienna (2006). Available at: www.cie.co.at/ (accessed on 18 June 2015).
7. Clarke FJJ and Parry DJ, Helmholtz reciprocity: its validity and application to reflectometry, *Lighting Res. Technol.* **17**, 1–11 (1985).
8. Nicodemus FE, Richmond JC, Hsia JJ and Ginsberg IW, *Geometrical Considerations on Nomenclature for Reflectance*, National Bureau of Standards, Washington, DC (October 1977).
9. Rössler G, Multigeometry color measurement of effect surfaces, *Die Farbe*, **37**, 111–121 (1990).
10. DIN Standard 6175-2, *Colour Tolerances for Automobile Lacquer Finishes, Part 2: Effect Lacquer Finishes* (2001). Available at: http://www.din.de/cmd?level=tpl-home&contextid=din&languageid=en (accessed on 18 June 2015).
11. ASTM E2539, *Standard Practice for Multi-angle Color Measurement of Interference Pigments*, ASTM International, West Conshohocken, PA, USA (2008).
12. Donaldson R, Spectrophotometry of fluorescent pigments, *Brit. J. Appl. Phys.*, **5**, 210–214 (1954).
13. Wyszecki G, Basic concepts of the colorimetry of fluorescent materials, *J. Color Appearance* **1**(5), 8–17 (1972).
14. Alman DH, Billmeyer FW Jr and Philips DG, A comparison of onemonochromator methods for determining the reflectance of opaque fluorescent samples, in *Proceeding 18th Session CIE (London, 1975)*, Bureau Central de la CIE, Paris, pp. 237–244 (1976)
15. Billmeyer FW Jr, Colorimetry of fluorescent specimens: a state-of-art report, *NBS-GCR*, 79–185 (1979).
16. Grum F, Colorimetry of fluorescent materials, in *Color Measurement*, Grum F, and Bartleson CJ, eds., Chap. 6, Academic Press, New York, pp. 235–288 (1980).
17. Leland J, Johnson N and Arecchi A, Principles of bispectral fluorescence colorimetry, *Proc. SPIE*, **3140**, 76–87 (1997).
18. Springsteen A, Introduction to measurement of color of fluorescent materials, *Anal. Chim. Acta*, **380**, 183–192 (1999).
19. ASTM E 991-98, *Standard Practice for Color Measurement of Fluorescent Specimens*, ASTM International, West Conshohocken, PA (2004).
20. ASTM E 2152-01, *Standard Practice for Computing the Colors of Fluorescent Objects from Bispectral Photometric Data*, ASTM International, West Conshohocken, PA (2004).
21. ASTM E 2153-01, *Standard Practice for Obtaining Bispectral Photometric Data for Evaluation of Fluorescent Color*, ASTM International, West Conshohocken, PA (2004).
22. Lerner JM and Thevenon A, *The Optics of Spectroscopy - A tutorial*, V2.0. Jobin Yvon-Spex, Instruments SA, Inc. (1988). Available at: http://www.horiba.com/us/en/scientific/products/optics-tutorial/ (accessed on 18 June 2015).
23. JCGM 200:2012, *International Vocabulary of Metrology – Basic and General Concepts and Associated Terms* (VIM) 3rd ed., 2008 version with minor corrections. Available at: http://www.bipm.org/en/publications/guides/#vim (18 June 2015).

24. Stearns EI, Influence of spectrometer slits on tristimulus calculations. *Color Res. Appl.*, **6**, 78–84 (1981).

25. Stearns EI and Stearns RE, An example of a method for correcting radiance data for bandpass error, *Color Res. Appl.*, **13**, 257–259 (1988).

26. Venable WH, Accurate tristimulus values from spectral data, *Color Res. Appl.*, **14**, 260–267 (1989).

27. Fairman HS, Results of the ASTM field test of tristimulus weighting functions, *Color Res. Appl.*, **20**, 44–49 (1995).

28. Oleari C, Spectral reflectance factor deconvolution and colorimetric calculations by local power expansion. *Color Res. Appl.*, **25**, 176–185 (2000).

29. Ohno Y, A flexible bandpass correction method for spectrophotometers. in *Proceedings AIC Colour 05*, Granada, pp. 697–700 (2005).

30. Gardner JL, Bandwidth correction for LED chromaticity. *Color Res. Appl.*, **31**:374–380 (2006).

31. Fairman HS, An improved method for correcting radiance data for bandpass error. *Color Res. Appl.*, **35** (5), 328–333 (2010).

32. Oleari C, Deconvolution of spectral data for colorimetry by second order local power expansion, *Color Res. Appl.*, **35**, 334–342 (2010).

33. ASTM E 308-96, Standard test method for computing the colors of objects by using the CIE System. in *Annual Book of ASTM Standards*. ASTM International, West Conshohocken, PA (1996).

34. Li C, Luo M R, Melgosa M and Pointer M R, Testing the Accuracy of Methods for the Computation of CIE Tristimulus Values Using Weighting Tables. *Color Res. Appl.*, Article first published online: 24 FEB 2015 | DOI: 10.1002/col.

35. Sève R and Duval B, Interpolation procedure: proposals and comments. CIE 152:2003, *Proc. CIE 25th Session* **1** D1-74-77 (2003).

36. JCGM 100:2008, *Evaluation of Measurement Data – Guide to the Expression of Uncertainty in Measurement* (GUM 1995 with minor corrections) (2008). Available at: http://www.bipm.org/en/publications/guides/gum.html (accessed on 18 June 2015).

37. Taylor BN and Kuyatt CE, Guidelines for evaluating and expressing the uncertainty of NIST measurement results, *NIST Technical Note* 1297 (1994). Available at: http://www.nist.gov/pml/pubs/tn1297/ (accessed on 18 June 2015).

38. Clarke PJ, Hanson AR and Verrill JF, Determination of colorimetric uncertainties in the spectrophotometric measurement of colour, *Anal. Chim. Acta*, **380**, 277–284 (1999).

39. Clarke PJ, Surface colour measurements, *Measurement Good Practice Guide* No. **96**, National Physical Laboratory, Teddington (2006).

40. Verrill JF, Malkin F, Larkin JA and Wardman RH, The BCRA-NPL Ceramic Colour Standards series II – master spectral reflectance data and thermochromism data, *J. Soc. Dyers Colour.*, **113**, 84–94 (1997).

41. Ohno Y, Spectral color measurement, in *Colorimetry, Understanding the CIE System*, Schanda J, ed., John Wiley & Sons (2007).

42. Nadal ME, Early EA, Weber W and Bousquet R, NIST 0:45 reflectometer, *Color Res. Appl.*, **33**(2), 94–99 (2008).

43. Nadal ME, Early EA, Weber W and Bousquet R, *0:45 Surface Color*. National Institute of Standards and Technology NIST Special Publication SP250-71 (2008)

44. Berns RS and Petersen KH, Empirical modelling of systematic spectro-photometric errors, *Color Res. Appl.*, **13**, 243–256 (1988).

45. Fairchild MD and Reniff L, Propagation of random errors in spectrophotometric colorimetry, *Color. Res. Appl.*, **16**, 360–367 (1991).

46. Verrill JF, Advances in spectrophotometric transfer standard at the National Physical Laboratory, in *Spectrophotometry, Luminescence and Colour, Science and Compliance*, Burgess C, and Jones, DG, eds., Elsevier, Amsterdam, pp. 49–63 (1995).

47. Berns RS and Reniff L, An abridged technique to diagnose spectrophotometric errors, *Color Res. Appl.*, **22**, 51–60 (1997).

48. Hanson AR and Clarke PJ, Determination of uncertainty in spectrophotometric surface color measurement, *Proc. SPIE* 4421, *9th Congress of the International Colour Association, 808*, June 6 (2002).

49. Hanson AR, Pointer MR and Clarke PJ, The uncertainty of surface colour measurements made with commercially-available diode-array instrumentation and recommended calibration protocols, *NPL Report* DQL-OR 005, National Physical Laboratory, Teddington (2004).

50. Gardner JL, Uncertainties in surface colour measurements, *Measurement Good Practice Guide* No. **95**, National Physical Laboratory, Teddington (2006).

51. Gardner JL, Uncertainty propagation for NIST visible spectral standards, *J. Res. Natl. Inst. Stand. Technol.*, **109**, 305–318 (2004).

13

Basic Instrumentation for Radiometry, Photometry and Colorimetry

13.1 Introduction

The purpose of this chapter is to explain generalities of radiometric, photometric and colorimetric basic instrumentation. The instruments are divided into instruments for the measurement of the light and of the optical modulation – reflectance factor and transmittance factor – and the same instruments are divided into instruments with spectral analysis and with spectral weighting. This distinction leads to differentiate between instruments on the basis of the geometry with which they collect the light to be measured. There are also instruments based on the visual comparison made by the operator. A further distinction exists between imaging and non-imaging instruments. This leads to the classification of instruments according to the scheme proposed in Tables 13.1 and 13.2.

Briefly, the non-imaging instruments listed in the table with their characteristics are:

- *Spectroradiometers*, for measuring the spectral density of the radiometric quantities.
- *Spectrophotometers*, for measuring ratios of spectral radiometric quantities, such as the reflectance, transmittance, reflectance factor, transmittance factor and radiance factor (Section 12.2.2).
- *Photometers*, for the measurement of light, such as illuminance (lx), light intensity (cd) and luminance (cd/m^2), which have a spectral response according to the luminous efficiency function $V(\lambda)$, that is, the measurement of the power at each wavelength weighted according to a standardized model of human visual brightness perception (Chapter 7).
- *Tristimulus colorimeters*, for measuring the tristimulus values, according to the standard observer colour-matching functions $\bar{x}(\lambda)$, $\bar{y}(\lambda)$ and $\bar{z}(\lambda)$, that is, for measuring the power at each wavelength weighted according to three standardized sensors representing human colour vision.
- *Subtractive visual colorimeters* operating by subtractive colour mixing.

The instruments for the measurement of optical modulation have an internal light source.

Standard Colorimetry: Definitions, Algorithms and Software, First Edition. Claudio Oleari.
© 2016 John Wiley & Sons, Ltd. Published 2016 by John Wiley & Sons, Ltd.

Table 13.1   *Classification of the non-imaging radiometric, photometric and colorimetric instruments.*

Quantity measured	Spectral analysis measurement	Spectral weighting measurement	Visual comparison measurement
Light	Spectroradiometers	• Photometers • Luxmeters • Luminance-meters • Tristimulus colorimeters	
Optical modulation	Spectrophotometers		
Colour		Tristimulus colorimeters	Subtractive mixing colorimeters

Table 13.2   *Classification of the imaging radiometric, photometric and colorimetric instruments.*

Quantity measured	Spectral analysis measurement	Spectral weighting measurement
Light	Hyperspectral cameras	Imaging photometers
Optical modulation	Hyperspectral cameras	Multispectral cameras
Colour	Hyperspectral cameras	• Imaging colorimeters • Multispectral cameras • Trichromatic cameras

Among the tools mentioned above, only spectroradiometers and spectrophotometers make *spectral density measurements*. In these instruments, the photometric and colorimetric properties are obtained from the spectral data with the calculation. In photometry and colorimetry, measurements are generally in the range of visible radiation from 380 to 780 nm, often less valuable instruments in the restricted range from 400 to 700 nm, while sometimes the range is from 300 to 830 nm. The analysis in the UV region is necessary to consider the fluorescence phenomena.

All other instruments make *spectral weighing measurements* by using photodetectors combined with appropriate optical filters in order to have spectral sensitivity as $V(\lambda)$, in the case of photometric measurements, or likewise $\bar{x}(\lambda)$, $\bar{y}(\lambda)$ and $\bar{z}(\lambda)$ in the case of colorimetric measurements.

A very important feature in these instruments is given by the type of photodetector used, the choice of which depends on the type of measurements to be made.

There are also instruments without a photodetector because there is an operator, who, performing a visual comparison with his/her visual system, replaces the photodetector.

There are no measures without uncertainty and the instruments do not all have the same class (Chapter 12). The quality of the calibration of spectrophotometers is evaluated by measurements on physical standards traceable to a metrology laboratory (Section 12.6).

The visual comparison of colour samples is common in colorimetric practice, and the comparison is meaningful only if made in a controlled visual situation, as it is required for consulting colour atlases (Chapter 14). With this purpose, there are properly defined and equipped *lighting cabinets* or light booths (Section 13.2).

In this chapter, the light cabin, a visual comparison tool, instruments for spectral weighing measurements, instruments for spectral analysis measurements and gloss-meters are presented. The last kind of instruments, the gloss-meter, does not regard colour but gloss (Section 13.6), a very important attribute of colour appearance.

13.2   Lighting Cabinet

Colour visual evaluation is the visual comparison of colour equality and colour difference between two objects under identical illumination conditions. This comparison of non-luminous objects requires controlled visual situations in the illumination geometry, in the vision geometry and in the choice of the light sources. The light cabinet is the place where such comparisons are possible, and therefore is an indispensable tool. The shape is that of a parallelepiped with one of the six faces open for observation. The ceiling of the booth usually consists of a set of light sources selectable by the observer. Often, but not always, the light of the lamps above is diffused by frosted glass.

The evaluation area in the booth has to be properly made. The background, on which the specimens are placed, and the surrounding area should be glossless and neutral in colour to avoid any colour misjudgement. A typical background is grey Munsell N7/ (Section 14.4.3). White backgrounds are recommended for evaluating low-gloss or light-colour samples, grey backgrounds for evaluating medium-gloss or intermediate-colour samples, and matte and flat black backgrounds for evaluating high-gloss or dark-colour samples. Sometimes light booths with a black interior are used. Black booths definitely give greater prominence to colours, but this visual situation is in reality infrequent.

A mixing of the ambient light, natural and artificial, with that of the booth should be avoided because any contamination could reduce the effectiveness of a standardized source. The evaluation area should always be shielded from any external light.

The best location for a light box is in a windowless room with neutral dark walls.

Some models have a tilting table top in order to have the vision axis orthogonal to the object and the illumination incident at 45° from the top.

Different kinds of lighting can be selected for approximating the standard illuminants:

1. The light source A – incandescent home lighting (Figure 13.1).
2. The daylight D65 – simulated by a fluorescent lamp (F7, FL3.15) (Figure 13.2) – for various applications with a correlated colour temperature (*CCT*) of 6504 K, reference for the industry.[1–3] Depending on the needs of the user, the lamp D50 – simulated by F8 – is given as an alternative or in addition to D65. It should be noted that the norm ISO: 3664[4] recommends that the observation of graphic art and photographic images be made under source D50 at 2000 ± 500 lux for proofs and prints and 1270 ± 320 cd/m^2 for slides.
3. The fluorescent lamp with a three-band spectrum, generally F11 with a *CCT* of 4000 K, used in supermarkets, teaching rooms and wide interior rooms.
4. The 'cool white' fluorescent lamp with a *CCT* of 4230 K, used for classical lighting in offices.

All these lamps are joined by a source emitting UV in the region 315–400 nm, which, when activated alone or in combination with another lamp, is used to assess fluorescence phenomena as the presence of whitening agents and fluorescent dyes.

The lamps deteriorate over time and a substitution is necessary after a certain lifetime, typical of any type of lamp, for example, after approximately 2500 hours of use.

The quality of the simulation of daylight is very important (Figure 13.2). The lamps used to simulate it have approximate spectral distributions because they are obtained as a sum of a broad-band spectrum with a line spectrum, and these spectra are very different from that of daylight (Sections 8.4, 8.8 and 11.4.8). The high value of the colour rendering index (Section 16.8.12) of the fluorescent lamps used in the simulation is not a total guarantee. The fluorescent lamp FL3.15 was conceived as a D65 simulator and has a very high colour rendering index, $R_a = 99$.

Daylight is simulated also with a halogen lamp and a set of filters (two or more suitable filters arranged in series). The light emitted by this kind of daylight simulator matches the spectral power distribution of daylight approximately with a good metamerism index.

The problem of the simulation of daylight is still open. The CIE merely provides the '*special metamerism index* for change in illuminant' algorithm for the assessment of the quality of a daylight simulator for

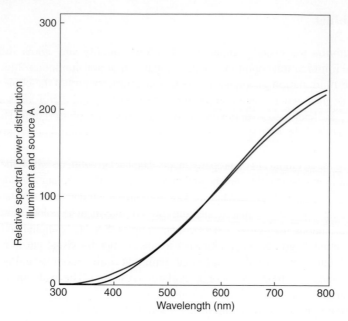

Figure 13.1   *Comparison of the spectral power distribution of a lamp in a lighting cabinet simulating the illuminant A (red line) and the standard illuminant A (black line).*

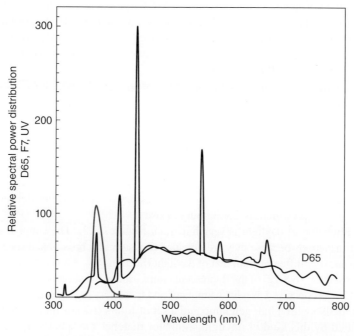

Figure 13.2   *Comparison of the daylight D65 (red line) with the fluorescent lamp F-7 (black line), used as a D65 simulator, and a UV lamp (blue line).*

colorimetry (Section 11.4.8). The purpose of this assessment is to quantify the suitability of a test source as a practical reproduction of the CIE standard illuminant considered for colorimetric tasks. The 'special metamerism index: change in illuminant' employs specified colour samples that are metameric matches for the standard daylight illuminant considered and the CIE 1964 standard colorimetric observer, and quantifies the mismatch resulting for the same observer when these samples are viewed under the illumination of the test source.[5-9]

Cards with pairs of metameric colours under daylight D65 are available in the market and these cards are useful for visually assessing the quality of the simulation of fluorescent lamps in use.

13.3 Visual Comparison Colorimeter

There are several instruments for visual comparison, but they are not classified as colorimetric instruments because they do not provide a colorimetric specification of the object in question, but are specific to particular laboratory measurements. In this section, we consider only one colorimeter with visual comparison, which is important for its high instrumental and historical value, the *Lovibond tintometer*®. The use of this instrument belongs to visual colorimetry.

This is a visual instrument for colour matching which uses three scales of subtractive filters – cyan, yellow and magenta- (Figure 13.3). Each scale consists of 250 values, graduated from the barely perceptible lighter colour to the most saturated ones, according to a rigorous arithmetic progression such that two identical filters labelled with 1.0 match the filter 2.0 combined with a colourless filter. (The colourless filter is necessary for taking into account the surface reflections that have to be equal in number.) The combination of three primary filters of equal value produces a neutral density filter. In this way, a series of grey filters are obtained, ranging from one that does not attenuate to one that fades to black with total absorption. The instrument is equipped with two light sources simulating the light of day from the north, close to the illuminant C. Colour matching is possible with colours of specimens seen in reflection and in transmission. The colour matching is described in the caption of figure 13.3.

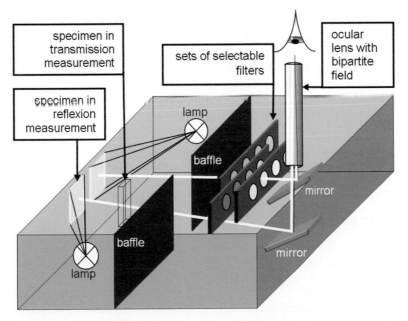

Figure 13.3 *Perspective view of Lovibond's tintometer. Two light sources illuminate two white grounds if the test sample is seen in transmission, while one ground is substituted by the test sample if this is seen in reflection. The absorptions of light by the test sample and by the filters in the two light paths are matched in colour.*

This instrument specifies the colour using a scale known as the *Lovibond notation*. In 1939 Schofield defined the units of the three reference colours as a function of the internal transmittance, and from these calculated the chromaticity of each individual filter and of all the combinations, thus putting in relation the Lovibond notation with the CIE notation.[10–13] Graphs of conversion between the two notations, called *nomograms*, were made for four different illuminants.

For this conception of visual instrument with the use of filters according to a proper colour-order system, the Lovibond tintometer can be considered as a colour atlas (Chapter 14).

13.4   Instruments with Power Spectral Weighting Measurement

As is known, in colour science, radiometry is concerned with the measurement of the power (watt) of electro-magnetic radiation – excluding radio waves, X- and γ-rays– and the radiometer is an instrument designed for this kind of measurement. In photometry and colorimetry, the common measurements are by the spectral power distribution or by the power spectral weightings, where the photodetectors are paired with suitable optical filters to have spectral response according to standardized sensors of human vision. Here the spectral weighting instruments are considered. The kind of filters used leads to the following distinction between the instruments:

1.   instruments for photometric measurements, in which a filter coupled to the photodetector is a device with a spectral responsivity equal to the luminous efficiency function $V(\lambda)$ and
2.   instruments for colorimetric measurements, in which three filters coupled to three photodetectors constitute three devices with spectral responsivities equal to the colour-matching functions $\bar{x}(\lambda)$, $\bar{y}(\lambda)$ and $\bar{z}(\lambda)$, respectively, or their linear combinations.

The type of photodetector strongly characterizes these instruments. The selenium photocells are devices of the past, even if some instruments make use of them today. Their limitation is in the memory effect, that is, indicating a value that can be influenced by previous measurements with high illuminance. The silicon photodetectors today are immune to this problem and are by far the most widely used. There are also photo-multiplier photodetectors, with very high sensitivity. Finally, the most widely used type of photodetector in photography is the photoresistor. The non-linearity between the electrical signal and the power absorbed by the photodetector is one of the possible sources of error, which can be corrected by a careful calibration of the instrument.

Two features are present in all of these types of measurements:

1.   the geometry used to collect the radiation to be measured and
2.   the dependence of the response from the direction of the incident radiation, distinguishing between response according to the cosine law and response conceived for unidirectional illumination.

13.4.1   Photometric Instruments

All photometric instruments have a spectral sensitivity equal to the photopic luminous efficiency $V(\lambda)$. Just the difference between the spectral sensitivity of the instrument and the function $V(\lambda)$ is the source of error, which cannot be reduced by careful calibration (Figure 13.4).

For high temperatures, above 50 °C, or for using these instruments for long periods at high levels of illumination, over 100 000 lux, the correction optical filter combined with the photodetector can be altered and the spectral sensitivity of the system has a significant deviation from the $V(\lambda)$ function. The advice is to avoid high temperatures and make intermittent use of the instrument avoiding long exposure of the photodetector with high continuous illumination.

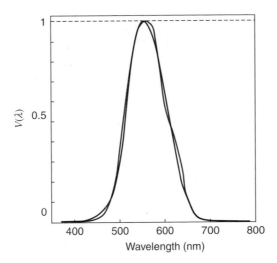

Figure 13.4 *Comparison between the CIE standard photopic observer curve (black line) and the typical curve of a photometric instrument (red line).*

The calibration of these instruments is made with the use of a standard source with a tungsten filament reproducing the standard illuminant A of 2850 ± 10 K.[14]

Generally, these instruments have auto-resetting active at each startup: covering the photodetector with the cap, the system associates the null value of the illumination at the dark signal of the sensor. The photometric quantity is proportional to the current minus the dark current.

Photometric equipment is designed to operate primarily in the field, and is simple to use, with easy portability, with great strength and internal power supply by battery. Often these instruments are constituted of two parts, linked together by a cable: one part is the actual instrument and the other is the photodetector equipped with appropriate optics depending on the kind of measurement to make. By changing this second part, the instrument performs a variety of functions. The photometric quantities to be measured are illuminance (lx), light intensity (cd) and luminance (cd/m²). To each one of these quantities corresponds an optical device to be coupled to the filtered photodetector used in the measurement. The illuminance meter and the luminance meter are here considered separately.

Luxmeter or Illuminance Meter

The luxmeter measures illuminance and, unless otherwise specified, has a photodetection system that follows the Lambert cosine law (Section 3.5). This is its main feature. In the measurement of illuminance, it is always assumed that the illuminated surface is flat and perpendicular to the average direction of the illuminating light. If the surface is rotated at an angle ϑ to the direction of the luminous flux, the illuminance is reduced in accordance with the cosine law, $E_v(\vartheta) = E_v(\vartheta = 0) \cos \vartheta$. But each photodetector, like all bodies, absorbs and reflects light (in a regular and diffused way) as a function of the incidence angle ϑ. This means that a photodetector hardly accurately measures the flow of light that strikes, giving a response in accordance with the cosine law. The deviation between the true quantity and that produced by the naked photodetector has to be corrected and this correction is made by placing a suitable opal diffusing filter, flat or dome-shaped, in front of the photodetector. This correction is termed *cosine correction* and the photodetector coupled to the cosine corrector filter is termed the *cosine receptor*. Figure 13.5 shows the dependence of the illumination angle on light incidence (black line) in the hemisphere placed in front of the illuminated surface: for ϑ equal to 0°,

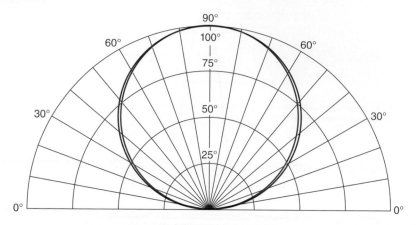

Figure 13.5   *Comparison of the cosine law (black line) and the behaviour of a typical luxmeter (red line).*

E_v is maximum (100%), while, for increasing $|\vartheta|$, E_v decreases until reaching a zero value for ϑ equal to ±90°. The agreement between the ideal cosine law (black line) and that of a typical light meter (red line) shows as an example a deviation of about 2% at angles of incidence of ±30°.

There are instruments in which the cosine correction is implemented by using an integrating sphere that has the function of collecting the light that strikes the input port. In this way and in the case of the ideal sphere, the photodetector placed on an output port always receives a Lambertian diffused flux in proportion to the flux entering the input port. In this case, the cosine law is respected on simple geometrical considerations. There are special cases where the front port of the sphere is equipped with a special diffusing filter suitable to reduce any deviation from the cosine law.

Luminance Meter

The measurement of luminance requires that the instrument collects light within a solid angle with a section lower than 4°, then in front of the sensor should be placed an optical device suitable to perform this function.

The single body instruments are usually equipped with an objective lens and an eyepiece so as to operate at a distance and to visually check the region in which the measurement is made. Figure 13.6 shows a spectroradiometer with an ocular Pritchard mirror combined with a spectrometer rather than with a sensor coupled to a filter, such as it is in luminance meters.

The photodetector, when separated from the body of the instrument, is behind a suitable radiance lens-barrel. In this case the tube can be designed to operate in contact with the source of the luminance to be measured, for example, a television/computer monitor.

13.4.2   Colorimetric Instruments

Colorimetric instruments that do not perform spectral analysis can be divided into

1. *Colorimeters for light sources* – colorimetrically specify light sources and are without an internal light source.
2. *Colorimeters for non-luminous colours* – colorimetrically specify non-luminous bodies and are with an internal light source.

Almost always these instruments are portable, robust, easy to use, with an internal power battery and low cost.

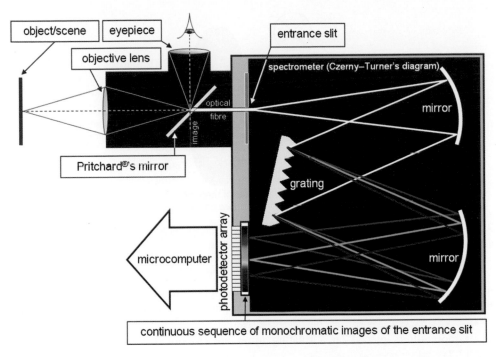

Figure 13.6 *Diagram of the optical system of a portable spectroradiometer. The mirror (known as Pritchard®'s mirror) placed at 45° with respect to the optical axis and coupled to an eyepiece allows the operator to visually examine the analysed region. An optical fibre inserted into the hole of the mirror leads the light to the spectrometer. The signals generated by the array of photodetectors suitably processed according to the calibration of the instrument provide the spectral power distribution of the radiation of the analysed object.*

Colorimeters for Light Sources

Colorimeters for the characterization of light sources or of illuminated objects in a scene consist of two parts:

1. An optical device to collect the light to be measured, that is, an objective lens or a mirror; the finest instruments collect light through a telescope with a Pritchard® mirror for a visual check on the measurement region.
2. An optoelectronic device for the colorimetric analysis of the collected light; originally, three glass filters combined with three photo-sensors mimicked the colour-matching functions of the CIE standard observer. Today some manufacturers have increased the number of filters and corresponding sensors up to 40, obtaining a spectral fitting of the responsivities of the CIE standard observer – abridged instrument –.

Consider a simple instrument with only three photo-sensors combined with three optical filters. The spectral responsivity $\varepsilon(\lambda)$ of three equal sensors is corrected by three filters with transmittances, $\tau_1(\lambda)$, $\tau_2(\lambda)$ and $\tau_3(\lambda)$. Three transducers are obtained with spectral sensitivities equal to the CIE colour-matching functions, $\bar{x}(\lambda) = \tau_1(\lambda)\varepsilon(\lambda)$, $\bar{y}(\lambda) = \tau_2(\lambda)\varepsilon(\lambda)$ and $\bar{z}(\lambda) = \tau_3(\lambda)\varepsilon(\lambda)$, or equal to their linear combinations. Consequently, the tristimulus values X, Y and Z are obtained from the signals

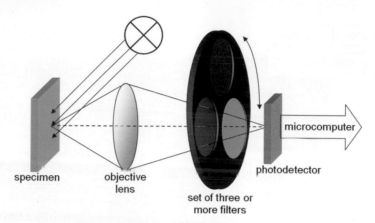

Figure 13.7   *Diagram of a tristimulus colorimeter with one photo detector combined with three coloured filters on a rotating wheel.*

generated by the photo-sensors combined with the corresponding filters. There are instruments termed *tristimulus colorimeters* and built according to different design patterns:

1. Three or more filters placed on a rotating turret are combined with a single sensor in sequence (Figure 13.7).
2. Three filters are combined with three sensors and operate simultaneously. The light coming from the object of measurement through the optical system crosses a protective filter and a lens, then it is collected by three optical fibres that lead to three different photo-sensors. Each photo-sensor is coupled to a filter in order to reproduce one of the standard colour-matching functions and the signal of the light incident on the photo-sensor is converted into an electrical signal. The digital signals, obtained by conversion from analogue signals, are processed by a computer to display the tristimulus values (X, Y, Z), the luminance L_v and the chromaticity coordinates (x, y).

Generally the choice of filters and the instrument calibration are made for obtaining the tristimulus values X, Y and Z from the signals of the photo-sensors by linear transformations. Moreover, the filters, classified as R, G and B − red, green and blue − and combined with the sensors constitute three transducers with spectral sensitivities $R(\lambda)$, $G(\lambda)$ and $B(\lambda)$. The filter G is chosen suitable to have the value of Y directly from its sensor, that is, the spectral response of the sensors combined with its filter $G(\lambda)$ is equal to the function $\bar{y}(\lambda)$ of the standard observer. The function $\bar{z}(\lambda)$ requires a scale factor with respect to the response $B(\lambda)$ of the short-wavelength channel B. Instead, the reproduction of a response equivalent to that of the function $\bar{x}(\lambda)$ generates some problems related to the difficulty of producing a filter whose transmittance satisfies the two maxima (bimodality) of the function $\bar{x}(\lambda)$ at the long and short wavelengths, respectively. It is clear that the degree of agreement between the CIE colour-matching functions and the responses produced by each photo-sensor with its filter influences the accuracy of the colorimeter. Figures 13.8a and 13.8b show typical examples of the existing agreement between the CIE colour-matching functions and the responses of the filter colorimeter. There are two ways to overcome the problem:

1. One is to use two filters, R_1 and R_2, each with single-band transmittance, so as to have two systems with photosensitive spectral responses $R_1(\lambda)$ and $R_2(\lambda)$, which, combined, reproduces $\bar{x}(\lambda)$, and then the tristimulus values are obtainable with the equations $X = a\,R_1 + b\,R_2$, $Y = G$ and $Z = c\,B$, where R_1, R_2, G and B are the signals produced by the four sensors (filters combined with detectors).

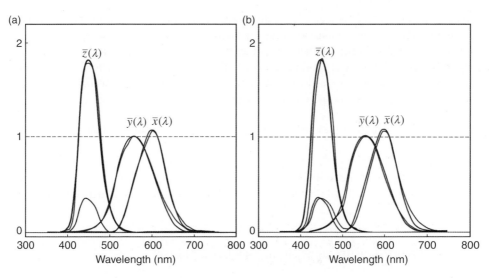

Figure 13.8 *Comparison between the 1931 CIE colour-matching functions (black line) and spectral responsivities of a typical tristimulus colorimeter with three (a) and four filters (b) (red line).*

2. The other method is to use one R filter with the peak transmittance at a long λ and consider the function in the range of short wavelengths proportional to the contribution of the B filter: the tristimulus values are $X = a\,R + b\,B$, $Y = G$ and $Z = c\,B$, where R, G and B are the signals produced by the three filters combined with three detectors.

This kind of instrument also by computation implements the measurement of the *CCT*. In this way, the characterization of the light source is complete.

An opal filter to be placed before the objective lens is used to make measurements of illuminance.

Colorimeter for Non-luminous Colours

Colorimeters for the measurement of non-luminous bodies are constituted of three parts:

- a light source;
- an optical device for the illumination and vision geometry; and
- an optoelectronic device for the analysis of the collected light.

The light source is usually

- a xenon-flash lamp filtered for reducing the content of UV light and for simulating the daylight, which has the advantage of a spectrum that is equally powerful over the whole visible range or
- a lamp with a tungsten filament suitably filtered to correct the power imbalance between long and short wavelengths and mimicking the obsolete illuminant C.

The optical device for the illumination and collection of the light to be analysed is made according to one of the diagrams recommended by the CIE (Section 12.2.2) and the kind of geometry characterizes the instrument.

The optoelectronic device for the analysis of light is similar to that present in the colorimeter for light sources and provides the percentage luminance factor Y and the chromaticity coordinates (x, y) with the display, giving a complete colorimetric characterization of the examined surface.

Tristimulus colorimeters with filters continue to be a viable alternative to the use of spectrophotometers and spectroradiometers, when fast and easy measurements, such as colour control in industrial processes, are required.

In order to have accurate measurements, it is necessary

- to use filters combined with a photodetector to mimic the CIE colour-matching functions as closely as possible;
- to use sources that reproduce a CIE illuminant (e.g., D65 illuminant);
- to use the measuring geometries recommended by the CIE; and
- to calibrate the instrument with a reflectance standard diffuser traceable to a metrology laboratory at least once a year.

13.5   Instruments for Measurements with Spectral Analysis

The instruments that implement measurements with spectral analysis can be divided into spectroradiometers and spectrophotometers. The analysis of the collected light is spectral and is obtained with different spectrometers:

- Spectrometer with dispersion by a rotating diffraction grating and only one photodetector – *monochromator* –, a typical configuration of high-class instruments for research laboratories; in some cases, the photodetector is of the photomultiplier type, thus offering very high sensitivity.
- Spectrometer with dispersion by a fixed diffraction grating and an array of photodetectors – *polychromator* –, configuration for laboratories and portable instruments.
- Spectrometer in which the collected light uniformly illuminates an array of photodetectors coupled to interference filters, whose passband is generally of ~15 nm – system termed *abridged* –, a typical configuration of portable robust instruments.

In some instruments, the photodetector or the array of photodetectors is cooled by the Peltier effect, significantly reducing the thermal noise of the signal.

13.5.1   Spectroradiometer

Spectroradiometers measure the spectral power distribution of light radiation, which typically is collected by means of optical devices:

1. An optical fibre – in this case the spectrum is represented in units of spectral density of flux (W/nm). The entrance of the optical fibre is equivalent to the entrance slit of the spectrometer. Moreover, in this case the instrument can be joined with optical devices for collecting the light in all possible ways.
2. A telescope with a Pritchard® mirror – in this case the spectrum is represented in units of spectral density of radiance (W/(m^2 sr nm)) (Figure 13.6).
3. A lens with a laser-beam pointing system – in this case the spectrum is represented in units of spectral density of radiance (W/(m^2 sr nm)).

In all cases, the chromaticity of light considered, the dominant wavelength, purity and colour-rendering index can be obtained by calculation. In cases (2) and (3) it is also possible to calculate the luminance and, with a cosine-correction accessory to be put before the lens, it is possible to measure the illuminance.

Often the specification is in arbitrary units, which is sufficient for many applications, except for the measurement of illuminance (lux) and of luminance (cd/m^2).

Figure 13.6 shows the diagram of the optical system of a portable spectroradiometer with Pritchard$^®$'s mirror, in which the path of the measured radiation is traced:

- the light, after passing through the lens, crosses a mirror, at the centre of which is a hole coincident with the region of measurement;
- the light entering the hole in the mirror continues its path within an optical fibre in which, thanks to various reflections, it becomes a uniform beam;
- the light is then collimated by a suitable lens – or mirror – on a diffraction grating that disperses its monochromatic components ordinately in different directions;
- the contributions of different wavelengths are focused by a further lens, or mirror, on an array of photodetectors; and
- the light incident on the remaining part of the mirror is reflected and sent to an eyepiece, by which the operator sees the measurement region that appears as a black ellipse in the centre of the field of view, while the remaining part corresponds to the background of the measurement area.

Despite the easy grip of portable instruments, it is recommended to always put the instrument on a stand for optimal use.

The calibration of spectroradiometers is considered in Section 12.3.

13.5.2   Spectrophotometer

Spectrophotometers measure the relative amount of a spectral radiant flux, which can be transmitted or reflected, for each wavelength of the spectral interval considered, according to the definitions of spectral reflectance factor and spectral transmittance factor given in Section 1.22.2. Spectrophotometer consist of three parts (Figures 12.1a, 12.1b and 13.6):

1. a lighting system, generally a xenon-flash lamp, or a lamp with a tungsten filament suitably filtered to correct the power imbalance between long and short wavelengths;
2. an optical apparatus for illuminating the surface under examination and collecting the light to the spectrometer-entrance slit, according to the geometries recommended by the CIE (Section 12.2.2); for some instruments this part is an accessory and for others it is an internal part of the instrument; and
3. one or more spectrometers for spectral measurements.

A computer, usually external to the instrument, manages the measurement and processes the raw data, providing the results of the measurement.

The spectrometric part is similar to that present in spectroradiometers.

Also in this case, there are fixed instruments for laboratory and portable instruments.

The high complexity of these instruments requires the calibration to be checked periodically. For instruments for measuring reflectance there are *Physical standards for calibration*, which are considered in Section 12.6.

13.5.3   Multiangle Spectrophotometer

Multiangle spectrophotometers are instruments offering many geometries of vision and of the illumination conceived to characterize particularly finishes with metallic pigments, producing highly non-Lambertian diffusion, and special effect pigments, such as mica pigments, producing interference colours. The visual perception of this kind of finish is goniodependent and is measured by multiangle geometries. Two different multiangle geometries have been standardized,[15–17] as described in Section 12.2.2.

In addition to the goniodependence of the colour appearance, other effects exist:

1. *Sparkle* – A sparkling or glitter impression, micro brilliance or glint, can be observed under directional illumination sunlight and depends on the illumination angle. Three different illumination angles are used, 15°, 45° and 75°. This phenomenon is influenced by
   - the pigment flake type and size,
   - the concentration level of the special effect pigment
   - the orientation of the special effect pigment, and
   - the application method.
2. *Graininess* – effect described as coarseness, observed under diffuse illumination and almost independent of the viewing angle. This visual graininess is influenced by the size and orientation of the pigment flakes.

These effects cannot be characterized in a spectrophotometric way and are quantified by a digital camera with resolution suitable to measure the size and the surface distribution of the pigment flakes. The direction of the camera vision is orthogonal to the specimen.

For a visual evaluation of all these goniodependent appearance effects, a tool, termed Gonio-Vision-Box, was developed by the company Merck and is manufactured under licence by BYK. This tool has the shape of a quarter of a sphere and on the semicircle of the vertical plane has a series of holes through which the colour sample is illuminated and observed (Figures 13.9a, 13.9b, and 13.9c). The position of the holes follows the standard geometries. Two samples with the edges in contact are visually comparable in all standard viewing directions (Figure 13.10). The practical utility of this simple tool is high. An image sensor provides information from three different lighting directions.

13.5.4   Fibre-Optic-Reflectance Spectroscopy (FORS)

The acronym FORS means 'Fibre-Optic-Reflectance Spectroscopy'. Commercial instruments for measurements of the reflectance or transmittance factor are usually tools in a unique body and, although they are lightweight and portable, are not always suitable for in-situ measurements, as, for example, on artworks. In recent years interesting miniaturized spectrometers have been developed, in which the light being measured enters through an optical fibre. The easy coupling of fibres allows a modular build-up of a system that consists of a light source, sampling accessories and a fibre optic spectrometer. Today, associated with these spectrometers, the market produces accessories for different measurement types: light sources with a light guide for illumination, tiny integration spheres, heads that illuminate and collect the reflected light. This technique began in the 1980s at the Institute IFAC-CNR in Florence,[18,19] where many varieties of tiny heads were designed for measuring works of art. Many proposals have been made, standard and non-standard, all of great interest. Two typical examples of probes are given below:

1. A probe-head is a dark hemisphere, 2.5 cm in diameter, terminating with a flat base and having three apertures on the dome. One aperture is at the top of the dome (for receiving the light reflected in the direction orthogonal to the sample), and the other two are placed at 45°, symmetrically with respect to the former, for illuminating the investigated area of the sample. This geometry could be defined as $(2 \times 45°/0°)$. The use of only one illuminating fibre is similar to the geometry $(45°/0°)$ (Figure 13.11).

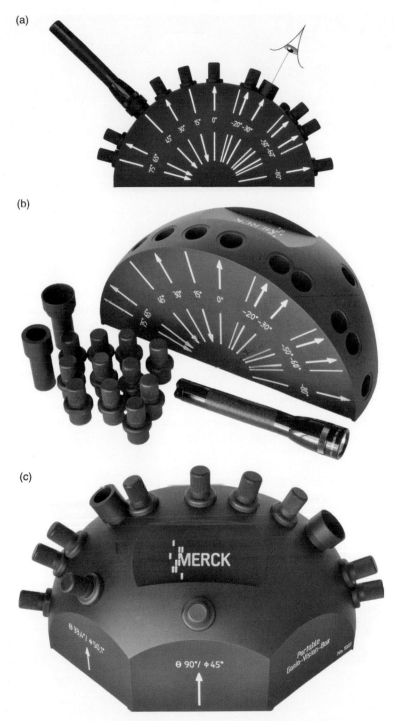

Figure 13.9 *Different views of the 'Gonio-Vision-Box'. All the standard geometries for illuminating and viewing are marked on the face of the instrument parallel to the plane containing the illumination and vision directions. Reproduced with permission from Werner Rudolf Cramer © 2015.*

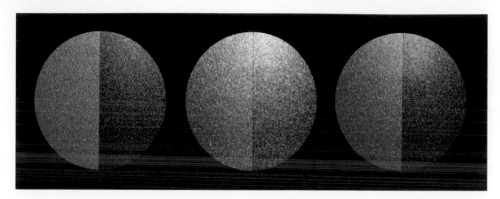

Figure 13.10   *Example of compared views by 'Gonio-Vision-Box' of three pairs of gonio-apparent specimens.*

2.   A flat-tip probe is a bifurcated bundle of fibres that are mingled at the front end and separate into two branches ending at the light source and the spectrometer, respectively. This kind of probe can be positioned with respect to the object being measured with different orientations. Typically the position is orthogonal and recalls the geometry (0°/0°). Figure 13.12 gives the section of the flat-tip of the probe, where the central fibre is for collecting reflected light to be analysed by the spectrometer and six external fibres are for the illumination. Probes with various arrangements of the fibres are used in different fields.

Advantages of fibre optic spectroscopy are the modularity and flexibility of the system, the speed of measurement and the low cost.

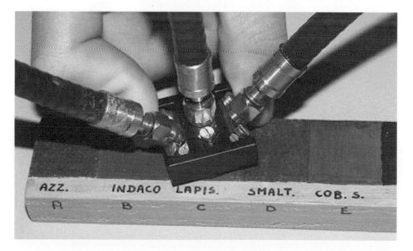

Figure 13.11   *Photography of the probe-head for FORS with two fibres set symmetrically at 45° for the illumination and an orthogonal fibre for carrying the light emerging from the specimen to the spectrometer. Reproduced with permission from Marcello Picollo © 2015.*

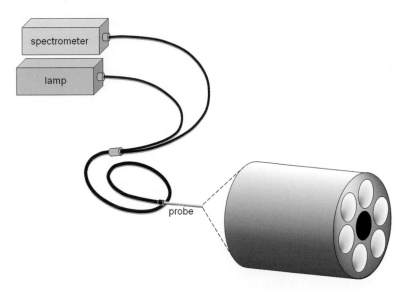

Figure 13.12   *Sketch of the FORS apparatus with a light source, a spectrometer and an optic fibre probe, where six external fibres are used for the illumination and the internal fibre collects the backlight reflected to the spectrometer.*

13.6   Glossmeter

> "*Gloss* (of a surface) – mode of appearance by which reflected highlights of objects are perceived as superimposed on the surface due to the directionally selective properties of that surface."[20]

Gloss is a surface property of bodies whose perception has an important and complex role in appearance. Gloss seems to be independent of colour and dependent only on the specular reflection. As is known, specular reflection strongly depends on the characteristics of the reflecting surface, such as its refractive index (Section 3.4), roughness and smoothness. R. S. Hunter and R. W Harold[21] spent much time studying these properties, and came to define five different kinds of gloss, distinguished according to the following scheme and gave a nominal distinction, which shows the definition of the quantity to be measured:

1. *Specular gloss* – the ratio of light flux specularly reflected to light flux incident at a defined angle.
2. *Sheen* – specular gloss at large angles of incidence and vision ($\vartheta_i = 85°$).
3. *Lustre, contrast gloss* – the ratio of the light flux reflected by diffusion from the test sample in the direction orthogonal to the surface $R_{45°:0°}$ to the light flux specularly reflected ($\vartheta_i = \vartheta_r = 45°$) from a reference – for example, black glass – $R_{45°:45°}$, with equal angles of incidence and reflection.
4. *Absence-of-bloom gloss* – measurement of the degree of absence of light diffused from the test sample in the direction adjacent to the specular direction.
5. *Distinctness-of-image gloss* – measurement of the minimum distance in which two neighbouring points of an extended light source reflected from the test sample appear resolved, that is, distinct, if observed in the specular reflection.

The *distinctness-of-image gloss* well represents the diffusion of the reflected light in the proximity of the specular reflection and is a good correlate of the quality of the perceived gloss.

In these definitions the relationship of the reflected light flux with the incident light flux is considered, but, in practice, the incident light flux is replaced by the light flux specularly reflected by a reference surface.

In practice, according to the existing standards, there are at least seven different definitions of *gloss-reflectance factor*, presented in more than 16 ASTM norms and recommendations (without considering other organizations for standardization):

$$G(\theta_i) = f \frac{S_S(\vartheta_i, \vartheta_r)}{S_R(\vartheta_i, \vartheta_r)}$$
(13.1)

where, in accordance with the various standards,

- ϑ_i is the angle of incidence, which is equal to the angle of specular reflection, $\vartheta_i = \vartheta_r$, which in the different standards assumes the values $\vartheta_r = (85°, 75°, 60°, 45°, 30°, 20°)$.
- $S_S(\vartheta_i, \vartheta_r)$ and $S_R(\vartheta_i, \vartheta_r)$ are the signals of the photodetectors relative to the sample (test) and the reference, respectively; the reference, according to the standard, is the ideal mirror, or black glass with a refractive index $n = 1.567$, or the black glass with a refractive index $n = 1.54$.
- f is a scaling factor that is set equal to 100 when the reference is reflective black glass – percent scale – and equal to 1000 in the case of an ideal mirror.

The gloss factor is measured by illuminating with white light C. The angles ϑ_i depend on the product category and the most usual values are

- 20° for paint in general (cars, appliances, etc.);
- 60° for almost all surfaces, except those with very high and very low gloss; and
- 85° for surfaces with very low gloss.

The scale of the gloss factor at the different angles is summarized in Table 13.3.

The geometry of illumination and of light collection is defined differently in the various standards. Referring to Figure 13.13, the various standards define the angles α, β, γ and δ related to pyramidal illuminance and vision.

Such a high number of definitions and standards means that the appearance of gloss is a complex phenomenon, which depends on the reflectance associated with each direction of illumination and on every point of view, that is, from the Bi-directional-Reflectance Distribution Function (BRDF) (Section 12.2.2). Unfortunately, tools for measuring the BRDF are now accessible only to a few research laboratories.

The glossmeters offered by the market operate at one or a few angles and meet general requirements. They are compact, lightweight and robust.

Table 13.3 *Gloss factor and measuring angle in cases with black-glass reference.*

Gloss-reflectance factor	Kind of gloss	Measuring angles
Greater than 70	Full gloss	All
Between 35 and 70	Semi-gloss	60°
Between 20 and 35	Eggshell	60°
Lower than 15	Flat or matte	85°

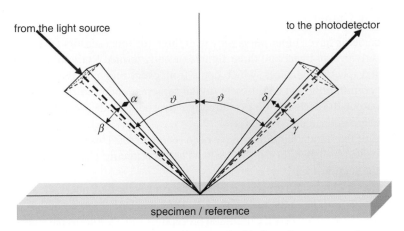

Figure 13.13   *Geometry of illumination and of light collection in a glossmeter.*

13.7   Imaging Instruments

In recent years, image sensors have had an impressive development and this has led to advanced cameras for photographic use and beyond. These cameras are not yet measuring instruments because the spectral sensitivities of image sensors only approximate the photometric sensitivity $V(\lambda)$ and the standard colour-matching functions $\bar{x}(\lambda)$, $\bar{y}(\lambda)$ and $\bar{z}(\lambda)$.

The usual instruments (spectroradiometers and spectrophotometers) make measurements on an area with a size of at least a few millimetres, and in practice many times it is necessary to measure the colour of a very small object (e.g., a thread in a fabric) or small areas in a complicated object where it is not possible to make contact measurements. Imaging instruments are useful to overcome these problems. The need to measure the luminance, colour, spectral radiance factor and spectral reflectance factor of the various points of a scene has led to the development of cameras that are considerable measuring instruments. The easiest solution was to adapt a digital camera to become an instrument for measuring the luminance corresponding to each pixel of the scene, which led to imaging photometers.

13.7.1   Imaging Photometer

The imaging photometer, also termed video-photometer, differs from the traditional photometer in that it employs an array of photodetectors, generally a CCD or CMOS array with 1024×1024 photosites, instead of just one. Here, too, the array of photodetectors is filtered so that the sensitivity is equal to the spectral luminous efficiency function $V(\lambda)$.

As in a photographic camera, an objective lens focuses an image of the scene on the plane of the photodetector array, then a photometric analysis of all the pixels of the captured image is possible with high practical value.

Often the array of photodetectors is cooled by the Peltier effect to eliminate the contribution of thermal noise from the signal and this leads to a digital signal with enhanced dynamics. A multi-exposure mode provides up to six orders of magnitude of dynamic range in a single image, with an absolute measurement range covering 0.0001 to 100 000 cd/m². Moreover, the dynamics of the instrument can be extended by using neutral optical filters.

This instrument works in combination with a computer and software provides comprehensive functions for a convenient analysis and documentation of results.

The shape of these photometers is that of a usual photographic camera or the most essential shape of an instrument.

13.7.2   Colorimetric Camera

Digital trichromatic cameras have colour *photosites* with three (or four) sensors (Section 15.2), which have different spectral sensitivities and generate three signals (usually termed RGB signals) that are the result of three different weighted spectral radiances, which come from every region of the scene. These RGB signals are *device-dependent coordinates*, that is, different digital cameras produce different RGB responses for the same scene and the output RGB signals do not directly correspond to the *device-independent coordinates* based on the CIE standard colorimetric observer.

The same responses of the standard observer have been searched for. This is a challenge, since commercial cameras are not conceived as laboratory instruments. The colorimetric characterization of a trichromatic camera is considered in Section 15.2.

Today the market offers trichromatic cameras capable of measuring the radiance of the scene weighted according to colour-matching functions close to those of the CIE 1931 standard observer. In the case of a trichromatic camera generating an RGB signal, powerful calibration procedures and algorithms have been implemented with this aim with good results, but the metamerism of the observer and of the instrument is an open problem: lights exist that are metameric for the human observer and are non-metameric for the instrument and vice versa.[22–24].

Today the market also offers cameras conceived for photometric and colorimetric imaging. Often these cameras can be used for both kinds of measurement. The camera has only one image sensor with uncoated photosites and the filtering is obtained by a suitable set of filters – three or four, or even more – mounted on one rotating wheel. The quality of the results is comparable with that of a luminancemeter and a tristimulus colorimeter. Several images with different filtering have to be shot. The measurement of the colorimetric and photometric quantities is made by proper software. Moreover, this kind of camera can be calibrated for specific use with the standard CIE D65 and A illuminants.

13.7.3   Multispectral and Hyperspectral Camera

Trichromatic cameras, only in the case of a very accurate calibration, measure the colour of objects illuminated by a particular light source and this quantity cannot be considered a measure of the spectral reflectance factor or of the radiance of a scene. Today, this limit is overcome by *spectral imaging* or *imaging spectroscopy*. This terminology includes the words 'multispectral' and 'hyperspectral':

- *Hyperspectral imaging* refers to a large number of contiguous and narrow wavelength bands, allowing the spectrum to be collected at high resolution.
- *Multispectral imaging* refers to systems where a small number – greater than three – of wavelength bands are recorded that can be contiguous, partially overlapping or separated.

Multispectral imaging is an expansion of the conventional trichromatic system. Each pixel contains information about the spectral reflectance or the spectral radiance of the scene with more channels (or bands) of responses than the normal three. The number of channels varies from four to several hundreds of bands for hyperspectral images.

Two main techniques are used for acquiring multispectral colour images, but others are possible:

1. One system commonly used consists of a monochrome digital camera coupled with a set of colour filters from a filter wheel or from a tuneable filter, in which the spectral transmittance can be controlled electronically. By using a single sensor array with high spectral sensitivity across a wide range of electromagnetic wavelengths, from infrared, through visible light and to ultraviolet, multichannel information is obtained by taking multiple exposures of a scene, one for each filter (Figure 13.14a).
2. The camera works as a scanner. The objective lens focuses the image of the scene on the plane of the entrance slit of a spectrometer, but only the light coming from the points of a horizontal strip of the scene,

(a)

(b)

Figure 13.14 *Description of image collection in hyperspectral imaging. (a) Collection of images of the whole scene obtained by illuminating with narrow spectral bands of light or filtering the reflected light by narrow band filters. (b) Image relative to a strip of the scene, where each vertical line represents the radiance spectrum of a pixel of the strip. The whole spectral information is given by the collection of the images of all the strips. The strip in the figure is overly large, but in reality the width is about 300 microns. (Reproduction of 'La schiava turca' by Parmigianino produced by G. Antonioli, F. Fermi and R. Reverberi. Reproduction with permission of the authors).*[25]

conjugated with the points of the entrance slit, enters the spectrometer. The spectrometer disperses the light on a plane normal to its optical axis and, for example, focuses the dispersed light on the sensor matrix with the blue light on the top and the red light on the bottom of the sensor (or vice versa). Each shot gives a picture where the spectral information is on vertical lines and each vertical line is in correspondence with a point of a horizontal strip of the scene (Figure 13.14b). All the scene is scanned strip by strip.

Once all the pictures are captured, an image-processing module for spectrum reconstruction is employed. The measured quantity can be the spectral reflectance factor or the spectral radiance at each pixel of the scene.

The quality of a multi-hyper spectral colour image acquisition system depends on many factors: the spectral sensitivity of the camera, the spectral sensitivity of each channel, the spectral radiance of the light source and the performance of the image-processing module.

Spectral imaging, for the complexity of the optical system and of the data processing, is one of the most actively researched areas in colour technology. Anyway, today several systems are available in the market.

References

1. ISO 3668:1998, *Paints and Varnishes – Visual Comparison of the Colour of Paints*, ISO Central Secretariat, Geneva (1998).
2. ASTM D1729 - 96(2009), *Standard Practice for Visual Appraisal of Colors and Color Differences of Diffusely-Illuminated Opaque Materials*, ASTM International, West Conshohocken, PA (2014). Available at: www.astm.org (accessed on 18 June 2015).
3. DIN 6173-2, *Colour Matching; Lighting Conditions for Average Artificial Daylight.* 10/01, DIN e.V., Am DIN-Platz, Berlin (1983). Available at: www.din.de (accessed on 18 June 2015)
4. ISO 3664:2009, *Graphic Technology and Photography – Viewing Conditions*, ISO Central Secretariat, Geneva (2009).
5. CIE 51:1981, *A Method for Assessing the Quality of Daylight Simulators for Colorimetry.* Commission Internationale de l'Éclairage, Vienna (1981). Available at: www.cie.co.at/ (accessed on 18 June 2015).
6. CIE 51.2:1999, *A Method for Assessing the Quality of Daylight Simulators for Colorimetry.* Commission Internationale de l'Éclairage, Vienna (1999). Available at: www.cie.co.at/ (accessed on 18 June 2015).
7. CIE 51.2 - 1999, A method for assessing the quality of daylight simulators for colorimetry, *Color Res. App.*, 26(3), 255 (2001).
8. CIE 135/3:1999, (Supplement 1-1999 to CIE 51:1981) *Virtual Metamers for Assessing the Quality of Simulators of CIE Illuminant D50.* Commission Internationale de l'Éclairage, Vienna (1999). Available at: www.cie.co.at/ (accessed on 18 June 2015).
9. CIE Publication Standard DS 012:2001, *Standard Method of Assessing the Spectral Quality of Daylight Simulators for Visual Appraisal and Measurement of Colour.* Commission Internationale de l'Éclairage, Vienna (2001). Available at: www.cie.co.at/ (accessed on 18 June 2015).
10. Schofield RK, *Colorimetry: Part 1, Lovibond System, Lovibond-Schofield System*, T. Sci. In8t. 16, 74, The Tintometer, Ltd., Salisbury (1939).
11. Schofield RK, *Colorimetry: Part 2, Lovibond System, Lovibond-Schofield System*, T. Sci. In8t. 16, 74, The Tintometer, Ltd., Salisbury (1939).
12. Haupt GW and Douglas FL, Chromaticities of lovibond glasses, *U. S. Department of Commerce, National Bureau of Standards, Research Paper* RP1808, **39**, July, 11–20 (1947).
13. Haupt GW and Douglas FL, Chromaticities of lovibond glasses, *J. Opt. Soc. Amer.*, **37**(9), 698–704 (1947).

14. DIN 5033, *Colorimetry - Part 1, 2, ..., 9*. DIN e.V., Am DIN-Platz, Berlin. Available at: www.din.de (accessed on 18 June 2015).

15. DIN 6175-2, *Tolerances for Automotive Paints - Part 2: Goniochromatic Paints*, DIN e.V., Am DIN-Platz, Berlin, Germany (2001). Available at: www.din.de (accessed on 18 June 2015).

16. ASTM E2194-14, *Standard Test Method for Multiangle Color Measurement of Metal Flake Pigmented Materials*, ASTM International, West Conshohocken, PA, (2014). Available at: www.astm.org (accessed on 18 June 2015).

17. ASTM E2539-08, *Standard Practice for Multiangle Color Measurement of Interference Pigments*, ASTM International, West Conshohocken, PA (2008). Available at: www.astm.org (accessed on 18 June 2015).

18. Bacci M, Fibre optics applications to works of art, *Sensors Actuators B: Chemical*, **29**(1–3), 190–196 (1995).

19. Aceto M, Agostino A, Fenoglio G, Idone A, Gulmini M, Picollo M, Ricciardi P and Delaney JK, Characterisation of colourants on illuminated manuscripts by portable fibre optic UV-visible-NIR reflectance spectrophotometry, *Roy. Soc. Chem. Anal. Meth.*, **6**, 1488–1500 (2014).

20. CIE Publication S 017/E:2011. *ILV: International Lighting Vocabulary*. Commission Internationale de l'Éclairage, Vienna (2011). Available at: www.eilv.cie.co.at/ (accessed on 18 June 2015).

21. Hunter RS and Harold RW, *The Measurement of Appearance*, John Wiley & Sons, New York (1987).

22. Hong G, Luo MR and Rhodes PA, A study of digital camera colorimetric characterization based on polynomial modelling, *Color Res. App.*, **26**, 76–84 (2001).

23. Pointer MR, Attridge GG and Jacobson RE, Practical camera characterisation for colour measurement, *Imaging Sci. J.*, **49**, 63–80 (2001).

24. Lee Hsien-Che, *Introduction to Color Imaging Science*, Cambridge University Press, Cambridge (2005).

25. Antonioli G, Fermi F, Oleari C and Reverberi R, Spectrophotometric scanner for imaging of paintings and other works of art, *CGIV 2004: The Second European Conference on Colour Graphics, Imaging and Vision, Aachen*, 219-224 (2004).

14

Colour-Order Systems and Atlases

14.1 Introduction

Consider the general definitions of 'colour-order system', 'colour atlas' and 'colour scale':

> "*Colour order system* – arrangement of samples according to a set of principles for the ordering and denotation of their colour, usually according to defined scales.
>
> NOTE – A colour order system is usually exemplified by a set of physical samples, sometimes known as a *colour atlas*. This facilitates the communication of colour but is not a prerequisite for defining a colour order system."[1]

> "*Colour (order) system* – general term used for any effort to systematically arrange colour experiences."[2]

> "*Colour atlas* – a collection of colour samples arranged and identified according to specified rules."[1]

> "*Colour atlas* is a systematic collection of colour chips or colour prints encompassing a large systematic range of possible experiences."[2]

> *Colour scale* – a scale in which the perceived colour stimuli of objects changes in a systematic manner usually in one attribute, i.e., as having equal differences of some attribute of colour sensation.

The term 'colour atlas' designates a collection of physical colour samples made up of the same material, which can be paper, fabric, plastic, gelatine, etc., achieved by a colour-order system. Colour-order systems represent a practical approach to colorimetry, which brings us back in time and still today in many areas arises in a complementary, if not alternative, way to standardized CIE colorimetry. Up to 1600 colour-order systems, impregnated with philosophical meanings, were patterns contained in the larger scheme of unity in the world. Graphical representation of the colours started in 1611 with Aron Sigfrid Forsius.[3] The value of the colour atlases is cognitive and practical. Their practical strengths are:

- the concreteness and immediacy of viewing a colour of a physical sample;
- the colour specification with a set of three numbers easily readable and usable; and
- the possibility of making visual comparisons with any other coloured body.

Standard Colorimetry: Definitions, Algorithms and Software, First Edition. Claudio Oleari.
© 2016 John Wiley & Sons, Ltd. Published 2016 by John Wiley & Sons, Ltd.

Different principles are at the basis of the various atlases, so it may be that other systems will be proposed over time and that today it is hard to foresee one order system/atlas becoming as unique and definitive. The principles that define a system for colour ordering are expressed by physical relationships, or psychophysical, or perceptive, or cognitive (*cognition* – the process of knowing with its dependence on awareness and judgement), or cultural and intrinsically contain the rules with which to build a *colour space* with a coordinate system. Consider the following definitions:

"*Colour space* – geometric representation of colour in space, usually of 3 dimensions."[1]

"*Colour space* – a three-dimensional cognitive model illustrating the relationships between colour percepts."[4] (Percept – a mental impression of something perceived by the senses, viewed as the basic component in the formation of concepts.)

"*Colour space* – systematic geometric representation of colours in three-dimensional space. The coordinates have different meanings – geometrical, perceptive, psychometric, etc. – according to different choices and give rise to various spacings within the solid."

Here the word 'space' should not be considered as a mathematical space with its associated operations, excluding certain partial and particular cases, but as an interpretation *graph* and a three-dimensional ordering of physical samples.

Below the generally most significant systems are schematically listed, in order to highlight these basic principles, which are related to *practical* and *cognitive* reasons[2,5]:

- systems based on *colorant mixture*;
- systems based on an *additive mixture of colour stimuli*:
  – *Ostwald system* (Section 14.3) and
  – *halftone printing* (Section 15.3);
- systems based on *colour subtractive filters*:
  – *Lovibond tintometer* (Section 13.3);
- systems defined only on *psychophysical* and *psychometric* experiments:
  – *Munsell order system* and *Book of Color* (1905) (Section 14.4) and
  – *uniform color-scale* system of *the Optical Society of America* (OSA-UCS) (1977) (Section 14.6);
- systems defined on *cognitive* and *psychometric* experiments:
  – *Swedish Natural Colour System* (NCS) (1969) (Section 14.7);
- systems defined on the psychophysical CIE space and according to perceived colour scales:
  – *DIN-Colour Chart* (1953) (Section 14.5).

An atlas based on the CIELAB system, spanned by cylindrical coordinates H_{ab}^*, L^* and C_{ab}^* (Section 11.3.5) and organized in constant hue pages H_{ab}^*, is also available in the market.[6] A view of the CIELAB system is given in the software attached to this book (Section 16.6).

Among all these systems, five stands out for historical reasons, knowledge and cognitive value: the Ostwald system, the Munsell system, the DIN-Colour Chart, the OSA-UCS and the NCS. All five systems are outstanding for the rigour of their construction.

All the atlases made according to these colour-order systems are materially constituted by collections of physical samples, whose peculiarities are only valid if the samples are observed under a specific illuminant, placed on a neutral background of defined lightness and according to a defined geometry of illumination and observation, thus avoiding any unwanted conditioning. Where necessary, all this belongs to the definition of the system and at the same time it is a constraint.

The use of a colour atlas is by direct visual comparison between the test-colour sample and the standard-colour samples of the colour atlas. The visual comparison must be made with the illuminant and viewing geometry defined for the atlas use. If the colour match is not obtained, the standard samples closest to the match are considered and then an interpolation is made based on the visual judgement.

Colour specification by visual comparison with a colour atlas can be considered as belonging to visual colorimetry:

> "*visual colorimetry* – colorimetry in which the eye is used to make quantitative comparison between colour stimuli."[1]

although traditionally visual colorimetry is a procedure of analytical chemistry, defined as:

> "*visual colorimetry* (analytical chemistry) – A procedure for the determination of the colour of an unknown solution by visual comparison to colour standards (solutions or colour-tinted disks)."[7]

Colour atlases have a limited stability over time; therefore, they need periodic replacement with new atlases.

There are other collections of colour samples commercially available that are arranged in various ways without referring to general psychophysical and psychometric principles. Two collections are particularly successful and appreciated, and therefore referred to here: RAL® and Pantone®.[6,8]

14.2   Colour Solid, Optimal Colours and Full Colours

Usually, the colour-order systems and atlases are conceived for non-luminous colours. Therefore, the treatment of the colour ordering has to be anticipated by new concepts related to the region of the non-luminous colours in the colour space. Particularly optimal colours, here defined, have an important role in the use of the Maxwell disk and in the atlases of Ostwald and NCS:

> "*Colour solid* – part of a colour space which contains surface colours."[1]

> "*Optimal colour stimuli* – object colour stimuli corresponding to objects whose luminance factors have maximum possible values for each chromaticity when their spectral luminance factors do not exceed 1 for any wavelength (i.e., no fluorescence).
> NOTE 1 – These stimuli correspond, in general, to objects whose spectral reflectance factors have values of either unity or 0, with not more than 2 transitions between them.
> NOTE 2 – The luminance factors and chromaticity coordinates of these stimuli define the boundaries of a colour solid corresponding to non-fluorescent objects.
> NOTE 3 – For a given luminance factor, these colour stimuli define the maximum purity possible for non-fluorescent objects."[1]

> *Full colour* – sample in the Ostwald system having the maximum purity for a particular hue.

> "*Purity* (of a colour stimulus) – a measure of the proportions of the amounts of the monochromatic stimulus and of the specified achromatic stimulus that, when additively mixed, match the colour stimulus considered.
> NOTE 1 – In the case of purple stimuli, the monochromatic stimulus is replaced by a stimulus whose chromaticity is represented by a point on the purple boundary."[1]

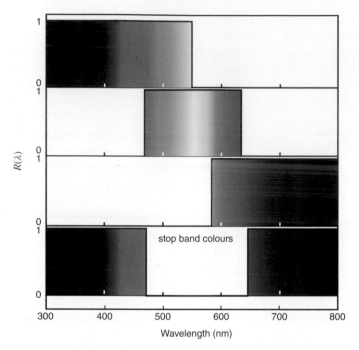

R(λ)

Wavelength (nm)

Figure 14.1 *Plot of spectral reflectance of optimal colours as defined by Schrödinger.*

The modern theory of the colour solid has been developed by Erwin Schrödinger.[9] All non-luminous colours lie within a finite, convex volume, termed *colour solid*, and Schrödinger's optimal colours make up the boundary of this solid. Its boundary is smooth, except for two points: the black and the white colours.

The colour solid can be represented in the tristimulus space, in CIELAB, CIELUV, etc.

The colour solid depends on the spectral power distribution of the illuminant.

For a mathematical definition of colour solid, the surface-colour reflectances have to be considered. Schrödinger defined *bivalent* reflectance factors as 0 or 1 at each wavelength, but may have any number of transitions between 0 and 1 over the visible wavelength range.

Schrödinger termed *optimal object colours* or *optimal colour*s the object colours having the maximum possible luminance factor for each chromaticity.

Therefore, *optimal colours* correspond to well-defined spectral reflectance factors: an optimal colour with dominant wavelength λ_d has the reflectance factor equal to 1 in a neighbourhood of λ_d and zero elsewhere (Figure 14.1). If the optimal colour is purple or with λ_d belonging to the extremities of the visible spectrum, the reflectance factor would be zero in a central interval of the spectrum and equal to 1 at the extremities (Figure 14.1). The shrinking of this region around λ_d increases the purity and decreases the luminance factor (Figure 14.2). This is stated by Schrödinger's:

"*Optimal Colour Theorem* – A … colour is optimal if and only if the values of its reflectance function are either 0 or 1, with at most two transitions between those two values."[9–12]

The *full colours* have the maximum distance from the achromatic axis of the Schrödinger colour solid; thus, they are the most colourful among optimal colours of defined dominant wavelength. Consider a plane containing the achromatic axis in the tristimulus space. The optimal colour vectors on this plane define a line that starts from the black colour B and ends in the white colour W (Figure 14.3).

Optimal pigments are called the pigments that produce optimal colour reflectances.

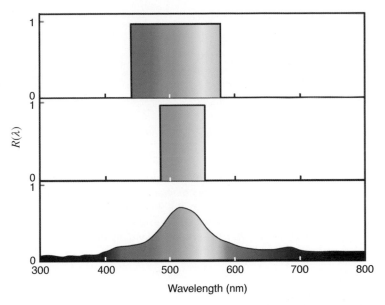

Figure 14.2 *Qualitative comparison between spectral reflectances of optimal colours with different purity and lightness (lightness increases and purity decreases with the extension of reflectance equal to 1) and of an actual colour with equal hue.*

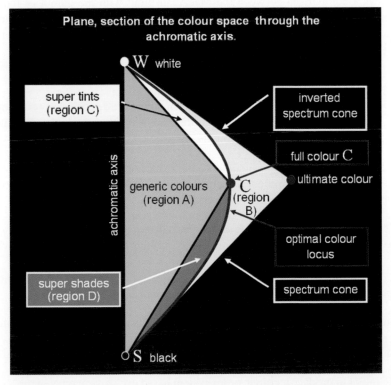

Figure 14.3 *Section of the colour solid containing the achromatic axis. The description and the names are given in the chapter.*

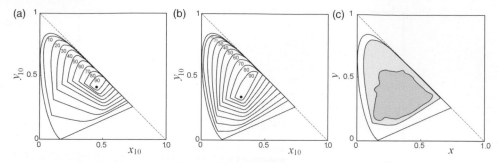

Figure 14.4 *(a) MacAdam limits related to the illuminant A and for percentage luminance factor Y = 10, 20, ...,80, 90 represented on the CIE 1964 chromaticity diagram. (b) MacAdam limits related to the illuminant D65 and for percentage luminance factor Y = 10, 20, ..., 80, 90 represented on the CIE 1964 chromaticity diagram. (c) MacAdam limit and Pointer's gamut of ink colours related to the illuminant C and to a percentage luminance factor Y = 20 represented on the CIE 1931 chromaticity diagram.*

14.2.1 MacAdam's Limit

> "*MacAdam limit* – the boundary of the optimal colour solid."

Since the specification of the surface colour depends on the considered illuminant, also the MacAdam limit depends on the illuminant. Figures 14.4a and 14.4b show the MacAdam limits for the sources A and D65 on the CIE (x, y) chromaticity diagram.

Figure 14.4c shows a comparison between the MacAdam limit with the gamut of the chromaticities of actual colours obtained with inks[13] for a percentage luminance factor $Y = 20$ and light source C. This figure shows that the MacAdam limit is a theoretical limit which is beyond the set of possible actual colours.

14.3 Ostwald's Colour-Order System and Atlas

> "*Ostwald Color System* – a color-order system in which colors are specified in terms of the attributes *hue*, *blackness*, and *whiteness*, and are spaced according to the results of spinning-disk mixing of specified amounts of ideal black, white, and maximally chromatic samples."[14]

Wilhelm Ostwald (1853–1932), as the first step before developing his colour-order system, considered related and unrelated colours and gave the following definitions[10] (compare these definitions with the corresponding ones given by the CIE and reported in Section 4.6):

- > "*Unrelated colours* are those that appear alone in an otherwise darkened visual field; we cannot judge their relation to the source of light; for unrelated colours there is no black, but only white and hue."

- > "*Related colours* are the colours of our surroundings, those that we see in relation to the light source; among related colours both blacks and greys appear."

For Ostwald, much trouble in understanding colour was because the physicist's unrelated colours do not match the artist's related colours. The scientist Ostwald privileged the scientific point of view, as it appears from the title of his book 'Physikalische Farbenlehre'.[15–26]

> Ostwald had the intent to produce a colour-order system based exclusively on Newton's centre-of-gravity rule, that is, on the colour matching obtainable with Maxwell's rotating disk (Section 4.3).

14.3.1 Ostwald's Hue Circle with Temperate Scale

For the construction of his atlas, Ostwald began the hue scale from the four Hering elementary hues, which were chosen in complementary positions as yellow and ultramarine, and red and sea-green – that is, these elementary hues were not exactly Hering's elementary hues –. The four intermediate hues were two complementary pairs, as orange and turquoise, and purple and leaf-green. With these eight colours, Ostwald constructed 24 colour hues with approximately equal spacing and numbered them from yellow upwards. For the equator of the solid, Ostwald developed a series of 100 hues, each colour as strong as could be obtained with pigments available at his time.

The ordered sequence of hues was arranged into a topological circle, without a metric; therefore, it was necessary to construct a 'temperate scale' with which to associate the dominant wavelengths with corresponding points on the colour circle with a uniform perceived scale (Figure 14.5). Ostwald pioneered a method to do this metrics based only on colour matching – that is, Newton's centre-of-gravity rule – and requiring no other visual judgement. Ostwald never completely explained this procedure, which he called the *principle of internal symmetry*. A reconstruction of this procedure has been tempted by Jan Koenderink.[27–29]

A possible reinterpretation of the *principle of internal symmetry* could be made in three steps on a chromaticity diagram, where the colour circle is a circle centred on the achromatic point:

1. Consider the straight lines radiating from the achromatic point and crossing the colour circle and the spectrum locus, creating in such a way a link between points of the colour circle and wavelengths of the spectrum locus.
2. Associate the wavelengths of the spectrum locus with the corresponding points on the colour circle as dominant wavelengths.
3. Once this correspondence is made, render the wavelength scale on the colour circle as uniform as possible by tilting the plane of the chromaticity diagram.

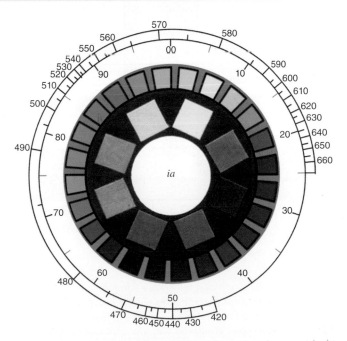

Figure 14.5 *Ostwald colour circle with a 'temperate scale' in correspondence with the wavelengths of spectral colours. The eight colours of the Ostwald original wheel are reproduced inside the circle.*

14.3.2 Ostwald's Semichrome

Ostwald introduced

1. the term *full colour* to denote a colour which permits the sensation of one single hue and is not 'tempered' by white or black and
2. the term *semichrome* for defining colours only by spectral reflectance – that is, only by a physical property –.

The colour circle traverses a wavelength-like domain composed of the spectrum locus with the line of purples. The Schrödinger optimal colours (Figure 14.1) are defined by dividing the colour-circle angle in two parts, one part with reflectance 1 and the other part with reflectance 0. The *semichrome* has the 1-to-0 transition points mutually complementary for a given white illuminant (Figure 14.6). Then Ostwald asserted the essential *semichrome theorem*, demonstrated by Schrödinger[9]:

Semichrome theorem –"The totality of colours (reflected by the colour sample where the reflectance is equal to 1) within half of the colour circle must be combined to obtain a saturated colour."[10] (Here the original words of Ostwald need an interpretation: 'saturated colour' means 'full colour' and 'half the colour circle' is a connected part of the spectrum whose end wavelengths are complementary relative to a given white.) That is, *full colours* are associated with pure semichromes.

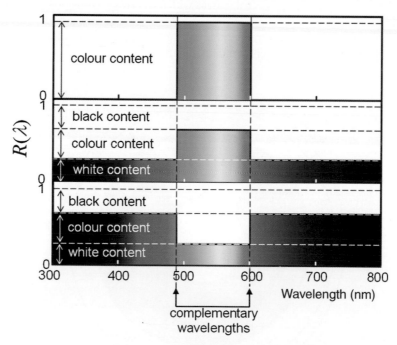

Figure 14.6 *From the top, spectral reflectance of a full-colour semichrome, a general semichrome with white and black content and its complementary semichrome. The abscissa axis can be associated with the colour circle on which the two points relative to complementary wavelengths are opposite points on the diameter and identify the steps of high-low reflectance passage. In this way, the reflectance has high value in a semicircle and low in the other semicircle.*

After these definitions, Ostwald termed *ideal pigments* as the pigments reflecting over a wavelength range of half of the spectrum – semichrome – and the corresponding *ideal complementary pigment* reflecting throughout the other half of the spectrum.

Moreover, Ostwald distinguished between

1.   *pure semichromes*, referring to ideal functions with transitions between 0 and 1, and
2.   *impure semichromes*, referring to reflectances achievable with real colorants.

14.3.3   Ostwald's Blackness, Whiteness and Purity

In the colour solid, consider a half-plane with a border constituted by the achromatic line. On this plane, consider the region delimited by the achromatic line and the bent line from white W to black B, which represents the optimal colours. The region inside these two lines represents non-luminous colours at a chosen dominant wavelength (and defined illuminant) (Figure 14.3). The Ostwald full colour is at the greatest distance from the grey axis.

Ostwald stated that black content, white content and full-colour content are the three elements of every non-luminous colour, and adopted these three elements as three independent dimensions of colour. The reproduction of these colours is made with the rotating disk by mixing the same contents of black, white and full colour (Figure 14.7). But, in this way the colours obtained by Maxwell disk mixing cover only the region 'A' of the surface colours 'A + C + D' (Figure 14.3). The colours in region 'C', termed *super-tints*, are not reproducible with the rotating disk because they have negative black content. Analogously those in region 'D', termed *super-shades*, are not reproducible because they have negative white content. Ostwald considered only the colours of region 'A' for his atlas.

The colours on a half-plane at a defined dominant wavelength are represented in a triangle, in which a set of three barycentric coordinates is defined according to the Ostwald-normalization assertion that "full chromatic colour *C* plus black *B* plus white *W* always resulted in unity":

$$C + B + W = 1 \tag{14.1}$$

where *C*, *B* and *W* mark only the quantity of full, black and white colours.

This equation regards also the three sectors of a Maxwell rotating disk (Figure 14.7), where the ideal reflectance is a bivalent function (Figure 14.6), the highest value of which, $(1-B)$, is lower than 1 and the lowest value *W* is higher than 0, as shown in Figure 14.6 (second and third part from the top). The reflectance is high in one half circle between two complementary wavelengths, and low in the second half circle. Then, equation (14.1) defines

*   the *degree of blackness*, represented by the black content *B*;
*   the *degree of whiteness*, represented by the white content *W*; and
*   the *colour purity*, represented by $C = [1 - (W + B)]$.

This definition of purity was given by Ostwald starting from its opposite, the *impurity* = $(W + B)$, that is, the degree of whiteness *W* plus the degree of blackness *B*.

The terms used to specify the colour in this organization of triangles with a defined hue are as follows (Figures 14.7a, 14.7b and 14.7c):

*   The colours of the outer row of samples, from white to full colour, belong to the *light clear series*. This row with the remaining rows parallel to it makes up the *isotones*, each row containing an increasing quantity of black. The lower member of each row is chromatic and the upper is grey.

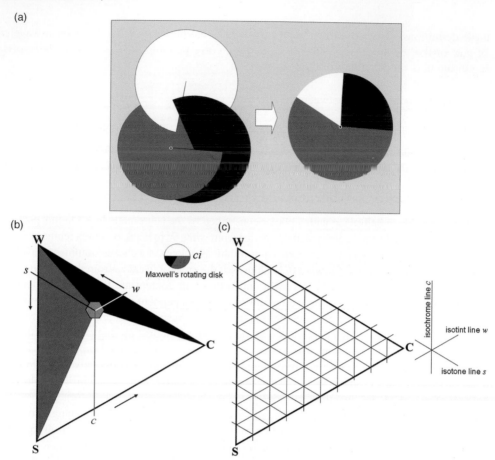

Figure 14.7 *(a) Use of the Maxwell disk (Figure 4.3b) made for producing a colour by mixing white, black and a full colour. The relative areas of the three sectors are equal to the content of white, black and full colour. (b) The colours produced by the rotating disk are set in correspondence to the points inside an equilateral triangle, where three barycentric coordinates are defined equal to the relative areas of white, black and a full colour in the rotating disk. (c) Isochrome, isotone and isotint lines drawn inside the triangle.*

- The colours of the outer row of samples from full-colour to black belong to the *dark clear series* and, with the rows parallel to it, form the group called *isotints*, which have a constant white content, increasing from row to row. The lower member in each row of this series is neutral while the upper member is chromatic.
- The colours of rows of samples parallel to the grey scale belong to a set of series of increased saturation, called the *shadow series*, or *isochromes*.

14.3.4 Ostwald's Atlas

The Ostwald atlas is a collection of non-luminous colours inside the triangles WCS, with corners in the white W, black S and full colours C on the planes at a defined dominant wavelength in the colour solid (Figure 14.3). These triangles are mapped into equal equilateral triangles. Opposite triangles correspond to complementary wavelengths. Each of these pairs of opposite triangles are represented in the atlas on the same page (Figure 14.8) and together are inscribed in a rhombus, in which the vertical diagonal line regards the grey colours. Figure 14.9 shows six pages of the Ostwald atlas. The spatial collection of these rhombuses is inscribed in a double cone.

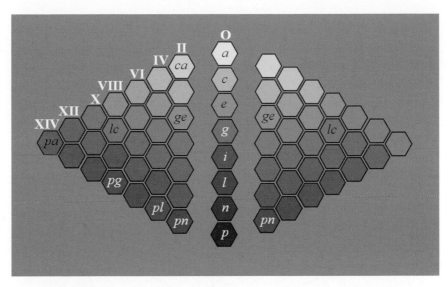

Figure 14.8 *Two pages of the Ostwald atlas with complementary colours (two hexagons are empty because their colours are not reproducible with an sRGB monitor).*

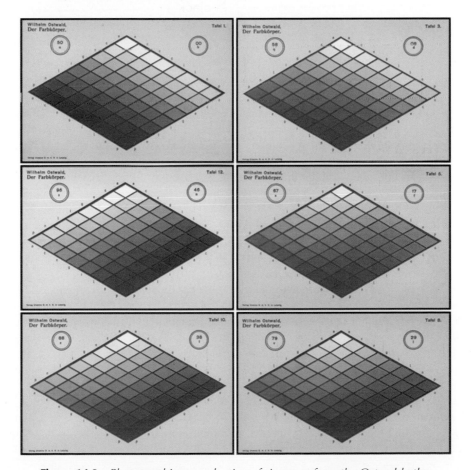

Figure 14.9 *Photographic reproduction of six pages from the Ostwald atlas*

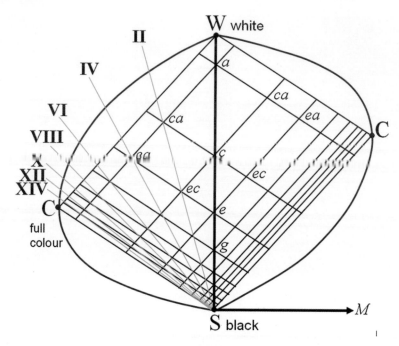

Figure 14.10 *Introduction of a log scale in the triangle WCS of the colour solid with the intention of obtaining uniform scales of perceived colours.*

The construction of the Ostwald atlas is independent of the illuminant, the observer and adaptation. The requirement is that the grey colours illuminated by the chosen illuminant appear achromatic.

In 1923 Ostwald rescaled the triangle WCS of the colour solid in term of whiteness and blackness according to the Weber-Fechner law in an attempt to obtain uniform perceptual scales (Figure 14.10).

The Ostwald atlas is presented visually in the software described in Section 16.5.1.

14.4   Munsell's Colour-Order System and Atlas

The usual glossaries present the Munsell system and atlas in a very operative way for colour specification:

> *Munsell colour system* – a colour system for describing the colour appearance of samples. The Munsell system uses matching against a set of samples and interpolation between them to arrive at a specification for the appearance of a given test sample, illuminated by daylight D65 and viewed by the observer CIE 1964 daylight adapted, in terms of three attributes, *hue*, *value* and *chroma*, using scales that are perceptually approximately uniform. The colour appearance in the Munsell system is characterized using sets of three symbols, for example, 2.5 YR 5/10, where 2.5 YR is the hue, 5/ the value – that is, lightness – and /10 the chroma.

> "*Munsell Book of Color* – Physical exemplification of the Munsell colour order system, consisting of about 1600 colour chips arranged in a cylindrical coordinate system of planes of constant Munsell hue on which Munsell value is displayed vertically and Munsell chroma horizontally."[14]

In 1905 Munsell started publishing the first edition of his book titled *A Color Notation*. Another seven editions followed. Geo H. Ellis Co. in Boston was the publisher of the first four editions (1905, 1907, 1913 and 1916) and the Munsell Color Company the publisher of the other four (1919, 1923, 1926 and 1936).

Munsell declared his work 'didactic' and its great clarity is evident to the reader. But the great value of the Munsell system and atlas is in the scientific rigour of the definitions and construction, and here this aspect is highlighted considering and citing the original Munsell book titled *A Color Notation* – 5th edition, 1919 –. In the subtitle "An illustrated system defining all colours and their relations by measured scales of Hue, Value, and Chroma, made in solid paint for the accompanying Color Atlas", the word "measured" draws attention to the fact that the construction is rigorous. Munsell himself wrote: "Every color can be recognized, named, matched, imitated, and written by its HUE, VALUE, and CHROMA", and used these three colour attributes to denote three independent coordinates in the colour space. Munsell gave a definition of these attributes, starting from the definition given by the 'Century Dictionary':

> "*hue* – Specifically and technically, distinctive quality of coloring in an object or on a surface; the respect in which red, yellow, green, blue, etc., differ one from another; that in which colors of equal luminosity and CHROMA may differ."

> "*value* – In painting and the allied arts, relation of one object, part, or atmospheric plane of a picture to the others, with reference to light and shade, the idea of HUE being abstracted."

> "*chroma* – The degree of departure of a color sensation from that of white or gray; the intensity of distinctive hue; color intensity."

and his concise definitions are:

> "*hue* is the quality by which we distinguish one color from another, as a red from a yellow, a green, a blue, or a purple."

> "*value* is the quality by which we distinguish a light color from a dark one."

> "*chroma* is the quality by which we distinguish a strong color from a weak one."

The terms 'hue', 'value' and 'chroma' with lowercase 'h', 'v' and 'c' represent colour attributes in general sense, while with uppercase 'H', 'V' and 'C' represent the corresponding quantities in Munsell's colour-order system and atlas, where Value and Chroma are specified by numbers of the Munsell scales. Moreover:

> "*Scale* is a graded system, by reference to which the degree, intensity, or quality of a sense perception may be estimated."

In 1920 the National Bureau of Standards published Technologic Paper No. 167, which states that from "an examination of the Munsell system, a revised edition of the Atlas' and 'A Color Notation', based upon the best present day methods of measurement and specification, would be a most important contribution to the science and art of chromatics generally". Five proposals were made:

1. standardization should be made of the value scale;
2. each colour should be specified in terms of physical measurement;
3. colorimetric and photometric specifications should accompany the atlas;
4. value measurements should be made with reference to a standard white; and
5. general agreement in the nomenclature should be obtained before issuing a revised publication.

Munsell died before this report was published. Anyway the 'Munsell Research Laboratory' (Baltimore, Maryland) continued the work in line with the proposals of the National Bureau of Standards and in 1929 published a revised edition of the *Atlas*, now called the *Munsell Book of Color*, and in 1933 published two papers in the *Journal of the Optical Society of America*.[30,31] Munsell's system became extremely popular, the *de facto* reference for American colour standards.

In the 1940s, the Optical Society of America made extensive measurements, and adjusted the arrangement of Munsell colours, issuing a set of 'renotations'. After a set of four systematic works on the Munsell system[32–34], a Subcommittee of the Optical Society of America on the spacing of the Munsell colours published a preliminary report.[35] In the same year D. Nickerson published the history and the scientific applications of the Munsell system.[36] After a long time, the Munsell atlas was measured and colorimetrically specified by national American laboratories[37,38] and in 1943 the Subcommittee of the Optical Society of America published the final report[39] with an important renotation of the Munsell system. This colorimetric specification is for the standard observer CIE 1931 and the standard illuminant C. The same renotated data are used in the software attached to this book (Section 16.5.2).

Today the commercial edition of the Munsell book is for the standard observer CIE 1964 and the illuminant D65.[40]

The Munsell system was built up starting with colours represented in a colour solid with a spherical shape, and afterword by enlarging this solid in a suitable way to contain permanent colours with chroma higher that that limited by the sphere. The irregularity of the colour solid is due to the colorant limitations and to the fact that the maximum saturations of colours vary from hue to hue and are reached at different value levels. Since each attribute within the solid is presented in measured steps of equally perceptible intervals, the boundary of the colour solid reveals these variations. The colour solid assumes a shape limited by the existing colorant but is open to a modification containing a new colorant with permanent colour.

14.4.1   Munsell's Instruments

Munsell used four instruments for quantitatively defining the Hue, Value and Chroma of the colour system and checking their self-consistency:

1.   a visual daylight photometer;
2.   a Maxwell rotating disk (Section 4.3);
3.   a sphere, spinning at high speed around any axis crossing its centre, with the surface subdivided into quadrilaterals by parallel and meridian lines, to be painted with paints of suitable hues, values and chromas; and
4.   a spectroscope.

The sphere, that is the spherical part of the Munsell colour solid, is a three-dimensional generalization of the bi-dimensional disk of Maxwell, with the function of mixing suitable amounts of colours in defined ratios.

14.4.2   Chromatic Tuning Fork

The set of all colours has to be organized in a rational way; therefore, a role is required for linking the colours together. First, a grey scale perceptually equispaced has to be constructed, then the hues with value and chroma such that two colours, which are in the colour solid in symmetrical positions with respect to the grey termed N5/, produce the grey N5/ if mixed in equal parts by the rotating disk. The grey N5/ is the "neutral centre and the balancing point for all colours, that a line through this centre finds opposite colors which balance and complement each other".

Figure 14.11 reproduces the sketch drawn by Munsell to represent the balancing between ordered colour pairs. The balancing is judged by the Maxwell disk. Figure 14.12 is a picture of the sphere, on which the

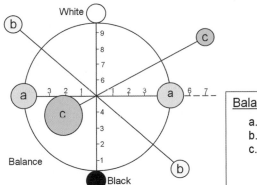

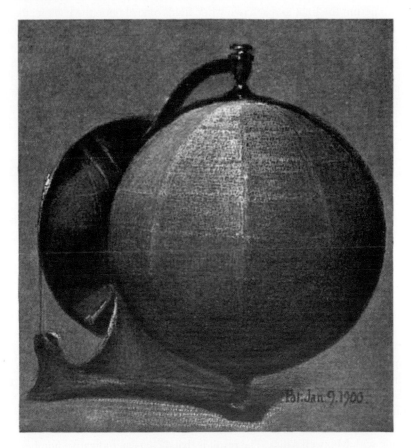

Figure 14.11 *Munsell's sketch showing the attempt at extending the centre-of-gravity rule to colour sample mixing by Maxwell's disk in a three-dimensional arrangement to obtain grey N5/. The word 'mass' represents the relative size of the sector in Maxwell's disk.*

A BALANCED COLOR SPHERE

Figure 14.12 *Munsell's coloured sphere as picture printed in the Munsell Book 'A Color Notation'. This coloured sphere is a three-dimensional rotating disk for testing that the mixing of the colours on the sphere, once the sphere is in rotation, is achromatic according to the Munsell achromatic scale.*

colours are ordered according to the rules given below. The idea of the sphere is so central in the construction of the Munsell system that the sphere of Figure 14.12 is painted on the page opposite to the frontispiece of the Munsell book 'A Color Notation'.

The exact judgements used to determine the construction of the colour samples according to the Munsell notations are not well documented; anyway, the ideas grasped from the book 'A Color Notation' are given in next sections.

14.4.3   Munsell's Value and Grey Scale

The measurement of the value is based on the use of the visual photometer. The description of this instrument, written by Munsell, explains its functioning, which helps to understand how to construct scales of values perceived as uniform:

"A photometer … measures the relative amount of light which the eye receives from any source, and so enables us to make a scale with any number of regular steps. The principle on which it acts is very simple. A rectangular box, divided by a central partition into halves, has symmetrical openings in the front walls, which permit the light to reach two white fields placed upon the back walls. If one looks in through the observation tube, both halves are seen to be exactly alike, and the white fields equally illuminated. A valve is then fitted to one of the front openings, so that the light in that half of the photometer may be gradually diminished. Its white field is thus darkened by measured degrees, and becomes black when all light is excluded by the closed valve. … One-half is thus said to be variable because of its valve, and the other side is said to be fixed. A dial connected with the valve has a hand moving over it to show how much light is admitted to the field in the variable half."

"A photometric scale of value places all colors in relation to the extremes of white and black, but cannot describe their hue or their chroma."

The first use of the photometer was to define a uniformly perceived grey scale. Consider a grey sample personally chosen with a supposed middle value of lightness. Munsell says:

"Let us now test one of these personal decisions about middle value. A sample replaces the white field – of the photometer – in the fixed half, and by means of the valve, the white field in the variable half is alternately darkened and lightened, until it matches the sample and the eye sees no difference in the two. The dial then discloses the fact that this supposedly middle value reflects only 42 per cent, of the light."

With this result it is possible to produce a scale of grey samples, denoted by $N1/$, $N2/$, $N3/$, $N4/$, $N5/$, $N6/$, $N7/$, $N8/$, $N9/$, $N10/$, where N means neutral and the numeral written to the right specifies the Value.

The photometer is used to associate each grey sample of the scale with its reflectance.

"Such a scale makes it easy to foresee the result of mixing light values with dark ones. Any two gray values in varying proportions unite to form a gray midway between them. Thus N4/ and N6/ being equally above and below the centre, unite to form N5/, as also N7/ and N3/, N8/ and N2/, or N9/ and N1/. But N9/ and N3/ unite to form N6/, which is midway between 3 and 9."

This check is made with Maxwell's disk. The result is that 42% is an indicative number. (Munsell does not consider this number an absolute but considers it a number relative to the midway grey.) This uniformly perceived grey scale is the first important part of the skeleton of the Munsell system, after the choice of the hues.

Although not clearly written, it was assumed that a coloured and a grey sample, compared by a photometer, have an equal value when the separation border is minimally distinct (Sections 6.5 and 7.3). This is the same assumption made for the definition of $V(\lambda)$.

Black and white as surface colours are ideal and cannot be used in a comparison with a photometer; therefore, they are not considered in the atlas.

A very good fitting of the luminance factor Y/Y_n of the Munsell grey scale as a function of the Munsell Value V is (Section 6.14)

$$100\left(\frac{Y}{Y_n}\right) = 1.22191V - 0.23111V^2 + 0.23951V^3 - 0.021009V^4 + 0.0008404V^5 \tag{14.2}$$

14.4.4 Munsell's Hue

The first classification of the colours is made visually and regards the distinction according to the hues.

Munsell started from five monolexemic hues that he considered as *principal hues* (the dominant wavelengths were measured by the National Bureau of Standards):

Red (R) ($\lambda_R = 612$ nm), Yellow (Y)($\lambda_Y = 585$ nm), Green (G)($\lambda_G = 508$ nm), Blue (B) ($\lambda_B = 488$ nm),

Purple (P) ($\lambda_P = 568_c$ nm)

then he added five intermediated hues:

blue – green (BG), purple – blue/ultramarine (PB), red – purple (RP), yellow – red/orange (YR), green – yellow (GY)

which

- are obtained by mixing contiguous principal hues and retain a suggestion of both and
- are opponent/complementary hues of the principal hues, in the same order: R is opponent to BG, Y to PB, etc.

In this way, Munsell decided to use a decimal system. These hues are called the 10 *major hues*, and are disposed in an equally spaced ordered sequence on the hue circle (Figures 14.13 and 14.14).

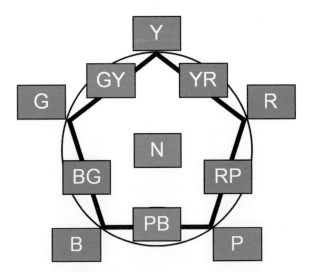

Figure 14.13 *The five principal hues of Munsell – red R, yellow Y, green G, blue B and purple P – sequentially situated on the corners of a pentagon and the corresponding complementary hues situated at the centre of the opposite sides –BG, PB, RP, YR and GY–. These hues together are the 10 major hues. N is the neutral grey.*

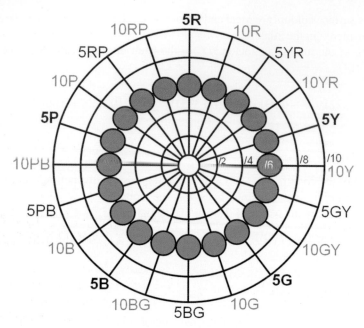

Figure 14.14 *20-colour Munsell circle. The colours in figure are from the renotated Munsell atlas and have Chroma equal to 6 and Value 5.*

Complementary hues unite in neutral grey. "But if, instead of mixing these opposite hues, we place them side by side, the eye is so stimulated by their difference that each seems to gain in strength; i.e., each enhances the other when separate, but destroys the other when mixed". (The mixing is made with the Maxwell disk.)

Munsell gave the list of pigments useful to produce paints with the major hues (Table 14.1).

Table 14.1 *Pigments used by Mansell to produce the 10 major hues.*

Colour complementary pairs	Pigments used	Chemical nature
Red	Venetian red	Calcined native earth
Blue-green	Viridian and cobalt	Chromium sesquioxide
Yellow	Raw sienna	Native earth
Purple-blue	Ultramarine	Artificial product
Green	Emerald green	Arsenate of copper
Red-purple	Purple madder	Extract of the madder plant
Blue	Cobalt	Oxide of cobalt with alumina
Yellow-red	Orange cadmium	Sulphide of cadmium
Purple	Madder and cobalt	See each pigment above
Green-yellow	Emerald green and Sienna	See each pigment above

Paints with pure pigments have various degrees of hue, value and chroma. Value and chroma can be stempered by additions of the neutral pigments (zinc white and ivory black) until each is brought to a middle value and tested on the value scale with a photometer. These paints with the major hues have to be balanced, and the balance is obtained according to the following sequence of operations:

1. define the five principal hues such that their mixing in equal amounts with the Maxwell disk produces grey N5/;
2. mix pairs of contiguous principal hues in equal parts and produce the intermediate hues;
3. check that the mixing of the five intermediate hues in equal proportions produce grey N5/;
4. by mixing the opposite pairs of hues (e.g., R-BG) (Figure 14.13), one principal and the other intermediate, quantify the difference in chroma with respect to grey N5/ and stemperate the principal hue with grey N5/ to obtain a match;
5. the 10 major hues are obtained with Value 5 and Chroma 5, denoted as R5/5, Y5/5, so on; and
6. the paints of all the major hues together, painted in appropriate spaces on the sphere equator, mixed by rotating the sphere quickly, give a visual result of grey N5/.

The original order of five principal hues is linked together by five intermediate steps (Figure 14.13). The balance is global and regards the Hues and the Value, fixed at 5, and the Chroma (although not yet defined, is fixed at 5).

This balancing of the hues with value and chroma is the second important part of the skeleton of the Munsell system, after the value scale.

Subsequently, the number of hues is grown by following the same logic. The 1929 edition of the *Book of Color* contains 20 instead of 10 hues and afterwards the number is increased to 40 (Figure 14.14).

14.4.5 Munsell's Value in Coloured Scales

This numbered scale of Values of grey samples:

> "serves not only to describe light and dark grays, but the value of colors which are at the same level in the scale. Thus R7/ (popularly called a tint of red) is neither lighter nor darker than the gray of N7/, … so that R1/, R2/, R3/, R4/, R5/, R6/, R7/, R8/, R9/, describes a regular scale of red values from black to white, while G1/, G2/, G3/, etc., is a scale of green values."

Moreover, by mixing equal amounts of a light yellow Y7/ with a dark red R3/ with the rotating disk, that is, two colours neighbouring in hue, but well removed in value, the line joining them centres at YR5/. In such a way, a link is generated among colours of different value and different hue.

14.4.6 Colour Sphere and Munsell's Colour Specification

The colour sphere is a convenient model to define chroma and illustrate the three colour qualities (hue, value, and chroma), and unites them by three measured and interconnected scales (Figures 14.15 and 14.16):

- The vertical axis of the sphere is spanned by a coordinate denoted by V, ranging between 0 and 10, and representing the Value. The north pole of the colour sphere is white, $V = 10$, and the south pole black, $V = 0$ (Figure 14.15).
- The colour Hues are spread around the equator of the sphere and represented by an angular variable H, not defined by number but by alphabetical acronyms (R, YR, Y, YG, G, BG, B, PB, P, RP) (Figures 14.13 and 14.14).

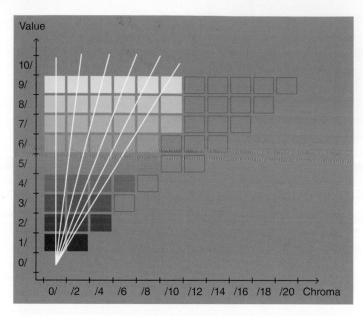

Figure 14.15   *Reproduction of a page from the Munsell Book of Colour, representing a constant Hue section spanned by Value and Chroma as orthogonal coordinates. Particularly, the constant saturation lines are drawn (white lines) to stress visually the difference between chroma and saturation (Figure 16.14).*

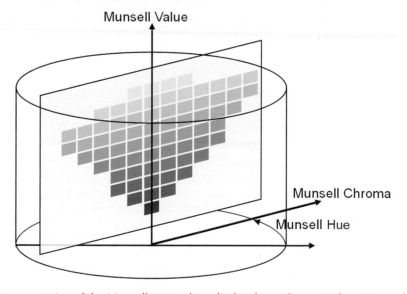

Figure 14.16   *Representation of the Munsell system by cylindrical coordinates, Value, Hue and Chroma. The Munsell sphere can be imagined as contained internally to this cylinder. The cylinder and sphere have a common Value axis.*

- The radial direction orthogonal to the Value axis V is the Chroma coordinate denoted by C. The scale of Chroma starts on the value axis, where $C = 0$ and represents an achromatic colour, and increases with the distance from this axis. The colours on a line beginning with neutral grey at the Value axis become more and more coloured until the line passes outside the sphere. The scale of the Chroma is defined by assuming that the Chroma of the balanced major hues used to define the colour circle on the equator of the sphere is $C = 5$. On this line, inward from the surface of the sphere to the neutral grey axis, the colours are greying with a loss of Chroma (Figures 14.14 and 14.15).

- The surface of the sphere is subdivided in quadrilaterals by parallel and meridian lines that are in correspondence with three coordinates, H, V and C, and a colour.
- The centre of the sphere is assumed the natural balancing point for all colours, then colour points equally removed from the centre must balance one another: thus white balances black, middle red balances middle blue-green, lighter red balances darker blue-green, etc. Every straight line through the centre indicates opposite qualities that balance one another (Figure 14.11). The colour points so found are said to be *complementary*.

Thus, the cylindrical coordinates associated with a sphere give a colour description (Figure 14.16), and each colour is specified by

$$\text{Hue Value/Chroma   (HV/C)} \tag{14.3}$$

where the Hue is represented by an alphabetical acronym, while the Value and the Chroma are numbers.

14.4.7   Munsell's Chroma

Chroma is the quality that distinguishes a strong colour from a weak one. A scale of Chroma has to be associated with any line perpendicular to the neutral axis of the sphere. On the plane of the equator, this scale is numbered 1, 2, 3, 4 and 5, from the centre to the surface of the sphere. The scale is equispaced according to visual perception. Two samples of opposite hues and an equal Value have different Chroma if, mixed in an equal ratio with the Maxwell disk, they do not produce a grey colour. The variation of the mixing ratio for obtaining grey quantifies the difference in Chroma. In this way, the scale of the Chroma is checked.

The best blue-green pigment has half the Chroma of the vermilion red pigment.

The mixture of two colours having an equal Value and Chroma and different non-contiguous Hues, for example, R and G, produces the hue Y with the same Value and lower Chroma (Figure 14.13).

14.4.8   Colour Tree

Some colours have the Chroma exceeding the numeric values defined by the sphere. A *complete* solid would contain all possible colours and a lot of colours are not contained in a sphere. The colour solid has an irregular spatial configuration, that is, a more complete model than the sphere. This configuration was called *colour tree* by Munsell.

The Chroma does not exceed the value of 16.

The scales associated with the three Munsell coordinates have been constructed in mutually independent ways and the steps on the three Munsell appearance scales are of different magnitudes: a step of 1 on the Value scale is designed to be perceptually equivalent to a step of 2 on the Chroma scale and a step of 3 on the Hue scale at Chroma 5. (These relationships between scales are between linear distances. The Hue is an angular variable and is not commensurable with the distances; therefore, it was chosen to evaluate the linear scale of the Hue on the equator of the sphere at Chroma 5.)

14.4.9   Munsell's System and CIE Chromaticity Specification

The observation of the Munsell colour atlas must be made by placing the samples on a grey background with reflected luminance equal to 20% that of equally lighted white.

It is customary to report the chromaticity of the samples of the Munsell book on the (x, y) CIE 1931 diagram and the chromaticity of samples with equal Chroma and Hue changes with the change of Value (Figure 14.17).

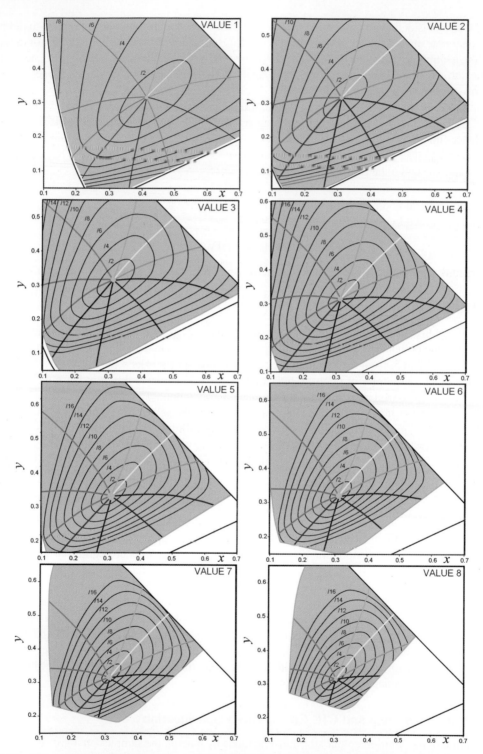

Figure 14.17 *Chromaticities of sections at a constant Value of the renotated Munsell atlas on the CIE 1931 chromaticity diagram. The constant Hue and constant Chroma lines are drawn. The constant Hue lines appear bent, proving that hue changes with decreasing purity approximately according to Abney's hue shift (Section 6.19).*

In addition, this change of chromaticity shows the difference between chroma and saturation: compare the constant Chroma loci of Figure 14.17 with the constant Sättigung loci of the DIN system of Figure 14.19 and Figure 16.14 (i.e. the chromaticity of colour samples with equal Sättigung and Farbetone does not change with the change of the Dunkelstufe).

Generally the loci of the chromaticity of the colour samples with an equal Hue (Figure 14.17) are in good agreement with empirical constant hue lines.

The 'Munsell Renotation System' is presented visually by the software described in Section 16.5.2.

14.4.10 Helmholtz-Kohlrausch's Effect and Abney's Hue Shift Phenomenon in the Munsell Atlas

Helmholtz-Kohlrausch's and Abney's phenomena are all evident in the Munsell atlas (Sections 6.15 and 6.19).

Figure 14.17 shows the constant Hue lines on the CIE 1931 chromaticity diagram that appear bent, proving that hue changes with decreasing purity approximately according to Abney's hue shift.

The Helmholtz-Kohlrausch effect is visually evident by observing the samples of the Munsell *Book of Color*. The samples of constant Munsell Value have been defined at a constant luminance factor; thus, Munsell chips of a given Hue and Value have a constant luminance factor while Chroma is changing. A visual analysis of these chips reveals that the chips with a higher Chroma appear brighter and that the magnitude of this phenomenon depends on the considered Hue and Value.

14.4.11 Munsell's Colour Atlas

The Munsell book is organized according to the Munsell order system and is presented on pages with a constant Hue (Figure 14.15), on which the Value increases from the bottom upwards and the Chroma from left to right. The intermediate colours are obtained by interpolation.

Today, the Munsell atlas presents a number of Hues higher than 10 of the major hues. Nine hues have been inserted between contiguous Hues, for which the notation is changed. The initials of the colours in the original notations have been preceded by the number 5, that is 5R, 5YR, 5Y, 5YG, 5G, 5BG, 5B, 5PB, 5P and 5RP, and the nine added intermediate Hues are 5R, 6R, 7R 8R, 9R, 10R, 1YR, 2YR, 3YR, 4YR, 5YR, etc.

Also finer subdivisions, represented by decimals, have been introduced: for example, 2.5YR, that fits between 2YR and 3YR.

The Munsell Value corresponds to the lightness and its scale, which ranges from 0 to 10, is equal with a good approximation to 1/10 of that of the CIELAB lightness L^* (Figure 6.36).

The Value is defined by heterochromatic matching with a visual photometer, based on the minimum distinct border technique; therefore, it does not take into account the Helmholtz-Kohlrausch effect and the scales are not strictly uniform.

The colours with a Munsell Chroma of less than or equal to 10 are called *weak colours*, and those with a higher Chroma are called *strong colours*.

A supplementary book is added to the Munsell book, which consists of pages with Hues inserted in between those listed above, with a sample Chroma greater than 11.

Other sets of samples arranged with the same logic and different purposes are[41]

- The *Munsell Neutral Value Scale* – a 37-step grey scale (glossy finish) with values of 0.5/ to 9.5/ and a 31-step grey scale (matte finish) with Values of 2.0/ to 9.5/.
- The *Munsell nearly neutrals collection*, whose colours have Chroma in the range 0–4 with step 0.5 and Value from 6 with pitch 0.5; the Value of 9.25 is added to this set of colours.
- The *Munsell soil color charts*.
- The *Munsell Plant Tissue Color Charts*.

14.5 DIN 6264's Colour-Order System and Atlas

DIN colour system – a three-dimensional colour system developed for the 'Deutsche Industrie Normung' (German Standardization Institute) upon which the *German standard colour chart* (DIN Farbenkarte) is based. The coordinates are Dunkelstufe *D* (darkness degree), Farbton *T* (dominant wavelength expressed on a perceptually equispaced scale) and Sättigung stufe *S* (purity on a basis yielding equal saturations).

> "*DIN color system.* Color order system developed for the Deutsche Industrie Normung (German Standardization Institute) to provide equality of visual spacing of colors in specified series, based on the attributes hue, saturation and relative darkness degree."[14]

The DIN colour system is the result of a practical compromise with the CIE 1931 system. It has been conceived for the CIE 1931 observer adapted to the illuminant C. The DIN colour system is spanned by a set of three cylindrical coordinates (*T*, *S*, *D*) (Figure 14.18), two of which, *T* and *S*, are directly readable on the CIE 1931 chromaticity diagram (Figure 14.19):

1. *T*, from *Farbton*, represents the dominant wavelength, or complementary wavelength for purple colours (approximately the hue). A set of 24 straight lines, associated with as many wavelengths, radiating from the C illuminant chromaticity (x_C, y_C), span the chromaticity diagram according to a uniform scale in the DIN system (later the illuminant D65 was used[42]). The hues on the colour circle for $S = 6$ and $D = 1$ show equal perceived colour differences. The choice of the dominant or complementary wavelength is a simplification that makes the transformation between tristimulus coordinates and DIN hue straightforward.[42,43]
2. *S*, from *Sättigung*, is the saturation spanned by a uniform scale in the DIN system; thus, also the DIN saturation of a sample may be calculated directly from its chromaticity coordinates.
3. *D*, from *Dunkelstufe*, is an inverse scale for lightness and can be called relative darkness, which is defined according to the logarithmic Fechner law.

$$D = 10 - 6.1723 \log_{10}\left[1 + 40.7\frac{Y}{Y_o}\right] \qquad (14.4)$$

where *Y* and Y_o are the luminances relative to the CIE illuminant C reflected by the colour sample and by the optimal colour (Section 14.2) with equal chromaticity, respectively ($D = 0$ for $Y = Y_o$). The loci of constant Y_o within the chromaticity diagram are shown in Figure 14.4b for standard illuminant D65 and the CIE 1931 observer.

Both Sättigung and Farbton scales were measured for a single lightness level, and the scales were then extended under the assumption that lines of constant Sättigung and the lines of constant Farbton are the same for all lightness levels.[42,43] The Dunkelstufe scale was determined for neutral colours. All these choices are convenient approximations that simplify the calculation of DIN coordinates.

The DIN atlas is organized in pages at defined *T* and on each of them the samples are arranged from left to right with increasing *S* and from top to bottom with *D* increasing from 0 to 10.

The constraint of *T* on lines of the CIE 1931 diagram implies that the pages of the atlas are not strictly with constant hue.

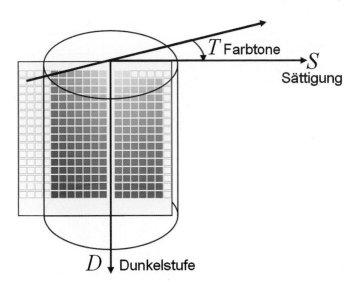

Figure 14.18 *Representation of the DIN system by cylindrical coordinates, Dunkelstufe, Farbton and Sättigung. The samples on vertical columns have equal chromaticity and equal saturation.*

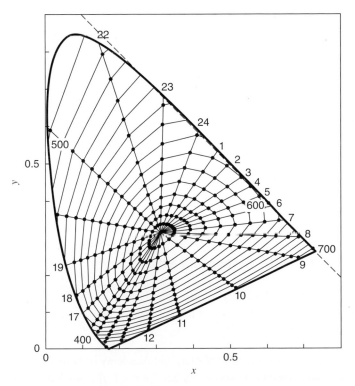

Figure 14.19 *CIE 1931 chromaticity diagram with the straight lines of constant hue T (equal dominant wavelength) and curves of constant Sättigung S of the DIN system. These lines are the same at any Dunkelstufe.*

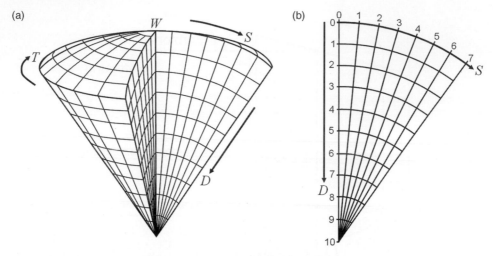

Figure 14.20   *(a) Schematic sphere-sector colour solid of the DIN colour system with coordinates (T, D, S) and the white point W. (b) Plane of constant hue taken from a schematic colour solid corresponding to the DIN colour-order system (a). A set of 24 such planes surrounds the common neutral axis with an equal angular separation of 15° and constitutes the whole schematic colour solid of the DIN system.*

Because S is independent of D, the distance, measured in *just noticeable difference* (jnd), between contiguous samples of equal D decreases with the increase in D, that is, the system has no uniform scales. The profound difference between chroma and saturation appears from a comparison of the DIN with the Munsell system (Figures 16.14 and 14.15).

For the DIN colour System, a sphere segment seemed to give a good description of the interrelations between the three colour coordinates (Figures 14.20a and 14.20b). The centre of the sphere is the ideal black with $D = 10$, and the radius is measured by $(10 - D)$, the equivalent of the lightness, which ends at optimal colours $D = 0$. The two spherical angles are represented by the degree of longitude T and the saturation S, that is, the degree of latitude measured from the north pole. The axis of rotation is the grey axis. Vertical planes containing the grey axis are planes of constant hue. Lines emerging from the ideal black are lines of constant saturation. The unit of hue angle is 15° and the unit of saturation angle has been chosen to be 5°.

The DIN system is presented visually in the software described in Section 16.5.3.

14.6   OSA-UCS's Colour-Order System and Atlas

"OSA-UCS color system – Optical Society of America Uniform Color Scales color order system based on equality of visual spacing, which uses the opponent-color scales $\pm L_{OSA}$ (Lightness), $\pm j$ (yellowness-blueness), and $\pm g$ (greenness-redness)."[14]

In 1947, the National Research Council of the National Academy of Sciences asked D. B. Judd at the US National Bureau of Standards whether he could suggest "some badly needed research program that would have nothing to do with armaments and war". Judd suggested a study of uniform perceived colour scales. A uniform colour scale is a linear sequence of colours that differ by perceptually equal amounts. The same year the Optical Society of America (OSA) formed a Committee on Uniform Scales, chaired by Judd. The project was ambitious and pioneering[44–61]:

"Color differences consisting solely of chromaticity differences without luminance differences are rare; there is no known way of evaluating the noticeability of combined luminance and chromaticity differences. The experimental clarification of this problem is one of the major programs yet to be completed in the field of colorimetric research."[44]

"The intention of the Committee was to make colors that correspond to about 500 lattice points of a regular-rhombohedral crystal in an *Euclidean color space* in which equal distances between points correspond to equal visual differences between the corresponding colors."

But the budget forced to study:

"a limited sample, consisting of 43 colors (made in the form of 5-cm-hexagonal, matte-finish, painted ceramic tiles, all having approximately the same luminous reflectance (30%) in the CIE D65 Standard daylight, and having chromaticities indicated to be equally different and triangularly arranged in a color space derived, tentatively, by modifying the Munsell renotations in ways indicated by ellipses of standard deviations of color matching)."[46]

In 1972 Judd died, before the end of the work, and he was certain that no colour space could have a Euclidean metric. The work was completed only in 1974 and, as the work failed in finding a Euclidean metric, it was generally considered a failure. Moreover, the OSA-UCS atlas is represented by sections (cleavage planes) of a three-dimensional lattice, whose use is not practical. This opinion was wrong because the content of knowledge in this system is very high. The concept of colour space is completely new because the space is spanned by a net of geodesic lines:

The *geodesic line* between two given colors is the line spanned by the lowest number of one-jnd steps.

"The OSA committee specified 424 colors that can be arranged in linear arrays (scales), that the committee intended would exhibit equally perceptible differences between all pairs of adjacent colors (that is, geodesic lines)."[60]

The OSA-UCS samples constitute "more than 400 uniformly spaced color scales" that "appear as linear arrays that pass in six different directions through each sample". The observers were asked only about the scales and "no particular color designation (as blue or red or green ...) was needed", that is, no explicit use of the concepts of colour appearance is made."[57]

"Every color would have 12 equally different nearest neighbours and would appear in six entirely different series (scales) of equally different colors."[46]

"The concept of uniform color scales is best served by a uniform lattice of points in a space that permits each sample of a scale to be a member of other scales in a highly coordinated manner."[53]

"The regular-rhombohedral system of color sampling adopted by the Optical Society Committee on Uniform Color Scales provides the maximum possible number and variety of uniform color scales and exhibits the maximum possible variety of relationships among colors."[46]

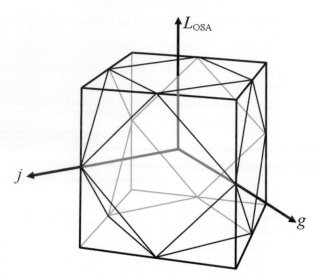

Figure 14.21 *Cube-octahedron lattice used for the spatial positioning of the OSA-UCS colour samples.*

The role played by the US National Bureau of Standards has been important and decisive in the construction of the system, but the literature presented in the books here considered does not describe in detail the construction to fully understand the OSA-UCS system:

"Five sets of those tiles were made and circulated among members of the committee, each of whom compared pairs of the tiles and recorded his estimates of the relative visual color differences, according to an experimental plan suggested by experts (?!) in the U. S. National Bureau of Standards (NBS). The tiles were surrounded and separated 3-4 mm from each other by a 30% reflectance gray card. The tiles and surround were illuminated with daylight (D65) of at least 500 lm/m^2."

"The judgment data were analyzed by use of appropriate psychometric techniques (?!), at the NBS."[46] Anyway, the scaling judgements used to define the space should be entirely those of colour difference.

The luminance factor of the surround produces a crispening effect (Section 4.7), which the OSA-UCS system takes into account.

The samples are arranged in a three-dimensional cube-octahedron lattice (Figure 14.21). The lattice consists of a succession of parallel planes, on which there are square lattices mutually shifted so that the points of each lattice, if projected on contiguous planes, are in the centre of the square patterns. In this way each sample is located at the centre of a cell consisting of 12 samples.

The Committee on Uniform Scales proposed a set of three orthogonal coordinates (L_{OSA}, j, g) to specify the colour samples arranged according the cube-octahedron lattice, which represent *lightness*, *yellowness* and *greenness*, respectively (g stands for 'green' and j for 'jeune', yellow in French).

The OSA-UCS system is specified for the CIE 1964 observers and D65 illuminant.

14.6.1 OSA-UCS's Lightness

The lightness L_{OSA} is positive if the luminance factor of the colour sample is greater that the luminance factor of the grey surround on which the colour sample is observed, zero if equal and negative if lower.

The lightness L_{OSA} takes into account the Helmholtz-Kohlrausch effect, according to which, in addition to luminance, chromaticity also contributes to the brightness (Figure 6.38b).

The lightness L_{OSA} takes into account the crispening effect, according to which the brightness of a sample is affected by the brightness of the grey surround, whose percentage luminance factor $Y_{10} = 30$ (Figure 6.36).

All this is described by a formula given by the Committee according to a modification of Semmelroth's formula[62,63]:

$$L_{OSA} = \left[5.9\left(Y_0^{1/3} - \frac{2}{3} + c\right) - 14.4\right]\frac{1}{\sqrt{2}} \tag{14.5}$$

where

$$c = \begin{cases} +0.042|Y_0 - 30|^{1/3} & \text{for} \ (Y_0 - 30) > 0 \\ -0.042|Y_0 - 30|^{1/3} & \text{for} \ (Y_0 - 30) \leq 0 \end{cases} \quad \text{with} \ Y_0 = Y_{10}F \tag{14.6}$$

$$F = 4.4934x_{10}^2 + 4.3034y_{10}^2 - 4.2760x_{10}y_{10} - 1.3744x_{10} - 2.5643y_{10} + 1.8103$$

and Y_{10} is a percentage luminance factor and (x_{10}, y_{10}) the chromaticity of the colour sample considered. This formula takes into account

- The crispening effect produced by the surround, whose percentage luminance factor $Y_{10} = 30$ (Figure 6.36).
- The effect of the Helmholtz-Kohlrausch by the factor F. Figure 6.38b shows the lines of equal F on the chromaticity diagram: for each value of F an ellipse exists centred at $(x_{10} = 0.3859, y_{10} = 0.4897)$, with a ratio between the axes equal to 1.4138 and with the major axis forming with the axis x_{10} an angle of 43.73°. The value of F for the D65 chromaticity is 1, that is, the illuminant D65 is assumed achromatic with $g = 0$ and $j = 0$.

14.6.2 OSA-UCS's (g, j) Coordinates

The sample coordinates of the system belong to the following ranges:

$$-7 \leq L_{OSA} \leq 5, \quad -6 \leq j \leq 12, \quad -10 \leq g \leq 6 \tag{14.7}$$

1. The coordinates g and j are zero for grey colours.
2. Positive values of j, with nearly zero values of g, indicate yellowish or brownish colours.
3. Negative values of j, with nearly zero values of g, indicate blues.
4. Positive values of g, with nearly zero values of j, indicate greens.
5. Negative values of g, with nearly zero values of j, indicate purples.

The coordinates g and j are obtainable from the tristimulus space by a two-step transformation proposed by the Committee:

1. First, a linear transformation is made in the tristimulus space:

$$\begin{pmatrix} R_{10} \\ G_{10} \\ B_{10} \end{pmatrix} = \begin{pmatrix} 0.7990 & 0.4194 & -0.1648 \\ -0.4493 & 1.3265 & 0.0927 \\ -0.1149 & 0.3394 & 0.7170 \end{pmatrix} \begin{pmatrix} X_{10} \\ Y_{10} \\ Z_{10} \end{pmatrix} \tag{14.8}$$

associated with the three basic vectors, with the following chromaticities:

$$(x_{10,R} = 0.747, y_{10,R} = 0.253)$$
$$(x_{10,G} = 2.920, y_{10,G} = -4.540)$$
$$(x_{10,B} = 0.171, y_{10,B} = 0.000)$$

(14.9)

– not too much different from the confusion points, with the exclusion of $(x_{10,G}, y_{10,G})$ – and the neutral vector with the chromaticity of the illuminant D65.

2. The second transformation is a combination of a non-linear transformation, made according to the Stevens power law, and a linear mixing with the aim of representing the perceived opponencies:

$$j = C\left(1.7R_{10}^{1/3} + 8G_{10}^{1/3} - 9.7B_{10}^{1/3}\right)$$
$$g = C\left(-13.7R_{10}^{1/3} + 17.7G_{10}^{1/3} - 4B_{10}^{1/3}\right)$$
$$C = 1 + \frac{c}{5.9\left(Y_0^{1/3} - 2/3\right)}$$

(14.10)

With regard to the spectrum locus, MacAdam[46] wrote: "The values of g and j can be computed for spectrum colours. However, because for any wavelength the values depend on Y_{10}, a specific value of Y_{10} must be assumed at each wavelength, in order to define the spectrum locus. However, admitting that the procedure is inconsistent, we show" in Figure 14.22 the spectrum locus for $Y_{10} = 30$. It is evident that the wavelength discrimination is represented incorrectly. Anyway, the Committee formulae are good for a colour specification in the region where the empirical data exist.

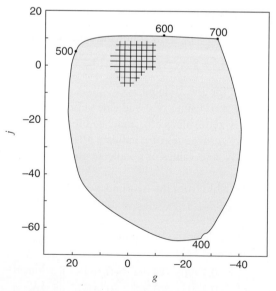

Figure 14.22   *OSA-UCS system spectrum locus at lightness $L_{OSA} = 0$, that is, $Y_{10} = 30$, as produced by the equations (14.5)–(14.10).*

14.6.3 OSA-UCS's Colour Difference Formula

"From colorimetric specifications of the tiles, formulas have been devised that define a three-dimensional Euclidean space in which the distances between the points that represent the colors correlate reasonably well with the reported magnitudes of the perceived color differences."[46]

MacAdam wrote:

"The committee strongly recommended that its formulae not be used to construct a formula for small colour differences, because the color differences used by the committee were all about 10 times just-noticeable differences."[61] Moreover MacAdam wrote without particular reason, "… Strictly interpreted, the formulas for L_{OSA}, j and g should not be used for evaluating large color differences either,…."[61]

14.6.4 OSA-UCS's Metrics

The points of the cube-octahedron lattice are crossed by six lines, on which the colours are monospaced. Each set of three lines of these belongs to a plane and the steps of an equal number of jnd in the six directions of the three lines draw a regular hexagon on this plane (Figure 14.23). The metrics are Euclidean if the sides of the hexagon measured in jnd have length equal to that of the half hexagon diagonals. This is not verified and therefore the metrics are not Euclidean. Particularly, in the case that the hexagon is on a constant lightness plane, Judd called this phenomenon the *hue superimportance effect*. This failure is an achievement, because now it is clear that any search for the Euclidean metrics is without success.

The project of a search for the Euclidean metric to express colour differences is clearly unrealistic, because if the formula has to represent the physiological process of colour comparison done by the network of neurons in the visual system, it is certain that the neuron system will not perform the mathematical operation of the

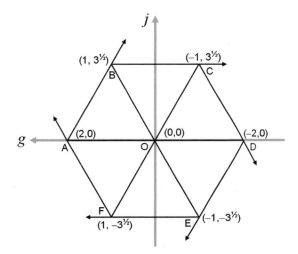

Figure 14.23 *Geometrical intuitive representation of the non-Euclidean metrics in OSA-UCS space. On a constant lightness plane a hexagon represents six colours, A, B, C, D, E and F, which are judged to be equally different from O. The differences between pairs of adjacent colours on the border, A – B, B – C, C – D, D – E, F – F and F – A are judged to be greater than the differences between the six colours and the centre O. The length of arrows on the border of the hexagon indicates the length that the sides of the hexagon should have for having Euclidean metrics.*[61]

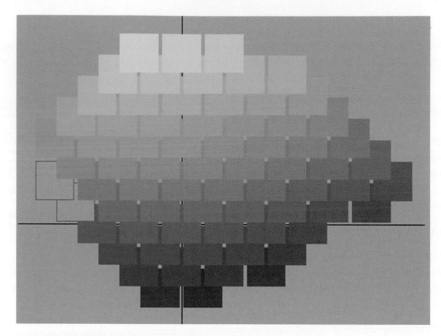

Figure 14.24   *Colour samples belonging to the cleavage $L_{OSA} - j = -4$. (Two rectangles are empty because their colours are not reproducible with an sRGB monitor.)*

Euclidean formula. The mathematical operations with this complexity are made by a conscious part of the brain. The evaluation of colour differences, when it will be understood, will be represented by a simple mathematical formula.

Although the OSA-UCS system has limitations that have been highlighted, its contribution of knowledge is very important, because it is based on a new concept of space spanned by a net of geodesic lines.

The OSA-UCS space was appreciated for gradations shown by combinations of samples belonging to cleavages not orthogonal to the coordinate axes. (Figure 14.24).

The OSA-UCS system is presented visually in the software described in Section 16.5.4.

14.7   NCS's Colour-Order System and Atlas

"*Natural Color System* – color-order system based on resemblances or colors to up to four of six 'elementary' colors – red, yellow, green, blue, black, and white –, in which the attributes of the colors are hue, chromaticness, and blackness." (Section 4.4)[14]

"NCS, Natural Colour System, is a way to describe and order, by means of psychometric methods, the characteristic relationships between all possible colour percepts of the 'surface mode'. By surface colour is meant, according to Katz (Section 6.1.3), a colour percept that appears as it were the surface layer of an object without appearing transparent or self-luminant. NCS is based on Hering's phenomenological analysis of the characteristic relationships of colour percepts and on the postulate of the six elementary colour sensations, the *NCS elementary colours*. These have no resemblance to each other – while all other colour percepts show more or less characteristic resemblance to them. These varying resemblances, which are possible to quantify, are called *NCS elementary attributes*."[66]

The locution *Natural Colour System* (NCS) was first used by Tryggve Johansson[64] and is a metric colour system based on Ewald Hering's opponent colour theory (Section 4.4). The word 'natural' is part of the name NCS because NCS deals with how humans experience and describe colour sensations. Today NCS is a proprietary perceptual colour model published by *NCS Colour AB* of Stockholm, Sweden.

The purpose of the NCS is to specify the appearance of an individual colour in terms of its resemblance attributes, not to define the difference between two colours; therefore, colour-difference scales are not equal everywhere in the NCS colour space.

14.7.1 NCS's Axioms

The axioms and criteria of NCS's system[65–68] are as follows:

1. Colours exist only as perceptions, and consequently normal observers are the only true colour measuring instruments. Observers are equal to each other.
2. Observers perceive and can describe the colour perception in the same way, provided that they have the same definitions of the colour describing words.
3. There exist pure perception qualities of six *elementary colours*, which cannot be defined perceptually in terms of other colours (e.g., the elementary red colour is only red, not a red with a little bit of yellow or a reddish-blue). The four chromatic elementary colours are yellow (Y), red (R), blue (B) and green (G), and the two non-chromatic elementary colours are white (W) and black (S) (Figure 14.25).
4. All other colours can be described in terms of their *degree of visual resemblance* to the elementary colours. These resemblances are the *elementary attributes* and at the same time the subjective *perceptive parameters* of *blackness* (s), *whiteness* (w), *yellowness* (y), *redness* (r), *blueness* (b) and *greenness* (g). These perceptive parameters are specified on a percentage scale.
5. Since the elementary attributes are based on resemblances, each perceptive parameter is graphically symbolized by a bipolar diagram (Figure 4.6), where the endpoints represent elementary colours, which are imaginary in the sense that it is not possible to produce colour stimuli that certainty can give rise to these pure colour percepts. Two elementary colour pairs, red-green and yellow-blue, have mutually opponent qualities (according to Hering and general empiricism), that is, at the same time, one observer perceives either redness or greenness, or yellowness or blueness.
6. A colour percept F can be specified with at most four perceptive parameters:

$$F = w + s + (y \text{ or } b) + (r \text{ or } g) = 100 \tag{14.11}$$

7. All the other perceptual colours are composite perceptions that can be defined in terms of the six perceptive parameters. This implies that the observers have the same reference system associated with these elementary colours and the appearances of a colour can be readily specified by NCS notation.

Figure 14.25 *NCS unique hues (NCS – Natural Colour System®©, property of and used on licence from NCS Colour AB, Stockholm, 2012. Reproduced with permission from NCS Colour AB © 2015.)*

14.7.2   NCS's Hue, Chromaticness and Nuance

The definitions of the other colour attributes follow from these axioms (Section 14.7.1).

> "*Hue* – attribute of a chromatic colour percept expressing the degree to which it resembles one or two chromatic elementary colours.
>   Note – colours with the same relation between the two chromatic elementary attributes are defined as having the same hue (constant hue)."[4]

The attribute NCS *hue*, defined in relation to the two contiguous chromatic perceptive parameters entering the percept *F*, written as indices, is specified by

$$\phi_{yr} = 100\frac{r}{y+r}, \phi_{rb} = 100\frac{b}{r+b}, \phi_{bg} = 100\frac{g}{b+g}, \phi_{gy} = 100\frac{y}{g+y} \tag{14.12}$$

and this numeric value enters the denotation surrounded by the letters for the two elementary colours between which the hue is situated.

> "*Chromaticness* – the degree of resemblance of a colour percept to the full chromatic colour of the same hue."[4]

According to this definition, chromaticness is derived from equation (14.11) as the sum of the chromatic elementary attributes of the colour:

$$c = (y \text{ or } b) + (r \text{ or } g) \tag{14.13}$$

Then the definition of a colour circle in the NCS follows (Figure 14.26):

> "*Colour circle (NCS)*, a circle which illustrates the relationship between chromaticness and hue."[4]

The complete colour specification in the NCS is given by the hue and the nuance, which is defined as follows (Figure 14.27):

> "*Nuance (in general)* – two-dimensional attribute that distinguishes amongst colours of the same hue.
>   Note – Colours of different hues but of the same nuance are also referred to as being 'equivalent' or 'corresponding'."[4]

> "*Nuance (NCS)* – two-dimensional colour attribute which (irrespective of hue) expresses the relationship between whiteness, blackness and chromaticness in a colour percept.
>   NOTE – colours of different hues but of the same nuance are also referred to as being 'equivalent' or 'corresponding'."[4]

From this definition, the NCS nuance is specified by the relationship between whiteness, blackness and chromaticness (*w/s/c*). Since $w + s + c = 100$, only two of the three values are sufficient for the specification of the nuance.

Notice that the lightness is not mentioned as a constituent factor of the NCS system. Lightness is a comparative magnitude, while whiteness and blackness are qualitatively descriptive attributes. There is no inner reference of absolute lightness, unless thereby one means white, and then the quantity is whiteness.

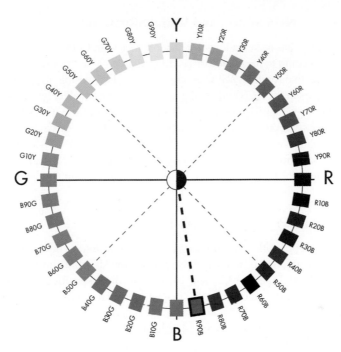

Figure 14.26 NCS hue circle. The hue R90B is highlighted. (NCS – Natural Colour System®©, property of and used on licence from NCS Colour AB, Stockholm, 2012. Reproduced with permission from NCS Colour AB © 2015.)

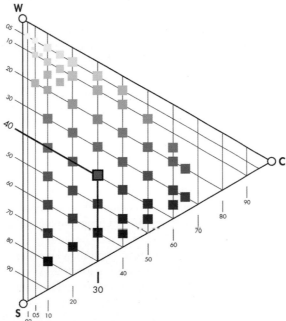

Figure 14.27 Triangular arrangement of the coordinates (w, s, c) specifying the nuances of equal hue R90B. The nuance 4030 is highlighted, where 40 is the blackness s and 30 is the chromaticness c. (NCS – Natural Colour System®©, property of and used on licence from NCS Colour AB, Stockholm 2012. Reproduced with permission from NCS Colour AB © 2015.)

14.7.3   Production of the NCS System and Visual Situation

The production of the NCS atlas has been made in strict accordance with Hering's theory, which says that a colour percept can be described according to the degree of *visual resemblance* to the six postulated basic colour sensations and that these six subjective resemblance phenomena were also possible to quantify. These resemblances are called elementary attributes in the NCS and are the attributes of which all colour percepts are composed.

The experimental conditions were performed in a light booth (Section 13.2) with simulated daylight at 5400 K and approximate illuminance of 1000 lx. The walls inside the booth were grey with a percentage luminance factor $Y = 54\%$ and the colour samples were judged on a white surround with $Y = 78\%$. The vision was orthogonal at a distance of 40 cm and the illumination of the colour samples was at 45°. The colour samples were specified colorimetrically according to the CIE 1931 standard observer and illuminant C in geometry (di/8°).

14.7.4   Psychophysics and Psychometrics for NCS

A preliminary work was the identification of samples with *elementary hues*. The task for observers was "Choose the sample you consider only red, that is, neither bluish nor yellowish", and similarly for the other hues.

Scaling experiments followed. Figures 14.26 and 14.27 show the disposition of the colour samples considered in the NCS Colour Circle and Colour Triangle, respectively. The steps of this scaling are:

1. For judgements of *whiteness* and *blackness*. 14 achromatic samples between the 'best' white and the 'best' black were considered.
    *   Question: "To what degree do you consider the colour of this sample to characteristically resemble, or remind you of, your own conception of the non-chromatic colours pure black (s) and pure white (w)?"
    *   Answer required was that the percentage specification of the blackness (s) and whiteness (w) was such that $s + w = 100$.
2. For assessment of *nuance* (chromaticness, blackness and whiteness). A total of 360 samples were considered: 6–8 samples within each of 24 preliminarily defined hues for the scales between white and closest to the pure chromatic colour (with $s \approx 10$) and an equal number between black and closest to the pure chromatic colour (with $w \approx 10$).
    *   Question: "To what degree do you consider the colour of this sample to characteristically resemble, or remind you of your own conception of pure black (s), pure white (w), and the most chromatic colour (c) you can imagine?"
    *   Answer required was that the percentage specification of the blackness (s), whiteness (w) and chromaticness (c) was such that $s + w + c = 100$.
3. For assessment of *hue*. A total of 72 samples were considered: 24 colour samples around the hue circle in three different preliminary nuance positions ($s/c \approx 1010$, $s/c \approx 7010$ and $s/c \approx 3050$).
    *   Question: "To what degree do you consider the colour of this sample to characteristically resemble, or remind you of your own conception of any of the chromatic colours pure yellow (Y), pure red (R), pure blue (B), and pure green (G)?"
    *   Answer required was that the percentage specification of the yellowness (y), the redness (r), the blueness (b) and the greenness (g) was such that $y + r + b + g = 100$. As expected, not one of the samples was judged to be simultaneously yellowish and bluish or reddish and greenish, and the result was either $y + r = 100$ or $r + b = 100$ or $b + g = 100$ or $g + y = 100$.

A statistical analysis on the psychophysical results and a colorimetric analysis on physical samples were made for all the samples considered. From all this follow the NCS system and the atlas.

Figure 14.28 shows the sequence of two bipolar diagrams for specifying colours by their resemblances, first by specifying grey by its resemblances to white and black, and then by the resemblances of the colour

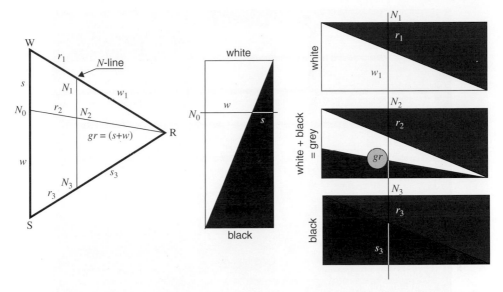

Figure 14.28 *The coordinates w, s and c used for specifying a nuance are three barycentric coordinates, but with a different meaning with respect to the barycentric coordinates in the Ostwald system. Here the barycentric coordinates do not refer to the content of black, white and full colour stimuli represented by Maxwell's disk, but to the content of resemblance with black, white and full colour, and designated by blackness s, whiteness w and chromaticness c. What is symbolized in the triangle (left) can be described in four bipolar diagrams (right). The caption of the original text[66]: "In the three colours N_1, N_2 and N_3, the redness is equal. This is illustrated by the equilateral triangle in which their symbols are situated on a line parallel with W-S axis for achromatic colours. The numeric value for the redness of the three colours may be written $r_1/(w_1+r_1)= r_2/[(w + s) + r_2]= r_3/(s_3 + r_3)$. If the elementary colour R is replaced by the general expression for the pure chromatic colour C, designating the hue, the distance to the N-line from the achromatic line designates the chromaticness c. This may be expressed in percent as $[100 \times c/(w + s + c)]$". (NCS – Natural Colour System®©, property of and used on licence from NCS Colour AB, Stockholm 2012. Reproduced with permission from NCS Colour AB © 2015.)*

sample with grey and full colour. An equal specification can be made by using barycentric coordinates, as those used in the Ostwald system and in the chromaticity diagrams, with the significant difference that here the coordinates represent the resemblance degree and not the intensities of the colour stimuli.

Figure 14.29 shows three scales of colours extracted from the NCS atlas in order to better grasp the perceptual meaning of whiteness, blackness and chromaticness.

14.7.5 Luminance Factor and NCS's Whiteness Scale

The NCS whiteness of the achromatic samples, estimated in a psychometrical way, can be related to the percentage luminance factor Y, and these points are well represented by the following fitting equation:

$$s = \frac{8736}{Y + 56} - 56 \text{ or } w = 100 - s = \frac{156Y}{Y + 56} \tag{14.14}$$

represented in Figure 6.36. The number 56 corresponds well with the mean percentage luminance factor of the entire adaptation field in the light booth used in the experimental stage, where the booth walls have a luminance factor equal to 54% and the surround to 78%.

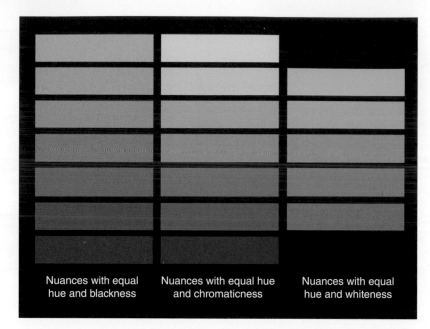

Figure 14.29   *Three scales of colours extracted from the NCS atlas in order to better grasp the perceptual meaning of whiteness, blackness and chromaticness also on black background. The three scales are at equal hue and equal whiteness, equal hue and equal blackness, and equal hue and equal chromaticness, respectively. (NCS – Natural Colour System®©, property of and used on licence from NCS Colour AB, Stockholm 2012. Reproduced with permission from NCS Colour AB © 2015.)*

In the triangle diagrams, loci with constant luminance factor and increasing chromaticness are well represented by straight lines converging on a point P outside the triangle. The position of P depends on the hue (Figure 14.30).[66]

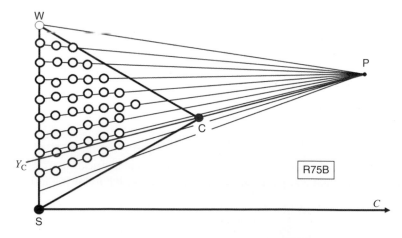

Figure 14.30   *In the triangular arrangement of the NCS colours at a defined hue, the colour samples with an equal luminance factor belong to straight lines, which intersect at a point P outside the triangle. Point P is different for the different hues.[66] (NCS – Natural Colour System®©, property of and used on licence from NCS Colour AB, Stockholm, 2012. Reproduced with permission from NCS Colour AB © 2015.)*

14.7.6 NCS's Atlas

The NCS atlas is organized into 40 pages arranged radially as shown in Figures 14.26 and 14.27. Each page is constant in hue and the samples are arranged according to the barycentric coordinates (resemblance degree expressed in %) of an equilateral triangle whose vertices are the white W on top, black S bottom and the pure colour C (devoid of white and black) right (Sections 4.3, 6.13.8 and 10.2). The samples placed on straight lines parallel to the sides of the triangle are at constant white content w, black content s and pure colour c, respectively, according to a resemblance scale (Figure 14.27). For each sample $w + s + c = 100$. By convention, the whiteness w is omitted. The complete specification of a colour sample is given by *sc-hue*, that in the case of the highlighted colour in Figure 14.27 is 4030-R90B.

The three-dimensional arrangement of the samples, although based on a resemblance scale, is not in uniform scales, because the uniform scales are based on the judgement of colour difference.

The NCS atlas is defined for the illuminant C.

From a comparison between the Munsell and NCS atlases[68,69], a general agreement appears with regard to the constant hue planes. The four elementary NCS colours, red, yellow, green and blue, correspond to the Munsell 5R, 5Y, 5G and 7.5B.

The NCS atlas is presented visually in the software described in Section 16.5.5.

References

1. CIE Publication, ILV CIE S 017/E:2011, *ILV: International Lighting Vocabulary*. Commission Internationale de l'Éclairage, Vienna (2011). Available at: www.eilv.cie.co.at/ (accessed 18 June 2015)
2. Kuehni RG and Schwarz A, *Color Ordered: A Survey of Color Systems from Antiquity to the Present*, Oxford University Press (2007).
3. Forsius SA, *Physica*, Codex Holmiensis D.6. Royal Library, Stockholm (1611)
4. NCS – Natural Color System. Available at: http://www.ncscolour.com/en/ncs/ (accessed 18 June 2015).
5. Billmeyer FW Jr, Survey of color order systems, *Color Res. App.*, **12**(4), 173–186 (1987).
6. RAL DESIGN System. Available at: http://www.ral-shop.com/information-for-users/the-ral-design-system/ (accessed 18 June 2015).
7. *McGraw-Hill Dictionary of Scientific & Technical Terms*, 6E, The McGraw-Hill Companies, Inc. (2003).
8. Pantone Color Formula guide. Available at: www pantone.com; www.xrite.com (accessed 18 June 2015).
9. Schrödinger E, Theorie der Pigmente von grösster Leuchtkraft, *Ann. Phys.*, **4**, 62 603–622 (1920). Available at: http://www.iscc.org/pdf/SchroePigments.pdf (accessed 18 June 2015).
10. Ostwald W, Neue Forschungen zur farbenlehre, *Physikalische Zeitschrift*, **XVII**, 322–332 (1916). Available at: http://www.iscc.org/pdf/OstwaldFarbenlehre.pdf (accessed 18 June 2015).
11. West G and Brill MH, Conditions under which Schrödinger object colors are optimal, *J. Opt. Soc. Am.*, **73** (9) 1223–1225 (1983).
12. Centore P and Zonohedral A, Approach to optimal colours (April 2011). Available at: https://sites.fas.harvard.edu/~cs278/papers/zone.pdf (accessed 18 June 2015).
13. Pointer MR, The gamut of real surface colours, *Color Res. Appl.*, **5**, 145–155 (1989).
14. ASTM E284 - 13b, *Standard Terminology of Appearance* (published on 11/01/2013 by ASTM International).
15. Ostwald W, *Die Farbenlehre*, Verlag Unesma G.m.b.H., Leipzig (1917) (1923).
16. Ostwald W, *Die Farbenfibel*, Verlag Unesma G.m.b.H., Leipzig (1916).
17. Ostwald W, *Der Farbatlas*, Verlag Unesma G.m.b.H., Leipzig (1917).
18. Ostwald W, *Physikalische Farbenlehre*, Verlag Unesma G.m.b.H., Leipzig (1919).

19. Ostwald W, *Der Farbkörper und seine Anwendung zur Herstellung farbiger Harmonien*, 12 Tafeln und Text, Text mit 9 Figuren, Verlag Unesma G.m.b.H. Leipzig (1919). Available at: http://home.arcor.de/dyck-berlin/Farbkoerper/Farbkoerper.htm (accessed 18 June 2015).

20. Ostwald W, *Farbkunde*, translated as *Colour Science* by J.Scott Taylor, Winsor & Newton Ltd, London, Volume I (1931), and Volume II (1933).

21. Bond ME and Nickerson D, Color-order systems, Munsell and Ostwald, *J. Opt. Soc. Am.*, **32**, 709–719 (1942).

22. Judd Ed, Symposium symposium on the Ostwald Color System, foreword, *J. Opt. Soc. Am.*, **34** (7), 353–354 (1944).

23. Zeishold H, Philosophy of the Ostwald Color System, *J. Opt. Soc. Am.*, **34** (7), 355–353 (1944).

24. Foss CE, Nickerson D and Granville WC, Analysis of the Ostwald Color System, *J. Opt. Soc. Am.*, **34** (7), 361–381 (1944).

25. Granville WC, Colorimetric specification of the *color harmony manual* from spectrophotometric measurements, *J. Opt. Soc. Am.*, **34** (7), 382–395 (1944).

26. Birren F, Application of the Ostwald color system to the design of consumer goods, *J. Opt. Soc. Am.*, **34** (7), 396–399 (1944).

27. Koenderink JJ and van Doorn AJ, Mensurating the colour circle: Ostwald's principle of internal symmetry, *Perception*, 29 ECVP Abstract Supplement (2000).

28. Koenderink J, *Color for the Sciences*, (New edition, 20 August 2010), MIT Press.

29. Koenderink JJ, Color atlas theory, *J. Opt. Soc. Am. A*, **4** (7), 1314–1321 (1987).

30. Munsell AEO, Sloan LL and Godlove IH, Neutral value scales. I. Munsell neutral scale, *J. Opt. Soc. Am.*, **23**, 394–411 (1933).

31. Godlove IH, Neutral value scales. II. Comparison of results and equations describing neutral scale, *J. Opt. Soc. Am.*, **23**, 419–425 (1933).

32. Judd DB, Foreword of the Munsell color system, *J. Opt. Soc. Am.*, **30**, 574 (1940).

33. Tyler JE and Hardy AC, An analysis of the original Munsell color system, *J. Opt. Soc. Am.*, **30**, 587–590 (1940).

34. Glenn JJ and Killian JT, Trichromatic analysis of the *Munsell Book of Color*, *J. Opt. Soc. Am.*, **30**, 609–616 (1940).

35. Newhall S, Preliminary report of the OSA Subcommittee on the spacing of the Munsell colors, *J. Opt. Soc. Am.*, **30**, 617–645 (1940).

36. Nickerson D, History of the Munsell color system and its scientific application, *J. Opt. Soc. Am.*, **30**, 575–586 (1940).

37. Kelly KL and Gibson KS, Tristimulus specification of the *Munsell Book of Color* from spectrophotometric measurements, *J. Opt. Soc. Am.*, **33**, 355–376 (1943).

38. Granville WC, Nickerson D and Foss CE, Trichromatic specifications for intermediate and special colors of the Munsell system, *J. Opt. Soc. Am.*, **33**, 376–385 (1943).

39. Sidney M, Newhall SM, Nickerson D and Judd DB, Final report of the O.S.A. subcommittee on the spacing of the Munsell colors, *J. Opt. Soc. Am.*, **33**, 385–418 (1943).

40. Munsell® Color Standard, X-Rite, Incorporated 4300 44th St. SE Grand Rapids, Minnesota 49512, USA.

41. Joy Turner Luke, Munsell™ Book of Color, nearly neutrals™ collection, *Color Res. Appl.*, **16**, 394–396 (1991).

42. Richter M and Witt K, The story of the DIN color system, *Color Res. Appl.*, **11**, 138–145 (1986).

43. Richter M, The official German standard color chart, *J. Opt. Soc. Am.*, **45**, 223–226 (1955).

44. Optical Society of America, Committee on colorimetry, *The Science of Color*, Optical Society of America, Washington; Crowell, New York (1953).

45. MacAdam D, Nonlinear relations of psychometric scales values to chromaticity differences, *J. Opt. Soc. Am.*, **53**, (1963).

46. MacAdam D, Uniform color scales, *J. Opt. Soc. Am.*, **64**, 1691–1702 (1974)

47. Nickerson D, Uniform color scales: Munsell conversion of OSA committee selection, *J. Opt. Soc. Am.*, **64**, 205–207 (1974).

48. Wyszecki G, Uniform color scales: CIE 1964 $U * V* W *$ conversion of OSA committee selection, *J. Opt. Soc. Am.*, **65**, 456–460 (1975).

49. Nickerson D, History of the OSA committee on uniform color scales, *Optics News*, Winter 1977, 8–17 (1977).

50. MacAdam D, Colorimetric data for samples of OSA uniform color scales, *J. Opt. Soc. Am.*, **68**, 121–130 (1978).

51. Davidson HR, Preparation of the OSA uniform color scales committee samples, *J. Opt. Soc. Am.*, **68**, 1141–1142 (1978).

52. Nickerson D, Munsell renotations for samples of OSA uniform color scales, *J. Opt. Soc. Am.*, **68**, 1143–1147 (1978).

53. Foss CE, Space lattice used to sample the color space of the Committee on Uniform Color Scales of the Optical Society of America, *J. Opt. Soc. Am.*, **68**, 1616–1619 (1978).

54. Nickerson D, Optical Society of America (OSA) uniform color scales samples, *Leonardo*, **12**, 206–212. (1979).

55. Davidson HR, Formulation for the OSA uniform color scales committee samples, *Color Res. Appl.*, **6**, 38–52 (1981).

56. Billmeyer FW Jr, On the geometry of the OSA uniform scales committee space, *Color Res. Appl.*, **6**, 34–37 (1981).

57. Nickerson D, OSA uniform color samples: a unique set, *Color Res. Appl.*, **6**, 7–33 (1981).

58. Billmeyer FW Jr and Taylor JM, Multidimensional scaling of selected samples from the Optical Society of America uniform color scales, *Color Res. Appl.*, **13**, 85–98 (1988).

59. MacAdam D and Man TM, Three-dimensional scaling of the uniform color scales of the Optical Society of America, *J. Opt. Soc. Am.*, *A*, **6**, 1218–138 (1989).

60. MacAdam D, Redetermination of colors for uniform scales, *J. Opt. Soc. Am.*, *A*, **7**, 113–115 (1990).

61. MacAdam DL, *Color Measurement*, Springer, Berlin (1985).

62. Semmelroth CC, Prediction of lightness and brightness on different backgrounds, *J. Opt. Soc. Am.*, **60**, 1685–1689 (1970).

63. Semmelroth CC, Adjustment of the Munsell Vaue and $W*$ scale to uniform lightness steps for various background reflectances, *Appl. Optics*, **10**, 14–18 (1971).

64. Johansson T, *Färg*, Lindtors Bokfortag AB, Stockholm (1937).

65. Hård A and Sivik L, NCS – Natural Color System; a Swedish standard for color notation, *Color Res, Appl.*, **6**, 129–138 (1981)

66. Hård A, Sivik L and Tonnquist G, NCS, Natural Color System – from concepts to research and applications, Part I, *Color Res. Appl.*, **21**, 180–205 (1996).

67. Hård A, Sivik L and Tonnquist G, NCS, Natural Color System – from concepts to research and applications. Part II, *Color Res. Appl.*, **21**, 206–220 (1996).

68. Judd DB and Nickerson D, Relation between Munsell and Swedish Natural Color System scales, *J. Opt. Soc. Am.* **65**, 85–90 (1975).

69. Billmeyer FW Jr and Bencuya AK, Interrelation of the Natural Color System and the Munsell color order system, *Color Res. Appl.*, **12**, 243–255 (1987).

15

Additive Colour Synthesis in Images

15.1   Introduction

References[1-4] are a general bibliography for this chapter.

Colour reproduction is one of the most important applications of colorimetry and among the different colour reproductions an important distinction has to be made between imaging and non-imaging colour reproduction.

Generally, in non-imaging colour reproduction, for example, in painting restoration, reflectance is to be reproduced in order to avoid any metamerism failure. In this case the reproduction concerns physics and the colour reproduction is made in a process of subtractive colour mixing, that is, the coloured surface subtracts part of the illuminating light in a spectrally dependent way. This process of subtractive colour mixing is physical and is considered in Section 3.9 dedicated to turbid media. The cases of structural colour – for example, iridescent colours, pearlescent colours, prismatic colours, interference colours, etc. – the physics of the phenomena involved renders the various cases so different that a generalised treatment for colour reproduction is almost impossible and every case has to be considered individually.

Colorimetry has a central role in image science – that is, the production, duplication, analysis, processing and visualization of images – and particularly in image-colour reproduction – for example, in movies (projection on a screen), television screens, computer monitors, halftone printing, ... –. Following RWG Hunt,[1] the colour reproduction is called *colorimetric* if the original and the reproduction have equal chromaticities and equal relative luminances. In this case the reproduction, if it is to be seen in reflection as in printed images, depends on the illuminant and a metamerism failure can occur at the change in illuminant.

Colorimetry enter the deep structure of images in breaking up the colour of each elementary area of the picture into suitable selected colours which are re-composed in an additive mixing adapted to the different reproduction media. The mechanisms of synthesis of colour are:

- *Time sequence additive synthesis*, obtained by mixing the different selected lights in various proportions in a time sequence. An example is given by the digital image projectors, termed Digital Light Processor - Digital Micromirror Device (LDP-DMD).

Standard Colorimetry: Definitions, Algorithms and Software, First Edition. Claudio Oleari.
© 2016 John Wiley & Sons, Ltd. Published 2016 by John Wiley & Sons, Ltd.

- *Mosaic-structure synthesis*, obtained by differently coloured dots, whose radiation is spatially integrated in the visual system of the observer. This integration can be considered a process of additive mixture. Examples are television/computer monitors, halftone printing.

The colour reproductions in additive and mosaic-structure syntheses are almost always metameric. Illuminant metamerism does not exist in self-luminous monitors, but the right illumination for halftone prints is important for a correct colour perception.

In this chapter, the mosaic-structure additive synthesis, that is the common technique used in electronic imaging and dye halftone printing, is considered.

This book deals with psychophysical and psychometric colorimetry, defined on colour-matching and colour-difference matching in a controlled visual situation, outside the context of typical daily vision of coloured objects. It is also known that colour appearance depends on the visual situation. Generally the visual context in which an image is viewed is different from the one in which the scene, reproduced in the image itself, is observed in reality. Moreover, generally in imaging no colorimetric quantity is reproduced with accuracy. The chromaticities and the relative luminances are such as to ensure equality of appearance and realistic reproduction. All this requires a new colorimetry, termed 'colour-appearance colorimetry'. The colour-appearance chapter is complex and rapidly changing, where research is in progress and the debate is lively. This chapter is not part of this book and the following treatise is strictly colorimetric.

15.2   Video Colour Image

The colour synthesis in television and computer monitors is additive with a mosaic structure and takes place on the retina of the observer, where the images of the different parts of the mosaic are overlapped. The monitor screen is made up of a mosaic of coloured light sources, that emit three types of lights, one red, one green and one blue – RGB –, the intensity of which is chosen according to the colour light that the element of the mosaic should emit. These three colours should constitute an RGB reference frame. The almost triangular shape of the chromaticity diagram requires that the choice of reference colours is RGB to get the widest range of colours by the additive mixing of only three coloured lights. The technology to use more than three kinds of light sources is in progress with the aim of having a wider gamut of colours. The minimum elements of the mosaic are also the minimum elements of the image that the screen shows and are termed pixels (or *pel*) – pixel means *picture element* –:

> "*Pixel* – the smallest element that is capable of generating the full functionality of a display."[4]
>   Therefore the pixel is a computer concept, which belongs

- to the software category, because its information content is a set of data that specifies the colour of the smallest detail of the image and
- to the hardware category, because it is the element of the screen emitting the colour light specified by the pixel, and therefore termed pixel again.

In the RGB reference frame of the tristimulus space the colours of an image are limited to the region of tristimulus space with the tristimulus values $0 \leq R, G, B \leq 1$, therefore the gamut of reproducible colours is limited.

The colorimetry of today's television and computer monitors was conceived at the origin of the television system. The starting point was the monitor, that was a *cathode-ray tube* (CRT), in which, according to the more widely used scheme, the three coloured light sources are obtained from three kinds of phosphors arranged in a mosaic on the monitor screen. Appropriate masks allow three electron beams to hit the three

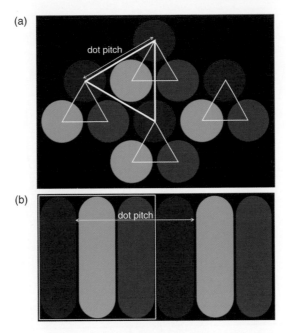

Figure 15.1 *Two of the most common types of pixel structure on a monitor screen: (a) 'delta' and (b) 'in-line'. Particularly, the in-line structure is used on LCD screens. 'Dot pitch' is the distance between two contiguous sources of equal colour and represents the spatial resolution of the monitors and defines the minimum distance of vision to avoid the perception of the internal structure of the pixel.*

different regions of the screen coated with the three different types of phosphors according to a mosaic structure. The most common kinds of mosaic are two (Figure 15.1):

- *delta shadow mask*, in which the mosaic is made by a set of three dots of the three reference colours – red, green and blue – arranged at the vertices of a Δ and
- *in-line shadow mask*, in which the screen is constituted by a succession of the three reference colours arranged in vertical line segments.

The three types of phosphors, hit by the three electron beams of varying intensity, are activated and then decay with their typical emission of light (Figure 15.2). The phenomenon is *cathode-luminescence*.

Every television standard is characterized

1. by the chromaticity of the light emitted by the three types of phosphors;
2. by the chromaticity of white radiation – obtained as a proper sum of the lights emitted by the three phosphors, corresponding to suitable intensities of the electron beams –; and
3. by the encoding of the video signal intended to drive the CRT.

Today the CRT monitor is considered obsolete and other types of monitors are available on the market – for example, liquid crystal display (LCD), plasma, Light Emission Diode (LED), Organic Light Emission Diode (OLED) and Digital Light Processor – Digital Micromirror Device (LDP-DMD) projector –. The subdivision of the screen into pixels within the three RGB sources persists, then here is considered what is also true for these monitors and projectors. Figure 15.2 shows the spectra of the reference lights and white light of an LCD monitor.

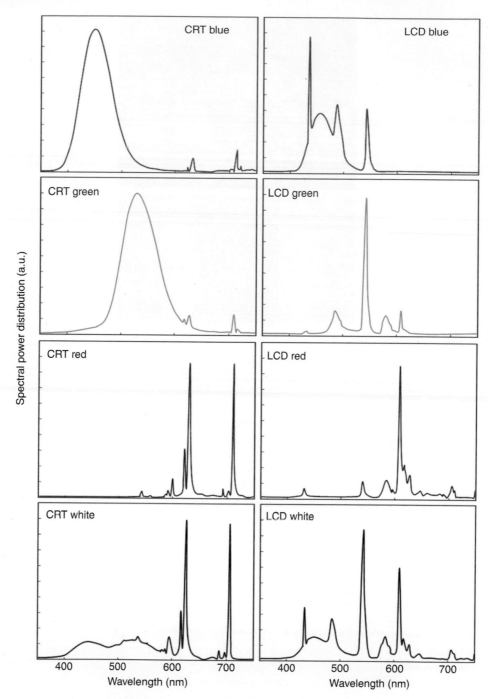

Figure 15.2   *On the left, emission spectra of the red, green and blue phosphors of common CRT monitors. On the right, emission spectra of the red, green and blue lights of common LCD monitors. White light is obtained as a mixing of the three coloured lights weighted in a suitable way to obtain the desired colour temperature.*

Table 15.1 *CIE 1931 chromaticity of the reference colours and white in the systems of television and digital imaging [the chromaticity of the white C is (x = 0.3101, y = 0.3162), and of D65 (x = 0.3127, y = 0.3290)].*

'Colour Space'	Colour Gamut	White	Chromaticities of the reference colours					
			x_R	y_R	x_G	y_G	x_B	y_B
HDTV (ITU-R BT.709 (1990)), sRGB (IEC 61966-2-1 (1999))	CRT	D65	0.64	0.33	0.30	0.60	0.15	0.06
NTSC (1953) (FCC1953, ITU-R BT.470 System M)	CRT	C	0.67	0.33	0.21	0.71	0.14	0.08
NTSC (1987) (SMPTE RP 145 'SMPTE C', SMPTE 170M)	CRT	D65	0.63	0.34	0.31	0.595	0.155	0.07
PAL/SECAM (1970) (EBU Tech. 3213, ITU-R BT.470 System B, G)	CRT	D65	0.64	0.33	0.29	0.60	0.15	0.06
Apple RGB	CRT	D65	0.625	0.34	0.28	0.595	0.155	0.07
ADOBE RGB 98	CRT	D65	0.64	0.33	0.21	0.71	0.15	0.06
ADOBE Wide Gamut RGB	Wide	D50	0.735	0.265	0.115	0.826	0.157	0.018
scRGB	Unlimited (signed)	D65	0.64	0.33	0.30	0.60	0.15	0.06
ROMM RGB	Wide	D65	0.7347	0.2653	0.1596	0.8404	0.0366	0.0001

Acronyms: ATSC, Advanced Television Systems Committee; CCIR, Comité consultatif international pour la radio, Consultative Committee on International Radio or International Radio Consultative Committee; CRT, cathode ray tube; DCI, digital cinema initiatives; EBU, European Broadcasting Union; HDTV, high definition television; IEC, International Electrotechnical Commission; ITU, International Telecommunications Union; NTSC, National Television System Committee; PAL, phase alternating line; PIMA, Photographic and Imaging Manufacturers Association, INC.; ROMM RGB, Reference Output Medium Metric RGB; SECAM, Système Électronique pour Couleur avec Mèmoire, or Sequential Couleur Avec Memoire; SMPTE, Society of Motion Picture and Television Engineers.

15.2.1 RGB Colorimetry

Colorimetry of television systems and computer monitors is usually defined for the CIE 1931 standard observer. The chromaticities of the reference stimuli, of the white of standard television systems and of the most widely used digital image systems are listed in Table 15.1. These reference stimuli should be the same stimuli emitted by the three phosphors of the monitor. The following three vectors, defined in the XYZ CIE 1931 reference frame (Figure 15.3), are the basis of the RGB reference frame

$$\hat{\mathbf{R}} = (X_r, Y_r, Z_r) = c_r(x_r, y_r, z_r) \text{ with } c_r = X_r + Y_r + Z_r$$
$$\hat{\mathbf{G}} = (X_g, Y_g, Z_g) = c_g(x_g, y_g, z_g) \text{ with } c_g = X_g + Y_g + Z_g \qquad (15.1)$$
$$\hat{\mathbf{B}} = (X_b, Y_b, Z_b) = c_b(x_b, y_b, z_b) \text{ with } c_b = X_b + Y_b + Z_b$$

whose length is chosen so that their sum is equal to the white stimulus (C or D65, according to the standard)

$$\mathbf{W} = \hat{\mathbf{R}} + \hat{\mathbf{G}} + \hat{\mathbf{B}} = (X_n, Y_n = 100, Z_n) \qquad (15.2)$$

(sometimes it is placed with $Y_n = 1$ instead of $Y_n = 100$, as in the case of the reference frame YIQ described below).

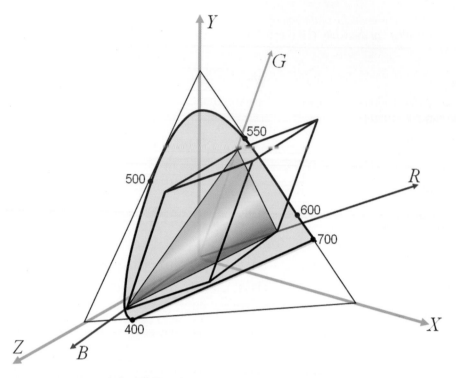

Figure 15.3   *Perspective view of the CIE 1931 tristimulus space in the XYZ reference frame with the unity plane of the chromaticity diagram and the parallelogram of the colours obtainable by mixing the RGB reference colours.*

The modulation of the intensities of the three stimuli **R**, **G** and **B** between 0 and 1, gives all the colours inside to a parallelogram in tristimulus space (Figure 15.3), that is a cube in the RGB reference frame (Figure 15.4).

Devices for reproduction of images – monitors, projectors – can reproduce only colour stimuli with the three non-negative tristimulus values in the RGB reference frame. This choice is a limitation, because only a part of the colour space can be considered, and, at the same time, a strong point, because a defined number of digits for a smaller space gives a more accurate colour specification.

Here the American television system (NTSC), proposed by the 'National Television Systems Committee' in 1951, is considered. The transformations between the XYZ and RGB reference frames are (Section 6.13.4)

$$
\begin{pmatrix} R \\ G \\ B \end{pmatrix} = \begin{pmatrix} 1.9099 & -0.5324 & -0.2882 \\ -0.9846 & 1.9991 & -0.0283 \\ 0.0583 & -0.1184 & 0.8979 \end{pmatrix} \begin{pmatrix} X \\ Y \\ Z \end{pmatrix}, \begin{pmatrix} X \\ Y \\ Z \end{pmatrix} = \begin{pmatrix} 0.6069 & 0.1735 & 0.2003 \\ 0.2989 & 0.5866 & 0.1144 \\ 0.0000 & 0.0661 & 1.1157 \end{pmatrix} \begin{pmatrix} R \\ G \\ B \end{pmatrix} \quad (15.3)
$$

obtained for reference RGB colour stimuli

$$
\begin{aligned}
\hat{\mathbf{R}} &= (X_r = 0.6069, Y_r = 0.2989, Z_r = 0.0000) \\
\hat{\mathbf{G}} &= (X_g = 0.1735, Y_g = 0.5866, Z_g = 0.0661) \\
\hat{\mathbf{B}} &= (X_b = 0.2003, Y_b = 0.1144, Z_b = 1.1157)
\end{aligned} \quad (15.4)
$$

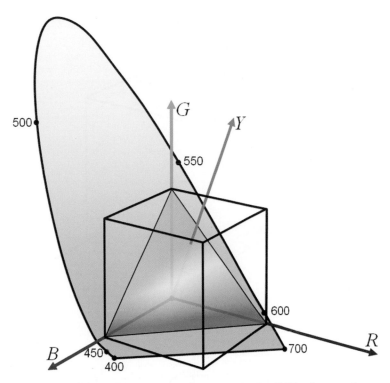

Figure 15.4 *Perspective view of the CIE 1931 tristimulus space in the RGB reference frame with the unity plane of the chromaticity diagram with the cube of the colours obtainable by mixing the RGB reference colours. The projection of the tristimulus vectors in the Y direction gives the luminance factor of the stimulus.*

with the condition that the unit vectors of the RGB reference frame have sum equal to the white stimulus C with luminance factor normalized to 1 (it corresponds to approximately 100–200 cd/m^2 in a CRT).

For colour management, we should consider the possible palettes at constant luminance, which are bounded by the intersection of the constant luminance planes with the cube (Figures 15.5 and 15.6). In particular, it is observed that the chromaticity palette begins to shrink with increasing luminance over a certain value until it is reduced to a single point at the stimulus white **W**.

The linear transformations between the XYZ and RGB reference frames considered here are calculated with the software presented in 16.8.6.

15.2.2 Video Signal and γ Correction

The luminance L_v emitted by the CRT monitor is in a non-linear relationship value of the electric signal E [volt] which drives the luminance, as would be desirable, but is in an exponential relationship (Figure 15.7)

$$\left(\frac{L_v}{L_{v,\max}}\right) = \left(\frac{E}{E_{\max}}\right)^{\gamma} \tag{15.5}$$

where $L_{v,\max}$ and $E_{\max}$ are the maximum values of L_v end of E, respectively, and γ is a positive number in the range 2-3.

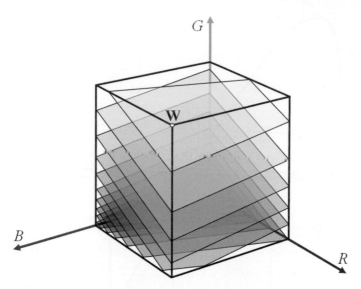

Figure 15.5 *Cube of the colours in the RGB reference frame of the CIE 1931 tristimulus space highlighting the planes of the stimuli with equal luminance. The colour palettes at different luminance are the intersection of these planes with the cube. The palettes have different size and shape at different luminances.*

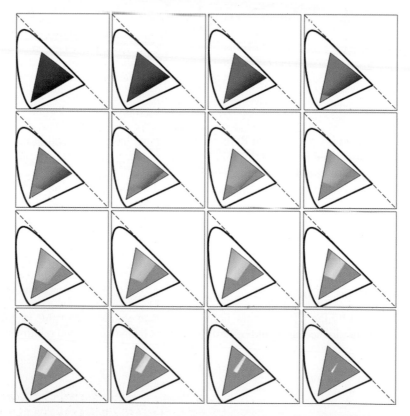

Figure 15.6 *Set of CIE 1931 chromaticity diagrams with the chromaticities of the palettes available at different luminance levels. All the chromaticities are possible only at low luminance and the palette is triangular, then it becomes quadrilateral and at the maximum luminance is a point, when the colour is white.*

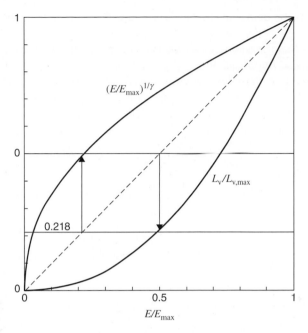

Figure 15.7 *Curve y = x^γ interpolating the normalized measurements of luminance ($L_v/L_{v,max}$) (black line) as a function of the normalized control signals (E/E_{max}) of the monitor. The upper curve (red line) represents the normalized signal subjected to correction to counteract the non-linear luminous emission of the phosphors (15.6).*

When television was conceived in the thirties of the twentieth century, the high cost of a television monitor with a circuit, which makes the linear relationship between the video signal and the luminance emitted, was avoided by a video-signal distortion before its broadcasting, known as *gamma correction* (Figure 15.7), and the number $\gamma = 2.2$ became a standard number. The 'corrected' signal E_M is

$$\left(\frac{E_M}{E_{M,max}}\right) = \left(\frac{E}{E_{max}}\right)^{1/\gamma} \tag{15.6}$$

In this way, the gamma-correction circuit of the broadcaster replaces the circuits of the single monitors with great economic savings. But since then, the evolution of television systems (colour, high definition, …) and the computerized management of the colour on the monitor (in video graphic stations, stations for ophthalmology diagnostics, …) has to deal with this choice, which often arises as a complication. For the right use of colour in a computer system it is necessary

- to know the chromaticity of the $\hat{R}$, $\hat{G}$, $\hat{B}$ and W stimuli of the monitor;
- to know the numbers γ_r, γ_g and γ_b related to the three reference colours of the monitor;
- to create the correspondence between (E_R, E_G, E_B), with which the videographic card works and the colour (R, G, B) made by the monitor, that, in the case of a card operating at 8 bits, is

$$E_R = 255(R/R_{max})^{1/\gamma_r}, \; E_G = 255(G/G_{max})^{1/\gamma_g}, \; E_B = 255(B/B_{max})^{1/\gamma_b} \tag{15.7}$$

where $(R_{max}, G_{max}, B_{max})$ is the colour of the monitor obtained for the maximum values of the signals that drive the electronic guns.

From all this, it appears that the additivity, which is colorimetrically significant for tristimulus values (R, G, B), is no longer meaningful for the quantities (E_R, E_G, E_B).

In practice, the number γ of the television monitor is 2.8 ± 0.3 and is different from the standard gamma correction $\gamma = 2.2$. This means that the image on the monitor is modified by a gamma $(2.8/2.2) = 1.273$ and therefore with increased contrast. This choice has two reasons: the image on the monitor is best read in the presence of diffused ambient light and also seems to be more acceptable by an average observer.

The colours obtainable by additive synthesis in a trichromatic RGB monitor are visually produced by the software presented in Section 16.10.1.

The relationship between luminance signal, given by Equation (15.5), is simplified if compared to the real one. An equation that best describes this relationship, termed *CRT transfer function*, is the following

$$\frac{L_v}{L_{v,max}} = \left[GAIN\left(\frac{E}{E_{max}}\right) + OFFSET \right]^{\gamma} \tag{15.8}$$

where *GAIN* and *OFFSET* are parameters operating on the brightness and contrast controls of the monitor, which satisfy the relation

$$GAIN + OFFSET = 1 \tag{15.9}$$

as appears for $E = E_{max}$ and $L_v = L_{v,max}$. The value of the *GAIN* is usually in the range of 1.0–1.1. Each phosphorus has its own *GAIN* and *OFFSET* parameters, which are distinguished by a subscript – 'r', 'g' and 'b', respectively –.

If the video-graphics card operates with 8-bit per channel, the correspondence between the colour (R, G, B) made by the monitor and the set of voltages (E_R, E_G, E_B), which operates the video graphics card, is

$$E_R = 255 \left\{ 1 + \left[\left(\frac{R}{R_{max}}\right)^{\frac{1}{\gamma_r}} - 1 \right] \frac{1}{GAIN_R} \right\}$$

$$E_G = 255 \left\{ 1 + \left[\left(\frac{G}{G_{max}}\right)^{\frac{1}{\gamma_g}} - 1 \right] \frac{1}{GAIN_G} \right\} \tag{15.10}$$

$$E_B = 255 \left\{ 1 + \left[\left(\frac{B}{B_{max}}\right)^{\frac{1}{\gamma_b}} - 1 \right] \frac{1}{GAIN_B} \right\}$$

In dealing with the CRT monitor, usually the CIE 1931 standard observer is considered, although the systematic error present in its definition below 460 nm leads to an incorrect specification of the luminance of the blue colours.

The software accompanying the book allows a complete characterization of the monitor in use (preferably a CRT) and this is presented in Sections 16.2.1 and 16.2.2.

Display technologies more modern than the CRT (LCD, Plasma, etc.) have a different native transfer function to that of CRT. Many modern monitors incorporate signal remapping that mimics the CRT transfer function and therefore the Equation (15.8) can be used as a good approximation.

15.2.3 Tristimulus Space and YIQ Reference Frame

The colour CRT is controlled by three signals E_R, E_G and E_B, which are in one to one correspondence with (R,G,B) tristimulus values of the colour to be displayed.

In 1938, Georges Valensi invented and patented a method for transmitting colour images suitable for reception by both colour and black-and-white television. All current television standards (NTSC, SECAN, PAL) and today's digital standards implement this idea of broadcasting by a composite video signal consisting of *luma* and *chrominance*.

- *Composite video signal* is the combination in only one signal of the video information required to recreate a colour picture, as well as line and frame synchronization pulses.
- *Luma* is the signal used to convey the brightness information of a picture (the 'black-and-white' or achromatic portion of the picture).
- *Chrominance* is the signal used to convey the colour information of the picture, which is usually represented as two colour-difference components in approximate correspondence with the perceived colour opponencies.

The corresponding electric signals are:

- E_Y, called luma or luminance signal, represents the luminance factor Y and is suitable for driving a black & white monitor;
- E_I and E_Q, called chrominance or signals of colour difference, contain the colour information.

The three signals E_Y, E_I and E_Q represent a set of three tristimulus values (Y, I, Q) defined in tristimulus space.

The present discussion ignores the gamma correction, that applies $\gamma = 1$, because any gamma value, 1 excluded, destroys the vector nature of the tristimulus space. This choice is made for a clearer understanding of the following explanation. In practice the tristimulus values have a power γ and are denoted as follows

$$R' = R^{1/\gamma}, \; G' = G^{1/\gamma}, \; B' = B^{1/\gamma} \tag{15.11}$$

and any quantity, that is a function of (R',G',B'), has the index ' ' '.

In this RGB reference frame with some approximation the three reference stimuli are assumed to be like spectral lights, therefore the relative luminance associated with the C white stimulus, represented by the vector $(\hat{\mathbf{R}} + \hat{\mathbf{G}} + \hat{\mathbf{B}})$, is $Y_r + Y_g + Y_b = 1$, where $Y_r \cong V(\lambda_r)$, $Y_g \cong V(\lambda_g)$ and $Y_b \cong V(\lambda_b)$. The luminance is relative, being compared to its maximum value, and the relative luminance of the stimulus (R, G, B) is defined by

$$Y = RV(\lambda_r) + GV(\lambda_g) + BV(\lambda_b) = R0.299 + G0.587 + B0.114 \tag{15.12}$$

that is the first of the three signals.

The other two signals, I and Q, are obtained from $(R - Y)$ and $(B - Y)$ by two scale factors, $(1.14)^{-1} \cong 0.8772$ and $(2.03)^{-1} \cong 0.4926$, and a rotation of 33°:

$$I = (\cos 33°)0.8772(R - Y) - (\sin 33°)0.4926(B - Y)$$
$$Q = (\sin 33°)0.8772(R - Y) - (\cos 33°)0.4926(B - Y) \tag{15.13}$$

The transformation between the coordinates RGB and YIQ is linear and writeable in matrix form

$$
\begin{pmatrix} Y \\ I \\ Q \end{pmatrix} =
\begin{pmatrix} 0.299 & 0.587 & 0.114 \\ 0.596 & -0.274 & -0.322 \\ 0.211 & -0.523 & 0.312 \end{pmatrix}
\begin{pmatrix} R \\ G \\ B \end{pmatrix},\
\begin{pmatrix} R \\ G \\ B \end{pmatrix} =
\begin{pmatrix} 1.0000 & 0.9563 & 0.6210 \\ 1.0000 & -0.2721 & -0.6474 \\ 1.0000 & -1.1070 & 1.70064 \end{pmatrix}
\begin{pmatrix} Y \\ I \\ Q \end{pmatrix} \tag{15.14}
$$

The Y's in the reference frames XYZ and YIQ could generate confusion

- in the XYZ reference frame, the axis associated to the Y component corresponds to non-physical stimuli and the Y component is the relative luminance;
- in the YIQ reference frame, the axis associated with the Y component corresponds to physical stimuli with the chromaticity of white C and the Y component is still the relative luminance.

By using transformations (15.3) and (15.14), it is possible to define on the (x, y) CIE 1931 diagram a system of barycentric coordinates in proportion to the tristimulus values (Y, I, Q) and referred to the YIQ triangle (Figure 15.8), whose vertices are associated with the chromaticities of the vectors $(Y \neq 0, I = 0, Q = 0)$, $(Y = 0, I \neq 0, Q = 0)$ and $(Y = 0, I = 0, Q \neq 0)$, respectively. It is observed that:

- the points $I = (x = -0.331, y = 0)$ and $Q = (x = 0.246, y = 0)$ are on the Alychne and represent non-physical colour stimuli with zero luminance;
- only the point $Y = (x = 0.3101, y = 0.3162)$ is within the triangle RGB and with the chromaticity of white C;
- the lines YI and YQ divide the diagram into four parts, in which, considering the barycentric coordinates referred to the triangle YIQ and associated with the vector (Y, I, Q), the Y coordinate is always positive, while I and Q assume both positive and negative values [in Figure 15.8 the sign of the I component is apparently reversed because in a spatial view of the XYZ system the vectors $(Y = 0, I \neq 0, Q = 0)$ intersect the plane of the chromaticity diagram $X + Y + Z = 1$ in the point I for the negative values of the I component];
- the points on the straight lines radiating from the point Y have defined ratio (I/Q), then a polar coordinate system centred on Y can be defined with the radial variable corresponding to the saturation and the angular variable to the hue.

In broadcast television, signals E_I and E_Q modulate the amplitude of two signals (I, Q) having the frequency of 4:43 MHz and their phase difference equal to $\pi/2$, that is, they are placed *in phase* and in *Quadrature* with respect to a reference oscillator and the signals are termed I and Q, respectively. The sum of these two signals originates a new signal having a phase shift with respect to the reference oscillator that represents the hue and is a function of the amplitudes of the signals E_I and E_Q. The main difference between NTSC and PAL systems is in the direction

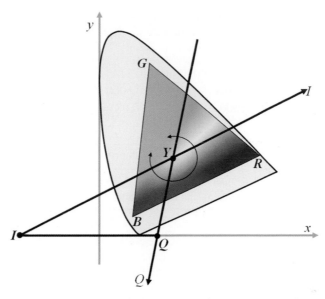

Figure 15.8   *CIE 1931 chromaticity diagram with triangles **RGB** and **YIQ**. Two distinct systems of barycentric coordinates are associated with these two triangles. The coordinates I and Q are on the I and Q axes. The angle centred at the **Y** point can be used to represent the hue.*

of rotation associated with the phase shift, which is fixed in NTSC while in PAL it is with *Phase Alternating Line*. This name describes the way that the phase of the video signal, that is part of the colour information, is reversed with each image line. This phase reversion has the effect that phase error, produced in the analogical transmission of the signal, display colour pairs that are shifted but on average correct. This correction is made at the expense of vertical frame colour resolution.

There are other systems of colour coding. In television the most important systems, in addition to the YIQ of the North American standard NTSC, already described, are

1.   the YUV, used in the European analogue PAL, with the RGB reference given in table 15.1

$$
\begin{pmatrix} Y \\ U \\ V \end{pmatrix} = \begin{pmatrix} 0.299 & 0.587 & 0.114 \\ -0.14713 & -0.28886 & 0.436 \\ 0.615 & -0.51499 & -0.10001 \end{pmatrix} \begin{pmatrix} R \\ G \\ B \end{pmatrix}, \quad \begin{pmatrix} R \\ G \\ B \end{pmatrix} = \begin{pmatrix} 1 & 0 & 1.13983 \\ 1 & -0.39465 & -0.58060 \\ 1 & 2.03211 & 0 \end{pmatrix} \begin{pmatrix} Y \\ U \\ V \end{pmatrix} \quad (15.15)
$$

where

$$
Y = 0.299R + 0.587G + 0.114B
$$
$$
U = \frac{0.436}{1 - 0.114}(B - Y), \quad V = \frac{0.615}{1 - 0.299}(R - Y) \quad (15.16)
$$

On the plane (U, V) a polar coordinate system centred on Y can be defined with the radial variable corresponding to the saturation and the angular variable to the hue.

2. The YC_bC_r of the CCIR 601-1 standard TV, today ITU-RBT.601 (International Telecommunication Union), used in digital PAL and in high-definition digital HDTV NTSC

$$
\begin{pmatrix} Y \\ C_b \\ C_r \end{pmatrix} = \begin{pmatrix} 0.2989 & 0.5866 & 0.1145 \\ -0.1688 & -0.3312 & 0.5000 \\ 0.5000 & -0.4184 & -0.0816 \end{pmatrix} \begin{pmatrix} R \\ G \\ B \end{pmatrix} \tag{15.17}
$$

where C_b, as U, is proportional to $(B-Y)$, and C_r, as V, is proportional to $(R-Y)$.

In addition, Xerox-YES, Kodak PhotoYCC® and TekHVC™ are current systems. It is evident that these systems are typical of the companies which have proposed them, and for the reader the original publications are referred to.[5–7]

15.2.4   sRGB System

The sRGB system is now the standard for colour management in computer monitors, scanners, digital cameras and the world-wide-web. This system was proposed in 1996 jointly by Hewlett-Packard hp® and Microsoft® with the aim of obtaining the simplest colour management in computerized systems. In fact, this standard remains just outside the high-quality printers, for which the range of possible colours with this system is insufficient. The sRGB standard considers the CRT monitor conceived for the high definition standard television ITU-R BT.709-2 1990 (International Telecommunications Union) as typical, whose colorimetric data are

$$
\begin{array}{ll}
\text{red} & (x_r = 0.6400, y_r = 0.3300) \\
\text{green} & (x_g = 0.3000, y_g = 0.6000) \\
\text{blue} & (x_b = 0.1500, y_b = 0.0600) \\
\text{white D65} & (x_{D65} = 0.3127, y_{D65} = 0.3290)
\end{array} \tag{15.18}
$$

Colour management starts from its specification in the XYZ CIE 1931 system and is carried out in three steps:

1. A linear transformation moves from the XYZ reference frame to the RGB one, in which the three components of the colour stimulus are denoted by R_{linear}, G_{linear} and B_{linear}, implemented assuming that the Y component of the stimuli is in the range $0 \le Y \le 1$, and that the stimulus of the white D65 is $(X_n = 0.9505, Y_n = 1.0000, Z_n = 1.0890)$

$$
\begin{pmatrix} R_{\text{linear}} \\ G_{\text{linear}} \\ B_{\text{linear}} \end{pmatrix} = \begin{pmatrix} 3.2410 & -1.5374 & -0.4986 \\ -0.9692 & 1.8760 & 0.0416 \\ 0.0556 & -0.2040 & 1.0570 \end{pmatrix} \begin{pmatrix} X \\ Y \\ Z \end{pmatrix}_{D65}
$$

$$
\begin{pmatrix} X \\ Y \\ Z \end{pmatrix}_{D65} = \begin{pmatrix} 0.4124 & 0.3576 & 0.1805 \\ 0.2126 & 0.7152 & 0.0722 \\ 0.0193 & 0.1192 & 0.9505 \end{pmatrix} \begin{pmatrix} R_{\text{linear}} \\ G_{\text{linear}} \\ B_{\text{linear}} \end{pmatrix} \tag{15.19}
$$

in which the initial stimulus, if a non-self-luminous colour, is considered due to the D65 illuminant, as the subscript denotes. In this transformation $0 \leq R_{\text{linear}}, G_{\text{linear}}, B_{\text{linear}} \leq 1$ and the numbers that are greater than 1 or less than 0 are set equal to 1 and 0, respectively.

2. In correspondence to the gamma correction ($\gamma = 2.2$) of high-definition standard television, a partly linear transformation and partly non-linear transformation, with $\gamma = 2.4$, is made, according to the sRGB standard,

$$C_{\text{sRGB}} = \begin{cases} 12.92 C_{\text{linear}} & \text{for } C_{\text{linear}} \leq 0.0031308 \\ (1+a)C_{\text{linear}}^{1/2.4} - a & \text{with } a = 0.055 \quad \text{for } C_{\text{linear}} > 0.0031308 \end{cases} \tag{15.20}$$

where C_{linear} is the generic R_{linear}, G_{linear} or B_{linear} and C_{sRGB} indicates the corresponding transformed value of R_{sRGB}, G_{sRGB} or B_{sRGB}; the standard inverse transformation is

$$C_{\text{linear}} = \begin{cases} \left(\dfrac{C_{\text{sRGB}} + a}{1+a} \right)^{2.4} & \text{for } C_{\text{sRGB}} > 0.04045 \\ \dfrac{C_{\text{sRGB}}}{12.92} & \text{for } C_{\text{sRGB}} \leq 0.04045 \end{cases} \tag{15.21}$$

with $0 \leq C_{\text{linear}} \leq 1$.

This transformation is chosen to be different from that of HDTV standard television to avoid that in the neighbourhood of $C_{\text{linear}} = 0$ the slope of the transformed function is too high. The best fit between the two transformed functions is for the constants 12.9232, 0.03928 and 0.00304, which are placed in the sRGB standard equal to 12.92, 0.04045 and 0.003108. Sometimes the optimum values are proposed instead of the default values, but the meaningful values are the standard ones.

---

3. The third step is the digitization of the signal according to an 8-bit encoding, which translates the numbers $0 \leq R_{\text{sRGB}}, G_{\text{sRGB}}, B_{\text{sRGB}} \leq 1$ into a set of three integers $R_{8\text{bit}}, G_{8\text{bit}}, B_{8\text{bit}}$ in the range 0 to 255: $C_{8\text{bit}} = \text{round} \, [255.0 \times C_{\text{sRGB}}]$, for the direct processing, and $C_{\text{sRGB}} = C_{8\text{bit}} / 255.0$ for the inverse transformation.

The digitization in digital television is carried out according to the formula

$$C_{8\text{bit}} = \left[(\text{WDC} - \text{KDC}) \times C_{\text{sRGB}} \right] + \text{KDC} \tag{15.22}$$

where WDC = 235 is the White Digital Count, and KDC = 16 is the Black Digital Count.

The sRGB standard not only defines the colorimetric system and the signal encoding for colour reproduction, but also defines the characteristics of the environment in which to edit the images and where to see the picture playback on a monitor. The standard monitor is a CRT with white D65 and a maximum luminance of 80 cd/m^2. The editing environment must have an illuminance of 64 lux due to a lamp of the D50 type and the objects must have an average reflected light equal to about 20% of that of the ideal reflecting diffuser. The image viewing environment must have an illuminance of 200 lux due to a lamp of the D50 type. It is necessary for the light reflected from the monitor to be always very low, in the order of 1% of the light reflected from the ideal diffuser in an editing room and 5% for use in a playback room.

15.2.5   Prints in the sRGB System

The environment of viewing printed pictures is standardized[8] and the illumination is defined by the illuminant D50 at 500 lux on a background with a reflectance factor equal to 20%. An sRGB image, conceived for the D65 illuminant, needs transforming before the printing process. Many chromatic-adaptation models have been proposed with different degrees of complexity. This point belongs to the debate on colour appearance. Here the transformation known as Bredford[9] is given

$$
\begin{pmatrix} X \\ Y \\ Z \end{pmatrix}_{D50} = \begin{pmatrix} 1.0479 & 0.0229 & -0.0502 \\ 0.0296 & 0.9905 & -0.0171 \\ -0.0093 & 0.0151 & 0.7517 \end{pmatrix} \begin{pmatrix} X \\ Y \\ Z \end{pmatrix}_{D65}
$$

$$
\begin{pmatrix} X \\ Y \\ Z \end{pmatrix}_{D65} = \begin{pmatrix} 0.9555 & -0.0231 & 0.0633 \\ -0.0283 & 1.0099 & 0.0021 \\ 0.0124 & -0.0206 & 1.3307 \end{pmatrix} \begin{pmatrix} X \\ Y \\ Z \end{pmatrix}_{D50}
$$

(15.23)

where the subscript denotes the illuminant.

15.2.6   Camera, Photo-Site and Pixel

Colour measurement is an operation that precedes its reproduction and occurs in the television system and in the photographic system by means of the camera. The camera, in complete analogy with the eye, is constituted by an objective lens and by a mosaic of three types of photo-transducers placed on its focal plane. These transducers are, depending on the cameras, of different nature, and, more frequently today, they are *Charge-Coupled Devices* (CCD) or *Complementary Metal Oxide Semiconductors* (CMOS). The photosite is the smallest space within an image sensor that transforms a luminous flux into an electric signal. The photosite is a physical place, belonging to the category of hardware. In literature, there are many similar definitions:

> *Photosite* – a set of one or more photo-transducer elements which are spatially organized in an adequate way to produce the colorimetric information necessary to define a pixel.

The mosaic spatial organization of the different types of photo-transducers present in the photosite is very important, because the signals generated by the various photo-transducers have to be put in relation to the pixels. Signal processing – a mathematical interpolation termed *demosaicing* or *demosaicking* – occurs from the signals generated by the various photo-transducers to the signals constituting the pixel, which depend on the spatial organization of the photo-transducers. The characteristics of the photosites allow us to understand, both from the electrical point of view and from the optical one, the way in which the pixels that form images are captured. The data produced directly by the photosites constitute a *raw image* representing a typical sampling of the image sensor. If the raw image data result from a capture made using a photosite with a *colour filter array* (e.g., the red, green and blue raw colour values are captured by a sensor array in which the individual photo-transducers have red, green or blue filters), a camera raw processing application is needed to create a viewable colour image. In some types of image sensors, – for example, colour Foveon type[10] – all the information related to a pixel is produced by overlapped photo-transducers. Instead other kind of image sensors, as in the Bayer pattern sensors[11] (Figure 15.9), the colour specification of each pixel is produced by a group of adjacent photo-transducers and depends on the signal processing used. The signal processing has a very important role in the definition of the quality of the image represented by the pixels.

The photosites of a typical colour image sensor have three spectral sensitivities in analogy with the fundamental colour-matching functions of the human visual system. Each image sensor has its own colour space, that should be built to be as close as possible to that of the CIE 1931 standard observer.

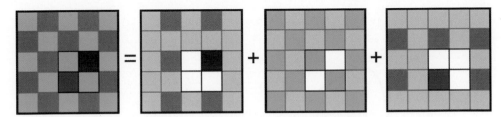

Figure 15.9   *Bayer pattern of arrangement of the colour sensors of the photosite (unveiled squares) constituting the sensors for the acquisition of digital images and its decomposition into three arrangements of sensors of the same type. (Bryce Bayer is the name of the Kodak researcher who first proposed this pattern.) In the unveiled squares, there are two sensors for the medium wavelength region – green – and only one for the short – blue – and long –red– regions. Since each type of sensor covers only a part of these units, an interpolation – demosaicing – is needed to obtain colour information at any point.*

Maxwell-Ives-Luther criterion – A conversion between the tristimulus space of the photo-camera and that of the standard observer is possible if the product of the spectral responsivity of the naked photo-sensor of the photo site and the spectral transmittance of the corresponding optical filters is a linear combination of the colour-matching functions of the CIE standard observer.[12]

If the spectral sensitivity functions of the three types of photo-sensor of a photosite can be written as a linear combination – *matrixing* – of the colour-matching functions of the standard observer, the electrical signals generated by the camera are

- in one to one correspondence with the colour stimuli represented in tristimulus space, and then the camera 'sees' as the standard observer;
- representable directly by tristimulus vectors in the basic reference frame of the camera, whose reference vectors S_R, S_G and S_B have the chromaticity placed at the vertices of a triangle with sides that are tangent to the spectrum locus.

This second property is a consequence of the fact that the spectral sensitivities of the photosites are always positive. The three stimuli S_R, S_G and S_B are said to be supersaturated because with saturation greater than 1.

In practice this does not occur exactly and an approximation, known as the *Compromise of Ives - Abney - Yule*, is considered. This approximation affects the colour reproduction entirely, from the monitor to the camera. The starting point is given by the standard reference stimuli $\hat{R}$, $\hat{G}$ and $\hat{B}$ of the monitor, which correspond to the chromaticities $R = (x_r, y_r)$, $G = (x_g, y_g)$ and $B = (x_b, y_b)$ (Figure 15.10). The colour-matching functions in the RGB reference frame (Figure 15.11) are all characterized by three large positive lobes equally spaced in wavelengths and joined by other lobes, positive and negative. The problem would be solved if the camera had as many light-sensitive elements as the number of lobes of these functions, and each one had the spectral sensitivity function equal to one lobe. The first approximation is to choose three types of photosensitive elements with spectral sensitivity very close to the big three positive lobes. The linear combinations of the colour matching functions that best approximate these spectral sensitivities correspond to three stimuli S_R, S_G and S_B, which can be considered as belonging to the fundamental reference frame of the camera (Figure 15.12). Their chromaticities (Figure 15.10) are the vertices of a triangle $S_R S_G S_B$, which contains all of the chromaticity diagram, and with good approximation are on the extension of the segments *WR*, *WG* and *WB*, where $W = (x_n, y_n)$ is the chromaticity of the white chosen by the standard system. The compromise is to use colour reproduction stimuli **R**, **G** and **B** instead of S_R, S_G and S_B. Since the stimuli **R**, **G** and **B** are obtained as the sum of the stimuli S_R, S_G and S_B with the white stimulus W, all the colours are reproduced with chromaticity shifted towards white, and then with a lower purity level. In practice, this desaturation is reduced

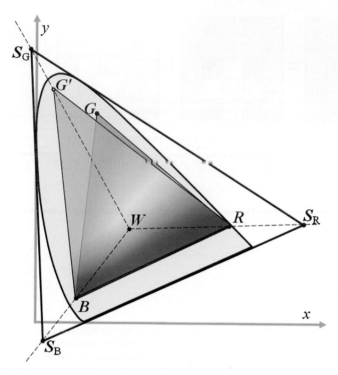

Figure 15.10   *CIE 1931 chromaticity diagram with the chromaticity of the stimuli S_R, S_G and S_B of a camera with the spectral sensitivity of Figure15.12. The construction of the stimuli S_R, S_G and S_B takes into account the chromaticity of the red and blue reference stimuli of the NTSC system but not the green stimulus, because this is not completely compatible. The Ives-Abney-Yule compromise considers the RGB monitor compatible with the S_R, S_G, and S_B stimuli of the camera if also the chromaticity of the green stimulus is within the segment WS_G (here represented by the point G').*

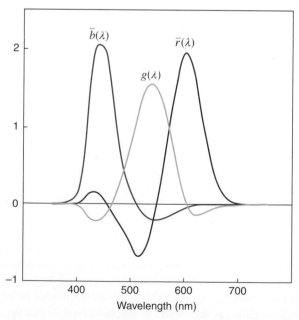

Figure 15.11   *Colour matching functions of the CIE 1931 observer in the monitor RGB reference frame.*

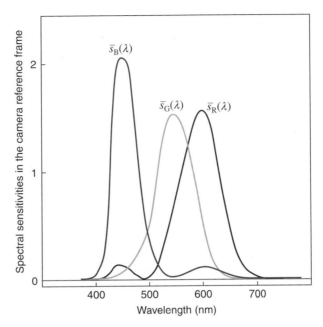

Figure 15.12 *Colour matching functions of the CIE 1931 observer in the $S_R S_G S_B$ reference frame.*

with an appropriate change in the number γ (cameras almost always allow more choices for the γ correction, among which is the standard $\gamma = 2.2$).

It is observed that in all cases, in which the matrixings of spectral sensitivities of the camera are different from the colour-matching functions of the standard observer, the problem of metamerism is twofold – metamerism for the human visual system and for the camera –, because there are metameric stimuli for the standard observer that are not metameric for the camera and vice-versa. An *instrumental metamerism* is added to the metamerism of the human visual system.

A similar problem arises in the construction of scanners.

15.2.7 Spectral Sensitivities of Digital Cameras

Each camera has its own colour space and the colorimetric characterization of the camera is given by its colour-matching functions. The most important reference frames in this space are two:

1. the fundamental reference frame, in which the colour-matching functions represent the spectral sensitivities of the three types of – R, G and B – filtered sensors of the photosites;
2. instrumental reference frame, in which the reference stimuli, for example **R**, **G** and **B**, are three suitable colours, as the reference colours of a reference monitor,

(generally in both reference frames, the reference stimuli and tristimulus values are denoted with the same letters often creating confusion, but the curves of the colour-matching functions dissolve any doubt. Anyway, here the quantities in the fundamental reference frame have the subscript 'F' and in the instrumental reference frame 'RGB').

The fundamental reference system is defined exactly as that for the standard observer, but with the difference that this system refers to the camera and not the human visual system – Equations (6.16)–(6.18) –. For example, the image sensors of the camera have quantum efficiencies similar to those of the eye. In a camera the activations of the photo-sensors are easily measurable, because they are electrical signals. ISO 17321[13] provides a standard methodology for the evaluation of the relative spectral sensitivity of digital cameras.

The instrumentation for the measurement of the spectral sensitivities of the camera consists of a tuneable monochromatic light source – that is, a light source with a monochromator –, an integrating sphere and a spectroradiometer.

The measurements are performed in correspondence with a defined set of wavelengths from 360 nm to 780 nm (or, better, 830 nm) with a step lower than 10 nm.

It is supposed that the response of the sensor is linear with the light intensity, conversely a linearization is required (the linearization is not considered here).

The operations are:

1. a monochromatic radiation is selected with the monochromator (actually, not a monochromatic light but a light with a spectral bandwidth of few nanometres, < 5 nm is selected);
2. the selected monochromatic light enters the integrating sphere;
3. the objective lens of the camera focusses the plane of the exit port of the integrating sphere on the image sensor, that is a Lambertian source of light uniformly distributed inside the port. A diffuser placed over the exit port is generally used. Attention must be paid for the image formed on the photosensors to be solely made of points within the port of the sphere;
4. the signal T_λ of the spectroradiometer generated by the light flux leaving another port of the integrating sphere is recorded;
5. once the value γ in the camera is pre-set equal to 1 and any automatic white balancing is stopped, the signals $S_{R,\lambda}$, $S_{G,\lambda}$ and $S_{B,\lambda}$ of the three elements of photosite of the camera are recorded. The exposure time and the lens aperture must be fixed for all the measurements such that the maximum response falls between 50% and 90% for the full range of the camera. These three signals should be the same for all the photosites, up to a statistical variation due mainly to noise. (Lens *vignetting* must be taken into account, i.e., the brightness of the image is not the same at all the points of the sensor. The falloff of the brightness of the image is approximated by the 'cos^4' or *cosine fourth law of illumination falloff*. The maximum value is at the centre of the image, at the point where the optical axis of the lens meets the image plane on the sensor, and decreases away from this point).

Assuming that the three elements of photosites of the camera generate a signal in proportion to the incident power, the spectral sensitivities of the camera in the fundamental reference frame are

$$r_F(\lambda) = \frac{S_{R,\lambda}}{T_\lambda}, \; g_F(\lambda) = \frac{S_{G,\lambda}}{T_\lambda}, \; b_F(\lambda) = \frac{S_{B,\lambda}}{T_\lambda} \qquad (15.24)$$

and this result depends on the pre-set gains on the three channels of the camera, which act as multiplicative factors on the three spectral sensitivities. In this reference frame, the spectral sensitivities are always positive. Figure 15.13 shows the spectral sensitivities of a commercial camera in the fundamental reference frame.[14]

The RGB instrumental reference frame needs a choice of three independent reference colours, suitable to define the reference frame itself. A convenient choice is that the red, green and blue colours are equal to the reference colours of the trichromatic standard monitor. The method of measurement is the Maxwell minimum saturation method in which a matching of white lights occurs on a bipartite field (Section 6.13.5). The instrumentation for the measurement consists of

-   a monochromatic light source, that is, a light source and a monochromator;
-   three reference light sources (A trichromatic monitor conveniently replaces the three reference sources and in this case the R, G and B reference colours of the monitor correspond to the stimuli of the reference frame of the colour space chosen to characterize of the camera. In this case, an appropriate software drives the reference lights of the monitor separately.);

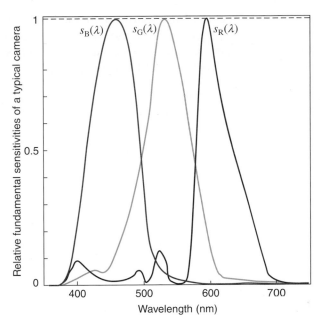

Figure 15.13 *Relative spectral sensitivity function of a typical commercial camera in its fundamental reference frame.*

- an integrating sphere;
- a calibrated photodetector.

There are three preliminary measurements:

(a) the objective lens of the camera focusses the plane of the exit port of the integrating sphere on the photo-sensor, that is a Lambertian source of monochromatic light, which is uniformly distributed inside the port. Attention must be paid for the image formed on the image sensor to be solely made of points within to the port of the sphere;

(b) the white light $\mathbf{W} = \mathbf{R} + \mathbf{G} + \mathbf{B}$ of the monitor enters the sphere and the three signals $S_{R,W}$, $S_{G,W}$ and $S_{B,W}$ of the camera are recorded;
  - if possible, by adjusting the gains of the three channels, the three signals are imposed to be equal (as in the previous case, the cosine fourth law of illumination falloff could produce a non-uniformity of the brightness of the image);
  - the exposure has to be chosen such that the three signals are approximately at 80% of the maximum of the full range of the camera;

(c) with the spectroradiometer the three fluxes r_W, g_W and b_W of the three reference colours exiting another port of the integrating sphere are measured.

The measurement is performed in correspondence with a defined set of wavelengths and the operations are:

1. a monochromatic radiation is selected with the monochromator (actually, not a monochromatic light is selected but a light with a spectral bandwidth of few nanometres, < 5 nm) with a step of 10 nm from 360 to 780 nm;

2. the selected monochromatic light in conjunction with the reference lights of the monitor enter the integrating sphere;

3. the intensities of the lights entering the integrating sphere are adjusted until the signals of the camera are the same as those recorded with white light $S_{R,W}$, $S_{G,W}$ and $S_{B,W}$; in this adjustment the intensity of one of the three reference lights goes to zero, and the reference light to which this happens depends on the spectral region in which the selected wavelength λ is located;

4. the signals of the spectroradiometer, produced by the flow of radiation leaving another port of the integrating sphere relative to the reference lights r_λ, g_λ and b_λ, and by the monochromatic light e_λ, are recorded (as in the previous case, the cosine fourth law of illumination falloff produces a non-uniformity of the brightness of the image).

The spectral sensitivities of the camera in the RGB instrumental reference frame of the monitor are

$$r_{RGB}(\lambda) = \frac{r_W - r_\lambda}{e_\lambda}, \quad g_{RGB}(\lambda) = \frac{g_W - g_\lambda}{e_\lambda}, \quad b_{RGB}(\lambda) = \frac{b_W - b_\lambda}{e_\lambda} \tag{15.25}$$

In this case, the spectral sensitivities take positive or negative values, depending on the spectral regions.

15.3   Principles of Halftone Printing

Start with the following fundamental definitions:

Ink is a liquid or paste that contains colorants (typically pigments and/or dyes) and is used to colour a surface, generally of paper, to produce a colour image.

Colour printing is a process for reproducing coloured images, typically with inks on paper.

Halftone is the printing technique that simulates colours in perception with the use of coloured dots, varying either in size, in shape or in spacing and randomly overlapped.

Dot is the smallest physical element of an image with a homogeneous colour. The dot shapes are not visible to the naked eye. A printed image is constituted by dot patterns of different inks, whose internal structure is not visible to the naked eye and which appear with a defined colour. The smallest dot patterns are the printed elements of an image corresponding to the software picture element, termed *pixel*.

The *pixel* (Section 15.2)

–   in digital imaging is the smallest controllable element of a picture to be represented on a screen or a print,
–   in a print is represented by a dot pattern visually perceived with a uniform colour.

(This definition of pixels is integrated with that given in Section 15.2.)

When printing with the halftone technique, colour is brought into a patchwork of dots of variable cross-section and arranged so as to avoid obvious geometric and unwanted configurations, termed *moiré patterns*.

The case in which inks are made with dyes is considered here. Inks, once dried on paper, are transparent coloured filters with a result that is independent of their overlapping order. The light diffusion is due to the cellulose fibres of the paper. Often the number of inks used is three – *tri-colour printing* denoted by the acronym CMY from cyan, yellow and magenta –. The overlaps of these inks produce other colours – red, green, blue and black –. The colours of these dots are eight (Tables 15.2 and 15.3), where the eighth colour is the white of the paper.

Table 15.2 *Colours obtained by superimposing CMY ideal inks on white paper.*

Paper not inked	One ink	Two inks overlapped	Three inks overlapped
White	Cyan	Red = magenta + yellow	Black = cyan + magenta + yellow
	Magenta	Green = yellow + cyan	
	Yellow	Blue = cyan + magenta	

Halftone colour production can be considered both as subtractive colour mixing and as additive colour mixing, it depends on the point at which the light begins to be considered:

1. *Subtractive colour mixing* is the phenomenon that starts with a light source, deals with the wavelength selective subtraction of light by materials (colorants, dyes, pigments, inks, paints, …) and ends with the light thus modulated entering the eye.

2. *Additive colour mixing* is the phenomenon that starts with the emission of coloured lights by illuminated materials and light sources, which are mixed in various proportions, and ends with the light thus mixed entering the eye.

A.C. Hardy and F.L. Wurzburg[15] wrote:

"In a three-color print, the colour of an area that is large in comparison with the size of the dots of the structured image can be regarded as an additive mixture of eight colours."

Inks can be considered as operating in subtractive mixing mode, while mosaic arrangement of coloured dots is according to additive mixing as when viewing a CRT or an LCD monitor (Section 15.2). The colorimetric description of the halftone-colour reproduction considering this phenomenon as additive synthesis is due to H. E. J. Neugebauer,[16] who used the additive mixing of tristimulus values. It is briefly reported here. Consider a paper element of unit area representing a pixel and this is also called a pixel. The pixel should be seen at a distance for which its interior appears indistinct. Following the analysis of M. E. Demichel,[17] the part of the pixel of area $c \leq 1$ is covered by cyan ink, the part $m \leq 1$ by magenta ink and the part $y \leq 1$ by

Table 15.3 *Tristimulus vectors associated to the unit printed area and the fractions the area, 'c', 'm' and 'y', related to the eight types of inking.*

Colour	Tristimulus vector associated to the unit printed area	Relative area (Demichel equations)
White	(X_1, Y_1, Z_1)	$a_1 = (1 - c)(1 - m)(1 - y)$
Cyan	(X_2, Y_2, Z_2)	$a_2 = c(1 - m)(1 - y)$
Magenta	(X_3, Y_3, Z_3)	$a_3 = (1 - c)m(1 - y)$
Yellow	(X_4, Y_4, Z_4)	$a_4 = (1 - c)(1 - m)y$
Red	(X_5, Y_5, Z_5)	$a_5 = (1 - c)m y$
Green	(X_6, Y_6, Z_6)	$a_6 = c(1 - m)y$
Blue	(X_7, Y_7, Z_7)	$a_7 = c m(1 - y)$
Black	(X_8, Y_8, Z_8)	$a_8 = c m y$

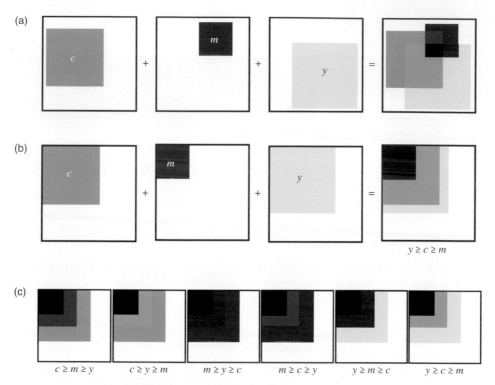

Figure 15.14 *(a) Pixel with partial coatings of cyan, magenta and yellow ideal inks with random overlapping and (b) in maximum overlapping. The shape of the spots of the three inks is square for graphic simplicity, but actually the spots are round. In the first case there are eight colours – white, black, cyan, yellow, magenta, red, green and blue – and in the second case only four – white, black, the colour of the most extensive ink and the colour of the overlapping of the two most extensive inks –. (c) The six possible cases in which the pixels have maximum overlapping of the inks inside them then produce only four colours.*

yellow ink (Figure 15.14). The pixel is divided into eight different parts, each uniform as regards to inking and corresponding to one of the eight colours listed in Table 15.2. The pixel, once lit, reflects a light flux in proportion to the illuminating one and decreased by the presence of the three inks that act as filters. The tristimulus values corresponding to the reflected luminous flux are obtained as a sum of eight stimuli, related to the eight colours considered above, and with intensity proportional to the compound probability that a photon passes through the surfaces with the *cmy* inking requested for a unit-area pixel: for example, the probability a_{red} of having the red colour is given by the probability that the incident photon crosses the surface area m filtered by the magenta ink, times the probability of crossing the area y filtered by the yellow ink and times the probability of crossing the area $(1 - c)$ devoid of cyan ink, that is, $a_{red} = my (1 - c)$. (Table 15.3)

A pixel totally inked with three CMY inks appears black in colour, and greyed out if it is only partially inked with equal areas inside the pixel.

The ideal inks operate as filters and their subtraction of light is independent of their position within the pixel, therefore we can consider as equivalent the cases in which the CMY dots are randomly arranged or stacked in an orderly manner, as shown in Figures 15.14a and 15.14b.

Figure 15.15 shows the spectral reflectance factor of ideal paper inked with ideal inks. The wavelength range is divided into three equal parts, 400–500, 500–600 and 600–700 nm. The reflectance factors of the ideal CMY inks are bivalent functions, equal to 1 or to 0 in the different parts of the spectrum, as the optimal colours (Section 14.2). Ideal white paper has a reflectance ideally equal to 1. The overlap of two inks

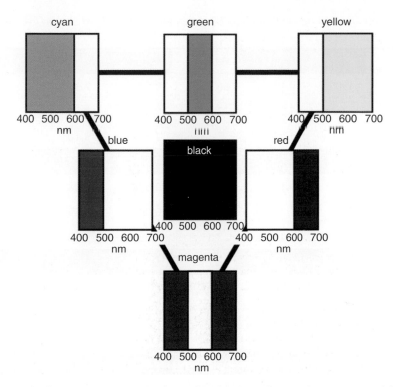

Figure 15.15 *Spectral reflectance of optimal colour inks of cyan, yellow and magenta, and their overlapping if the pixel is completely coated.*

produces a third colour with bivalent reflectance too and the overlap of all three inks a reflectance ideally equal to zero. All this is represented in Table 15.2. The colour specifications of the standard papers and inks are given in Table 15.4 (N.B. the standard quantities are specified in CIELAB units, while here are given the *XYZ* tristimulus values because they are the same used in the vector equations of Neugebauer). Spectral reflectance factors of three actual inks, measured on the Agfa 'Postscript Process Colour Guide'[18] are shown in Figure 15.16 for a comparison with the ideal inks of Figure 15.15.

> In graphic arts and photography, ISO standards[19] require
>
> - the illuminant D50 at 500 lux on a background with reflectance factor equal to 20% (ISO 3664:2009 Graphic technology and photography - Viewing conditions) and
> - the colour-measurement geometry (45°/0°) or (0°/45°) on a black plane.

The equations given in Table 15.3, which define the relative areas $a_1, a_2, \ldots, a_8$, are known as Demichel's equations. The Neugebauer equations specify the colour represented by the vector (X, Y, Z) and obtainable by a linear combination of the eight vectors (X_i, Y_i, Z_i), with $i = 1, 2, \ldots, 8$, relative to stimuli due to the pixel surface entirely coated by the i-th colour

$$(X, Y, Z) = \sum_{i=1}^{8} a_i (X_i, Y_i, Z_i), \quad \sum_{i=1}^{8} a_i = 1 \tag{15.26}$$

Table 15.4 Colours obtained using the flexographic process and ISO standard inks on five different substrates of standard paper.[20]

Paper 1	White	Cyan	Magenta	Yellow	YM_Red	YC_Green	CM_Blue	Black
X	80.00	14.47	30.12	66.74	29.67	7.58	4.51	2.02
Y	82.97	21.97	15.27	72.07	16.02	17.59	2.99	2.10
Z	71.80	50.41	14.45	7.35	2.08	5.10	13.76	1.73
Paper 2	White	Cyan	Magenta	Yellow	YM_Red	YC_Green	CM_Blue	Black
X	77.82	14.47	30.12	66.74	29.67	7.58	4.51	2.02
Y	80.70	21.97	15.27	72.07	16.02	17.59	2.99	2.10
Z	69.81	50.41	14.45	7.35	2.08	5.10	13.76	1.73
Paper 3	White	Cyan	Magenta	Yellow	YM_Red	YC_Green	CM_Blue	Black
X	67.00	15.19	29.59	59.65	27.20	7.34	4.61	2.88
Y	70.01	22.93	15.27	64.07	14.54	16.79	3.25	2.99
Z	54.90	47.54	13.69	6.21	2.14	5.15	13.65	2.47
Paper 4	White	Cyan	Magenta	Yellow	YM_Red	YC_Green	CM_Blue	Black
X	77.82	19.60	35.91	67.64	32.54	11.64	10.16	6.51
Y	80.70	25.96	21.97	72.07	20.14	20.14	9.01	6.65
Z	69.87	51.21	19.05	11.71	6.85	10.70	18.57	5.29
Paper 5	White	Cyan	Magenta	Yellow	YM_Red	YC_Green	CM_Blue	Black
X	69.49	20.06	33.10	64.22	31.45	10.16	8.58	6.51
Y	72.07	27.03	20.14	67.99	19.27	17.59	7.54	6.65
Z	53.70	46.60	15.79	9.97	5.54	9.15	15.08	5.09

The global compound probability, represented by the last equation, is equal to 1 and this is verifiable specifying the a_i's as a function of c, m and y. The set of Equations (15.26) is non-linear and with c, y and m three unknown quantities.

Here a solution to the set of Equations (15.26) is proposed. There are six separate cases represented by six triangles in the chromaticity diagram (Figure 15.17): **CWB**, **BWM**, **MWR**, **RWY**, **YWG**, **GWC**. Placed at random the areas c, m and y coated by the three inks, there may be up to eight regions with the eight colours of Table 15.3 (Figure 15.14a), while if the areas c, m and y are set in an orderly way, such that the greater contains the intermediate and this the smaller, the pixel contains only four colours (Figure 15.14b):

(a) the white appears on the part of the paper given by the difference between the entire pixel and the area coated by the ink with the greatest extension;

(b) the colour of the most extended ink appears with an area equal to the difference between its coated area and the area of the intermediate ink;

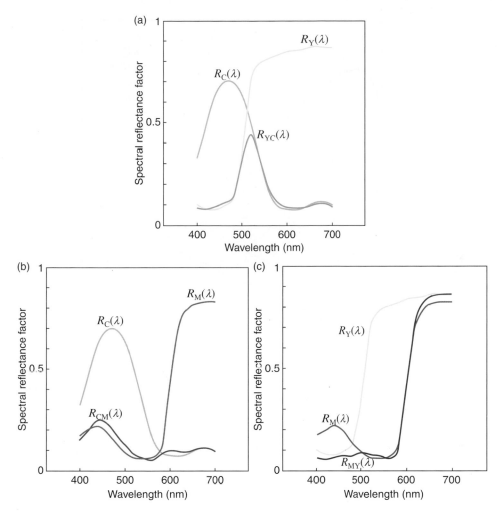

Figure 15.16 *Spectral Reflectance factors of three actual inks, measured on the 'Postscript Process Colour Guide' by Agfa 1993, given for a comparison with the ideal optimal colour inks of Figure15.15: (a) spectra of the yellow and cyan inks and their overlapping producing green; (b) spectra of the cyan and magenta inks and their overlapping producing blue; (c) spectra of the magenta and yellow inks and their overlapping producing red.*

(c) the colour of the overlap of the two most extensive inks appears with an area equal to the difference between the areas of the less extensive inks; and

(d) the black appears with an area equal to that of the less extensive ink (call the black area k).

The four colours change in correspondence to the change in the order of the extensions of the three inks. The six cases are (Figure 15.14c and 15.17):

1. For $c \geq m \geq y$, there are the colours white, cyan, blue and black to the sum of which corresponds a colour with chromaticity within the triangle **CWB**.

2. For $m \geq c \geq y$, there are the colours white, magenta, blue and black to the sum of which corresponds a colour with chromaticity within the triangle **BWM**.

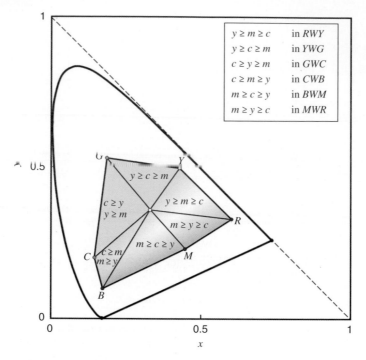

Figure 15.17   *CIE 1931 chromaticity diagram with the chromaticity of the colours obtained by trial inks and D65 illuminant. The hexagon of the available chromaticity **CBMRYG** is divided into six triangles, to which correspond the six cases considered for solving the vector-equations system (15.26).*

3.   For $m \geq y \geq c$, there are the colours white, magenta, red and black to the sum of which corresponds a colour with chromaticity within the triangle **MWR**.

4.   For $y \geq m \geq c$, there are the colours white, yellow, red and black to the sum of which corresponds a colour with chromaticity within the triangle **RWY**.

5.   For $y \geq c \geq m$, there are the colours white, yellow, green and black to the sum of which corresponds a colour with chromaticity within the triangle **YWG**.

6.   For $c \geq y \geq m$, there are the colours white, cyan, green and black to the sum of which corresponds a colour with chromaticity within the triangle **GWC**.

As an example, consider the case 1., where the Equations 15.26 become

$$\begin{cases} (X,Y,Z) = a_1(X_1,Y_1,Z_1) + a_2(X_2,Y_2,Z_2) + a_7(X_7,Y_7,Z_7) + a_8(X_8,Y_8,Z_8) \\ a_1 = 1 - c \\ a_2 = c - m \\ a_7 = m - y \\ a_8 = y \end{cases} \qquad (15.27)$$

and the vector equation is equivalent to a set of three equations, that are linear with three unknowns. Its solution is immediate. In a similar way the solutions for the other cases are obtained. The triangle, to which the chromaticity of the stimulus (X, Y, Z) belongs, defines the case among the six.

This analysis shows that the colours can be represented by four stimuli, two coloured stimuli in addition to the white and black stimuli. The sum of white and black stimuli is equal to a grey stimulus. If black is not obtained with the superposition of the three inks, but with a fourth ink black in colour, it has *four-colour process,* denoted with the acronym CMYK (the 'K' in CMYK stands for 'Key', because cyan, magenta and yellow printing plates are aligned with the 'black Key plate'). When printing colour images by combining multiple ink colours, the coloured inks usually do not contain much image detail. The key plate, which is impressed using black ink, provides the contrast of the image increasing the quality of the details. Black ink used to add details and darkness in shadowed areas is called a *Skeletal Black*. In 1725 Jakob Christoffel Le Blon first understood the important role of the black ink in printing and invented the four-colour-process.[21] In practice different amounts of coloured inks are substituted by black ink:

1. *Grey component replacement* (GCR) is the complete replacement of CMY inks with *full black* ink in the k area.
2. *Under colour removal* (UCR) is a partial replacement of CMY inks with *full black* ink in an area smaller than the k area, which is obtained starting from the CMY colour specification by reducing the overlapping area of the CMY inks in an equal amount and inking an equal area with a black ink.
3. *Under colour addition* (UCA) is a partial replacement of *black* ink with an overlapping of CMY inks in an area smaller than the k area, which is obtained starting from the CMYK colour specification of the GCR by reducing the black ink area and inking an equal area with an overlapping of the CMY inks.

Photographic reproduction also occurs with very similar colour synthesis to that of the halftone printing. In this case the section of the elements of the mosaic is in the range between 0.5 and 1.5 µm, depending on the type of film used. The colorimetric basis of colour photography obtained by chemical and colour printing inks is considerably similar.

The software presented in subsection 16.10.2 displays the palettes of the colour obtainable with three inks, yellow, cyan and magenta on the monitor.

15.4 Towards the Colorimetry of Appearance

Colour reproductions by the self-luminous monitor or by halftone printing concern the colour within an image and pose three main problems:

1. the techniques that produce images have a smaller set of colour stimuli than that of the actual scenes of which the images are taken;
2. sets of colour stimuli available in various media are limited and different and this poses the problem of passing from one medium to another, for example from the set of tristimulus values (R, G, B) of a monitor to the set (c, m, y), or (c, m, y, k), of a printer; and
3. the visual context in which the images are comfortably viewed must be defined, because the perceived colours depend on the visual context.

The concern of points 1 and 2, known as *gamut mapping*, has no solution for all images without compromise. Since different devices have different gamuts of perceived colours, the transformation from one device to another cannot occur without losing colours. At most, the transformation could be related to the subset of common colours, which is the intersection of the gamuts and is considered too small. An image has a change in colour in the passage from one reproduction technique to another. In the past, the criteria for these transformations were non-colorimetric, but were required to produce realistic images in all technical achievements. The constraints are now classics:

- lightness concerns all the colours of the image as a whole and must be realistic and
- chromaticity involves two typical kinds of colours: the colours of human skin and the green of the football field, which must be recognized as such.

The chromaticity of the sky, though important, has not such a crucial role for a realistic appearance.

Point 3. requires a step beyond psychophysical colorimetry (Chapter 9) and psychometric colorimetry (Chapter 11). It requires the construction of a new colorimetry, addressed to the measurement of *colour appearance* in relation to the visual context. Colour appearance in the transition between different devices and different visual situations should be unchanged or little changed, while the colour stimuli change. Similar problems are present in the photographic shooting. Colour appearance is a complex chapter in evolution, where research is in progress. Particularly open problems have unsatisfactory and non-convincing solution. The debate is open. Therefore no colour appearance model is considered in this book.

Today, these problems have a more rigorous definition, although not definitive. The places for the use of images in the various media are defined (e.g., in Section 15.2 the sRGB system indicates the place in which to observe a picture on a monitor). The International Color Consortium (ICC) (www.color.org) and national and international institutions for standards (ISO) have an important role in this field. The problem of gamut mapping is considered by introducing compromises, called *rendering intent*. All that is not considered here and the reader is referred to the specific bibliography, for example [22–30].

References

1. Hunt RWG, *Reproduction of Colour*, 6th ed. The Wiley-IS&T Series in Imaging Science and Technology (2004).
2. Lee HC, *Introduction to Color Imaging Science*, Cambridge University Press, Cambridge UK (2005).
3. MacDonald LW and Luo MR, *Color Imaging*, John Wiley & Sons, Chichester (1999).
4. CIE Publication CIE S 017/E:2011, *ILV: International Lighting Vocabulary*. Commission Internationale de l'Éclairage, Vienna (2011). Available at: www.eilv.cie.co.at/ (18 June 2015).
5. *Color Encoding Standard*, XNSS 288811, Xerox Corp., Sunnyvale, CA (1989).
6. *Kodak Photo CD System: A Planning Guide for Developers*, Eastman Kodak Company, Rochester, NY (1991).
7. *TekColor™ Color Management System*, Chap. 2, Tektronix, Inc. Beaverton, OR (1990).
8. ISO 3664:2009, *Graphic Technology and Photography – Viewing Conditions*
9. Lam KM, *Metamerism and Colour Constancy*, Ph.D. Thesis, University of Bradford (1985).
10. Foveon http://www.foveon.com/article.php?a=74
11. *Bayer Color Imaging Array*, B.E. Bayer, U.S. Patent 3971065 A, July 20, 1976
12. Luther RTD, Aus dem Gebiet der Farbreizmetrik, *Zeitschrift für technische Physik*, **8**, 540–558 (1927).
13. ISO 17321 WD 4, *Colour Characterisation of Digital Still Cameras Using Colour Targets and Spectral Illumination*, International Organization for Standardization, Geneva (1999).
14. Sigernes F, Dyrland M, Peters N, Lorentzen DA, Svenøe T, Heia K, Chernouss S, Sterling DC and Kosch M, The absolute sensitivity of digital colour cameras, *Optics Express* **17** (22), 20211–20220 (2009).
15. Hardy AC and Wurzburg FL Jr Color correction in color printing, *J. Opt. Soc. Amer.*, **38** (4), 300–307 (1948).
16. Neugebauer, HEJ, Die theoretische grundlagen der mehrfarbendrucks, *Z. wiss. Photogr.*, **36**, 73–89 (1937).

17. Demichel ME, *Procédé* **26**, 17-21 e 26–27 (1924).

18. *Agfa "Postscript Process Colour Guide"*, Agfa Gevaert NV (1993)

19. ISO 13655: 2009, *Graphic Technology – Spectral Measurement and Colorimetric Computation for Graphic Arts Images*, International Organization for Standardization, Geneva (2009). http://www.iso.org/iso/home/standards.htm

20. ISO 12647-6:2012, *Graphic Technology – Process Control for the Production of Half-Tone Colour Separations, Proofs and Production Prints – Part 6: Flexographic Printing*, International Organization for Standardization, Geneva (2012). http://www.iso.org/iso/home/standards.htm

21. Le Blon JC, *Coloritto: or the Harmony of Coloring in Painting*. London (1725). Reprinted from British Library by ECCO,Medicine Science and Technology, Lighting Source Ltd, London (2011)

22. Fairchild MD, *Color Appearance Models*, 3rd ed., The Wiley-IS&T Series in Imaging Science and Technology (2013).

23. Kipphan H, *Handbook of Print media*, Springer, Berlin (2000).

24. ISO 12647-1:2013; ISO 12647-2:2012; ISO 12647-3:2013; ISO 12647-4:2014; ISO 12647-5:2015; ISO 12647-6:2012; ISO 12647-7:2013; ISO 12647-8:2012: *Graphic Technology — Process Control for the Production of Half-Tone Colour Separations, Proof and Production Prints*, International Organization for Standardization, Geneva, Switzerland. http://www.iso.org/iso/home/standards.htm

25. AllenE and TriantaphillidouS, eds., *The Manual of Photography*, Focal Press, Oxford and Elsevier, Burlington, MA (2011).

26. ISO 22028-1:2004; ISO 22028-2:2013; ISO 22028-3:2012; ISO 22028-4:2012: *Photography and Graphic Technology — Extended Colour Encodings for Digital Image Storage, Manipulation and Interchange*, International Organization for Standardization, Geneva. http://www.iso.org/iso/home/standards.htm

27. *Color Management: Understanding and Using ICC Profiles*, PhilGreen, ed., The Wiley-IS&T Series in Imaging Science and Technology (2010).

28. Ebner M, *Color Constancy*, The Wiley-IS&T Series in Imaging Science and Technology (2007).

29. Motovič J, *Color Gamut Mapping*, The Wiley-IS&T Series in Imaging Science and Technology (2008).

30. Giorgianni EJ and Madden TE, *Digital color Management*, 2nd ed., The Wiley-IS&T Series in Imaging Science and Technology (2008).

16

Software

(Software developed by Gabriele Simone)

16.1 Introduction to the Software

Colorimetry is a science with high visual value and with a strong connection with measurements and computations. Measurements, colorimetric computations, monitor visualizations of the colours, and today colour management in image acquisitions and reproductions and colorant formulations, are made by a computer. These practical reasons induced us to collect all the standard definitions of the colorimetric quantities, the algorithms, the atlases and to present all that in ready to use software, named 'Colorimetric eXercise' – shortened with the acronym CX –. The software and this chapter are mutually linked through a common index. The main purposes are:

1. to support teachers and students in learning colorimetry and
2. to provide the lab technician with an useful tool for computations.

The software presents 32 toolboxes subdivided into ten sections (Table 16.1)

All the colorimetric computations are made following the updated algorithms of colorimetry given by the CIE.[1]

Once we launch the CX program, the toolboxes listed in Table 16.1 are presented with a dialogue menu, as shown on Figure 16.1. By clicking on the various buttons on the menu, we open the corresponding toolboxes.

At the beginning, software installation and monitor calibration are the operations to carry out in sequence. The colorimetric characteristics of the monitor are understandable only once the user has an adequate knowledge of the colorimetry, presented in Sections 15.2.1–15.2.2. Anyway the user can work with no monitor setting, because the software, by default, works assuming the monitor is an sRGB monitor and in this case the colours are reproduced in a significant although approximate way.

16.1.1 Software Installation

The CX software runs with the operating system Windows 7/8/10. CX requires framework .NET 4.0 or higher and a minimum resolution of 1366x768. The CX software can also run with the old operating system Windows XP/Vista if the latest version of .NET is installed. To download the appropriate framework related to the language, the user can visit the web site: www.microsoft.com.

Standard Colorimetry: Definitions, Algorithms and Software, First Edition. Claudio Oleari.
© 2016 John Wiley & Sons, Ltd. Published 2016 by John Wiley & Sons, Ltd.

Table 16.1   *Ordered list of the sections and toolboxes of the CX program.*

Sections	Number	Toolbox name
Monitor	1	Monitor setup
	2	Visual evaluation of the gamma
Colour-Vision test	3	Colour-Vision test
	4	Achromatic spectral stimulus detection
	5	Nagel-like anomaloscope
Visual contrast phenomena	6	Brightness contrast and crispening
	7	Brightness contrast in colour scales
	8	Brightness and chromatic contrast
	9	After image
Colour atlases	10	Ostwald Atlas
	11	Munsell Atlas
	12	NCS Atlas
	13	DIN Atlas
	14	OSA-UCS Atlas
CIE UCS systems	15	CIELAB
	16	CIELUV
Tristimulus and cone activation	17	Tristimulus and cone activation
Colorimetry	18	CIE colour specification
	19	CIE systems
	20	Chromaticity diagrams
	21	Fundamental systems
	22	Dominant wavelength and purity
	23	Tristimulus space transformation
	24	Colour difference ΔE
	25	Colour rendering index
Black body and daylight spectra	26	Black body and daylight spectra
Colour synthesis	27	RGB monitor colour additive mixture
	28	Halftone CMY printing
	29	Two-pigment mixture
	30	Four-pigment mixture
Tools	31	Spectral data view and download – illuminant-observer weights
	32	Saved file opening

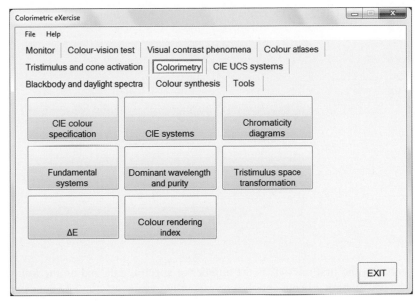

Figure 16.1 *Graphic interfaces collected in only one figure for opening by clicking on the toolboxes related to the 'monitor', 'Colour-vision test', 'Visual contrast phenomena', 'Colour atlases', 'Tristimulus and cone activation', 'Colorimetry', 'CIE UCS systems', 'Reference illuminant spectra', 'Color synthesis' and 'Tools'.*

The CX program is installed by double clicking on the icon of the file 'setup.exe'.

The numeric decimal separator is the dot '.'.

For a complete use of the CX program, the user has to be an administrator.

The correct colour management is obtained once the characteristics of the monitor used is set in CX program (preferably the CRT monitor). For a correct color reproduction it is also recommended to set "Windows Style" on "Windows Classic" from the Control Panel Menu. Graphic effects such as transparency may lead to altered color perception.

Any image obtained from CX may be used but the source must be cited.

16.1.2 Data Files

In the CD, in addition to CX program, is the folder 'Colour files' containing all the files in the original format used in the program and useful for any other calculations that the user wishes to make. The files are written in ASCII format and have the .TXT extension. The files, grouped into different areas, are:

- *Luminous efficiency function V(λ), folder 'Vlambda':*

CIE24_V.TXT	$V(\lambda)$ CIE 1924 photometric photopic observer (Section 7.3)
CIE51_Vp.TXT	$V'(\lambda)$ CIE 1951 photometric scotopic observer (Section 7.6)
CIE88_VM.TXT	$V_M(\lambda)$ CIE 1924 photometric photopic Modified observer (Section 7.4)
CIE05_V10.TXT	$V_{10}(\lambda)$ CIE 2005 10° photometric photopic observer (Section 7.7)
CIE_VF2.TXT	$V_{F2}(\lambda)$ CIE 2° Fundamental photometric observer (Section 7.8)
CIE_VF10.TXT	$V_{F10}(\lambda)$ CIE 10° Fundamental photometric observer (Section 7.8)

These files, for example, CIE24_V.TXT for $V(\lambda)$ CIE 1924, are written as follows

```
41
380,.000039
390,.00012
400,.000396
410,.00121
420,.004
...
540,.954
550,.995
560,.995
...
770,.00003
780,.000015
```

where the number in the first row, 41, is the number of spectral data and on the following 41 lines the numbers are the wavelength λ in nm and the corresponding value of $V(\lambda)$. The numbers are separated by a comma.

- *Colour matching functions* (CMF), folder 'CMF':

CIE31_xyz.TXT	CMF of the 2° CIE 1931 observer in XYZ
CIE64_xyz.TXT	CMF of the 10° CIE 1931 observer in $X_{10}Y_{10}Z_{10}$
CIEF2_xyz.TXT	CMF of the 2° CIE fundamental observer in $X_FY_FZ_F$
CIEF2_lms.TXT	CMF of the 2° CIE fundamental observer in LMS
CIEF10_xyz.TXT	CMF of the 10° CIE fundamental observer in $X_{F,10}Y_{F,10}Z_{F,10}$
CIEF10_lms.TXT	CMF of the 10° CIE fundamental observer in $L_{10}M_{10}S_{10}$
VOS_xyz.TXT	CMF of the Vos observer in X'Y'Z'
VOS_lms.TXT	CMF of the Vos observer in L'M'S'
CIE_SDO.TXT	CMF of the standard deviate observer

These files, for example CIE31_xyz.TXT for $\bar{x}(\lambda)$, $\bar{y}(\lambda)$ and $\bar{z}(\lambda)$, are written as follows

```
81
380,   0.0014,   0.0000,   0.0065
385,   0.0022,   0.0001,   0.0105
390,   0.0042,   0.0001,   0.0201
395,   0.0076,   0.0002,   0.0362
400,   0.0143,   0.0004,   0.0679
...
545,   0.3597,   0.9803,   0.0134
550,   0.4334,   0.9950,   0.0087
555,   0.5121,   1.0000,   0.0057
560,   0.5945,   0.9950,   0.0039
...
```

```
760,   0.0002,   0.0001,   0.0000
765,   0.0001,   0.0000,   0.0000
770,   0.0001,   0.0000,   0.0000
775,   0.0001,   0.0000,   0.0000
780,   0.0000,   0.0000,   0.0000
```

where the number in the first row, 81, is the number of spectral data and on the following 81 lines the numbers are the wavelength λ in nm and the corresponding values of $\bar{x}(\lambda)$, $\bar{y}(\lambda)$ and $\bar{z}(\lambda)$. The numbers are separated by a comma. In the case of the file CIE_SDO.TXT, the three numbers after the wavelength are $\Delta\bar{x}(\lambda)$, $\Delta\bar{y}(\lambda)$ and $\Delta\bar{z}(\lambda)$ according to Equation (9.21).

- *Illuminants*, folder 'Illuminants':

A.TXT	CIE illuminant A
B.TXT	CIE illuminant B
C.TXT	CIE illuminant C
D50.TXT	CIE illuminant D50
D55.TXT	CIE illuminant D55
D65.TXT	CIE illuminant D65
D75.TXT	CIE illuminant D75
ID50.TXT	CIE illuminant indoor D50
ID65.TXT	CIE illuminant indoor D65
D_S0S1S2.TXT	Principal components of daylight spectra $S_0(\lambda)$, $S_1(\lambda)$ and $S_2(\lambda)$
F_1. TXT, …	Fluorescent illuminant F1, F2, … , F12
FL3-1. TXT, …	new set of fluorescent illuminant FL3.1, FL3.2, … , FL3.15
HP_1.TXT, …	gas discharge illuminants HP1, HP2, … , HP5
G_1.TXT, …	gas discharge illuminants G1=NaLP, G2=NaHP, G3=MB, G4= MBF, G5=MBTF, G6=HMI, G7=Xenon

These files, for example, A.TXT for A standard illuminant are written as follows

```
107
300,      0.93
305,      1.13
310,      1.36
315,      1.62
320,      1.93
...
550,     92.91
555,     96.44
560,    100.00
565,    103.58
570,    107.18
...
820,    258.07
825,    259.86
830,    261.60
```

where the number in the first row, 107, is the number of spectral data and on the following 107 lines the numbers are the wavelength λ in nm and the corresponding radiometric value of $S(\lambda)$. The numbers are separated by a comma.

The file D_S0S1S2.TXT, principal components of daylight spectra (Section 8.4), is written as follows

```
300,      0.04,      0.02,      0.00
305,      3.02,      2.26,      1.00
310,      6.00,      4.50,      2.00
315,     17.80,     13.45,      3.00
320,     29.60,     22.40,      4.00
...
550,    104.40,      1.90,     -0.30
555,    102.20,      0.95,     -0.15
560,    100.00,      0.00,      0.00
565,     98.00,     -0.80,      0.10
570,     96.00,     -1.60,      0.20
...
800,     61.00,     -9.70,      6.40
805,     57.15,     -9.00,      5.95
810,     53.30,     -8.30,      5.50
815,     56.10,     -8.80,      5.80
820,     58.90,     -9.30,      6.10
825,     60.40,     -9.55,      6.30
830,     61.90,     -9.80,      6.50
```

where on each line the numbers are the wavelength λ in nm and the corresponding radiometric values $S_0(\lambda)$, $S_1(\lambda)$ and $S_2(\lambda)$ defined in Section 8.4.

- *Munsell's standard for colour-rendering index*, folder 'CRI':

The files are 14 and named CRI_01.TXT, CRI_02.TXT, …and CRI_14.TXT. For example, the file CRI_01.TXT is written as follows

```
400,.256
420,.244
440,.230
460,.220
...
640,.451
660,.451
680,.455
700,.462
```

where on each line the numbers are the wavelength λ in nm and the reflectance factor value $R(\lambda)$

The tables used in the CX software, and downloadable from the folder COLORFILES, are free published data or are reproduced with the permission of the CIE, Central Bureau, Kegelstrasse 27, A-1030 Wien (ciecb@ping.at). Most of these tables are downloadable from the CIE website www.cie.co.at/ and from the CVRL website www.cvrl.ucl.ac.uk/

16.2 Monitor

A colorimetrically correct use of a monitor needs a *colour calibration* for a properly coded input colour signal to produce a colour response on the monitor, whose measure is equal to the colour specification of the signal. Colour calibration is a requirement for all the devices linked to a computer, such as monitors, printers, scanners, cameras, ... The colour calibration of a monitor is made by a colorimeter or, better, a spectroradiometer, and dedicated software. The market offers many solutions with different costs. Colour calibration is necessary for a correct view on a monitor of a digital photograph obtained by a camera or scanner. In this case the observer is the CIE 1931, the image format is sRGB, the white of the monitor is D65 and the adaptation illuminant for the observer is D50. The CX software is not conceived for displaying images, thus no colour calibration is needed, but is conceived for correctly reproducing single colours on a monitor, that are numerically specified. The correct reproduction of these colours needs some operations explained below. The inexpert user may skip this section and work with the default settings, and come back here to carry out these operations, once he knows their reasons.

16.2.1 Monitor Setup

The characteristics of the CRT monitor are (Section 15.2)

1. the chromaticities of the three reference colours – red, green and blue – and of white and
2. the gamma of the monitor response (Section 16.2.2).

The chromaticities of the three reference colours of the monitor should be supplied by the monitor producer, who, optimistically, supplies only the chromaticity of the reference colours for the CIE 1931 observer. Otherwise these data, if unknown, are obtainable from the spectral power distribution of the red, green, blue and white lights measured by a spectroradiometer and computation by the 'CIE colour specification' program of the 'Colorimetry' toolbox of CX program for all the observers: the CIE 1931, CIE 1964, Vos and the CIE fundamentals for 2° and 10°. The chromaticities of the reference colours of the monitor can be written in the

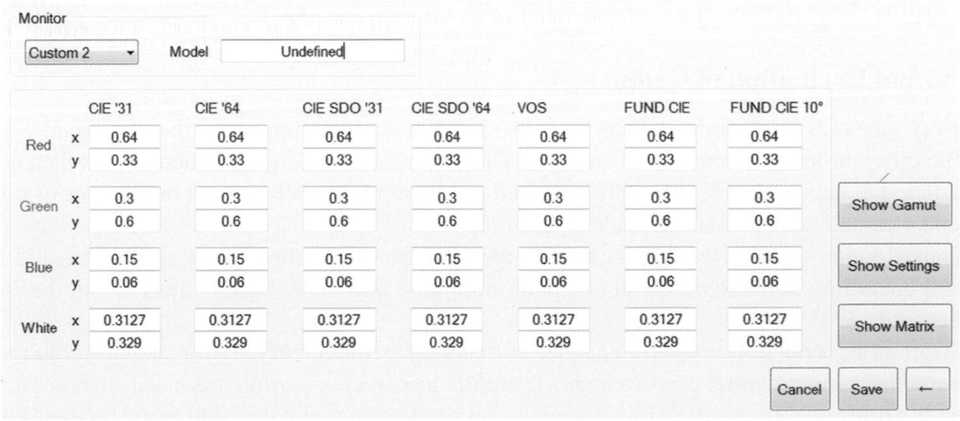

		CIE '31	CIE '64	CIE SDO '31	CIE SDO '64	VOS	FUND CIE	FUND CIE 10°
Red	x	0.64	0.64	0.64	0.64	0.64	0.64	0.64
	y	0.33	0.33	0.33	0.33	0.33	0.33	0.33
Green	x	0.3	0.3	0.3	0.3	0.3	0.3	0.3
	y	0.6	0.6	0.6	0.6	0.6	0.6	0.6
Blue	x	0.15	0.15	0.15	0.15	0.15	0.15	0.15
	y	0.06	0.06	0.06	0.06	0.06	0.06	0.06
White	x	0.3127	0.3127	0.3127	0.3127	0.3127	0.3127	0.3127
	y	0.329	0.329	0.329	0.329	0.329	0.329	0.329

Figure 16.2 *'Monitor setup' dialogue window of the 'monitor' toolbox. Chromaticities of the reference red, green, blue and white lights of the monitor specified according to the CIE 1931, CIE 1964, CIE fundamentals 2° and 10°, Vos observers and Standard Deviate Observers. Here the numbers are related to an sRGB monitor and by default are the same for all observers. The user should write the chromaticities of his/her monitor.*

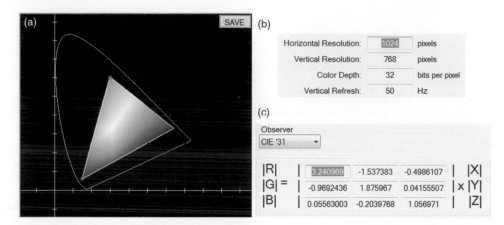

Figure 16.3   *(a) Chromaticity diagram CIE 1931 with the triangle representing the reference RGB colours of the monitor for the standard CIE 1931 observer, shown in the window 'Show gamut'. (b) Current monitor setting shown in the window 'Show settings' (c) Transformation matrix from the XYZ CIE 1931 reference frame to the reference frame associated with the RGB colours of the monitor shown in the window 'Show matrix'.*

dialogue window 'monitor setup' (Figure 16.2) of the 'monitor' toolbox. Figure 16.3 shows the content of the windows 'Show Gamut', 'Show Settings' and 'Show Matrix'.

The correct colour management needs the measurement of the 'gamma' and of the 'gain' of the monitor. All these quantities can be evaluated visually by using the 'Visual evaluation of the gamma' program of the 'monitor' toolbox (Section 16.2.2).

The CX program runs correctly for the standard CIE 1931, CIE 1964, Standard Deviate Observers, CIE fundamental observers and for the Vos observer after the colorimetric specifications of the reference colours of the monitor are written in the 'monitor' window.

When no setting is made, the program runs supposing that the monitor is an sRGB monitor. The consequence is that, if the monitor is a true sRGB monitor, the colours are correctly reproduced only for the CIE 1931 observer. If the monitor is not a true sRGB monitor, nothing can be said about the quality of the reproduction. Generally, the reproduction is approximate for all the observers.

16.2.2   Visual Evaluation of Gamma (γ)

A cathode ray tube (CRT) monitor converts the video signal to emitted luminance in a nonlinear way (Section 15.2.2). The other monitors mimic a CRT monitor. The *gamma correction* signal neutralizes such non-linearity. In practice the display devices have different values of gamma, which depend on the type of display and on the luminance and contrast setting made by the user.

The standard video signal used in personal computers depends on the Operating System, and today the video-signal encoding, generally used in computer monitors, scanners, digital cameras and the Internet, is sRGB (Section 15.2.4).

This CX program is not conceived for encoded colour management, but only for showing colours obtained by computations or instrumental measurement. Therefore a correct colour display needs the measurement of the gamma distortions of the three colour channels and a colour signal correction made on their knowledge. This program proposes visual evaluation of the gammas. This γ correction is evaluated by visual comparison of two contiguous squares with the same reference colour – red, green and blue –. One right square emits light on all its surface, while the left square emits only on alternate lines. The user can modulate the intensity of the right square by moving a slider with the mouse. In correspondence with the fixed luminance of this second

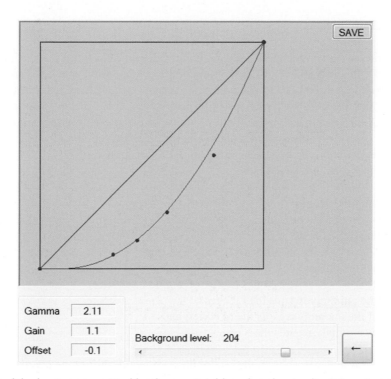

Figure 16.4   *Plot of the luminance emitted by the monitor blue phosphor (ordinate) against the signal producing such a luminance (abscissa). The blue dots are empirical points and the blue line is the best interpolating line. The values of the parameter – Gamma, GAIN and OFFSET – regard the interpolating line.*

square, the signal driving the right square is modified until it appears equal to the left one (the squares have to be viewed at a distance such that the alternate lines are undistinguishable.)

Once the averaged luminances are matched, the luminous lines of the left square have a luminance equal to the double luminance of the right square, because emitting areas of the two squares are in a ratio of 1 to 2. Then, clicking on "next matching" both squares are driven with the signal defined for right and the user search for a new match. By repeating this operation four or five times, a correspondence between the signals and luminances is generated. By clicking on 'Gamma computation' the program evaluates the best parameters – γ, *GAIN* and *OFFSET* – of the unknown fitting line (Figure 16.4). These matches are made also for the white light for a check. By clicking on 'update gamma record', these parameters combined with the chromaticity of the reference RGB lights and of the White light are recorded in the database. Any time the program is launched, these parameters are retrieved and consequently the colour signals are modified in order to reproduce the colours correctly.

In practice, the colorimetry of the CRT monitors is defined for the CIE 1931 observer, although the systematic error of this observer in the short wavelength region below 460 nm does not reproduce the blue-phosphor luminance correctly. The CX program reproduces the colours correctly for all the observers – the CIE 1931, CIE 1964, Standard Deviate Observers, CIE fundamental 2° and 10° and VOS – if the reference lights of the monitor have been specified according to all the observers and their chromaticities inserted in the 'Monitor Setup' toolbox.

In practice, LCD monitors have an emission of light that is not Lambertian, that is, dependent on the direction of light emission and therefore on the viewing position of the observer. This phenomenon makes gamma correction difficult and generally viewing quality not completely satisfactory, but the quality of LCD monitors is increasing.

16.3   Colour-Vision Tests

The explanation of these colour-vision tests needs a short recall of the physiology of the human colour vision system (Sections 5.4, 5.5 and 6.13.7) and a description of the colour-vision anomalies and deficiencies.

Colour vision is based on three kinds of photosensitive cones, which are activated by the number of photons absorbed. The three kinds of cones have typical spectral sensitivity functions – that is, colour-matching functions –.

The observers with different spectral sensitivities to those of the standard observer are named *anomalous trichromats*; the observers with only two kinds of cones are *dichromats* (or *daltonics*); the observers with only one kind of cones are *monochromats* and have no chromatic discrimination.

Dichromats are subdivided into three kinds:

1.   *protanopes*, lacking in L cones;
2.   *deuteranopes*, lacking in M cones; and
3.   *tritanopes*, lacking in S cones.

Analogously, three kinds of *anomali*:

1.   *protanomali*, if the anomaly regards the L cones;
2.   *deuteranomali*, if the anomaly regards the M cones; and
3.   *tritanomali*, if the anomaly regards the S cones.

The severity of the anomaly depends on the size of the wavelength shift of the photopigment absorbance. The causes of almost all the deficiencies have a molecular genetic explanation.

The vision of dichromats and anomalous trichromats allows a normal life and visual deficiencies are often shown only by specific tests.[2–10]

Because dichromats have only two kinds of cones, their colour sensations are represented in a two-dimensional space (Section 6.13.7). The comparison of this space with that of a trichromat shows that sets of equal colours for one kind of dichromat correspond to sets of different colours for the trichromats. The points representing the chromaticities of these colours belong to straight lines on the chromaticity diagram, named *confusion lines* (Figures 6.32 and 16.6). Any kind of dichromats has its own confusion lines. The confusion lines of any kind of dichromat have a common point named *confusion point* (Figures 6.32 and 16.6). Particular spectral radiations appear as *achromatic* to the dichromats and the wavelength of this neutrality characterizes the different kinds of dichromats (Table 16.2).

Luminance, which is a photometric quantity related to the luminosity sensation, depends on the activations of the M and L cones. Dichromats with one kind of these cones lacking or anomalous trichromats have different luminous sensations to those of the standard observer. This phenomenon is very small and not discriminating.

The tests presented in this software diagnose the anomalies of the visual system of the observer, although in a non-decisive way.

At the end, diagnosis is accomplished by comparing the results of the considered observer and the standard observer and by combining the results of the different tests.

The set of tests begins with considering 15 colour patches, equiluminant for the CIE 1931 standard colorimetric observer and for the CIE 1924 standard photometric observer, and is organized in two steps, both considerable as tests:

Luminance Matching of Heterochromatic Colours

The matching of the luminances of the 15 samples perceived by the individual observer, obtained by changing the luminances until minimum is the distinction of the borders between samples and background (*minimum*

Table 16.2 *Human population and colour vision anomalies and deficiencies.*

Observers		♂ %	♀ %	Wavelength with $V(\lambda)_{max}$ [nm]	λ [nm] Neutrality	Cones
Normal Trichromat		~92	~99.5	555		Three kinds of cones (*L, M, S*) with normal spectral sensitivities
Anomalous Trichromat	Protanomalus	1.0	0.02	540		Three kinds of cones with spectral sensitivities with maximum shifted with respect to the normal.
	Deuteranomalus	4.9	0.38	560		
Dichromat	Protanope	1.0	0.02	540	494	Without L cones
	Deuteranope	1.1	0.01	560	499	Without M cones
	Tritanope	0.002	0.001	555	570 400	Without S cones
S cones Monochromat				~440		With only S cones
Rods Monochromat		0.003	0.002	507		With only rods

distinct border technique) (Figure 16.5a). Once the heterochromatic matching is obtained for all the colours, the comparison between the user and the standard observer is made by clicking on the corresponding toolbox (Figure 16.5b). The differences between the standard observer and individual observers are small: tritanopes have normal photometric vision and the others have the contribution to the luminance given by M or L cones, whose spectral differences are small. Normal observers, with the crystalline lenses of the eyes yellowish and darkened by age, have filtering of the light before the light crosses the retina. This filtering could produce a different luminous sensation with respect to a young observer, partially minimized by an adaptation phenomenon. A limitation of this test is due to the number of 8 bits used to specify the colours, in accordance with the sRGB standard. This test has a low discrimination power and the diagnosis is difficult and not decisive. The 8-bit colour coding is a very strong limitation

Ordering of Heterochromatic Colours

This test regards the chromatic ordering of 13 of the 15 colour samples, previously matched in luminance and recalls the Farnsworth test. The ordering starts from a fixed sample and follows on by choosing the closest perceived colour sample from the samples not yet chosen until the whole sample set is exhausted (Figure 16.6a). Once the colour ordering is obtained for all the colours, the comparison between the user and the standard observer is made by clicking on the corresponding toolbox (Figure 16.6b). Figures 16.6a, 16.6b and 16.6c are related to a deuteranopic observers while 16.6 (d) to a normal observer. This test is significant.

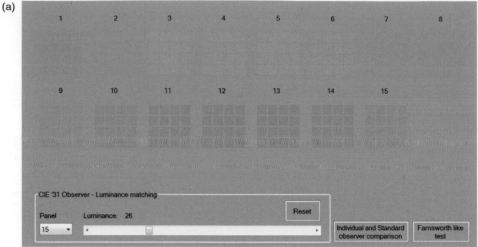

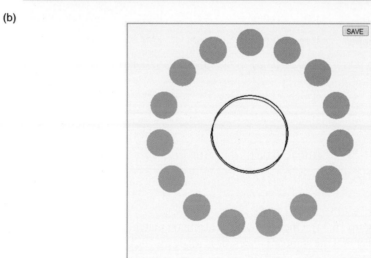

Figure 16.5   *(a) Fifteen colour patches with an internal fragmentation on a uniform grey. The luminance of these patches and of the ground is the same for the CIE 1924 standard photometric observer. The user must modify the luminance of the colour patches with that of the ground by minimizing the perceived border between the colour patches and the background. The colour patch is selected by clicking on the patch or via the panel window. The luminance is modulated by moving the luminance slider with the mouse. (b) Plot of the luminance variations made in the experiment of Figure16.5a represented by a circular red line in comparison with a black circle associated with the standard photometric observer. Any increment in the luminance is represented by a distance from the centre greater than that of the circle and, conversely, any decrement by a minor distance. The difference between the red circular line and the black circle is representative of the difference between the user and the standard photometric observer. The origin of such a difference may be either a yellowing of the crystalline lens, or anomalies, or dichromatism. The difference between the standard observer and individual observers is small, as in the case shown here. The 8-bit colour coding is a very strong limitation.*

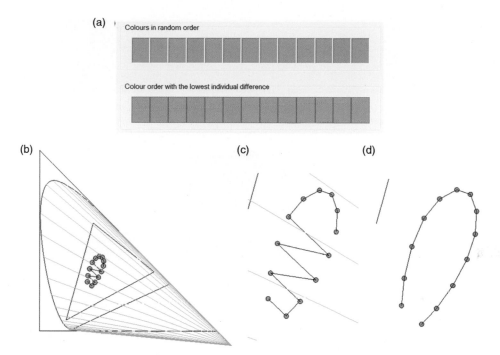

Figure 16.6 *(a) Ordering made by a deuteranope of 13 colour patches proposed in random order. The user must order these colour patches in such a way that any colour has the minimum perceived difference with respect to the previous one. The choice of the colour sample to put in the ordered sequence is made by clicking on the colour sample and, to correct an incorrect movement, by clicking on the moved sample. (b) Chromaticity diagram CIE 1931 with a black line connecting the colour samples of Figure 16.6a in the perceived order produced by a deuteranope. The green lines are typical deuteranopic confusion lines, to be compared in direction with the jumps of the black line connecting the colour samples. (c) Enlarged part of the chromaticity diagram (b) related to the ordering of Figure16.6a. (d) Enlarged part of the chromaticity diagram related to the ordering of a normal trichromat.*

Achromatic Spectral Stimuli Detection

The third test regards the detection of perceived achromatic colours on the colour circle. Dichromats have two achromatic points on the colour circle, of which one is easier to detect. At the top of the screen, the colour of the colour circle is shown on a strip [Figure 16.7a]. At the centre of the screen is a grey square (grey selectable from black to white) with a coloured disk inside it, whose colour is one of those in the colour circle. The observers change the colour in the disk searching for an achromatic colour, if they see one. The observer has to record the detected achromatic colour by clicking on the 'Record first/second achromatic appearance' button. After making the detection by clicking on the 'Show dominant wavelengths' button, a window will appear showing the wavelengths associated with the achromatic points [Figure 16.7b]. The protanopes have an achromatic point around 490–500 nm, the deuteranopes around 500–505 nm and both have a second achromatic point in the magenta hue region (Table 16.2). The tritanopes have one achromatic point around 565–575 and a second one around 400 nm (Table 16.2). Since the colour presented by the monitor is not monochromatic but obtained from broad band lights, the achromaticity is perceived with some uncertainty, mainly in the magenta hues. Anyway, the method is well selective for dichromats.

(a)

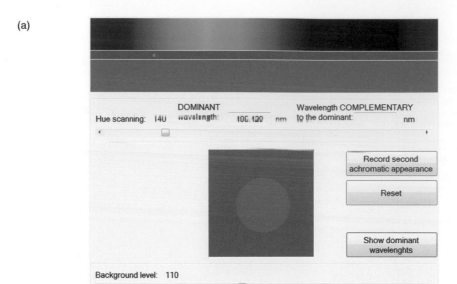

(b)

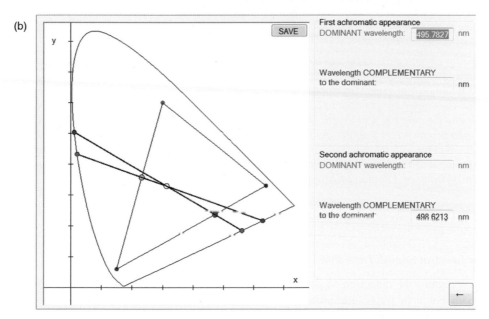

Figure 16.7 *(a) Window for achromatic spectral stimuli detection. (b) Chromaticity diagram with the wavelengths that appear achromatic to the observer. In this deuteranopic case, the wavelength appearing as achromatic is ~502 nm, and the magenta hue appearing as achromatic corresponds to a complementary wavelength ~499, as expected for a deuteranope.*

Red-Green Mixture-Yellow Matching

The fourth test regards the matching of a red-green colour mixture with yellow colour, and recalls the use of the Nagel anomaloscope and the Rayleigh match equation. In this anomaloscope, the light sources are red, monochromatic with wavelength $\lambda_R = 670\text{–}671$ nm, green, $\lambda_G = 546\text{–}549$ nm and yellow, $\lambda_Y = 589$ nm, and are represented by colour stimuli with chromaticities lying on the confusion line common to the protanopes and deuteranopes.

Consider a yellow radiance $L_{Y,\lambda}$ obtained as sum of the radiances $L_{R,\lambda}$ and $L_{G,\lambda}$, of the red and green lights of the monitor, respectively and suppose as an approximation that these radiances do not activate the S-cones. The activation of the L- and M-cones produced by these lights when separated are

$$
L_R = \int_{380}^{780} L_{R,\lambda}\,\overline{l}(\lambda)\,d\lambda, \;\; L_G = \int_{380}^{780} L_{G,\lambda}\,\overline{l}(\lambda)\,d\lambda
$$
$$
M_R = \int_{380}^{780} L_{R,\lambda}\,\overline{m}(\lambda)\,d\lambda, \; M_G = \int_{380}^{780} L_{G,\lambda}\,\overline{m}(\lambda)\,d\lambda
\tag{16.1}
$$

and when together are

$$
L_Y = \int_{380}^{780} \left(L_{R,\lambda} + L_{G,\lambda} \right)\overline{l}(\lambda)\,d\lambda = L_R + L_G
$$
$$
M_Y = \int_{380}^{780} \left(L_{R,\lambda} + L_{G,\lambda} \right)\overline{m}(\lambda)\,d\lambda = M_R + M_G
\tag{16.2}
$$

A modulation of the radiance of the red light by a factor R and that of the green light by G, with the sum leaving the activations L_Y and M_Y unchanged is represented by the equations

$$
\begin{cases} L_R + L_G = L_Y \\ M_R + M_G = M_Y \end{cases} \rightarrow \begin{cases} RL_R + GL_G = L_Y \\ RM_R + GM_G = M_Y \end{cases}
\tag{16.3}
$$

This set of two equations with two parameters, R and G, considered as unknowns, has only one solution, that regard the trichromats and is $R = G = 1$. For the protanpopes and deuteranopes, the two equations of the set have to be considered separately and it follows that R and G have a linear relationship and the parameters of these equations are of the protanopes and deuteranopes, respectively,

$$
\text{deuteranopes } RL_R + GL_G = L_Y \;\; \rightarrow R + G\frac{L_G}{L_R} - \left(1 + \frac{L_G}{L_R} \right) = 0
$$
$$
\text{protanopes } \;\; RM_R + GM_G = M_Y \rightarrow R + G\frac{M_G}{M_R} - \left(1 + \frac{M_G}{M_R} \right) = 0
\tag{16.4}
$$

The test proposed by the software comprises searching for a different colour match to $R = G = 1$ (Figure 16.8a). If such a match exists for the observer, he or she is not a trichromat and the parameters of the fitting line characterise the kind of dichromatism (Figure 16.8b).

This test is approximate because the chromaticities of the red and green colours of the monitor do not belong to a confusion line common to the protanopes and deuteranpes, although not too distant. For correctness, the observer should look through an optical high-pass filter with cut-off at the 550 nm wavelength to be sure of the S-cone exclusion.

The software considers the colour stimuli in the RGB laboratory reference frame, where the tristimulus value is represented by 8-bit numbers. In this case, the two stimuli displayed in the two halves of the bipartite field are $(R, G, 0)$ and $(Y, Y, 0)$, where $0 \le R, G, Y \le 255$. These stimuli specified in the fundamental reference frame become (Section 6.13.4)

$$
\begin{pmatrix} L \\ M \\ S \end{pmatrix} = \mathbf{T} \begin{pmatrix} R \\ G \\ 0 \end{pmatrix} = \begin{pmatrix} L = T_{11}R + T_{12}G \\ M = T_{21}R + T_{22}G \\ S = T_{31}R + T_{32}G \end{pmatrix}, \quad \begin{pmatrix} L_Y \\ M_Y \\ S_Y \end{pmatrix} = \mathbf{T} \begin{pmatrix} Y \\ Y \\ 0 \end{pmatrix} = \begin{pmatrix} L_Y = T_{11}Y + T_{12}Y \\ M_Y = T_{21}Y + T_{22}Y \\ S_Y = T_{31}Y + T_{32}Y \end{pmatrix} \tag{16.5}
$$

where $\mathbf{T}$ is the transformation matrix. Colour matching exists between the two halves of the bipartite field if

$$
\begin{cases} L = L_Y = T_{11}R + T_{12}G = T_{11}Y + T_{12}Y \\ M = M_Y = T_{21}R + T_{22}G = T_{21}Y + T_{22}Y \\ S = S_Y = T_{31}R + T_{32}G = T_{31}Y + T_{32}Y \end{cases} \tag{16.6}
$$

In the Nagel anomaloscope $S = S_Y = 0$, for which the set of two equations has to be considered. In the case of dichromates, protanopes and deuteranopes, the equation system becomes a single equation for each of the two types of dichromatism, that represent a straight line (Figure 16.8b) for the L cones of deuteranopes

$$
\frac{R}{G} = -\frac{T_{12}}{T_{11}} + \left(1 + \frac{T_{12}}{T_{11}}\right) \frac{Y}{G} \tag{16.7}
$$

and analogously for the M cones of protanopes

$$
\frac{R}{G} = -\frac{T_{22}}{T_{21}} + \left(1 + \frac{T_{22}}{T_{21}}\right) \frac{Y}{G} \tag{16.8}
$$

Stimuli $(R, G, 0)$ and $(Y, Y, 0)$, matched for deuteranopes or protanopes, satisfy the Equations (16.7) or (16.8), respectively. For normal trichromat observers, the two Equations (16.7) and (16.8) have to be satisfied simultaneously, and this happens for the intersection point of the two straight lines. For normal trichromat observers there is only one colour match. For protanopes and deuteranopes all the points of the corresponding straight lines represent good colour matches. The identification of the lines classifies the type of colour blindness.

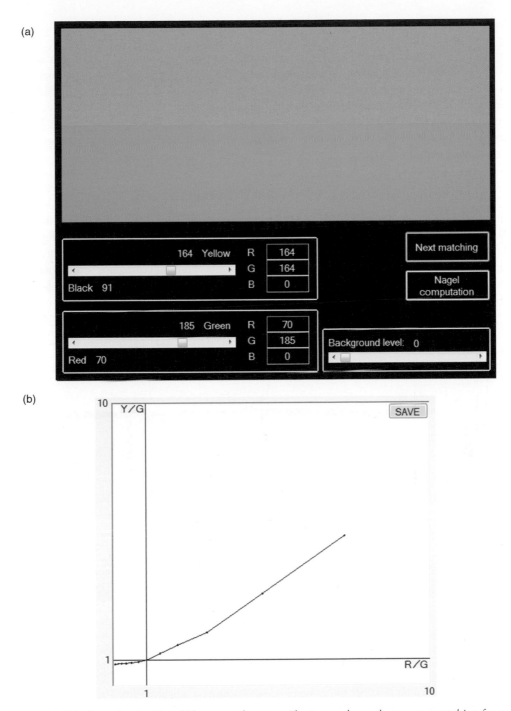

Figure 16.8 *(a) Windows for the Nagel-like anomaloscope. The two colours shown are matching for a deuteranope. (b) Pairs of (R/G) and (Y/G) data related to a set of matches are plotted and show the typical straight trend of a dichromate as requested by* Equations (16.7) *and* (16.8), *that in this case is a deuteranope.*

16.4   Visual Contrast Phenomena

This section of the CX software is didactical and presents three visual phenomena requiring an active position of the observer. Other visual static phenomena are presented in Section 4.7. All these phenomena, belonging to the set of the colour-appearance phenomena, have almost no rule in colorimetry, if we consider many colour-appearance phenomena external to the classical CIE colorimetry, anyway they are presented here in order to know the borders of the colorimetry.

This software considers in the next sections examples of *simultaneous brightness contrast*, of *simultaneous chromatic contrast* and of *crispening*.

16.4.1   Simultaneous Brightness Contrast and Crispening

This section presents phenomena related to the brightness induced by contiguity of different fields.

The 'Brightness contrast and crispening' program shows two equal sets of rectangles in grey scales, from the black to the white of the monitor (Figure 16.9a). The difference between the two sets is produced by the separating line, that originally is set black in the left set, while the right rectangles are mutually touching. The consequence is that the left rectangles appear internally uniform and the right rectangles show brightness increasing from top to bottom.

After launching the 'Brightness contrast and crispening' program, the observer can modify the luminance of the left background and, the brightness contrast crispening appears: the rectangles brighter than

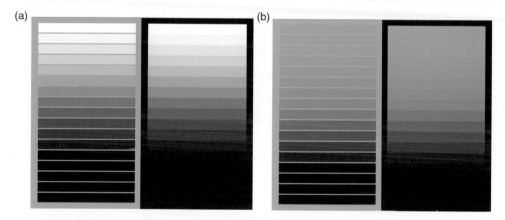

Figure 16.9   *(a) Simultaneous brightness contrast and brightness crispening in achromatic figure. This figure is the same presented by the 'Brightness contrast and crispening' program in an achromatic and static situation. Two phenomena are represented in this figure. Two exactly equal grey scales are represented. The strips of the left side scale are on a uniform grey background while on the right side contiguous strips are mutually touching. The left side strips appear uniform and the right side strips appear with a non-uniform luminosity, that is higher where a strip is touching a darker strip and is lower where is touching a lighter strip. The observer can isolate a left rectangle by using two uniform and equal cards and so verify the internal uniformity of the rectangle. It appears that the left and right rectangles have equal luminance. This phenomenon is the simultaneous lightness contrast. Moreover, the left side strips that are darker than the background appear darker and the left side strips that are lighter than the background appear lighter. This phenomenon is the brightness contrast crispening. Any variation in the background luminance shifts this discontinuity line. (b) Simultaneous brightness contrast and brightness crispening in chromatic figure.*

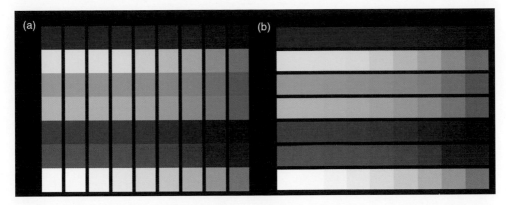

Figure 16.10 *Simultaneous brightness contrast in chromatic figure. Two figures are represented, related to situations 3 and 4 presented in the 'Brightness contrast in colour scales' program. These two figures are constituted by equal coloured squares, that are ordered in equal hue lines with decreasing lightness from left to right. Black lines are separating the coloured squares: in the left figure the black lines separate the squares of different lightness and in the right part separate the squares with different hues. In the right figure, the contact between equal hue and different luminance squares induces a lightness simultaneous contrast such that the squares appear non-uniform, darker at left and lighter at right. In the left figure, the contact between different hues produces a non-visual effect and the squares appear uniform as they are physically. This does not mean that chromatic interaction does not exist, but means only that in this situation does not induce visible colour gradients.*

the background appear with increased brightness while the darker ones appear with lowered brightness. The effect is particularly evident during the continuous modification of the background.

The observer can modify the hue of the rectangles and verify that the phenomenon is hue independent (Figure 16.9b).

16.4.2 Simultaneous Brightness Contrast in Colour Scales

The 'Brightness contrast in colour scales' program shows a matrix of colour samples, in which any row has defined hue and the brightness increases from right to left (Figure 16.10). The colour samples can be seen in four different ways with different appearances:

	Structure of the scene	Appearance
1	The colour samples are separated by a black line vertically and horizontally	The colour samples appear internally uniform
2	No black line separates the colour samples	The colour samples appear internally non-uniform and bended as a cylindrical surface with a vertical axis.
3	Only the rows are separated by black lines	The colour samples appear internally non-uniform and bended as a cylindrical surface with a vertical axis.
4	Only the columns are separated by black lines	The colour samples appear internally uniform

The experiment shows only the dependence of the brightness on the luminance of the contiguous fields. The phenomenon has an explanation in the physiology with the receptive fields.

Figure 16.11   *Example of an image obtained by the 'Brightness and chromatic contrast': an open ring defined by additive mixing of Red = 180, Green = 178 and Blue = 65, is over to two rectangles as backgrounds, defined by Red = 201, Green = 116 and Blue = 65, and Red = 184, Green = 220 and Blue = 155, respectively. The ends of the open ring are on different backgrounds and the upper end appears lighter and slightly more greenish while the lower end appears darker and slightly reddish. These appearance differences are known as brightness contrast and chromatic contrast, respectively.*

16.4.3   Brightness and Chromatic Contrast

Simultaneous contrast is presented in Section 4.7. In this section a case of simultaneous contrast is considered, where a more complex figure is guided by the user.

The 'Brightness and chromatic contrast' program shows an open ring (Figure 16.11), whose background is constituted by two contiguous fields with different colours and with a separation region between the two ends of the open ring. The observer can continuously modify the contents of the reference colours – Red, Green and Blue – that additively constitute the colour of the open ring. Generally, the colour of the open ring appears different in the two parts, going from one colour to the other continuously. The perceived difference of the two parts of the open ring can be a lightness difference and/or a chromatic difference. The observer, by modifying the colours of the open ring and of the two backgrounds, can investigate almost all the possible situations of simultaneous contrast and detect the situations where the phenomenon regards the lightness, hue and colourfulness.

16.4.4   After Image

After-image is the typical phenomenon produced by *successive contrast* (Section 4.7) and happens when a viewer sees an imaginary coloured field in the area of view after having intently viewed an object – coloured pattern – for a while and then moved their gaze to a different and contrasting area. The 'After image' program shows this effect explained in the caption of Figure 16.12.

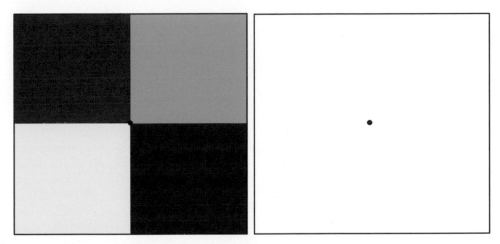

Figure 16.12   *The 'After image' program displays the left coloured pattern. The observer has to stare at it for approximately 15 seconds (better with one eye) with attention fixed on the black dot in the centre of the pattern. Then by clicking quickly the pattern changes and a figure equal to the right square follows. The observer must continue to be with attention fixed on the black dot in the centre of the square. The four coloured squares of the previous image immediately appear in complementary colours. This afterimage is due to a local colour adaptation of the retina. Moreover the afterimage has smooth edges, revealing mutual interaction between colour stimuli related to contiguous coloured regions. By way of explanation, this phenomenon is due to the reduced sensitivities of the cones produced by constant stimulation for a long time.*

16.5   Colour Atlases

WARNING – This section is dedicated to the monitor reproduction of the most important colour atlases. A warning must be given to the user. There are so many different monitor screens displaying different colours, also approximately calibrated. The accuracy of colour vision on monitors is limited. Therefore the view of atlas colours on monitor is for screening and didactical purposes only, not for substituting the use of real atlases. Moreover, the atlases have to be seen under proper controlled lighting, to which the observer must be adapted. The practical use of colour atlases is made for the visual colour specification of a real object or of a colour to be used in an object in a project. It is used by taking the colour samples off the atlas and comparing their contact edge with the object of study under controlled illumination.

This section considers only colour atlases presented in Chapter 14, therefore, before dealing with this program, the user should read that chapter.

The Background of any view of the atlas is grey lit by the illuminant used and its luminance factor is chosen by the user.

Clicking on each sample colour atlas opens a window with all the colorimetric data and presents the selected colour on three backgrounds, one white, one black and one grey with the chromaticity of the considered illuminant, whose luminance factor is variable. By varying this factor, the operator can identify the situation of minimal border distinction between the sample and background. In this situation, the colour sample and the background should have the same luminance factor. The numerical difference between the background level and the luminance factor of the sample is due to many factors: observer, 8-bit colour digitization, approximate monitor calibration, Helmholtz-Kohlrausch effect, etc. This is a very useful exercise.

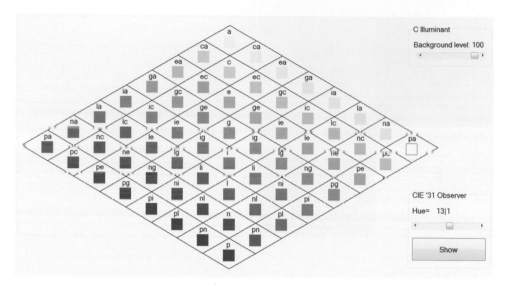

Figure 16.13   *Page at defined hue 13|1 of Ostwald's atlas, corresponding to the unique yellow (right) and its complementary hue (left). The yellow page has to be compared with the corresponding one of the NCS atlas at the same unique yellow (Figure 16.17).*

16.5.1   Ostwald's Atlas

Ostwald's Atlas is described in Section 14.3.

No illuminant and no observer are defined for viewing and specifying this colour atlas. Simply the illuminating light must be achromatic. The atlas here presented is specified for the illuminant C and the observer CIE 1931 (Figure 16.13).

This colour specification of the Ostwald Atlas is published[11] (Section 14.3).

16.5.2   Munsell's Atlas

Munsell's Atlas is described in Section 14.4.

This program shows the whole Munsell system of colours renotated by Newhall, Nickerson and Judd under the standard C illuminant and for the CIE 1931 standard colorimetric observer.

The comparison with the DIN system shows the deep difference between chroma and saturation (Figures 16.14, 16.15 and 14.15).

All the numerical data related to the Munsell renotation, used for the reproduction of this atlas on computer monitors, are published[12] (Section 14.4).

The Munsell Atlas on the market is different than the renotated published version and used here (different observer and illuminant).

The Munsell Atlas is sold by 'X-Rite, Incorporated 4300 44th St. SE Grand Rapids, MN 49512, USA,' (http:/www.xrite.com/top_munsell.aspx?action=products).

16.5.3   DIN's Atlas

DIN's Atlas is described in Section 14.5.

The DIN 6264 system of the *Deutsche Institut für Normung* (DIN) is defined by a compromise on the colour appearance and the CIE 1931 system, by which any colour sample is easily specified.

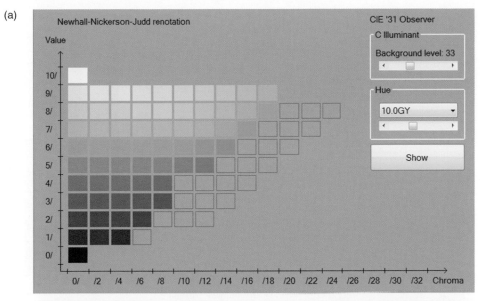

(a)

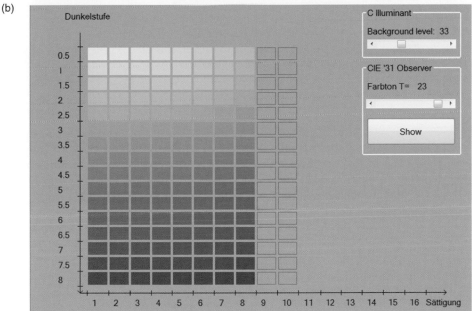

(b)

Figure 16.14 *(a) Organization of the 10GY constant hue page of Munsell's colour system as shown by the program. The lines radiating from the black point are loci with equal saturation and equal chromaticity (Figure 14.15). (b) DIN's page at defined Farbtone T = 23. All the colours at equal Sättigung have the same chromaticity. Compare Figure 16.14a with Figure 16.14b and Figure 14.15.*

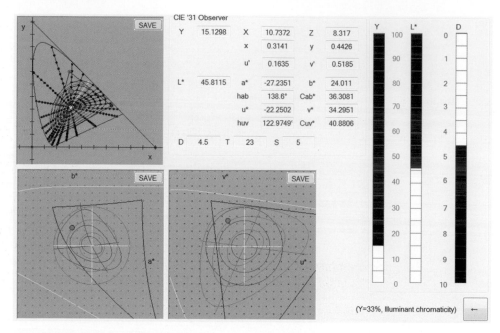

Figure 16.15   *DIN's colour sample (T = 23, D = 4.5, S = 5). Chromaticity of the DIN colours on the CIE 1931 Chromaticity diagram and constant lightness plane in CIELAB and CIELUV. The chromaticities at defined Farbtone and Sättigung are the same for different values of the Dunkelstufe. This window is open by clicking on a colour sample of the DIN atlas and gives any colorimetric information on the selected colour.*

Figure 16.14b shows a page of the DIN atlas with a hue close to the chosen one for Figure 16.14a of the Munsell Atlas. The comparison of these two figures shows the difference between saturation and chroma, the two quantities spanning the abscissa in the two atlases.

The colour specification of this atlas is for the C illuminant and for the CIE 1931 standard colorimetric observer.

All the numerical data used for the reproduction of this atlas on a computer monitor are published[13] (Section 14.5).

The 'Farbsystem DIN 6164' atlas is sold by 'Munster-Schmidt Abt. Verlag, Brauweg 26a, Postfach 2741, D 37073 Göttingen.' (http://www.unitycolor.com/Color-Sample-Store/Color-Cards-others/Germany/DIN-6164-color-ystem::206.html)

16.5.4   OSA-UCS' Atlas

OSA-UCS's Atlas is described in Section 14.6.

This system has been conceived for the CIE 1964 standard colorimetric observer and the colour specification is for the standard D65 illuminant.

In OSA-UCS system are particularly interesting palettes made by the various sections of the colour space (Figure 16.16), for which the system was considered interesting for artists.

The numerical data used for the reproduction of this atlas on computer monitor are published.[14]

The OSA-UCS Atlas is sold by 'Optical Society of America, 2010 Massachusetts Avenue, NW – Washington, DC 20036-1023, USA.' (http://www.osa.org/en-us/home/)

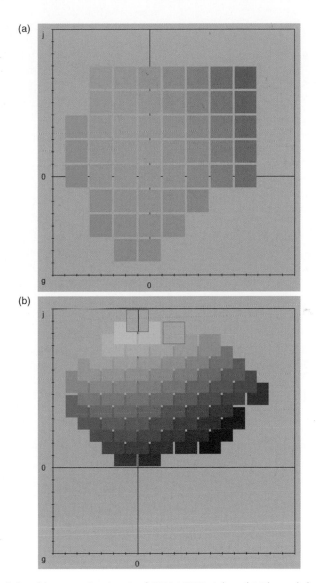

Figure 16.16   *(a) Plane defined by equation L = 0 of OSA-UCS's Atlas. (b) Plane defined by equation L − j = −8 of OSA-UCS's Atlas. The colour samples represented by an empty rectangle cannot be shown by the monitor used. The background has a luminance factor equal to 30%.*

16.5.5   NCS' Atlas

NCS's Atlas is described in Section 14.7.

Colour samples in the NCS system are seen under C illuminant and the colour specification here used is for both the CIE 1931 and the CIE 1964 standard observer (Figure 16.17).

The Natural Color System®© (NCS®©) is a proprietary perceptual colour model published by *NCS Colour AB* of Stockholm Sweden. All the numerical data needed for the reproduction of this atlas on a computer monitor are used with permission from the NCS Colour AB, P.O. Box 49022, SE-100 28 Stockholm Sweden, www.ncscolour.com.

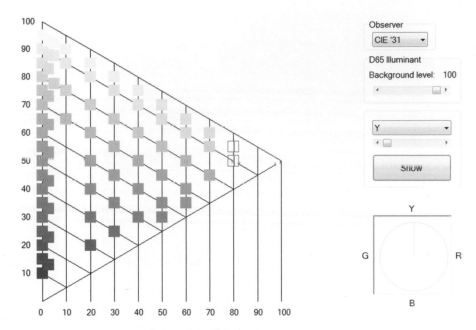

Figure 16.17 *NCS's page at defined Y hue, the NCS unique yellow. This page has to be compared with the corresponding one of the Ostwald Atlas at unique yellow (Figure16.13).*

16.6 CIE 1976 CIELUV and CIELAB Systems

CIELUV and CIELAB systems are described in Sections 11.3.4 and 11.3.5, respectively.

This program shows these systems on a computer monitor for both CIE observers and for different illuminants. The colours are shown according to the cylindrical coordinates (L^*, C^*_{uv}, h_{uv}) or (L^*, C^*_{ab}, h_{ab}) in two kind of tables: tables at defined hue angle and tables at constant lightness (Figures 16.18 and 16.19).

The user has to choose

- The colour system to be represented, CIELAB or CIELUV.
- Sections of the space,
    1. constant lightness planes;
    2. defined hue-angle planes.

If CIELAB is chosen, the user must select

- an observer, CIE 1931 or CIE 1964,
- an illuminant, from all the standard and recommended CIE illuminants; and
- the way to represent contiguous colours, with or without separating line (chromatic contrast appears in the rows and the simultaneous brightness contrast between contiguous rows).

If the user choses the section of the cylindrical coordinates at constant hue angle h_{ab}, the plot has the Chroma C^*_{ab} on the abscissa and lightness L^* on the ordinate. Any table at defined hue angle h_{ab} is shown combined with the "complementary" hue-angle page ($h_{ab} \pm 180°$) (Figures 16.18a and 16.18b).

If the user chooses the section at defined lightness L^*, the plot has a^* on the abscissa and b^* on the ordinate. Moreover the user can choose to plot either only the colours of the system, or only the MacAdam limit of the optimal colours, or both (Figures 16.19a and 16.19b). Figure 16.19a shows colours outside the MacAdam limit, which are possible because the monitor is self-luminous, but are not representative of non-self-luminous colour samples.

The CIELAB atlas is sold by www.RAL-Colours.de.

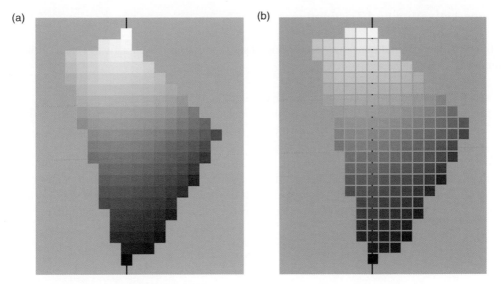

Figure 16.18   *(a) This figure represents two pages at constant and "complementary" hue angles – 0° and 180° – of the CIELAB system for the CIE 1931 observer and D65 illuminant. The coloured squares are physically uniform. The lightness L* increases from bottom to top and the chroma C_{ab}^* increases moving away from the achromatic central vertical line. The lightness simultaneous contrast appears between contiguous samples with different lightness while simultaneous chromatic contrast appears between contiguous colours with equal lightness. The right side of the squares appears desaturated tingeing with the opponent green hue, and analogously the left side of the squares appears desaturated tingeing with the opponent red hue. The grey squares of the central line appear tinged with the green hue on the right side and with the red hue on the left. These grey colour samples appear achromatic when covering all the other colour samples with white papers. This phenomenon does not regard the squares at the border. Only the colours reproducible by the monitor used are represented. (b) The same two pages of Figure 16.18a but related to the D75 illuminant. Contiguous colour samples are with a separation line and the contrast phenomena are absent. Only the colours reproducible by the monitor used are represented.*

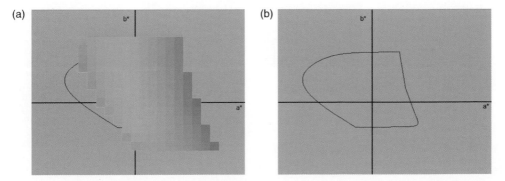

Figure 16.19   *(a) Plane at defined L* = 67 of the CIELAB space for the CIE 1931 observer and the illuminant D75 with plot of MacAdam's limit and colour samples. Only the colours reproducible by the monitor used are represented. (b) The same plane of Figure 16.19a with only plot of MacAdam's limit.*

16.7   Cone Activation and Tristimulus

This section is didactical and considers the sequential steps of colour-vision activation and of the tristimulus computation. In sequence are considered

1.   The spectral power distribution $S(\lambda)$ of a CIE standard and non-standard illuminant.
2.   The spectral reflectance factor $R(\lambda)$ of a set of five colour samples of the Munsell atlas, with equal Value $V = 7$, equal Chroma $C = 8$ and different Hue: 5R/7/8, 5Y/7/8, 5G/7/8, 5B/7/8, 5P/7/8.
3.   The spectral fundamental sensitivities $\bar{l}(\lambda), \bar{m}(\lambda), \bar{s}(\lambda)$ of the three kinds of cones.

   These three individual quantities and their product are shown by graphs (Figure 16.20). This representation is a simulation of the cone activation process. It explains the amount of activation L, M, S of the three kinds of cones and the resulting colour sensation, represented by the following integrals (or summations):

$$L = K \int_{380}^{780} S(\lambda)R(\lambda)\bar{l}(\lambda)\,\mathrm{d}\lambda = K \sum_{\lambda_i=380}^{780} S(\lambda_i)R(\lambda_i)\bar{l}(\lambda_i)\Delta\lambda_i$$

$$M = K \int_{380}^{780} S(\lambda)R(\lambda)\bar{m}(\lambda)\,\mathrm{d}\lambda = K \sum_{\lambda_i=380}^{780} S(\lambda_i)R(\lambda_i)\bar{m}(\lambda_i)\Delta\lambda_i$$

$$S = K \int_{380}^{780} S(\lambda)R(\lambda)\bar{s}(\lambda)\,\mathrm{d}\lambda = K \sum_{\lambda_i=380}^{780} S(\lambda_i)R(\lambda_i)\bar{s}(\lambda_i)\Delta\lambda_i$$

$$K = 100 / \int_{380}^{780} S(\lambda)\bar{y}(\lambda)\,\mathrm{d}\lambda = 100 / \sum_{\lambda_i=380}^{780} S(\lambda_i)\bar{y}(\lambda_i)\Delta\lambda_i$$

$$(16.9)$$

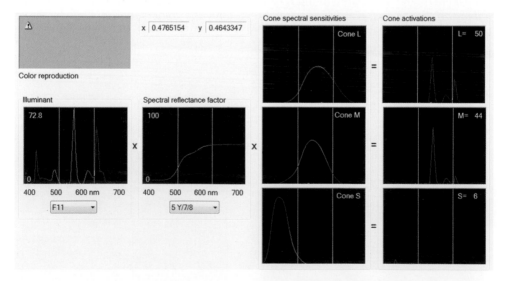

Figure 16.20   *Cone activation process produced by the illuminant F11 and the colour sample 5 Y/7/8 of the Munsell atlas. The colour stimulus of the colour sample is not reproducible by the monitor used and an exclamation mark is over printed.*

where $\Delta\lambda_i = 1$ nm and the spectral data are with a 1 nm step. The computation of the constant K is made with the requirement that the perfect reflecting diffuser (Section 3.5), whose spectral reflectance factor is $R(\lambda) = 1$, is characterized by a percentage luminance factor $Y_n = 100$.

The user can understand the origin of the colour sensation from a mutual comparison of the functions $S_\lambda, R(\lambda), \overline{l}(\lambda), \overline{m}(\lambda), \overline{s}(\lambda)$

The colours non-reproducible by the monitor used are simulated and a warning appears.

16.8   CIE Colorimetry

This Section has Chapters 7, 8, 9 and 11 as reference.

This Section considers the colour specification for seven observers defined in Chapter 9:

- Standard colorimetric observer CIE 1931 for 2° visual field (mainly used in imaging).
- Standard colorimetric observer CIE 1964 for 10° visual field (mainly used in industrial colorimetry).
- Standard deviate observer for 2° visual field.
- Standard deviate observer for 10° visual field.
- Vos observer for 2° visual field (used in physiology).
- Stockman and Sharpe 2°observer (used in physiology).
- Stockman and Sharpe 10° observer (used in physiology).

Two CIE colour specifications are defined for the CIE standard observers (Sections 9.2 and 9.3) and standard deviate observers (Section 9.4), one psychophysical and one psychometric (Chapter 11). The Vos observer and the two Stockman and Sharpe observers are defined only with the psychophysical colour specification and their chromaticity diagram is that defined by MacLeod-Boynton (Sections 9.5 and 9.6).

The colour specification of any colour considered is made in all the ways (psychophysical and psychometric) defined by CIE, in order to have a comparison between the different specifications. A click on any displayed colour opens a window with all the specifications of this colour, and by clicking again it is possible to have the quantities specifying the colour represented on all the CIE diagrams – (x, y) CIE chromaticity diagram and CIELAB and CIELUV planes at defined lightness –. Other quantities are given, as *whiteness* and *tint*. Before the description of the computer tools of this section, it is necessary to define the whiteness and tint.

Whiteness and Tint

A typical characteristic of paper is the whiteness and the tint and CIE recommended formulas for the evaluation of these two quantities, adequate for commercial use: whiteness, W or W_{10}, and tint, T_w or $T_{w,10}$, for the 2° and 10° CIE observers,[1] respectively. These formulae are defined only for the CIE standard illuminant D65. These quantities are defined as a function of the chromaticities, therefore depend on relatively 'big' coefficients (16.10), that render the use of the formulae rather unsteady. Therefore CIE says:

"The application of the formulae is restricted to samples that are called 'white' commercially, that do not differ much in colour and fluorescence, and that are measured on the same instrument at nearly the same time."

The formulas are:

$$\text{CIE } 1931 \quad W = Y + 800(x_n - x) + 1700(y_n - y) \qquad\qquad T_W = 1000(x_n - x) - 650(y_n - y)$$
$$\text{CIE } 1964 \quad W_{10} = Y_{10} + 800(x_{n,10} - x_{10}) + 1700(y_{n,10} - y_{10}) \quad T_{W,10} = 900(x_{n,10} - x_{10}) - 650(y_{n,10} - y_{10}}$$
$$(16.10)$$

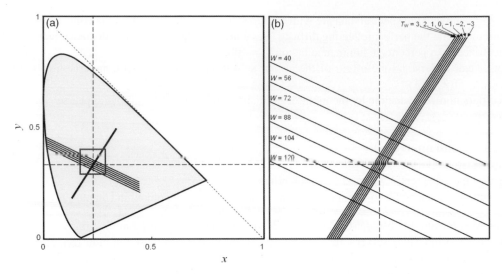

Figure 16.21   *(a) Chromaticity diagram with constant whiteness and constant tint lines. (b) Achromatic region of the chromaticity diagram enlarged.*

where, for the CIE 1931 standard colorimetric observer,

Y is the percentage luminance factor value of the sample,
(x, y) is the chromaticity of the sample,
(x_n, y_n) is the chromaticity of the perfect diffuser lit by D65 illuminant,

and similarly Y_{10}, (x_{10}, y_{10}) and $(x_{n,10}, y_{n,10})$ for the CIE 1964 standard colorimetric observer.

These formulae are the equations of lines on the CIE chromaticity diagrams (Figure 16.21). The whiteness and the tint have values in the ranges

$$40 < W < 5Y - 280, \quad -4 < T_W < +2$$
$$40 < W_{10} < 5Y_{10} - 280, -4 < T_{W,10} < +2$$

(16.11)

The ideal diffuser has $W = 100$ and $T_W = 0$ for the CIE 31 observer, and $W_{10} = 100$ and $T_{W,10} = 0$ for the CIE 64 observer.

The CX program computes the whiteness and tint all the times that this quantity can be computed. Anyway the user has to be careful to apply these formulae in the restricted situation defined by CIE.

16.8.1   CIE Colour Specification

Colour specification of a colour stimulus is a representation of the colour stimulus in terms of operationally defined values, such as three tristimulus values (psychophysics, Chapter 9), or as three attributes of the colour perception (psychometrics, Chapter 11).

Conventionally, the colour specification of light sources and of non-self-luminous surfaces is made with a normalization: the *percentage luminance factor Y* is used, which assumes the value $Y = 100$ for the radiance reflected by the *perfect reflecting diffuser* (Section 9.7 and Chapter 11). This colour specification is independent of the luminous existence of the light source and of illuminance of the specimens. The spectral reflected radiance is proportional to the *relative spectral power distribution* of the illuminant $S(\lambda)$ times the *spectral reflectance factor* $R(\lambda)$, that is, $L_e\lambda \propto S(\lambda)R(\lambda)$.

This program computes the colour specification of light sources and of non-luminous colours starting from the relative spectral power distribution of the light sources and from the spectral reflectance factor of the non-luminous colour. These spectral data are inputs for the program and must be written in an appropriate format.

Colour Specification of Light Sources

All the psychophysical computations are made for all the observers – CIE 31, CIE 64, standard deviate, Vos and Stockman-Sharpe –.

The spectral power distribution of the selected light source is read or hand drawn. If it is read from file, the file must be an ASCII file, named with the extension .TXT, and written as the illuminant files of the Folder 'Colour files/Illuminants' (Section 16.1.2).

Generally, the spectral data are supposed as obtained by measuring the light reflected from the perfect reflecting diffuser with reflectance factor equal to 1 (see Section 12.3.2). In practice, a standard tile certified by a metrological laboratory, or a working standard, is used and the spectral radiance data have to be divided by the spectral reflectance factor of the standard, or working standard, tile at equal wavelength. By clicking 'Reflected light from a standard white tile', the file of the reflectance factor of this white tile is selected and then automatically used in the computation. The reflectance factor file is analogously written as the illuminant file. A trial file is selectable.

The spectral power distribution can be entered manually by selecting points on the diagram for spectral power distribution against wavelength. Once the points are entered, it must be determined whether the interpolation involves negative interpolated values. If there are negative values, it is necessary to define the spectrum with a greater number of points.

These spectral data are supposed measured by a spectroradiometer with a bandwidth lower than 1 nm, therefore they do not need any deconvolution. These spectral data are interpolated by Lagrange formula at a step of 1 nm – Equation (12.43) – as required by CIE (Figure 16.22a). The illuminant obtained can be recorded by clicking 'Save Lagrange interpolated data' button and used in other computations.

The tristimulus values are computed executing the summations of the formulae (9.36).

If the spectral data S_λ are given by an abridged spectroradiometer, that is, with spectral data in a range shorter than 380–780 nm and generally with a step longer than 1 nm (e.g., 10 nm or 20 nm), the user has to choose the key 'abridged spectrum' (differently 'non-abridged spectrum') and the summations (9.36) are made in the range of the spectral data. In the case of 'abridged spectrum', the colour specification is approximate.

Clicking on 'show data spectra' opens a window where the data of spectral power distribution of the chosen light source are plotted and their interpolation and extrapolation are drawn. If the extrapolation is clearly meaningless, the user must select 'abridged' and click on 'enter data' again. In this case, the colour specification is approximate, but meaningful.

Once the tristimulus vector (X, Y, Z) is computed, all the other specifications are obtained by the known formulae (Figure 16.22b).

Colour Specification of Non-luminous Colours

All the colorimetric computations are made for the CIE 31 and CIE 64 observers, while for Vos and Stockman-Sharpe observers only the psychophysical computations are made.

The user has to choose the observer and the illuminant. If the illuminant is not in the list and has to be read from file, the file must be written as the illuminant files of the folder 'Colour files/Illuminants' (Section 16.1.2).

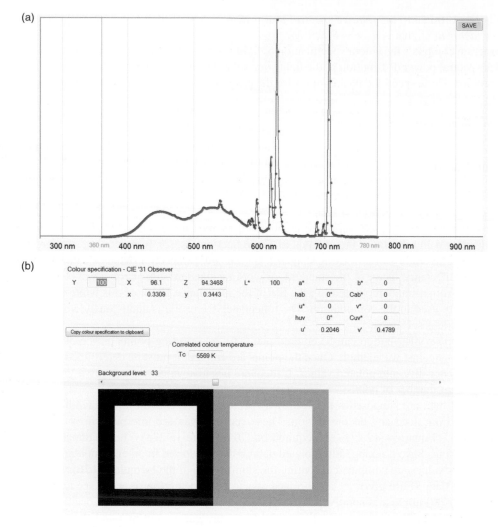

Figure 16.22   *(a) Relative spectral power distribution of the white light emitted by a Cathode-Ray-Tube Monitor, and (b) corresponding colour specification and visualization. Clicking on 'Copy colour specification to clipboard' all the data related to the colour specification are recorded on the clipboard for their use in MS Excel, in other mathematical computation programs or in a word processor.*

The computation of the non-luminous colour specification needs the knowledge of the spectral reflectance factor (or the spectral transmittance), entered by file or hand drawn. The file can also be written on a % scale (the software should automatically recognise the kind of scale), and recorded on an ASCII file, named with the extension .TXT, and written as follows:

```
15
375.2, 34.8
390.7, 54.2
405.1, 62.9
425.8, 70.2
```

```
470.2,  65.8
494.5,  62.1
538.7,  55.7
547.0,  50.3
570.8,  60.8
591.0,  70.1
610.3,  62.1
642.0,  48.2
673.1,  39.5
721.4,  33.8
745.0,  32.1
```

where the number in the first row, 15, is the number of spectral data and on the subsequent lines the numbers are the wavelength in nm and the corresponding spectral percentage reflectance factor $R(\lambda)$.

The spectral reflectance factor can be entered manually by selecting points on the diagram for spectral reflectance factor against wavelength. Once the points are entered, it must be determined whether the interpolation involves negative interpolated values. If there are negative values, it is necessary to define the spectrum with a greater number of points.

These spectral data are supposed measured by a spectroradiometer with a bandwidth lower or equal to 1 nm, therefore they do not need any deconvolution. These spectral data are interpolated by Lagrange formula at a step of 1 nm, as required by CIE. In this case, the computation needs the choice of an illuminant $S(\lambda)$ and of an observer – CIE 31, CIE 64, Standard Deviate, Vos and Stockman-Sharpe –. The tristimulus values are computed executing the summations of the following formulae (9.36) or (9.37) or (9.38).

This colour specification is relative to the perfect reflecting diffuser, with luminance factor $Y = 100\%$, and not to the absolute power of the light source.

In the case of an abridged instrument, that gives measurement of the reflectance factor $R(\lambda)$ in a range shorter than 380–780 nm (e.g. 400–700 nm with a step of 5 nm, 10 nm or 20 nm), the user has to choose an 'abridged' spectrum (differently 'non-abridged' spectrum). In the case of an abridged spectrum, the extrapolation is made with the Lagrange formula and the extrapolated spectrum is plotted, but the colorimetric computation is restricted to the range where $R(\lambda)$ is known. In this case, the colour specification is approximate (Figure 16.23).

Once the tristimulus vector (X, Y, Z) is computed, all the other specifications are obtained by the known formulae.

Clicking on 'show data spectra' opens a window where the spectral reflectance data are plotted and their interpolation and extrapolation are drawn. If the extrapolation is clearly meaningless, the user must select 'abridged' and click 'enter data' again (Figure 16.23a).

Clicking on 'output data' opens a windows (Figure 16.23b) where all the colorimetric data are written and the colour is reproduced in three squares with three different backgrounds – one black, one of the colour of the illuminating light with the luminance factor $Y = 100\%$, and the intermediate square with $Y = 33\%$ –. The user can modify the 'background level', that is, the luminance factor of the intermediate background, with the slider and check whether at the same Y of the colour sample the border distinctness between the square and the background appears minimum. This is also to check on the consistency of the colorimetric computations.

Clicking on 'show diagrams' opens a window (Figure 16.24) where all the colorimetric data are written again and are plotted on the standard diagrams. These diagrams show circles of the constant luminance CIELAB and CIELUV planes in all the diagrams. In this way, the distortion between the different spaces appears. In particular, this distortion depends on the chosen illuminant that defines the origin of the a^* and b^* coordinates in CIELAB and of u^* and v^* coordinates in CIELUV.

Clicking on 'Monitor RGB specification' opens a window with the reproduction of the colour and the colour specification according to the sRGB standard where the numbers are specified with 8 bits.

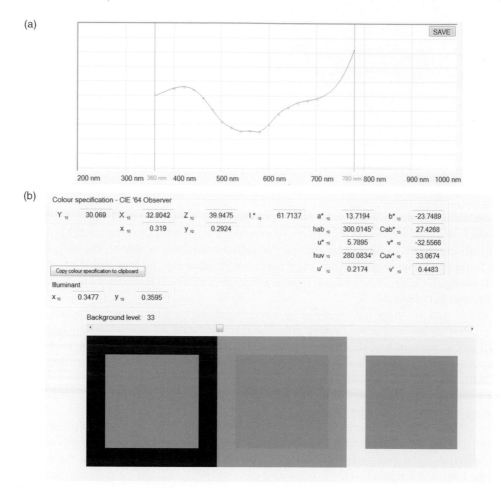

Figure 16.23 *(a) Spectral reflectance factor of the 7th colour sample used for the evaluation of the colour rendering index lit by D50 illuminant. The observer considered for the colour specification is the CIE 1964. The colorimetric computation should be made in an approximate way by choosing 'abridged' mode, since no measurement exists below 400 nm and over 700 nm. The Lagrange extrapolation in these spectral regions is meaningless. (b) Colour specification and visualization of the considered colour sample of Figure 16.23a. The colour is displayed on three backgrounds, one white, one black and one grey with the chromaticity of the considered illuminant, whose luminance factor is variable. By varying this factor, the operator can identify the situation of minimal border distinction between the sample and the background. In this situation, the colour sample and the background should have the same luminance factor. The numerical difference between the background level and the luminance factor of the sample is due to many factors: observer, 8-bit colour digitization, approximate monitor calibration. In this case, the numerical values are 30.0690 and 33. Clicking on 'Copy colour specification to clipboard' all the data related to the colour specification are recorded on the clipboard for their use in MS Excel, in other mathematical computation programs or in a word processor.*

16.8.2 CIE Systems

The program makes the transformations between all the colorimetric systems of CIE, with regard to all the observers, the illuminants (standard and considered): (X, Y, Z), (Y, x, y), (L^*, a^*, b^*), (L^*, h_{ab}, C_{ab}^*), (L^*, u^*, v^*), (L^*, h_{uv}, C_{uv}^*) (Chapters 9 and 11). Moreover the test colour is reproduced on the monitor. (Figures 16.25, 16.26a and 16.26 b)

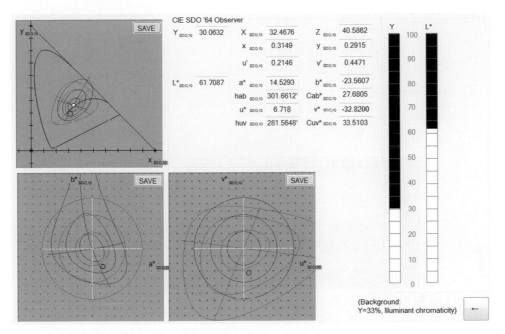

Figure 16.24   *Graphical representation of the colour specification of the colour sample, whose spectral reflectance factor is shown in16.23a and colour specification in16.23b. In this case the observer is the CIE 1964 Standard Deviate Observer and the illuminant D50. At right, two columns compare the luminance factor to the CIE 1976 lightness. Planes of CIELAB and CIELUV systems with equal lightness are compared. Particularly on these three planes circles of one plane is transformed in the corresponding closed lines of the other showing the relative distortion of one space with respect to the other.*

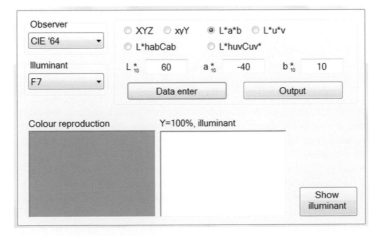

Figure 16.25   *Example of input data and colour visualization in the toolbox 'CIE systems'.*

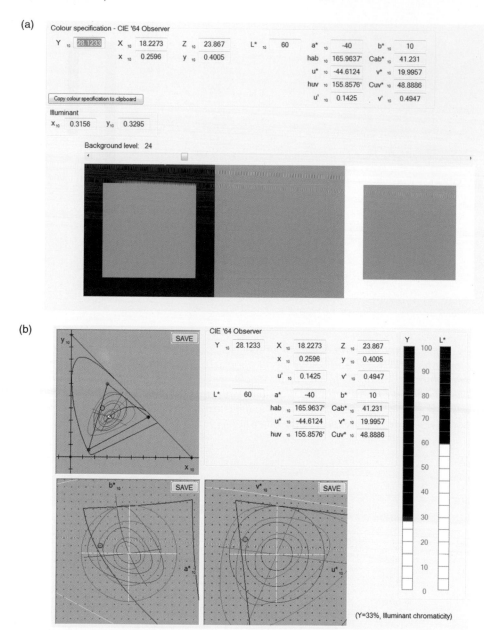

Figure 16.26   *(a) Complete CIE colour specifications and colour visualization of the colour selected in Figure 16.25. The selected colour is displayed on three backgrounds, one white, one black and one grey with the chromaticity of the considered illuminant, whose luminance factor is variable. By varying this factor, the operator can identify the situation of minimal border distinction between the sample and background. In this situation the colour sample and the background should have the same luminance factor. The numerical difference between the background level and the luminance factor of the sample is due to many factors: observer, 8-bit colour digitization, approximate monitor calibration, etc. This is a very useful exercise. Clicking on 'Copy colour specification to clipboard' all the data related to the colour specification are recorded on the clipboard for their use in MS Excel, in other mathematical computation programs or in a word processor. (b) Graphical representation of the colour specification of the input colour sample considered in 16.26a.*

16.8.3 Chromaticity Diagrams

Over time, many attempts have been made to produce chromaticity diagrams with uniform scales, that is, with Euclidean metrics. The MacAdam ellipses (Section 6.17) are considered the most important empirical data set to represent the chromatic discriminability for the foveal vision adapted to the C illuminant (recall that these ellipses represent one standard deviation in colour-matching and the jnd is assumed three times the standard deviation). This program considers the 11 most meaningful chromaticity diagrams related to the CIE 1931 observer produced in the research for scale uniformity.

With this program it is possible, on the chosen diagram, to plot the MacAdam ellipses (enlarged 10 times), the Planckian locus, the triangle of the RGB chromaticities of the monitor used. The xy net of the CIE 1931 diagram is reproduced on all the considered diagrams for a comparison. The new coordinates are generally denoted by (a, b) with exclusion of two cases.

The diagrams are:

- (r, g) *CIE (1931)*

$$\begin{pmatrix} r \\ g \end{pmatrix} = \frac{\begin{pmatrix} 2.36458+0.46809 & -0.896509+0.46809 & -0.46809 \\ -0.51515-0.08875 & 1.4264-0.08875 & 0.08875 \end{pmatrix}\begin{pmatrix} x \\ y \\ 1 \end{pmatrix}}{(1.85464-0.62987)x+(0.51548-0.62987)y+0.62987} \tag{16.12}$$

- *Judd (1935)*

$$\begin{pmatrix} a \\ b \end{pmatrix} = \frac{\begin{pmatrix} 2.7760 & 2.1543 & -0.1192 \\ -2.9446 & 5.0322 & 0.8283 \end{pmatrix}\begin{pmatrix} x \\ y \\ 1 \end{pmatrix}}{-x+6.3553y+1.5405} \tag{16.13}$$

- *Breckenridge-Schaub (1939)* (Figure 16.27a)

$$\begin{pmatrix} a \\ b \end{pmatrix} = \frac{\begin{pmatrix} -0.74030 & -1.35203 & 0.70001 \\ 3.19700 & -1.55045 & -0.54884 \end{pmatrix}\begin{pmatrix} x \\ y \\ 1 \end{pmatrix}}{x-7.05336y-1.64023} \tag{16.14}$$

- *Farnsworth (1944)*

$$\begin{pmatrix} a \\ b \end{pmatrix} = \frac{\begin{pmatrix} 0.6600 & 0.0000 & 0.0000 \\ 0.0000 & 1.2666 & 0.0000 \end{pmatrix}\begin{pmatrix} x \\ y \\ 1 \end{pmatrix}}{y+0.2666} \tag{16.15}$$

(a)

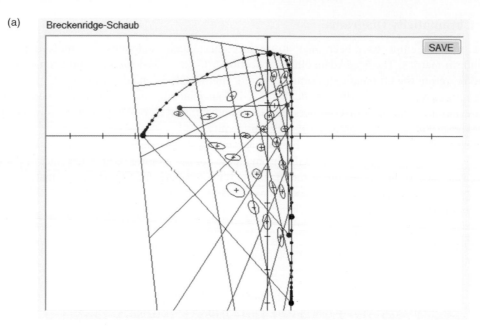

(b)

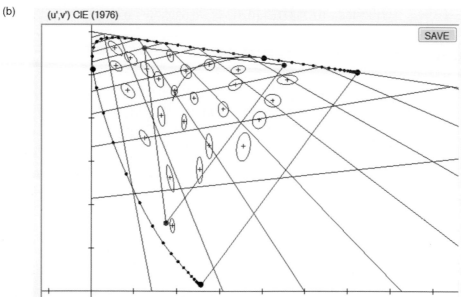

Figure 16.27   *(a) Breckenridge-Schaub chromaticity diagram with MacAdam's ellipses, monitor RGB triangle and Planckian locus. This diagram is used in glass factories since it has scales with high uniformity in the achromatic region. (b) CIE 1976 (u', v') chromaticity diagram with MacAdam's ellipses, RGB triangle of the monitor used and Planckian locus. This diagram is very important because it is at the core of the CIELUV system.*

- *OSA (1963)*

$$
\begin{pmatrix} a \\ b \end{pmatrix} = \frac{\begin{pmatrix} 0.4661 & 0.1593 & 0.0000 \\ 0.0000 & 0.6581 & 0.0000 \end{pmatrix} \begin{pmatrix} x \\ y \\ 1 \end{pmatrix}}{-0.1574x + y + 0.2424}
\tag{16.16}
$$

- *Sugiyama - Fukuda (1959)*

$$
\begin{pmatrix} a \\ b \end{pmatrix} = \frac{\begin{pmatrix} 0.230 & 0.000 & 0.000 \\ 0.000 & 0.600 & 0.000 \end{pmatrix} \begin{pmatrix} x \\ y \\ 1 \end{pmatrix}}{-0.038x + y + 0.163}
\tag{16.17}
$$

- (u', v') *CIE (1976)* (Figure 16.27b)

$$
\begin{pmatrix} u' \\ v' \end{pmatrix} = \frac{\begin{pmatrix} 4 & 0 & 0 \\ 0 & 9 & 0 \end{pmatrix} \begin{pmatrix} x \\ y \\ 1 \end{pmatrix}}{-2x + 12y + 3}
\tag{16.18}
$$

This diagram is the CIE 1976 uniform-chromaticity-scale diagram spanned by the orthogonal coordinates v' against u', defined in Section 11.3.1–11.3.3. This diagram is a modification of the (u, v) CIE 1960 UCS diagram, and supersedes it.

- *Hunter (1941)*

$$
\begin{pmatrix} a \\ b \end{pmatrix} = \frac{\begin{pmatrix} 2.4266 & -1.3631 & -0.3214 \\ 0.5710 & 1.2447 & -0.5708 \end{pmatrix} \begin{pmatrix} x \\ y \\ 1 \end{pmatrix}}{x + 2.2633y + 1.1054}
\tag{16.19}
$$

- *MacAdam (1937)*

$$
\begin{pmatrix} a \\ b \end{pmatrix} = \frac{\begin{pmatrix} 4 & 0 & 0 \\ 0 & 6 & 0 \end{pmatrix} \begin{pmatrix} x \\ y \\ 1 \end{pmatrix}}{-2x + 12y + 3}
\tag{16.20}
$$

16.8.4   Fundamental Observers

This program considers the Vos observer (Section 9.5) and the CIE fundamental observers for 2° and 10° (Section 9.6).

This program makes the transformation between the different coordinate systems (X_F, Y_F, Z_F), (Y_F, x_F, y_F), (L, M, S), (Y_F, l, s). The coordinates (x_F, y_F) and (l, s) are plotted and the colour is reproduced on the monitor (Figure 16.28).

These observers are mainly used by physiologists and are generally represented in the fundamental reference frame (L, M, S).

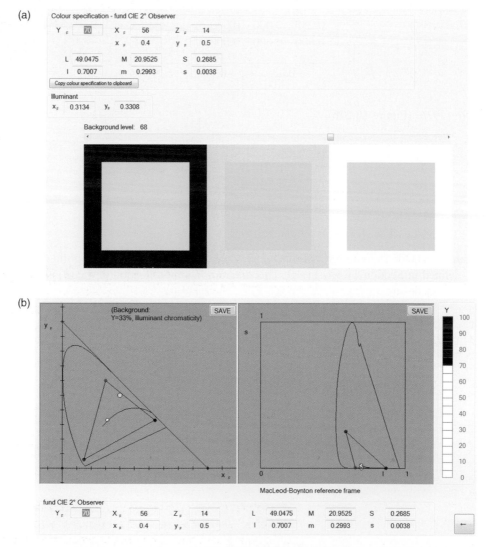

Figure 16.28   *(a) Output data and colour visualization in the toolbox 'fundamental observer'. Clicking on "Copy colour specification to clipboard" all the data related to the colour specification are recorded on the clipboard for their use in MS Excel, in other mathematical computation programs or in a word processor. (b) Chromaticity specification shown on the CIE (x_F, y_F) and MacLeod-Boynton chromaticity diagrams.*

The chromaticity coordinates (l', m', s') of the Vos observer, introduced by MacLeod-Boynton, are obtained by transformations (9.24) and are defined by (9.25)

$$l' = \frac{L'}{L' + M'}, \ m' = \frac{M'}{L' + M'}, \ s' = \frac{S'}{L' + M'}, \ l' + m' = 1 \tag{16.21}$$

The plane of this chromaticity diagram has constant luminance, according to the hypothesis that the S-cones have no contribution to the luminance, while L- and M-cones have equal contribution.

An analogous computation can be made with the 2° and 10° fundamental observers. In this case the transformations are given by (9.28) and (9.31), respectively, and the equations for the MacLeod-Boynton diagram are (9.32)-(9.35).

As an exercise, with this program the user has the opportunity to use the coordinate $(l, m = 1-l, s)$ and to observe the great difference in scale between these coordinates and the other $(x_F, y_F, z_F = 1 - x_F - y_F)$. The input $(Y_F; x_F, y_F)$ is easy. The input (Y_F, l, s) needs some exercise if the input data are not obtained by computation but are trial numbers.

16.8.5   Dominant Wavelength and Purity

This program computes the Helmholtz coordinates, which are defined in Section 6.13.8 for all the observers, purities, dominant and complementary wavelengths of a colour stimulus.

Recall that

- The excitation purity is defined on the chromaticity diagram and the colorimetric purity on the luminances.
- The purity is a pure number in the range 0–1.
- The achromatic stimulus has purity 0 and the colour stimulus with chromaticity on the border of the chromaticity diagram has 1.

The user should become familiar with wavelength dominancy and complementarity for chromaticities in the different regions of the chromaticity diagram, particularly in the regions of the green and magenta hues, where the complementarity regards non-spectral lights. An example of computation is given in Figure 16.29.

16.8.6   Tristimulus Space Transformations

Tristimulus space is a three dimensional linear vector space (Section 6.13), in which any set of three independent vectors can be chosen as reference frame (Section 6.13.4). Three vectors are independent if none of the three vectors can be written as a linear combination of the other two.

Consider two reference frames in the tristimulus space, XYZ and ABC, which is specified by the chromaticities of the reference stimuli **A**, **B**, **C** and of the white $\mathbf{W} = \mathbf{A} + \mathbf{B} + \mathbf{C}$ on the (x, y) chromaticity diagram: (x_A, y_A), (x_B, y_B), (x_C, y_C) and (x_n, y_n).

The reference frame XYZ may be any reference frame, and thus different from the reference frames so denoted.

The normalization in the transformed reference frame is such that the vector with the chromaticity (x_n, y_n) is equal to $(A = 1, B = 1, C = 1)$.

The program gives the matrix transformations between XYZ and any chosen ABC reference frames, the Exner coefficients for the luminance (Section 6.13.6), the graphs of the chromaticity diagrams in the two reference frames and the graph of the colour-matching functions. As default example of computation, the

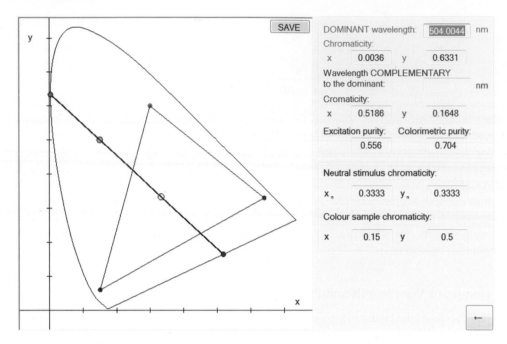

Figure 16.29 *Example of computation of the dominant wavelength and its complementary wavelength, of the colorimetric and excitation purity related to the chromaticity (x = 0.15, y = 0.50) and the neutral point (x_n = 0.3333, y_n = 0.3333) of the CIE 1931 observer.*

coordinates reference frame of the sRGB system (Sections 6.13.4 and 15.2.4) is considered and the result is given in Figures 16.30a 16.30b, 16.30c.

16.8.7 Colour-Difference Formulae ΔE

Over time many colour-difference formulae have been proposed (Section 11.4) and here the most important, today still in use, are recalled.

The program computes the colour-difference applying all the formulae considered here starting from the colour specification of a set of colour samples or from a set of spectral reflectance (or transmittance) factors. Moreover the user can define the values of the parametric factors present in these formulae.

The formulae are (Section 11.4):

- CIE 1976 $L^*a^*b^*$ colour-difference (ΔE^*_{ab})
- CMC(l:c) colour-difference formula (ΔE_{CMC})
- CIE 1994 colour-difference formula (ΔE^*_{94})
- CIEDE2000 total colour-difference formula (ΔE_{00})

CIE recommends the use the CIEDE2000 formula whenever in the past the CIE 94 or CMC formula were used and such a recommendation is in agreement with the persons who developed the CMC formula.

Figures 16.31a and 16.31b give a comparison between two transmittances and the corresponding colour-difference values, computed with all the considered colour-difference formulae. A visual comparison of the reproduced colours is only indicative for perception because it depends on the monitor calibration and, in the case of printed colour as in this figure, on the printing profile. In any case a colour difference judged on the monitor can be considered approximately meaningful.

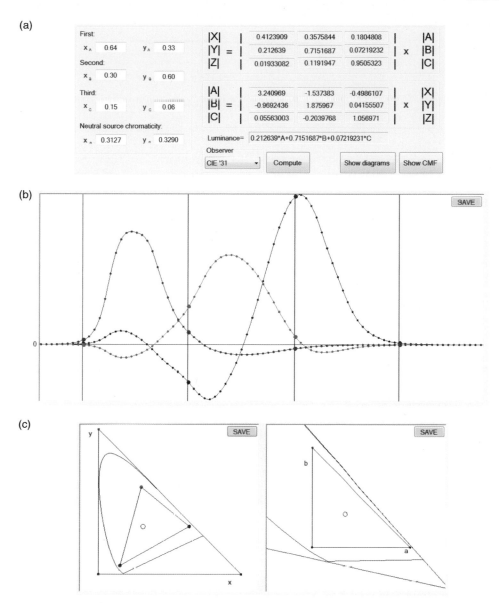

Figure 16.30 *(a) Computation of the transformation matrices between the reference frames CIE XYZ and RGB of the sRGB system, and Exner's coefficients of the 'Helligkeit' equation (luminance) of Exner-Schrödinger. (b) Colour matching functions of the CIE 1931 standard observer in sRGB reference frame. (c) Chromaticity diagrams related to the reference frames CIE XYZ and RGB of the sRGB system.*

In the case of input from file, the user must consider whether the computation is abridged or not, following the procedure given in Section 16.8.1.

16.8.8   CIE 1974 Colour Rendering Index R_a

> "*Colour rendering of a light source* – effect of an illuminant on the colour appearance of objects by conscious or subconscious comparison with their colour appearance under a reference illuminant."[15]

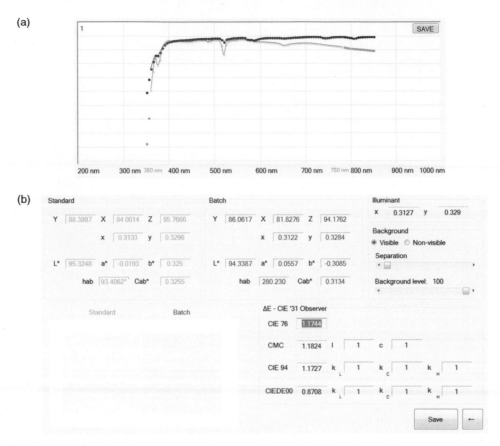

Figure 16.31   *(a) Comparison between transmittances of two glasses for colour-difference computation. (b) Corresponding colour-difference computation. The two colours in comparison are displayed on a background of variable grey and with a variable separation. In this way, the user can judge the colour difference in different situations. The colour difference is visually enhanced by reducing the mutual distance of the colour reproductions and brightening the background.*

"*Colour rendering index* [R] – measure of the degree to which the psychophysical colour of an object illuminated by the test illuminant conforms to that of the same object illuminated by the reference illuminant, suitable allowance having been made for the state of chromatic adaptation."[15]

"*CIE 1974 general colour rendering index* [Ra] – mean of the CIE 1974 special colour rendering indices for a specified set of eight test colour samples."[15]

"*CIE 1974 special colour rendering index* [Ri] – measure of the degree to which the psychophysical colour of a CIE test colour sample illuminated by the test illuminant conforms to that of the same sample illuminated by the reference illuminant, suitable allowance having been made for the state of chromatic adaptation."[15]

Commonly, the locution 'colour rendering index' is shortened with the acronym 'CRI'.

	Munsell specification	Appearance under D65	swatch
1	7.5 R 6/4	Light greyish red	
2	5 Y 6/4	Dark greyish yellow	
3	5 GY 6/8	Strong yellow green	
4	2.5 G 6/6	Moderate yellowish green	
5	10 BG 6/4	Light bluish green	
6	5 PB 6/8	Light blue	
7	2.5 P 6/8	Light violet	
8	10 P 6/8	Light reddish purple	
9	4.4 R 4/13	Strong red	
10	5 Y 8/10	Strong yellow	
11	4.5 G 5/8	Strong green	
12	3 PB 3/11	Strong blue	
13	5 YR 8/4	Light yellowish pink	
14	5 GY 4/4	Moderate Olive green	

Figure 16.32 *Set of colours from the Munsell atlas, specified for the D65 illuminant and CIE 1931 observer, used for the CIE 1974 colour rendering index.*

According to the definition, the *CIE 1974 general colour rendering index*[16] of an illuminant is the arithmetic average of the first *eight special colour rendering indices* R_i associate with the conventional set of (8+6) colour samples, with reflectances equal to colour samples from the Munsell atlas (Figure 16.32). The first eight colour samples have the same Munsell value $V = 6$ and medium chroma, while four of the other six have almost unique hues and high Munsell chroma. The last two – skin tone and leaf green –are very meaningful for their role in the daily life.

The special colour rendering indices R_i, $i = 1$-8, and the general colour rendering index R_a are computed by the following sequence of operations:

1. Test and reference illuminants must have the same colour temperature. For test illuminants with a correlated colour temperature $T_{cp} < 5000$ K, the Planckian radiator is used as a reference illuminant, while for $T_{cp} \geq 5000$ K the daylight illuminant is used.

2. Once the correlated colour temperature T_{cp} of the test illuminant is computed and the reference illuminant is defined, the computation begins for the colour specifications with the CIE 64 UVW system (Section 11.3.2) of the conventional colour samples ($i = 1, \ldots, 8, \ldots, 14$), $(U^*_{t,i}, V^*_{t,i}, W^*_{t,i})$ and $(U^*_{r,i}, V^*_{r,i}, W^*_{r,i})$ lit by the test and reference illuminants (subscript "t" for "test" and "r" for "reference"), respectively.

3. The colour specifications $(U^*_{t,i}, V^*_{t,i}, W^*_{t,i})$ and $(U^*_{r,i}, V^*_{r,i}, W^*_{r,i})$, $i = 1, \ldots, 8, \ldots, 14$, do not take into account the chromatic adaptation to the test and reference illuminants. A chromatic adaptation correction is necessary and its computation is schematically presented here. This correction operates only on the

chromaticities (u_t, v_t) of the test illuminant and $(u_{t,i}, v_{t,i})$ of the colour samples. The adaptation correction is schematically reported here for implementation in a computer program. For a thorough understanding of this correction, the reader is invited to study the original text.[16] The steps of this correction are:

(i)   Two new coordinates are introduced, that for the test illuminant are

$$\left\{ \begin{array}{l} c_t = \dfrac{4 - u_t - 10v_t}{v_t} \\[3mm] d_t = \dfrac{0.404 - 1.481u_t + 1.708v_t}{v_t} \end{array} \right. \tag{16.22}$$

and for the colour samples i lit by the test illuminant are

$$\left\{ \begin{array}{l} c_{t,i} = \dfrac{4 - u_{t,i} - 10v_{t,i}}{v_{t,i}} \\[3mm] d_{t,i} = \dfrac{0.404 - 1.481u_{t,i} + 1.708v_{t,i}}{v_{t,i}} \end{array} \right. \tag{16.23}$$

(ii) These coordinates are chromatically corrected by a von Kries type transformation (Section 6.20.3), transforming the chromaticity of the test source to that of the reference illuminant, as follows

$$\left\{ \begin{array}{l} c'_{t,i} = c_{t,i} \dfrac{c_r}{c_t} \\[3mm] d'_{t,i} = d_{t,i} \dfrac{d_r}{d_t} \end{array} \right. \tag{16.24}$$

(iii) After this correction the CIE 1960 chromaticity coordinates are

$$\left\{ \begin{array}{l} u'_{t,i} = \dfrac{10.872 + 0.404c'_{t,i} - 4d'_{t,i}}{16.518 + 1.481c'_{t,i} - d'_{t,i}} \\[3mm] v'_{t,i} = \dfrac{5.520}{16.518 + 1.481c'_{t,i} - d'_{t,i}} \end{array} \right. \quad \text{and} \quad \left\{ \begin{array}{l} u'_{r,i} = u_{r,i} \\[2mm] v'_{r,i} = v_{r,i} \end{array} \right. \tag{16.25}$$

$$\left\{ \begin{array}{l} u'_t = u_r \\[2mm] v'_t = v_r \end{array} \right. \quad \text{and} \quad \left\{ \begin{array}{l} u'_r = u_r \\[2mm] v'_r = v_r \end{array} \right. \tag{16.26}$$

(iv). Finally the coordinates to be used for the colour differences and the colour rendering indices are

$$\begin{cases} W^*_{t,i} = 25Y^{1/3}_{t,i} - 17 \\ U^*_{t,i} = 13W_{t,i}*(u'_{t,i} - u'_t) \\ V^*_{t,i} = 13W_{t,i}*(v'_{t,i} - v'_t) \end{cases} \text{ and } \begin{cases} W^*_{r,i} = 25Y^{1/3}_{r,i} - 17 \\ U^*_{r,i} = 13W_{r,i}*(u'_{r,i} - u'_r) \\ V^*_{r,i} = 13W_{r,i}*(v'_{r,i} - v'_r) \end{cases} \tag{16.27}$$

where $Y_{t,i}$ and $Y_{r,i}$ are the percentage luminance factors of the *i*-th test tile lit with the test and reference illuminant, respectively.

4. The colour difference for the *i*-th colour sample lit with the test and reference illuminants is computed by applying the following Euclidean formula

$$\Delta E_i = \sqrt{(U^*_{r,i} - U^*_{t,i})^2 + (V^*_{r,i} - V^*_{t,i})^2 + (W^*_{r,i} - W^*_{t,i})^2}, \quad i = 1-14 \tag{16.28}$$

The corresponding CIE 1974 special colour rendering index is defined by the formula

$$R_i = 100 - 4.6\Delta E_i \tag{16.29}$$

and the CIE 1974 general colour rendering index as the arithmetic average of the first eight special rendering indices

$$R_a = \frac{1}{8}\sum_{i-1}^{8} R_i \tag{16.30}$$

The computer program 'colour-rendering index' first computes the correlated colour temperature T_{cp} of a test illuminant, then computes the special R_i and the $R_{a,8}$ and $R_{a,14}$ (average of the $14 = 8 + 6$ special colour rendering indices). The spectral power distribution of the test illuminant can be entered manually or by file.

The file must be recorded on as ASCII file named with extension .TXT, and written according to the rules for the illuminants in the folder 'Colour files/Illuminants' shown in Section 16.1.2.

The spectral power distribution of the test illuminant can be entered manually by selecting points on the diagram for spectral power distribution against wavelength. Once the points are entered, it must be determined whether the interpolation involves negative interpolated values. If there are any negative values, it is necessary to define the spectrum with a greater number of points.

The adaptation correction is very complex, therefore it is not often implemented, although not declared. This computer program makes both computations and the results are given for a comparison. Generally the difference is very small. The results for the CIE standard and recommended illuminants are equal to those given by CIE.

The last six special indices are related to very meaningful colour samples and therefore have to be considered individually, anyway the average of all 14 special indices is given.

Figure 16.33 gives an example of the output of a CRI computation.

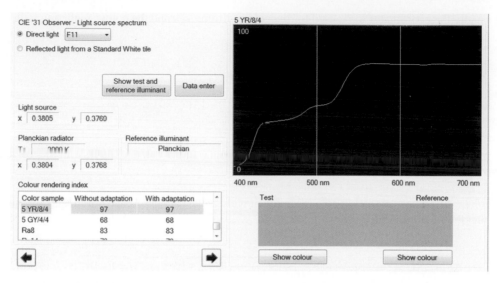

Figure 16.33   *Example of colour-rendering-index computation related to the F11 fluorescent illuminant. Colorimetric data and CRI for any considered reference specimen used in the CRI computation and colour visualization of the selected specimen illuminated by the standard and test illuminants.*

The user is invited to correlate the shape of the illuminant spectra, the region where the spectral power density is higher, the position of the narrow bands, with the special CRIs, taking into account the spectral reflectance factors of the surface colour samples. In this way he or she can understand the meaning of colour rendering.

Today, with the new LED light sources, the formula here presented seems no longer completely useful and the debate for an updated formula is very hot.

16.9   Black Body and Daylight Spectra and Other CIE Illuminant Spectra

In colorimetry and photometry, an important distinction has to be made between light source and illuminant (Section 8.1.), although in everyday English, the term illuminant is not restricted to this sense, but is also used for any kind of light falling on a body or scene.

Generally illuminants are CIE considered or standard. The use of the word illuminant not referred to a CIE considered or standard illuminant means that we are considering the numerical spectral power distribution of a known or theoretical light source.

The light sources/illuminant important for colorimetry are presented in Chapter 8.

Here self-luminous objects, such as monitors, that are important in colorimetry, but not in photometry, are not considered.

The first quantity that characterizes these illuminants is the colour temperature or the correlated colour temperature, if it is definable.

Since colour rendering index of an illuminant (Section 16.8.8) is computed by using a reference illuminant of equal correlated colour temperature T_{cp}, and the reference illuminants defined by CIE are the Planckian radiators for $T_{cp} < 5000$ K and the daylight illuminants for $T_{cp} \geq 5000$ K, these two kinds of illuminants (defined in Chapter 8) are considered here and compared.

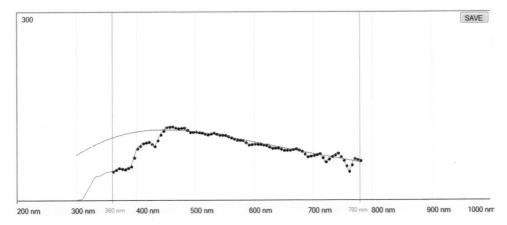

Figure 16.34 *Comparison of the Planckian radiator (red line) and daylight spectra (black dotted line) given at the correlated colour temperature of 6500 K.*

Colorimetry and photometry consider the relative spectrum of the daylight illuminants normalized to 100 at 560 nm.

Black Body and Daylight Spectra

The 'Black Body and Daylight spectra' program shows a comparison by plots on the scale of the spectra of Planckian radiator and daylight at equal correlated colour temperature (Figure 16.34). It is also possible to download the file of these spectra.

It is interesting to compare blackbody radiation and daylight in relation to the correlated colour temperature: below 4000 K the two curves are well overlapped, over 4000 K a difference below 450 nm begins and increases with the temperature, while the spectra over 450 nm are always well overlapped.

Other CIE Illuminant Spectra

CIE is the body responsible for publishing all of the standard and recommended illuminants, which have spectral power distribution published by CIE. Each illuminant is denoted by a capital letter or by a letter-number combination (Chapter 8). The program 'Spectral data view and download' of the toolbox 'Tools' shows all the illuminant considered in the book and the user can add his or her own files. These files can be seen by selecting 'illuminants'. The selection of 'handplot' among the illuminants shows the last handplot illuminant recorded in any of the toolboxes with this option.

16.10   Additive Colour Synthesis

Colorimetry for colour reproduction in images is presented in Chapter 15.

Two methods of colour reproduction are considered and comprise additive mixing based on the mosaic structure.

The programs for colour mixing presented in this section are useful for the user as training tools and for understanding the most widely used techniques of image reproduction.

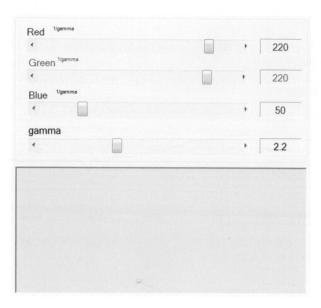

Figure 16.35   *Visualization of the colour produced by additive mixing of three RGB lights represented by tristimulus values R = 220, G = 220 and B = 50 in the RGB reference frame of the sRGB system. The user has only to move the three sliders freely with the mouse and colour mixing automatically follows. In this case the mixture of R and G produces yellow while any increment of B produces a desaturation of the yellow. This visualization takes into account the monitor γ that the user can modify.*

16.10.1   RGB Monitor, Additive Colour Mixture

Before entering this section, the user should read Section 15.2.

This toolbox visualizes on the computer monitor a colour specified by a set of three reference stimuli (R, G, B), constituting a trichromatic system. The user chooses freely the intensities of the red, green and blue reference stimuli, proper of the monitor configuration (Sections 15.2.1-15.2.2 and 15.2.4).

The visual phenomenon associated with this vision on trichromatic monitor is an *additive mixture of colour stimuli* combined on the retina producing a visual action in such a manner that they cannot be perceived individually. The view of an RGB monitor through an enlarging lens shows that the monitor surface has a mosaic structure (Section 15.1), whose RGB elements are so small that theirs lights are not visible with naked eye and the RGB lights are mixed additively only on the retina.

The additive RGB colour spaces corresponding to the monitor displays can be treated as *device-dependent colour spaces* having a simple functional relationship to CIE colorimetry, thus they are categorized as colorimetric colour spaces.

The aim of this toolbox is to familiarize the user with the additive mixture of colour stimuli (Figure 16.35), that is often counterintuitive for people accustomed to the more everyday subtractive colour system produced by mixtures of pigments, dyes and other colorant substances, which produce colour of objects to be seen in reflection or in transmission.

16.10.2   Halftone CMY Printing

Before entering this section, the user should read Section 15.3.

Screen plate printing and *halftone printing* are techniques for printing generally on paper and reproducing a wide colour gamut with three coloured inks, generally cyan, yellow and magenta (CMY). Often a fourth

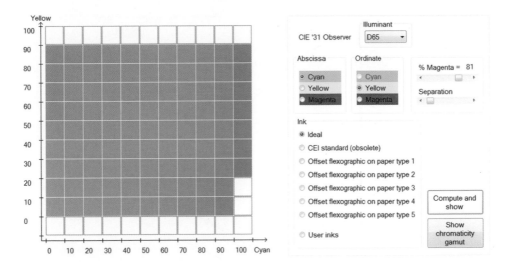

Figure 16.36 *Halftone printing simulation on a monitor. The magenta ink is equal for all the samples and covers the 81% area. The percentage areas covered by cyan and yellow inks are on the abscissa and on the ordinate, respectively, with steps of 10%. The colour samples represented by an empty square cannot be shown by the monitor used.*

ink, black (K), is added with many advantages. The main reason to use black ink is that the overlapping of the three inks does not produce a pure black, but only a dull colour. In this case the set of four inks is denoted by CMYK, where K means 'Key black' and the word 'Key' is due to the printing technique.

The view of a halftone printed image through an enlarging lens shows that the printed surface has a mosaic structure, whose elements are small coloured dots that can be considered as light sources invisible with the naked eye and these coloured lights are mixed additively only on the retina.

This toolbox shows (Figures 16.36 and 16.37)) the colours obtainable on the monitor according to Neugebauer equations by using

– the standard inks of the 'Comité Européen d'Imprimerie' (CEI), today obsolete, under the illuminant D65;
– the ideal inks under all the CIE illuminants (standard and considered);
– the standard ISO inks for flexography (ISO 12647-6: 2012) under the illuminant D50 (ISO 13655 : 2009); and
– User inks.

Colour samples are shown in tables of 11 columns and 11 rows according to a system of Cartesian coordinates. The percentage areas covered by two inks are on the abscissa and on the ordinate, respectively, in steps of 10%. The covering percentage of the third ink is equal for all the samples and can be chosen with a step of 1%. The colour samples, which are not displayed, cannot be represented by the monitor used. The observer is the CIE 1931 standard observer. The user chooses:

• two inks, whose percentage covering areas are on the abscissa and ordinate;
• the percentage covering area of the third ink;
• the kind of inks: ideal, standard CEI, standard ISO 12647-6: 2012, or user inks;

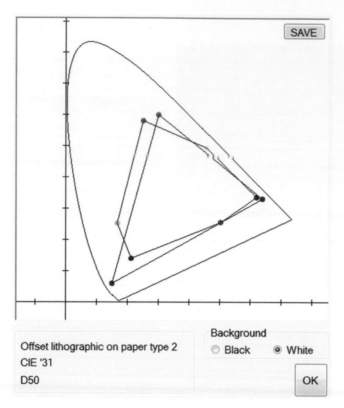

Figure 16.37   *CIE 1931 chromaticity diagram with comparison of the gamut's of colours possible for the monitor used and of the inks selected. It is obvious that not all the colours defined of the RGB gamut can be mapped into the CMY gamut and vice versa.*

- the illuminant, in the case of ideal inks; and
- the way to propose contiguous colours, with or without a separating line (in this way, simultaneous contrasts disappear or appear).

The ideal inks have optimal colours due to bivalent reflectances (Sections 14.2 and 15.3) (Figure 15.15):

- Cyan ink has reflectance equal to 1 for λ <500 nm and 0 elsewhere.
- Yellow ink has reflectance equal to 1 for λ >400 nm and 0 elsewhere.
- Magenta ink has reflectance equal to 0 for $400 < \lambda$ <500 nm and 1 elsewhere.

16.11   Subtractive Colorant Mixing

Before entering this section, the user should read Sections 3.9.1-3.9.3.

The daily experience leads to be familiar with the mixing of coloured pastes and does not approach to the phenomenon of mixing of coloured lights, simply because the first exercise is done consciously by hands, while the latter is almost never done consciously. The visual result of the mixing of coloured pastes can be understood only with physical models that consider the elementary phenomena of absorption and scattering of light, and colorimetry is only applied. Here we apply the model of Kubelka-Munk-Saunderson.

The program 'Colour synthesis' has two Toolboxes – 'Two pigment mixture' and 'Four pigment mixture' – with different purposes. With a mixture of two pigments it is observed the non-linear effect on colour due to the variation of the percentages of mixing only two pigments. In particular the desaturating role of white pigment and darkening role of black pigment can be observed. 'Four pigment mixture' considers almost all the complexity of the reproduction of a paint.

16.11.1 Two Pigment Mixture

The aim of this toolbox is only didactical and works in a particular approximate way:

- The observer is CIE 1931.
- The illuminant D65 is fixed.
- This program gives the colour subtractive mixing for colour layers with spectral reflectance obtained by a two-flux Kubelka-Munk model, that considers only the absorption and mono-dimensional scattering internal to the painting layer.
- The surface reflection considered by Saunderson's correction is ignored.

Anyway, this representation is enough

(i) To understand the effect of any percentage variation in a mixture of two pigments.
(ii) To see that often the resulting colours are external to the colour gamut of the computer monitor used.
(iii) To understand the important role of the white pigment.
(iv) To see on the chromaticity diagram the difference between colour additive mixing and subtractive colorant mixing.
(v) To compare the reflectance with the function $\log[S(\lambda)/K(\lambda)]$.
(vi) The user has to choose a pigment pair from a set of 28 pigments (the colours shown in the choice windows are obtained by mixtures at 50% with the white pigment, since the pure pigment is generally too dark).
(vii) Eleven colour samples are obtained by mixing two pigments in variable ratios with 10% steps.
(viii) The spectral reflectance, the quantity $\log[S(\lambda)/K(\lambda)]$ and the chromaticity (x, y) are plotted.
(ix) An exclamation mark is shown for the colours reproduced on the monitor in an approximate way.

Two examples are given in Figures 16.38 and 16.39.

The user is advised to compare the light mixing of the program 'RGB monitor additive mixing', for example, $R = 255$, $G = 255$ and B variable, with the light reflected by the mixtures of the pigment Yellow b and Blue a. In the first case an increment of the blue light produces a desaturation of the Yellow colour due to the R and G mixture. In the second case, a scale of green hues is obtained. The user should freely make comparison like this as training for visually judging colours produced with different techniques.

16.11.2 Four Pigment Mixture

The colorant-formulation problem is considered in the mixture of four pigments, of which two pigments are placed in the mix for creating the chromatic content of the colour, in addition to the black and white pigments. Here, too, the effect due to the change of the percentage by which the pigments enter the mixture is highly non-linear.

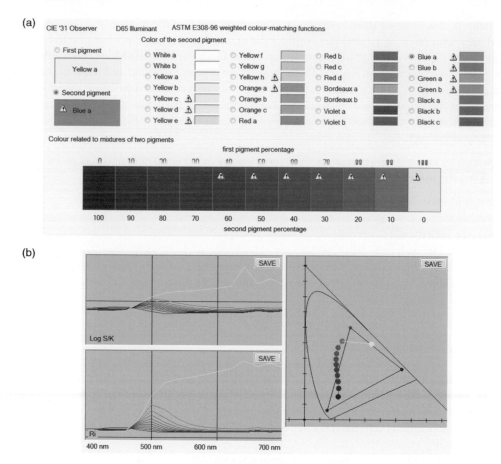

Figure 16.38   *(a) Mixture with 'Yellow b' and 'Blue a' pigments. Only pure blue and pure yellow pigments produce blue and yellow appearances, while the mixtures give a variable green hue. An exclamation mark at any concentration c means that this mixture is not reproducible with the monitor in use and this is shown on the chromaticity diagram, (b) where many mixtures are outside the triangle of the monitor gamut. Moreover, there is a jump in colour appearance between the pure yellow pigment and the mixture constituted by 90% yellow and 10% blue. This phenomenon has its origin in the spectral reflectance, which shows a regular variation with the variation of the mixing ratio in a wide range from 100% to 10% of blue pigment and only between 10% and 0% it shows a great variation. In (b) CIE 1931chromaticity diagram, spectral reflectances and log(S/K) are shown for the different mixtures considered. The mixture of these two pigments shows that all the reflectances cross the same point at a particular wavelength, which is independent of the pigment concentrations.*

This section considers colour subtractive mixing for painting layers with spectral reflectance obtained by two-flux Kubelka-Munk model and Saunderson correction (Figures 16.40 and 16.41). The user chooses four pigments (the colours shown in the choice windows are obtained by mixtures at 50% with the white pigment). The colour related to the chosen four pigments is shown on a three-component diagram in correspondence with any fixed concentration of the fourth pigment.

For a more analytical view of the possible colours obtainable by changing the relative percentage of the pigments chosen, it is convenient to use the lightest pigment as fourth, which can be selected in 1% steps from

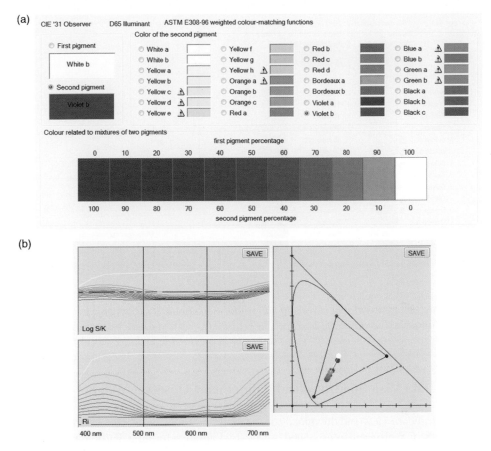

Figure 16.39 *(a) Mixture with 'White b' and 'Violet b' pigments. In (b) CIE 1931 chromaticity diagram, spectral reflectances and log(S/K) are shown. The mixture of these two pigments shows that log(S/K) curves are almost parallel with exclusion of the extremities and analogously the reflectances are regularly changing with the concentrations. The greatest changes happen when a coloured pigment is present with a percentage lower than 10%.*

0% to 100%. Anyway, a change of the order of the selected pigments allows a finer view of the effect of the fourth pigment in the mixture (1% steps).

The colour reproduction depends on the following other choices made by the user:

- The standard observer (CIE 1931 or CIE 1964).
- The illuminant.
- The refraction index of the vehicle containing the pigment grains (*vehicle* is a liquid that holds pigments together without dissolving them. It can make paint dry more slowly or appear more translucent and sometimes acts as a binder).
- The measurement geometry of the spectral reflectance factor (that is supposed measured with integration sphere and may with specular component be included or excluded).

The observer can visually verify the effect of variation on measurement geometry of the illuminant and of the refraction index of the vehicle.

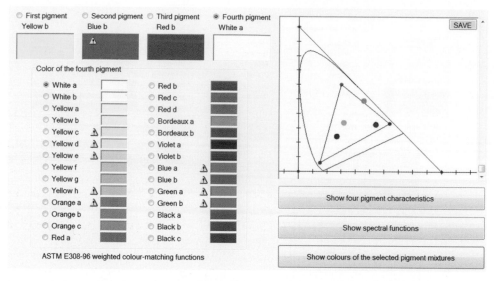

Figure 16.40 *Selection of four pigments: 'Yellow b', 'Blue b', 'Red b' and 'White a'. On the right side, the CIE 1931 chromaticity diagram with the chromaticities of these pigments is shown and by clicking on 'Show spectral functions' the spectral plots of the functions K(λ) and S(λ) related to the four pigments appear. An exclamation mark at any colour means that this colour is not reproducible with the monitor in use.*

16.12   Spectral Data View and Download – Illuminant-Observer Weights

The program 'Spectral data view and download' of the toolbox 'Tools' allows the user to plot and download all the files used by the CX program interpolated at 1 nm by Lagrange formula. Particularly all the illuminants can be shown and downloaded, as already said in section 16.9.2. Moreover it is possible to compute, plot and download the weights for the calculation of tristimulus values at defined illuminant and observer according to Equations (12.42), for steps of 10 and 20 nm in the spectral reflectance factor data. These weights take into account the Lagrange interpolation at 1 nm [Equations (12.43)–(12.45)]. No deconvolution is considered. These weights correspond to those defined by ASTM Table 5 (Section 12.4).

Figure 16.42 shows the weights for the CIE 1964 observer and the illuminant F11.

16.13   Save File Opening

This toolbox allows the user to open '.cvc' files (own software format) relative to screenshots obtained by clicking on the SAVE button on the figures shown by the program.

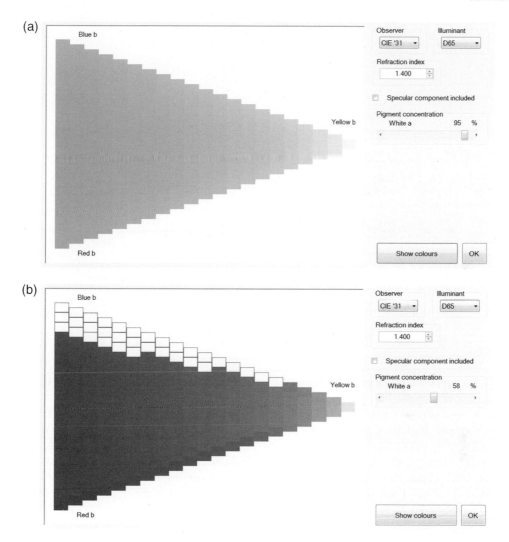

Figure 16.41   *Colours obtained by mixing the pigments selected in Figure 16.40 under illuminant D65 for the observer CIE 1964. The computation of the reflectance factor of this painting is computed by the Kubelka-Munk-Saunderson model with refraction index selected for the vehicle equal to 1.4 and supposing that the reflectance factor is measured with the specular component excluded. The content of the 'White a' pigment is 95% in (a) and 58% in (b) and the content of the other pigments is defined by the position of the colour sample in the triangular plot where barycentric coordinates are defined.*

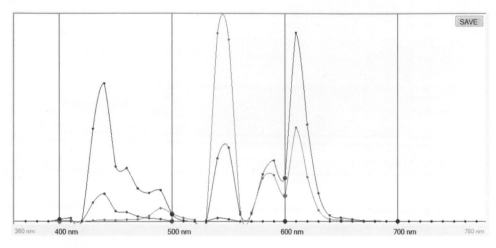

Figure 16.42   *Weights for 10 nm wavelength steps and for the CIE 1964 observer and the illuminant F11 obtained by Lagrange interpolation:* $W_{X,CIE64,F11,\lambda}$ *(red line),* $W_{Y,CIE64,F11,\lambda}$ *(green line),* $W_{Z,CIE64,F11,\lambda}$ *(bue line).*

References

1. CIE 15:3, *Colorimetry*, 3rd ed., Commission Internationale de l'Éclairage, Vienna (2004). Available at: www.cie.co.at/ (accessed on 18 June 2015).
2. Linksz A, *An Essay on Color Vision and Clinical Color-Vision Tests*. Grune & Stratton (1964).
3. Rubin ML and Barbara Cassin, Sheila solomon, *Dictionary of Eye Terminology*. Triad Pub. Co., Gainesville, Fla (1984).
4. Shinobu Ishihara, *Ishihara's Tests for Colour Blindness*, Kanehara & Co. LTD (1985).
5. Mertens HW, *Validity of clinical color vision tests for air traffic control specialists*. (SuDoc TD 4.210:92/29) Office of Aviation Medicine, U.S. Federal Aviation Administration. Available to the public through the National Technical Information Service (1992).
6. Kaiser PKand Boynton RM, *Human Color Vision*, Optical Society of America, Washington DC (1996).
7. Hilbert D and Byrne A, *Readings on Color*, MIT Press, Cambridge, Mass (1997).
8. McIntyre John D, *Colour Blindness: Causes and Effects*, Dalton Publishing. Chester (2002).
9. Shevell SK, *The Science of Color*. Elsevier, Amsterdam (2003).
10. Wang K and Wang X, *Color Vision Test Plates*. People's Medical Publishing House (2008).
11. Granville WC, Colorimetric specification of the *color harmony manual* from spectrophotometric measurements, *J. Opt. Soc. Amer.*, **34** (7), 382–395 (1944).
12. Sidney M, Newhall SM, Nickerson D and Judd DB, Final report of the O.S.A. subcommittee on the spacing of the munsell colors, *J. Opt. Soc. Amer.*, **33**, 385–418 (1943).
13. Richter M, The official german standard color chart, *J. Opt. Soc. Amer.*, **45**, 223–226 (1955).
14. MacAdam D, Colorimetric data for samples of OSA uniform color scales, *J. Opt. Soc. Amer.*, **68**, 121–130 (1978).
15. ILV CIE S 017/E:2011, *ILV: International Lighting Vocabulary*, Commission Internationale de l'Éclairage, Vienna (2011). Available at: www.eilv.cie.co.at (accessed on 18 June 2015).
16. CIE Publication 13:3, *Methods of Measuring and Specifying Colour Rendering Properties of Light Sources*, Commission Internationale de l'Éclairage, Vienna (1995). Available at: www.cie.co.at/ (accessed on 18 June 2015).

Index

Note: Bold numbers indicate that the reference in the text is in a title of a section or subsection.
(Fig.) means that the term is related to a figure on the page.
(file) means that the term is related to one data file of the set described in section 16.1.2 and belonging to the folder 'Colour files' of the CD attached to the book.
(sw) means that the term is related to a software toolbox presented on the page.

Standard Colorimetry: Definitions, Algorithms and Software, First Edition. Claudio Oleari.
© 2016 John Wiley & Sons, Ltd. Published 2016 by John Wiley & Sons, Ltd.